Graduate Texts in Physics

Graduate Texts in Physics

Graduate Texts in Physics publishes core learning/teaching material for graduate- and advanced-level undergraduate courses on topics of current and emerging fields within physics, both pure and applied. These textbooks serve students at the MS- or PhD-level and their instructors as comprehensive sources of principles, definitions, derivations, experiments and applications (as relevant) for their mastery and teaching, respectively. International in scope and relevance, the textbooks correspond to course syllabi sufficiently to serve as required reading. Their didactic style, comprehensiveness and coverage of fundamental material also make them suitable as introductions or references for scientists entering, or requiring timely knowledge of, a research field.

More information about this series at http://www.springer.com/series/8431

Edouard B. Manoukian

Quantum Field Theory II

Introductions to Quantum Gravity, Supersymmetry and String Theory

 Springer

Edouard B. Manoukian
The Institute for Fundamental Study
Naresuan University
Phitsanulok, Thailand

ISSN 1868-4513   ISSN 1868-4521   (electronic)
Graduate Texts in Physics
ISBN 978-3-319-81617-3   ISBN 978-3-319-33852-1   (eBook)
DOI 10.1007/978-3-319-33852-1

Printed on acid-free paper

This Springer imprint is published by Springer Nature
The registered company is Springer International Publishing AG Switzerland

Preface to Volume II

My motivation in writing this second volume was to have a rather introductory book on quantum gravity,[1] supersymmetry,[2] and string theory[3] for a reader who has had some training in conventional quantum field theory (QFT) dealing with its foundations, with abelian and non-abelian gauge theories including grand unification, and with the basics of renormalization theory as already covered in Vol. I *Quantum Field Theory I: Foundations and Abelian and Non-Abelian Gauge Theories*. This volume is partly based on lectures given to graduate students in theoretical and experimental physics, at an introductory level, emphasizing those parts which are reasonably well understood and for which satisfactory theoretical descriptions have been given.

Quantum gravity is a vast subject,[4] and I obviously have to make a choice in this introductory treatment of the subject. As an introduction, I restrict the study to two different approaches to quantum gravity: the perturbative quantum general relativity approach as the main focus and a non-perturbative background-independent one referred to as "loop quantum gravity" (LQG), where space emerges from the theory itself and is quantized. In LQG we encounter a *QFT in a three-dimensional space*.

---

[1] For more advanced books on quantum gravity that I am familiar with, see the following: C. Kiefer (2012): Quantum Gravity, by Oxford University Press, T. Thiemann (2007): Modern Canonical Quantum Gravity, C. Rovelli (2007): Quantum Gravity, as well as of the collection of research investigations in D. Oriti (2009): Approaches to Quantum Gravity, by Cambrige University Press.

[2] For more advanced books on supersymmetry that I am familiar with, see the following books: H. Baer & X. Tata (2006): Weak scale supersymmetry: from superfields to scattering events, M. Dine (2007): Supersymmetry and string theory - beyond the stadard model, S. Weinberg (2000): The Quantum theory of fields III: Supersymmetry, by Cambridge University Press, and P. Binetruy (2006): Supersymmetry, experiments and cosmology by Oxford University Press.

[3] For more advanced books on string theory that I am familiar with, see the following books: K. Becker, M. Becker & J. H. Schwarz (2006): String theory and M-theory - a modern approach, M. Dine (2007): Supersymmetry and string theory - beyond the standard model, and J. Polchinski (2005) : Superstring theory I & II by Cambridge University Press.

[4] See the references given above on quantum gravity.

Some unique features of the treatment given are:

- No previous knowledge of general relativity is required, and the necessary geometrical aspects needed are derived afresh.
- The derivation of field equations and of the expression for the propagator of the graviton in the linearized theory is solved with a gauge constraint, and a constraint necessarily implies that not all the components of the gravitational field may be varied independently—a point which is most often neglected in the literature.
- An elementary treatment is given of the so-called Schwinger-DeWitt technique.
- Non-renormalizability aspects of quantum general relativity are discussed as well as of the renormalizability of some higher-order derivative gravitational theories.
- A proof is given of the Euler-Poincaré Characteristic Theorem which is most often omitted in textbooks.
- A uniqueness property of the invariant product of three Riemann tensors is proved which is also most often omitted in textbooks.
- An introductory treatment is provided of "loop quantum gravity" with sufficient details to get the main ideas across and prepare the reader for more advanced studies.

Supersymmetry is admittedly a theory with mathematical beauty. It unites particles of integer and half-integer spins, i.e., with different spins, but with equal masses in symmetry multiplets. Some important aspects in the treatment of the subject are the following:

- A fundamental property of supersymmetric theories is that the supersymmetry charge (supercharge) operator responsible for interchanging bosonic and fermionic degrees of freedom obviously does not commute with angular momentum (spin) due to different spins arising in a given supermultiplet. This commutation relation is explicitly derived which is most often omitted in textbooks.
- The concept of superspace is introduced, as a direct generalization of the Minkowski one, and the basic theory of integration and differentiation in superspace is developed.
- A derivation is given of the so-called Super-Poincaré algebra satisfied by the generators of supersymmetry and spacetime transformations, which involves commutators and anti-commutators[5] and generalizes the Poincaré algebra of spacetime transformations derived in Vol. I.
- The subject of supersymmetric invariance of integration theory in superspace is developed as it is a key ingredient in defining supersymmetric actions and in constructing supersymmetric extensions of various field theories.
- A panorama of superfields is given including that of the *pure* vector superfield, and complete derivations are provided.

---

[5]Such an algebra is referred to as a graded algebra.

- Once the theory of supersymmetric invariant integration is developed, and superfields are introduced, supersymmetric extensions of basic field theories are constructed, such as that of Maxwell's theory of electrodynamics; a spin 0–spin 1/2 field theory, referred to as the Wess-Zumino supersymmetric theory *with* interactions; the Yang-Mill field theory; and the standard model.
- There are several advantages of a supersymmetric version of a theory over its non-supersymmetric one. For one thing, the ultraviolet divergence problem is much improved in the former in the sense that divergences originating from fermions loops tend, generally, to cancel those divergent contributions originating from bosons due to their different statistics. The couplings in the supersymmetric version of the standard model merge together more precisely at a high energy. Moreover, this occurs at a higher energy than in the non-supersymmetric theory, getting closer to the Planck one at which gravity is expected to be significant. This gives the hope of unifying gravity with the rest of interactions in a quantum setting.
- Spontaneous symmetry breaking is discussed to account for the mass differences observed in nature of particles of bosonic and fermionic types.
- The underlying geometry necessary for incorporating spinors in general relativity is developed to finally and explicitly derive the expression of the action of the full *supergravity* theory.

In string theory, one encounters a *QFT on two-dimensional surfaces* traced by strings in spacetime, referred to as their worldsheets, with remarkable consequences in spacetime itself, albeit in higher dimensions. If conventional field theories are low-energy effective theories of string theory, then this alone justifies introducing this subject to the student. Some important aspects of the treatment of the subject are the following:

- In string theory, particles that are needed in elementary particle physics arise naturally in the mass spectra of oscillating strings and are not, a priori, assumed to exist or put in by hand in the underlying theory. One of such particles emerging from closed strings is the evasive graviton.
- With the strings being of finite extensions, string theory may, perhaps, provide a better approach than conventional field theory since the latter involves products of distributions at the same spacetime points which are generally ill defined.
- Details are given of all the massless fields in bosonic and superstring theories, including the determination of their inherited degrees of freedom.
- The derived degrees of freedom associated with a massless field in D-dimensional spacetime, together with the eigenvalue equation associated with the mass squared operator associated with such a given massless field, are consistently used to determine the underlying spacetime dimensions D of the bosonic and superstring theories.
- Elements of space compactifications are introduced.
- The basics of the underlying theory of vertices, interactions, and scattering of strings are developed.
- Einstein's theory of gravitation is readily obtained from string theory.
- The Yang-Mills field theory is readily obtained from string theory.

This volume is organized as follows. In Chap. 1, the reader is introduced to quantum gravity, where no previous knowledge of general relativity (GR) is required. All the necessary geometrical aspects are derived afresh leading to explicit general Lagrangians for gravity, including that of GR. The quantum aspect of gravitation, as described by the graviton, is introduced, and perturbative quantum GR is discussed. The so-called Schwinger-DeWitt formalism is developed to compute the one-loop contribution to the theory, and renormalizability aspects of the perturbative theory are also discussed. This follows by introducing the very basics of a non-perturbative, background-independent formulation of quantum gravity, referred to as "loop quantum gravity" which gives rise to a quantization of space and should be interesting to the reader. In Chap. 2, we introduce the reader to supersymmetry and its consequences. In particular, quite a detailed representation is given for the generation of superfields, and the underlying section should provide a useful source of information on superfields. Supersymmetric extensions of Maxwell's theory, as well as of Yang-Mills field theory, and of the standard model are worked out, as mentioned earlier. Spontaneous symmetry breaking, and improvement of the divergence problem in supersymmetric field theory are also covered. The unification of the fundamental couplings in a supersymmetric version of the standard model[6] is then studied. Geometrical aspects necessary to study supergravity are established culminating in the derivation of the full action of the theory. In the final chapter, the reader is introduced to string theory, involving both bosonic and superstrings, and to the analysis of the spectra of the mass (squared) operator associated with the oscillating strings. The properties of the underlying fields, associated with massless particles, encountered in string theory are studied in some detail. Elements of compactification, duality, and D-branes are given, as well as of the generation of vertices and interactions of strings. In the final sections on string theory, we will see how one may recover general relativity and the Yang-Mills field theory from string theory. We have also included two appendices at the end of this volume containing useful information relevant to the rest of this volume and should be consulted by the reader. The problems given at the end of the chapters form an integral part of the books, and many developments in the text depend on the problems and may include, in turn, additional material. They should be attempted by every serious student. *Solutions to all the problems are given* right at the end of the book for the convenience of the reader. We make it a point *pedagogically* to derive things in detail, and some of such details are sometimes *relegated* to appendices at the end of the respective chapters, or worked out in the problems, with the main results *given* in the chapters in question. The very detailed introduction to QFT since its birth in 1926 in Vol. I,[7] as well as the introductions to the chapters, provide the motivations

---

[6]The standard model consists of the electroweak and QCD theories combined, with a priori underlying symmetry represented by the group products $SU(2) \times U(1) \times SU(3)$.

[7]*Quantum Field Theory I: Foundations and Abelian and Non-Abelian Gauge Theories*. I strongly suggest that the reader goes through the introductory chapter of Vol. I to obtain an overall view of QFT.

and the pedagogical means to handle the technicalities that follow them in these studies.

This volume is suitable as a textbook. Its content may be covered in a 1 year (two semesters) course. Short introductory seminar courses may be also given on quantum gravity, supersymmetry, and string theory.

I often meet students who have a background in conventional quantum field theory mentioned earlier and want to learn about quantum gravity, supersymmetry and string theory but have difficulty in reading more advanced books on these subjects. I thus felt a pedagogical book is needed which puts these topics together and develops them in a coherent introductory and unified manner with a consistent notation which should be useful for the student who wants to learn the underlying different approaches in a more efficient way. He or she may then consult more advanced specialized books, also mentioned earlier, for additional details and further developments, hopefully, with not much difficulty.

I firmly believe that different approaches taken in describing fundamental physics at very high energies or at very small distances should be encouraged and considered as future experiments may confirm directly, or even indirectly, their relevance to the real world.

I hope this book will be useful for a wide range of readers. In particular, I hope that physics graduate students, not only in quantum field theory and high-energy physics but also in other areas of specializations, will also benefit from it as, according to my experience, they seem to have been left out of this fundamental area of physics, as well as instructors and researchers in theoretical physics.

Edouard B. Manoukian

Contents

Notation

- Latin indices $i, j, k, \ldots$ are generally taken to run over 1,2,3, while the Greek indices $\mu, \nu, \ldots$ over $0, 1, 2, 3$ in 4D. Variations do occur when there are many different types of indices to be used, and the meanings should be evident from the presentations.
- The Minkowski metric $\eta_{\mu\nu}$ is defined by $[\eta_{\mu\nu}] = \text{diag}[-1, 1, 1, 1] = [\eta^{\mu\nu}]$ in 4D.
- The charge conjugation matrix is defined by $\mathscr{C} = i\gamma^2\gamma^0$.
- Unless otherwise stated, the fundamental constants $\hbar$, c are set equal to one.
- The gamma matrices satisfy the anti-commutation relations $\{\gamma^\mu, \gamma^\nu\} = -2\eta^{\mu\nu}$.
- $\bar{\psi} = \psi^\dagger\gamma^0, \bar{u} = u^\dagger\gamma^0, \bar{v} = v^\dagger\gamma^0$. A Hermitian conjugate of a matrix M is denoted by $M^\dagger$, while its complex conjugate is denoted by M^*.
- The Dirac, the Majorana, and the chiral representations of the γ^μ matrices are defined in Appendix I at the end of this volume.
- γ^μ matrices are defined in other dimensions in Appendix I as well.
- The step function is denoted by $\theta(x)$ which is equal to 1 for $x > 0$ and 0 for $x < 0$.

Chapter 1
Introduction to Quantum Gravity

All particles, whether massive or massless, experience the gravitational interaction due to their energy content.[1] Although the gravitational coupling is much smaller than other couplings such as the electromagnetic coupling, the Fermi coupling, the QCD couplings, and so on, the incorporation of gravity in quantum field theory interactions seems important. For one thing, we have seen that in grand unified theories[2] that the effective couplings of various theories merge at high energies at which gravitation may play an equally important role as the other interactions. This will also lead to the ultimate goal of developing a unified theory for all the fundamental interactions, from which the various interactions become distinguishable at limiting low energy limits of such a unified theory. Unification of the interactions in Nature is a major theme in fundamental physics. Even Einstein tried to unify gravity and electrodynamics many years ago. It is expected that gravitation would play, in general, a fundamental role in the ultraviolet divergence problem in quantum field theory when considering theories at small distances.

A quantum gravity (QG) theory as such is needed in early cosmology for the description of the origin of the universe, as well as in black hole physics. It also has to deal, in general, with singularities that may arise in a classical treatment, and problems at small distances, or equivalently at high energies. In particular, it is of interest to provide a unified description of Nature which is applicable from microscopic to cosmological distances. Fundamental constants for a unit of length and a unit of mass expected to be relevant to this end are, respectively, the Planck length and the Planck mass. Out of the fundamental constants of quantum physics $\hbar$, of relativity c, and the Newtonian gravitational one G_N, these units of length and

---

[1]For an overall view of quantum field theory since its birth in 1926 see Chap. 1 of Vol. I [43]. The present introduction is partly based on the latter.

[2]See, e.g., Chap. 6 of Vol. I [43].

© Springer International Publishing Switzerland 2016
E.B. Manoukian, *Quantum Field Theory II*, Graduate Texts in Physics,
DOI 10.1007/978-3-319-33852-1_1

mass, respectively, relevant to quantum gravity, are given by the following

$$\ell_P = \sqrt{\frac{\hbar G_N}{c^3}} \simeq 1.616 \times 10^{-33} \, \text{cm}, \quad m_P = \sqrt{\frac{\hbar c}{G_N}} \simeq 1.221 \times 10^{19} \, \text{GeV}/c^2.$$

In units $\hbar = 1$, $c = 1$, dimensions of physical quantities may be then expressed in powers of mass ($[\text{Energy}] = [\text{Mass}]$, $[\text{Length}] = [\text{Mass}]^{-1} = [\text{Time}], \ldots$). Since gravitation has a universal coupling to all forms of energy, one may hope that it may be implemented within a unified theory of the four fundamental interactions, as mentioned earlier, with the Planck mass providing a universal mass scale. Unfortunately it is difficult in practice to investigate quantum properties of gravitation as one has to work at such high energies that are not accessible experimentally.

A key observation of Einstein, referred to as the principle of equivalence, in developing his general theory of gravitation, is that at any given point in space and any given time, one may consider a frame in which gravity locally, at the point in question, is wiped out. For example, in simple Newtonian gravitational physics, a test particle placed at a given point inside a freely falling elevator on its way to the Earth, remains at rest, relative to the elevator, for a very short time, depending on the accuracy being sought, and, depending on its position relative to the center of the Earth. The particle eventually moves, in general, from its original position in a given instant.[3] Einstein's principle of equivalence applies only locally at a given point of space and at a given time. At the point in question, in the particular frame in consideration, gravity is wiped out and special relativity survives. The reconciliation between special relativity and Newton's theory of gravitation, then readily leads to Einstein's General Theory of Relativity (GR), where gravity is accounted for by the curvature of spacetime and its departure from the flat spacetime of special relativity one has started out with, through the application of the principle of equivalence. As a consequence of this, a geometrical description arises to account for the role of gravity. By doing this, one is able to enmesh non-gravitational laws with gravity via this principle.

GR predicts the existence of Black Holes.[4] Recall that a black hole (BH) is a region of space into which matter has collapsed and out of which light may not escape. It partitions space into an inner region which is bounded by a surface, referred to as the event horizon which acts as a one way surface for light going in but not coming out. The sun's radius is much larger than the critical radius of a BH which is about 2.5 km to be a black hole. We will see in Appendix E of this chapter by examining a spherically symmetric BH of mass M that this critical radius is given by $R_{BH} = 2G_N M/c^2$.

---

[3]The corresponding details will be given in Sect. 1.1 vis-à-vis Fig. 1.1.

[4]Here it is worth recalling that gravitational waves have been detected from the merger of two *black holes* 1.3 billion light-years from the Earth via the Laser Interferometer Gravitational Wave Observatory (LIGO). See B. P. Abbott et al.: Phys. Rev. Lett. 116, 061102 (2016), Astrophys. J. Lett. 818, L22 (2016).

One may argue that the Planck length may set a lower limit spatial cut-off. The following formal and rough estimates are interesting. Suppose that by means of a high energetic particle of energy E, $\langle E^2 \rangle \sim \langle p^2 \rangle c^2$, with $\langle p^2 \rangle$ very large, one is interested in measuring a field within an interval of size δ around a given point in space. Such form of energy acts as an effective gravitational mass $M \sim \sqrt{\langle E^2 \rangle}/c^4$ which, in turn, distorts space around it. The radius of the event horizon of such a gravitational mass M is given by $r_{BH} = 2G_N M/c^2$. Clearly we must have $\delta > r_{BH}$, otherwise the region of size δ that we wanted to locate the point in question will be hidden beyond a BH horizon, and localization fails. Also $\langle p^2 \rangle \geq \langle (p - \langle p \rangle)^2 \rangle \geq \hbar^2/4\delta^2$. Hence $M \geq \hbar/2c\delta$,

$$\delta > \frac{2G_N M}{c^2} \geq \frac{G_N \hbar}{c^3 \delta},$$

which gives $\delta > r_{BH} = \sqrt{G_N \hbar/c^3} = \ell_P$.

Hawking[5] has shown that a BH is not really a black body, it is a thermodynamic object, it radiates and has a temperature (Appendix E of this chapter) associated with it.[6] As a consequence of which the entropy[7] of a BH is given by (Appendix E)

$$S_{BH} = \frac{c^3 k_B}{4\, G_N \hbar} A = k_B \frac{A}{4\,\ell_P^2}, \qquad A = 4\pi \left(\frac{2G_N M}{c^2} \right)^2,$$

referred to as the Bekenstein-Hawking Entropy formula[8] of a BH, where A is the surface area of the BH horizon, and k_B is the Boltzmann constant. This result is expected to hold in any consistent formulation of quantum gravity, and shows that a BH has entropy unlike what would be naïvely expect from a BH with the horizon as a one way classical surface through which information is lost to an external observer. The proportionality of the entropy to the area rather than to the volume of a BH horizon should be noted. It also encompasses Hawking's theorem of increase of the area with time with increase of entropy.

From the geometrical description of gravitation given earlier, one may introduce a gravitational field to account for the departure of the curved spacetime metric from that of the Minkowski one, and make contact with the approaches of conventional field theories, dealing now with a field permeating an interaction between all dynamical fields. The quantum particle associated with the gravitational field, the so-called graviton, emerges by considering the small fluctuation of the metric, associated with curved spacetime of GR about the Minkowski one, as the limit of the full metric, where the gravitational field becomes weaker and the particle becomes

---

[5]Hawking [31, 32].

[6]Particle emission from a BH is formally explained through virtual pairs of particles created near the horizon with one particle falling into the BH while the other becoming free outside the horizon.

[7]Recall that entropy S represents a measure of the amount of disorder with information encoded in it.

[8]Bekenstein [14].

identified. This allows us to determine the graviton propagator in the same way one obtains, for example, the photon propagator in QED, and eventually carry out a perturbation theory as a first attempt to develop a quantum theory of gravitation, starting from the Lagrangian density of the action of GR, referred to as the Einstein-Hilbert action.

In units of $\hbar = 1, c = 1$, Newtons gravitational constant G_N, in 4 dimensional spacetime, has the dimensionality $[G_N] = [mass]^{-2}$, which is a dead give away of the non-renormalizability of a quantum theory of gravitation based on GR. The non-renormalizability of the theory is easier to understand by noting that the degree of divergence of a graph, in general, in the theory turns out to increase with the number of loops of integrations without a bound, implying the need of an infinite number of parameters are needed to be fixed experimentally,[9] indicating that perturbative quantum general relativity is not of practical value, in general. Also correspondingly, new interactions Lagrangians need to be added[10] to the theory indicating that the theory is far from being complete. Here we may pose to recall that in QED, for example, only two parameters may be fixed experimentally, the charge and the mass of the electron. Also the additional terms to be added to the original lagrangian density in doing so, have the same structure as the original terms in the original Dirac-Maxwell Lagrangian density.

The Lagrangian density of the action of GR involves two derivatives. Some higher order derivatives theories turn out to be renormalizable[11] but violate, in a perturbative setting, the very sacred principle of positivity condition of quantum theory. Unfortunately, such a theory involves ghosts in a perturbative treatment, due to the rapid damping of the propagator at high energies faster than $1/k^2$, and gives rise, in turn, to negative probabilities.[12]

It is generally believed that one is trying to use general relativity beyond its limit of validity, at energy scales where a more fundamental quantum gravity will be involved. In this sense, general relativity is expected to emerge as a low energy limit of a more fundamental theory, as the former has been quite successful in the low energy classical regime. As a matter of fact the derivatives occurring in the action, in a momentum description via Fourier transforms, may be considered to be small at sufficiently low energies. In view of applications in the low energy regime, one then tries to separate low energy effects from high energy ones even if the theory has unfavorable ultraviolet behavior such as in quantum gravity.[13] Applications of such an approach have been carried out in the literature as just cited, and, for example, the modification of Newton's gravitational potential at long distances has

---

[9]Manoukian [37], Anselmi [2].

[10]This fact is already revealed by going up to the two-loop contribution to the theory: 't Hooft and Veltman [61], Kallosh, Tarasov and Tyutin [34], Goroff and Sagnotti [30], van de Ven [65], Barvinsky and Vilkovisky [13].

[11]Stelle [59].

[12]Stelle [59]. Unitarity (positivity) of such a theory in a non-perturbative setting has been elaborated upon by Tomboulis [63].

[13]Donoghue [25–27]; Bjerrum-Bohr et al. [15, 16].

been determined to have the structure

$$U(r) = -\frac{G_N m_1 m_2}{r}\left[1 + \alpha\frac{G_N(m_1 + m_2)}{c^2 r} + \beta\frac{G_N\hbar}{c^3 r^2}\right],$$

for the interaction of two spin 0 particles of masses m_1 and m_2. Here α, β, are dimensionless constants,[14] and the third term represents a quantum correction being proportional to $\hbar$. We will consider the low energy behavior of quantum GR briefly later in Sect. 1.8.

Conventional quantum field theory is usually formulated in a fixed, i.e., in, a priori, given background geometry such as the Minkowski one. This is unlike the formalism of "Loop Quantum Gravity" (LQG) also called "Quantum Field Theory of Geometry". The situation that we will encounter in this approach is of a *quantum field theory in three dimensional space*, which is a non-perturbative background independent formulation of quantum gravity. The latter means that no specific assumption is made about the underlying geometric structure and, interestingly enough, the latter rather emerges from the theory. Here by setting up an eigenvalue equation of, say, an area operator, in a quantum setting, one will encounter a granular structure of three-dimensional space yielding a discrete spectrum for area measurements with the smallest possible having a non-zero value given to be of the order of the Planck length squared: $\hbar G_N/c^3 \sim 10^{-66}$ cm^2.[15] The emergence of space in terms of "quanta of geometry", providing a granular structure of space, is a major and beautiful prediction of the theory.

The 3 dimensional space is generated by a so-called time slicing procedure of spacetime carried out by Arnowitt, Deser and Misner.[16] The basic field variables in the theory is a gravitational "electric" field, which determines the geometry of such a 3 dimensional space and naturally emerges from the definition of the area of a surface in such a space, and its canonical conjugate variable referred to as the connection. By imposing equal time commutation relation of these two canonically conjugate field variables, the quantum version of the theory arises, and the fundamental problem of the quantization of geometry follows. Loop variables are defined in terms of the connection, and corresponding spin-network states are introduced to describe the underlying geometry of three space (Sect. 1.9). Here the spin-network states correspond to microscopic degrees of freedom. The basic idea goes to Penrose [47] whose interest was to construct the concept of space from combining angular momenta. It is also interesting that the proportionality of entropy and the surface area of the BH horizon in the Bekenstein-Hawking Entropy formula has been derived in loop quantum gravity.[17] For general references on LQG, see also [50, 51, 62].

---

[14]Recent recorded values are $\alpha = 3$, and $\beta = 41/10\pi$ [15].

[15]Ashtekar and Lewandoski [7], Rovelli and Vidotto [51], Rovelli and Smolin [54].

[16]Arnowitt et al. [3].

[17]See, e.g., [1, 44].

We begin this chapter by developing the general geometric notion of spacetime in general relativity from first principles. No previous knowledge of general relativity is required to follow the development. From this, the concept of a gravitational field is obtained and the graviton propagator and its inherited polarization aspects are obtained as done in conventional quantum field theory. Quantum fluctuations about a background metric, satisfying Einstein's field equation, are described and renormalizability aspects of quantum general relativity as well of more general quantum gravities with higher order derivatives are considered. In particular, to study the one-loop divergence contribution to quantum GR, we develop an elegant method, referred to as the Schwinger-DeWitt technique.[18] In this respect, we also prove two important theorems related to the "Euler-Poincaré Characteristic" and to the "Invariant Products of so-called three Riemann tensors". In the remaining part of the chapter, we provide an introductory presentation of loop quantum gravity.

1.1   Geometrical Aspects, Structure of Spacetime and Development of the General Theory of Relativity

In a geometrical context, gravity is accounted for by the curvature of spacetime and the departure of the latter from that of the flat Minkowski spacetime of special relativity. With gravity, one associates several geometrical terms to describe the underlying geometry of spacetime, and in this sense gravity and these geometrical terms become simply interchangeable words for the same thing. The structure of spacetime is then held "responsible" for the motion of a particle due to gravity without introducing a gravitational field as a dynamical variable as such. By such a geometrical description, one is able to enmesh non-gravitational physical laws with gravity via the principle of equivalence, to be discussed below, in a straightforward manner. In turn, starting from a consistent geometrical formalism, a gravitational field may be introduced, as a dynamical variable, permeating an interaction between all dynamical fields solely due to their energy-momentum content in the same way that the Maxwell field permeates the interaction between charged particles. Unlike the Maxwell field, however, which carries no charge, the gravitational field, due to its energy-momentum content, generates a direct self-interaction as well.

I present a simple treatment of this geometrical description in such a way that a reader who has never been exposed to general relativity may, hopefully, be able to follow.

A rather elementary and clear way to start and understand how Einstein's theory of gravitation arises is to consider, in Newtonian gravitational theory, a classic thought experiment of an elevator in free-fall in the Earth gravity, as shown in Fig. 1.1, neglecting, for simplicity, the Earth rotation. To account only for the

---

[18]Schwinger [55], DeWitt [21, 24]. See also [21] for the pioneering work on the description of fluctuations about an arbitrary spacetime background.

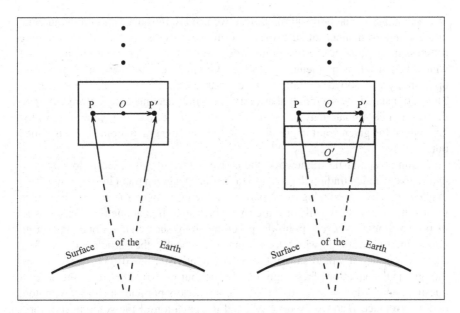

Fig. 1.1 A particle placed at the point O will remain at rest inside the elevator in free-fall, while from outside the elevator the particle accelerates with the gravitational force. A particle placed at P or at P$'$ will remain at rest inside the elevator, only momentarily, and will eventually move toward the center O due to the attraction of a particle to the center of the Earth. This leads to the basic concept that at every point in spacetime, way before a particle falls to the surface of the Earth, a coordinate system may be set up in which locally, and only locally, i.e., only at the point in question a particle is at rest with the gravitational force wiped out

gravitational force due to the Earth one would, of course, neglect other forces. By Newtonian gravitational theory, one usually means weak gravitational force and slowly moving particles. In free-fall, the elevator on its way to the earth, its enter O will move to a point, say, O' assumed to be tracing a line directed to the center of the earth.

We are interested in investigating the role of gravitation, due to the Earth, on the trajectory of a particle put, *in turn* at points O, P, P$'$, in the elevator in free-fall, from the point of view of what may seem to be happening inside the elevator, and what is perceived from outside of the elevator.

A particle set at point O will remain there, in reference to the elevator in free fall, with the gravitational force wiped out at that point, while from outside of the elevator the particle is seen to accelerate in the Earth gravity. On the other hand, a particle placed at point P or point P$'$ the situation is different. Inside the elevator, the particle will eventually move toward the center O due to the attraction of a particle toward the center of the Earth. For a very short time, however, depending on the accuracy being sought, the particle will be considered to be at rest at the point in question inside (i.e., relative to) the elevator, indicating, momentarily, the absence of a gravitational field, while from outside of the elevator the particle again accelerates in the Earth gravity. By considering the elevator, described by a coordinate system, in which a particle is momentarily at rest, we need to introduce an infinite number, in a continuous manner, of such local coordinate systems, as we move, indicating progress in time, along the line going from the point O to the point O', in each of which the particle is momentarily at rest, while in a general coordinate system set up in space, way above the Earth surface, a particle accelerates in the Earth gravity. This is translated, by saying, that at every point in spacetime, way before falling to the Earth, a coordinate may be set up in which a particle, locally and only locally, i.e., only at the given point in question in spacetime, the gravitational force is wiped out.

Within a relativistic framework, the above is formulated in the following way. One is interested in finding the role of gravitation at a given point in spacetime. This may be done by introducing a test particle at the point in question. As the particle moves in spacetime, it will trace a curve which may be parametrized in terms of its proper time τ. At every point of spacetime along such a curve, one may set up a local inertial frame, in which locally, and *only* locally, the particle has zero four acceleration, i.e., it would satisfy the special relativistic law $d(dX^{\mu}/d\tau)/d\tau = 0$, thus giving the equation of a straight line for the four velocity $dX^{\mu}/d\tau$ *at* the point in question, where $d\tau^2 = -\eta_{\alpha\beta}\, dX^{\alpha}\, dX^{\beta}$. For a massless particle, such as the photon, one may replace τ above by $q = X^0$, and the equation of the straight line, for a massless particle, in the local Lorentz coordinate system becomes $d(dX^{\mu}/dq)/dq = 0$, with $\eta_{\alpha\beta}\, (dX^{\alpha}/dq)\, (dX^{\beta}/dq) = 0$. In the sequel, we use, in general, the notation λ for such parameters.

The role of gravitation on the particle is then described by the comparison of all such inertial frames and by the elucidation on the way they *relate* to one another. Intuitively, in a geometrical sense, gravitation would imply a departure of a particle's path from a straight line, as defined in a so-called flat local Lorentz frame, to a curved one attributed to an underlying curved geometric structure as will be seen below.

This brings us to what is called the principle of equivalence in a more general context: At every point in spacetime, one sets up a local Lorentz frame, such that locally in it, and only locally, the laws of physics, not involving gravitation, may be formulated by the application of special relativity, and the role of gravitation is then taken into account by the comparison of such local Lorentz frames and by the way they are infinitesimally related to one another.

From a pure geometrical picture, the above means that for sufficiently small regions such as on a curved surface, these regions may be considered to be flat. In a limiting sense, at every point on such a curved surface, one may then set up a coordinate system corresponding to a completely *flat* space, in which special relativity applies at the point in question. To account for gravity, then one, clearly, needs a structure to tell us how such coordinates may be arranged relative to each and how the origin of one coordinate system is related to the origin of an infinitesimally close one and hence also give us the relation between the local Lorentz coordinates and of the underlying spacetime. This structure is referred to as the *connection*.

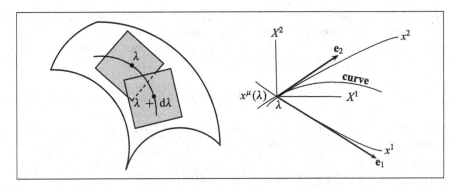

Fig. 1.2 The connection allows us to compare the tangent spaces shown in *grey*, on the *left-hand side* of the figure, at infinitesimally separated points on the *curve* parametrized by λ. At every point $x^\mu(\lambda)$ on the curve parametrized by λ, a Lorentz coordinate system is set up with coordinates $X^1, X^2, \ldots$

Figure 1.2 shows how Lorentz frames (flat spaces), referred to as tangent spaces here, may be set up at various points and arranged in a region of a curved space, with their origins falling, say, on a curve in spacetime parametrized by some parameter λ. The connection would then allow us to compare how a pair of such tangent spaces, at infinitesimally separated points, are arranged relative to each other. It is defined in terms of the concept of parallel transport to be discussed below. It is important to know that it is not necessary to consider such a curved structure to be embedded in a higher dimension. It just helps one to visualize the situation.

At every point $x(\lambda)$, on such a curve, in a general curvilinear coordinate description of curved spacetime, one sets up a Lorentz coordinate system with coordinate basis vector fields $e^\alpha{}_\mu(x)$, where α is a Lorentz index, i.e., it refers to the local Lorentz coordinate system, and the μ indices refer to the generalized coordinates. In a globally everywhere flat Minkowski space, the curved lines, specified by $x^1, x^2, \ldots$, originating from the origin of the local Lorentz coordinate system, on the right-hand side of Fig. 1.2, will straighten up and lie along the axes $X^1, X^2, \ldots$ and the basis vectors will reduce simply to $e^\alpha{}_\mu = \delta^\alpha{}_\mu$. In the more general case with a curved spacetime, and only within this context of the comparison of two such systems, it has been customary to use indices from the beginning of the Greek alphabet $\alpha, \beta, \ldots$ for the Lorentz ones, and indices from about the middle of the alphabet $\ldots, \mu, \nu, \ldots$ (and beyond) for the generalized coordinate ones. We use this notation in the present section to avoid any confusion. Clearly, the orientation of the axes specified by $X^1, X^2, \ldots$ are arbitrary and amounts to a freedom of carrying a local Lorentz transformation to re-orient these axes.

A vector field $\mathbf{V}(\lambda)$, with components $V^\alpha(x)$, $\alpha = 0, 1, 2, 3$, in a local Lorentz coordinate system set up *at point* λ, on a curve parametrized by λ, with coordinate label $x^\mu(\lambda)$ in curved spacetime, may be expressed as

$$\mathbf{V}(\lambda) = V^\mu(x)\,\mathbf{e}_\mu(x), \qquad \mathbf{e}_\mu(x) = \{e^\alpha{}_\mu(x)\}. \tag{1.1.1}$$

The fields $e^{\alpha}{}_{\mu}(x)$ are called tetrad or vierbein fields. Thus the index μ specifies the different vector fields in the curvilinear coordinate system, while an index α specifies the αth component of any of these vectors in the local Lorentz coordinate system set up at x, as mentioned above. Hence

$$V^{\alpha}(x) \;=\; V^{\mu}(x)\,e^{\alpha}{}_{\mu}(x), \qquad\qquad (1.1.2)$$

expressed as functions of the generalized coordinates. One may define the scalar product

$$\mathbf{V}(\lambda)\cdot\mathbf{V}(\lambda) \;=\; V^{\mu}(x)V^{\nu}(x)\,\mathbf{e}_{\mu}(x)\cdot\mathbf{e}_{\nu}(x), \qquad\qquad (1.1.3)$$

thus introducing a metric $g_{\mu\nu}(x) = g_{\nu\mu}(x)$ $\left(g^{\mu\nu}(x)\right)$ to lower (raise) the coordinate indices μ, ν, and define scalar products of two vectors $\mathbf{V}_1$, $\mathbf{V}_2$ in terms of the vector components V_1^{μ}, V_2^{ν}

$$\mathbf{e}_{\mu}(x)\cdot\mathbf{e}_{\nu}(x) \;=\; g_{\mu\nu}(x), \qquad\qquad (1.1.4)$$

in curved spacetime, in a curvilinear coordinate system,

$$\mathbf{V}_1\cdot\mathbf{V}_2 \;=\; V_1^{\mu}(x)V_2^{\nu}(x)\,g_{\mu\nu}(x) = V_1^{\mu}(x)V_{2\mu}(x), \quad V_{2\mu}(x) \equiv V_2^{\nu}(x)g_{\mu\nu}(x). \qquad (1.1.5)$$

Due to the symmetric nature of the product in (1.1.4), $g_{\mu\nu}$ is taken to be symmetric.

In the local Lorentz coordinate system in question set up at $x^{\mu}(\lambda)$, the scalar product in (1.1.3) then reduces to the familiar scalar product in the local Lorentz coordinate system at the point in question

$$\mathbf{V}(\lambda)\cdot\mathbf{V}(\lambda) \;=\; V^{\alpha}(x)V^{\beta}(x)\eta_{\alpha\beta}, \qquad\qquad (1.1.6)$$

expressed as functions of the generalized coordinates, where $\eta_{\alpha\beta}$ is the Minkowski metric, needed for lowering one of the Lorentz indices α, β.

Let us consider just for a moment a global flat Minkowski spacetime. In it we define the parallel transfer, or parallel transport, of basis vectors along a curve parametrized by a parameter λ, where they are literally moved parallel to themselves, by

$$\frac{d\,\mathbf{e}_{\mu}(x(\lambda))}{d\lambda} \;=\; \partial_{\nu}\,\mathbf{e}_{\mu}(x)\,\frac{dx^{\nu}(\lambda)}{d\lambda} \;=\; 0. \qquad\qquad (1.1.7)$$

Hence from the general relation in (1.1.1), we have in a global Minkowski space, with the parallel transport of the basis vectors as given above,

$$\frac{d\mathbf{V}(\lambda)}{d\lambda} \;=\; \frac{dV^{\mu}(\lambda)}{d(\lambda)}\,\mathbf{e}_{\mu}(x). \qquad\qquad (1.1.8)$$

In curved space, this parallelism is taken over by introducing, in the process, the concept of a covariant derivative, or equivalently by taking into account of the way the basis vectors $e^{\alpha}{}_{\mu}(x(\lambda))$ turn as λ is made to vary, to make up for the difference in such a geometrical context in a straightforward manner.

To find the derivative of $\mathbf{V}(\lambda)$, with respect to λ, we need to know how these tangent spaces at λ, $\lambda + d\lambda$ arrange themselves with respect to each other. That is, we need to know how the basis vectors change as we move infinitesimally when λ is made to vary. This is done by introducing the concept of parallel transport. It is here where we need a structure, referred to as the connection, to quantify this change. To this end, we note that $\mathbf{e}_{\mu}(x(\lambda + d\lambda)) - \mathbf{e}_{\mu}(x(\lambda))$, must vanish for $dx^{\mu}(\lambda) \rightarrow 0$. Also at λ, it may be expanded in terms of $\mathbf{e}_{\sigma}(x(\lambda))$. That is, the derivative of $\mathbf{e}_{\mu}(x(\lambda))$ with respect to λ at this point, may be written as

$$\frac{d}{d\lambda}\, \mathbf{e}_{\mu}(x(\lambda)) = \Gamma_{\nu\mu}{}^{\sigma}(x)\, \mathbf{e}_{\sigma}(x)\, \frac{dx^{\nu}(\lambda)}{d\lambda}, \tag{1.1.9}$$

where the totality of the *expansion* coefficients $\{\Gamma_{\nu\mu}{}^{\sigma}(x)\}$ is called the connection. This equation may be also rewritten as

$$\left(\partial_{\nu}\, \mathbf{e}_{\mu}(x) - \Gamma_{\nu\mu}{}^{\sigma}(x)\, \mathbf{e}_{\sigma}(x)\right) \frac{dx^{\nu}(\lambda)}{d\lambda} = 0, \tag{1.1.10}$$

which should be compared with the globally flat space case on the right-hand side of (1.1.7), and generalizes the concept of parallel transport of the basis vectors along the curve in question to that of a curved space thanks to the introduction of the connection as a result of the turning that the basis vectors go through to achieve this in spacetime.

Accordingly, the derivative of a vector $\mathbf{V}(\lambda)$ field, with respect to the parameter λ, follows from (1.1.1), (1.1.9) to be simply

$$\frac{d\mathbf{V}(x)}{d\lambda} = \left(\frac{dV^{\sigma}(x)}{d\lambda} + \Gamma_{\nu\mu}{}^{\sigma}(x)\, V^{\mu}(x)\, \frac{dx^{\nu}}{d\lambda}\right) \mathbf{e}_{\sigma}(x). \tag{1.1.11}$$

The components of the vector field $d\mathbf{V}(x(\lambda))/d\lambda$ are then given by

$$\frac{dV^{\sigma}(x)}{d\lambda} + \Gamma_{\nu\mu}{}^{\sigma}(x)\, V^{\mu}(x)\, \frac{dx^{\nu}}{d\lambda} \equiv \frac{DV^{\sigma}(x)}{d\lambda}, \tag{1.1.12}$$

where we have used the notation $DV^{\sigma}/d\lambda$ for a component in order not to confuse it with the first term $dV^{\sigma}/d\lambda$ on the left-hand side of (1.1.12).

To reconcile with the fact that the just mentioned components are indeed components of a vector field, a rule of transformation for the connection automatically follows. Under a coordinate transformation $x \rightarrow x'$, the relation between the components $V^{\mu}(x)$, $V'^{\mu}(x')$ is, by definition, given by the chain rule, to be

$$V^{\sigma}(x) = \frac{\partial x^{\sigma}}{\partial x'^{\nu}}\, V'^{\nu}(x'). \tag{1.1.13}$$

Accordingly, we must also have $DV^\sigma/d\lambda = (\partial x^\sigma/\partial x'^\nu) DV'^\nu(x')/d\lambda$ which is easily shown from (1.1.11) to give the following transformation rule for the connection

$$\Gamma'_{\rho\nu}{}^\gamma = \frac{\partial x'^\gamma}{\partial x^\sigma} \frac{\partial x^\mu}{\partial x'^\rho} \frac{\partial x^\kappa}{\partial x'^\nu} \Gamma_{\mu\kappa}{}^\sigma + \frac{\partial x'^\gamma}{\partial x^\sigma} \frac{\partial^2 x^\sigma}{\partial x'^\rho \partial x'^\nu}. \tag{1.1.14}$$

Due to the second term on the right-hand of (1.1.14), the connection is not a tensor but has just the right transformation property so that the covariant derivative of $V^\mu(x)$, defined below, is a tensor. On the other hand,

$$\left(\Gamma'_{\rho\nu}{}^\gamma - \Gamma'_{\nu\rho}{}^\gamma\right) = \frac{\partial x'^\gamma}{\partial x^\sigma} \frac{\partial x^\mu}{\partial x'^\rho} \frac{\partial x^\kappa}{\partial x'^\nu} \left(\Gamma_{\mu\kappa}{}^\sigma - \Gamma_{\kappa\mu}{}^\sigma\right), \tag{1.1.15}$$

and the combination $\left(\Gamma_{\mu\kappa}{}^\sigma - \Gamma_{\kappa\mu}{}^\sigma\right)$ is a tensor as it satisfies the correct transformation law.

With covariant derivatives properly introduced, theories may be then developed in terms of such derivatives which are invariant under general coordinate systems. That is, they would lead, self-consistently, to theories which are invariant in such a curved spacetime due to gravity.

Upon writing $dV^\sigma(x)/d\lambda = \partial_\nu V^\sigma(x) (dx^\nu/d\lambda)$, we may use (1.1.12) to introduce the covariant derivative of $V^\sigma(x)$

$$\nabla_\nu V^\sigma(x) = \partial_\nu V^\sigma(x) + \Gamma_{\nu\mu}{}^\sigma V^\mu(x) \equiv V^\sigma{}_{;\nu}(x), \tag{1.1.16}$$

where $V^\sigma{}_{;\nu}(x)$, $\nabla_\nu V^\sigma(x)$ are standard notations for the covariant derivative.

A scalar field, under a coordinate transformation $x \to x'$, by definition, satisfies the relation $\phi'(x') = \phi(x)$, and hence $\partial_\mu \phi(x)$ transforms as a vector field

$$\left(\frac{\partial \phi'(x')}{\partial x'^\mu}\right) = \frac{\partial x^\nu}{\partial x'^\mu}\left(\frac{\partial \phi(x)}{\partial x^\nu}\right), \tag{1.1.17}$$

In the literature, especially in the earlier one, components $V^\mu(x)$ of a vector field are referred to as contravariant components, while the components $V_\mu(x) = g_{\mu\nu}(x)V^\nu(x)$ are referred to as covariant components. Using the fact that $V^\mu(x)V_\mu(x)$ is a scalar, the covariant derivative of the components $V_\mu(x)$ is easily found from the fact that $V^\mu V_\mu$ is a scalar field, and hence $\partial_\nu(V^\mu V_\mu)$ is a vector field. In detail

$$\partial_\nu(V^\mu V_\mu) = (\partial_\nu V^\mu)V_\mu + V^\mu \partial_\nu V_\mu = \left(V^\mu{}_{;\nu} - \Gamma_{\nu\sigma}{}^\mu V^\sigma\right)V_\mu + V^\mu \partial_\nu V_\mu$$
$$= V^\mu{}_{;\nu} V_\mu + V^\mu \partial_\nu V_\mu - \Gamma_{\nu\sigma}{}^\mu V^\sigma V_\mu = V^\mu{}_{;\nu} V_\mu + V^\mu\left(\partial_\nu V_\mu - \Gamma_{\nu\mu}{}^\sigma V_\sigma\right), \tag{1.1.18}$$

where in writing the last expression we have simply interchanged the indices $\sigma \leftrightarrow \mu$ in $\Gamma_{\nu\sigma}{}^\mu V^\sigma V_\mu$. The above equation is a tensor equation, which means that the coefficient of V^μ in the last term defines a tensor as well. Thus the covariant derivative of V_μ is defined by

$$V_{\mu;\nu}(x) = \partial_\nu V_\mu(x) - \Gamma_{\nu\mu}{}^\sigma(x) V_\sigma(x) \equiv \nabla_\nu V_\mu. \tag{1.1.19}$$

At every point in spacetime we may set up locally a Lorentz coordinate system, in which the laws of special relativity hold such as, for example, that the partial derivative of a vector is a tensor locally and hence the connection vanishes. We cannot, however, conclude that the connection vanishes at every point in a general coordinate system since the connection is not a tensor as we have seen in (1.1.14). The combination $\left(\Gamma_{\mu\kappa}{}^\sigma - \Gamma_{\kappa\mu}{}^\sigma\right)$, on the other hand, as we have seen in (1.1.15) is a tensor. We may thus conclude from the transformation law in (1.1.15), that *at any given point* the latter combination must be zero *as* we consider the transformation from a local Lorentz coordinate system to a general coordinate system *at the given point* in question. That is, the connection is symmetric $\Gamma_{\mu\kappa}{}^\sigma = \Gamma_{\kappa\mu}{}^\sigma$.

By considering the products $V_1^\mu V_2^\nu$, $V_{1\mu} V_{2\nu}$, $V_1^\mu V_{2\nu}$, covariant derivatives of tensors, with two indices (second rank tensors), are easily obtained to be given by

$$\nabla_\sigma T^{\mu\nu} = \partial_\sigma T^{\mu\nu} + \Gamma_{\sigma\rho}{}^\mu T^{\rho\nu} + \Gamma_{\sigma\rho}{}^\nu T^{\mu\rho}, \tag{1.1.20}$$

$$\nabla_\sigma T_{\mu\nu} = \partial_\sigma T_{\mu\nu} - \Gamma_{\sigma\mu}{}^\rho T_{\rho\nu} - \Gamma_{\sigma\nu}{}^\rho T_{\mu\rho}, \tag{1.1.21}$$

$$\nabla_\sigma T^\mu{}_\nu = \partial_\sigma T^\mu{}_\nu + \Gamma_{\sigma\rho}{}^\mu T^\rho{}_\nu - \Gamma_{\sigma\nu}{}^\rho T^\mu{}_\rho, \tag{1.1.22}$$

$$\nabla_\sigma T_\mu{}^\nu = \partial_\sigma T_\mu{}^\nu + \Gamma_{\sigma\rho}{}^\nu T_\mu{}^\rho - \Gamma_{\sigma\mu}{}^\rho T_\rho{}^\nu, \tag{1.1.23}$$

with obvious extensions for higher rank tensors.

In view of developing theories invariant under general coordinate transformation and to account for the role of gravitation, we have learnt how to define covariant derivatives of tensor, vector and scalar fields. We have also learnt, in particular, how to introduce covariant derivatives of vector fields components $V^\mu(x)$, $V_\mu(x)$ in a general coordinate system. In Sect. 2.4, of Chap. 2 dealing with supersymmetry, we will see how the covariant derivative of a Dirac field is defined and further generalizations will be carried out in Sect. 2.5 needed to develop supergravity. This will not be needed in the present chapter.

In developing invariant theories involving such fields under general coordinate transformations to account for the role of gravitation, one needs also to introduce an invariant definition corresponding to a volume element in spacetime, which would be needed in defining an action integral. To this end, we note that under a general coordinate transformation $x \rightarrow x'$, a spacetime volume element (dx) changes to $|\det(\partial x^\nu/\partial x'^\mu)| (dx')$ via the Jacobian. On the other hand, under such a transformation, the basis vectors $e^\alpha{}_\mu(x)$ transform as covariant components, with general coordinate indices μ, as follows [see (1.1.9)]

$$e'^\alpha{}_\mu(x') = \frac{\partial x^\nu}{\partial x'^\mu} e^\alpha{}_\nu(x), \quad \det[e^\alpha{}_\nu(x)] = \det\left(\frac{\partial x'^\mu}{\partial x^\nu}\right) \det[e'^\alpha{}_\rho(x')]. \tag{1.1.24}$$

That is, $(\mathrm{d}x)\,|\det[\,e^\alpha{}_\nu(x)\,|\,]$ is an invariant. From (1.1.4), we also have

$$e^\alpha{}_\mu(x)\,\eta_{\alpha\beta}\,e^\beta{}_\nu(x) = g_{\mu\nu}(x), \quad -\big(\det e^\alpha{}_\nu(x)\big)^2 = \det[g_{\mu\nu}(x)] \equiv g(x) < 0,$$
$$(1.1.25)$$

since $\det\eta_{\alpha\beta} = -1$. Hence for a volume element of curved spacetime, one makes the replacement

$$(\mathrm{d}x) \;\Rightarrow\; \sqrt{-g(x)}\,(\mathrm{d}x), \qquad\qquad (1.1.26)$$

and the right-hand side is an invariant under general coordinate transformations.

The parallel transfer $V^\mu_\parallel(x + \mathrm{d}x)$ of the vector *components* $V^\mu(x)$ from a point x to a point infinitesimally close point $x + \mathrm{d}x$, in reference to some given curve, may be explicitly defined through the equation

$$V^\mu{}_{;\nu}(x)\,\mathrm{d}x^\nu \;=\; \Big(V^\mu(x + \mathrm{d}x) - V^\mu_\parallel(x + \mathrm{d}x)\Big), \qquad (1.1.27)$$

and hence from (1.1.16) by

$$V^\mu(x) \;\to\; V^\mu(x) - \Gamma_{\nu\rho}{}^\mu(x)\,V^\rho(x)\,\mathrm{d}x^\nu \;\equiv\; V^\mu_\parallel(x + \mathrm{d}x). \qquad (1.1.28)$$

The covariant derivative allows one to define the covariant derivative of a vector field $V^\mu(x)$ along the tangent vector $\mathrm{d}x^\nu(\lambda)/\mathrm{d}\lambda$ of a curve parametrized by a parameter λ, such as the proper time, with coordinate labels $x^\nu(\lambda)$, in a curvilinear coordinate system, by

$$\nabla_\nu V^\mu(x)\,\frac{\mathrm{d}x^\nu(\lambda)}{\mathrm{d}\lambda} \;\equiv\; \nabla_U V^\mu, \qquad U^\nu = \frac{\mathrm{d}x^\nu(\lambda)}{\mathrm{d}\lambda}. \qquad (1.1.29)$$

A vector field with components $V^\mu(x)$ is then said to be parallel transferred along the curve parametrized by λ and coordinate label $x^\nu(\lambda)$ if $\nabla_U V^\mu = 0$. A privileged curve is one for which its tangent vector is parallel transferred to itself, that is $\nabla_U U^\mu = 0$, which from (1.1.16), (1.1.29) it leads to

$$\frac{\mathrm{d}^2 x^\mu}{\mathrm{d}\lambda^2} + \Gamma_{\sigma\nu}{}^\mu(x)\,\frac{\mathrm{d}x^\sigma}{\mathrm{d}\lambda}\frac{\mathrm{d}x^\nu}{\mathrm{d}\lambda} = 0. \qquad (1.1.30)$$

It is referred to as the geodesic equation. Here we note that, in curved spacetime, in a curvilinear coordinate system, one has $\mathrm{d}\tau^2 = -g_{\mu\nu}(x)\,\mathrm{d}x^\mu\,\mathrm{d}x^\nu$, for a massive particle, and $g_{\mu\nu}(x)\,(\mathrm{d}x^\mu/\mathrm{d}\lambda)\,(\mathrm{d}x^\nu/\mathrm{d}\lambda) = 0$ for a massless particle.

What is special about a geodesic equation? Again coming back to describing the role of gravitation at a given spacetime point, one may consider a test particle at the point $x^\mu(\lambda)$ in question at which a local Lorentz coordinate has been set up. Locally, at the point $x^\mu(\lambda)$, and only at this point, with labeling $X^\alpha(\lambda)$ pertaining to the latter coordinate system, set up at $x^\mu(\lambda)$, the principle of equivalence states that

$$\frac{\mathrm{d}X^\beta(\lambda)}{\mathrm{d}\lambda} \frac{\partial}{\partial X^\beta} \frac{\mathrm{d}X^\alpha(\lambda)}{\mathrm{d}\lambda} = \frac{\mathrm{d}^2X^\alpha(\lambda)}{\mathrm{d}\lambda^2} = 0, \qquad (1.1.31)$$

i.e., $\mathrm{d}X^\alpha(\lambda)/\mathrm{d}\lambda$ satisfies the equation of a straight line at the point in question. The geodesic equation in (1.1.30) is the corresponding generalized straight line in curved spacetime, with the mathematical statement of the principle of equivalence that the connection $\Gamma_{\sigma\,\nu}{}^\mu(x)$ vanishes, locally, in the local Lorentz coordinate system *at* its origin with coordinate label $x^\mu(\lambda)$. Thus a geodesic in curved spacetime is the equation of a straight line in the local Lorentz coordinate system set up at the point in question.

By setting up local Lorentz coordinate at spacetime points in curved spacetime, and using the principle of equivalence, one is able to enmesh physical laws, as developed in special relativity, with gravity. In such a context, gravitation is visualized as a geometric property of spacetime. With it one associates such geometrical terms as the metric, involved in measuring intervals in spacetime and hence obviously appears in defining an invariant measure of a spacetime volume element as seen in (1.1.26), the connection which tells us how coordinate basis vectors rotate, or equivalently as how pairs of infinitesimally close tangent spaces are arranged relative to each other in investigating the parallel transfer of a vector along a curve such as a geodesic. With gravitation one also associates a structure of most importance called the Riemann curvature discussed next. With such geometric notions, information on gravity is obtained by probing the geometry of spacetime.

How does the Riemann curvature arise and what is its significance? As in the investigation of the non-abelian gauge fields, associated with internal degrees of freedom,[19] it arises due to the non-commutativity of the components of the covariant derivative. Moreover, properties based on commutation relations are statements concerning measurements. On may obtain the explicit expression for the Riemann curvature by considering the parallel transport of a vector along a closed loop in spacetime. A far more physically interesting way to investigate the nature of the Riemann curvature, is by comparing infinitesimally close paths (worldlines) of two test particles, referred to as the method of geodesic deviations. This is discussed next. The investigation of the non-commutativity of covariant derivatives will follow this discussion.

On the left-hand side of Fig. 1.3, consider the relative acceleration of the two test particles at P and P′, as now perceived from *outside* the elevator in free-fall, in Newtonian physics. On the other hand, on the right-hand side, two infinitesimally close geodesics are shown in curved spacetime. Points on the two geodesics are identified and compared at the same given values of a parameter λ, which may be

---

[19]It is worth comparing this with the non-abelian gauge theories cases investigated in Vol. I [43]. See (6.1.15) of Chap. 6 of Vol. I.

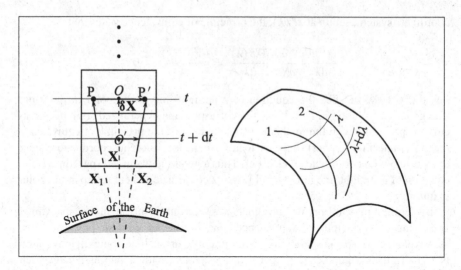

Fig. 1.3 Description of the relative acceleration of test particles, placed at P and P′, as perceived from the outside of the elevator as depicted on the *left-hand side* of the figure in Newtonian physics. Two close geodesics, marked by 1 and 2 in curved spacetime are shown on the *right-hand side* of the figure

chosen to coincide with the proper time of a massive particle or another parameter for massless particle as previously discussed.

In the Newtonian physics case, neglecting the interaction between the test particles, let $\varphi(\mathbf{X})$ denote the gravitational potential at a point specified by the vector $\mathbf{X}$, hence

$$\frac{d^2\mathbf{X}_1}{dt^2} = -\partial_1\varphi(\mathbf{X}_1), \quad \frac{d^2\mathbf{X}_2}{dt^2} = -\partial_2\varphi(\mathbf{X}_2), \tag{1.1.32}$$

which gives for the relative acceleration for the two test particles, in terms of an infinitesimal separation δX^i between the particles, as perceived from outside of the elevator in free-fall

$$\frac{d^2\delta X^i}{dt^2} = -\left(\partial^i\,\partial_j\varphi(\mathbf{X})\right)\delta X^j. \tag{1.1.33}$$

In the curved spacetime case, in a curvilinear coordinate description as shown in Fig. 1.4, we consider two close geodesics, denoted by 1 and 2, satisfying the equations

$$\frac{d^2x^\mu}{d\lambda^2} + \Gamma_{\nu\sigma}{}^\mu(x)\frac{dx^\nu}{d\lambda}\frac{dx^\sigma}{d\lambda} = 0, \tag{1.1.34}$$

$$\frac{d^2(x^\mu+\delta x^\mu)}{d\lambda^2} + \Gamma_{\nu\sigma}{}^\mu(x+\delta x)\frac{d(x^\nu+\delta x^\nu)}{d\lambda}\frac{d(x^\sigma+\delta x^\sigma)}{d\lambda} = 0. \tag{1.1.35}$$

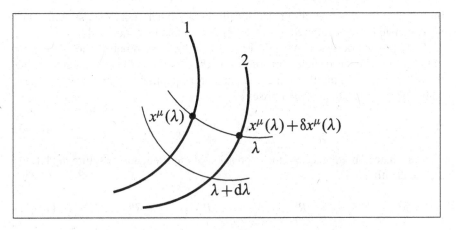

Fig. 1.4 $\mathbf{P}(\lambda) = \delta x^{\mu}(\lambda)\,\mathbf{e}_{\mu}(x(\lambda))$ denotes the displacement vector between two points $x^{\mu}(\lambda)$ and $x^{\mu}(\lambda) + \delta x^{\mu}(\lambda)$, respectively, on two neighboring geodesics marked by 1 and 2

Upon carrying a Taylor expansion of the connection in the latter equation about point x, and subtracting the first equation from the second, we obtain

$$\frac{\mathrm{d}^2 \delta x^{\mu}}{\mathrm{d}\lambda^2} + \partial_{\rho}\Gamma_{\nu\sigma}{}^{\mu}(x)\,\delta x^{\rho}\,\frac{\mathrm{d}x^{\nu}}{\mathrm{d}\lambda}\frac{\mathrm{d}x^{\sigma}}{\mathrm{d}\lambda} + 2\,\Gamma_{\nu\sigma}{}^{\mu}(x)\,\frac{\mathrm{d}x^{\nu}}{\mathrm{d}\lambda}\frac{\mathrm{d}\delta x^{\sigma}}{\mathrm{d}\lambda} = 0. \tag{1.1.36}$$

Now let $\mathbf{P}(\lambda) = \delta x^{\mu}\,\mathbf{e}_{\mu}(x(\lambda))$ denote the displacement vector between the two neighboring geodesics. The relative acceleration of the two particles is then given by taking the derivative of $\mathbf{P}(\lambda)$ with respect to λ twice and using, in the process, (1.1.10)/(1.1.11) and (1.1.36) to obtain

$$\frac{\mathrm{d}^2}{\mathrm{d}\lambda^2}\mathbf{P}(\lambda) = \frac{\mathrm{D}^2 \delta x^{\mu}}{\mathrm{d}\lambda^2}\,\mathbf{e}_{\mu}(x(\lambda)), \tag{1.1.37}$$

with components

$$\frac{\mathrm{D}^2 \delta x^{\mu}}{\mathrm{d}\lambda^2} = R^{\mu}{}_{\nu\rho\sigma}(x)\,\frac{\mathrm{d}x^{\nu}}{\mathrm{d}\lambda}\frac{\mathrm{d}x^{\rho}}{\mathrm{d}\lambda}\,\delta x^{\sigma}, \tag{1.1.38}$$

which should be compared with (1.1.33), defining, in the process, the Riemann curvature tensor

$$R^{\mu}{}_{\nu\rho\sigma} = \partial_{\rho}\,\Gamma_{\nu\sigma}{}^{\mu} - \partial_{\sigma}\,\Gamma_{\nu\rho}{}^{\mu} + \Gamma_{\rho\kappa}{}^{\mu}\,\Gamma_{\nu\sigma}{}^{\kappa} - \Gamma_{\sigma\kappa}{}^{\mu}\,\Gamma_{\nu\rho}{}^{\kappa}. \tag{1.1.39}$$

If the latter is not equal to zero, it provides evidence of the presence of gravitation, as a manifestation of the spacetime curvature, by the observed deviation of two neighboring geodesics. It is important to note that a tensor which vanishes in one coordinate system vanishes in every coordinate system. Accordingly, although gravitational forces may be eliminated locally by a suitable choice of a coordinate system, the non vanishing of the tensor $R^{\mu}{}_{\nu\rho\sigma}$ in a given coordinate system implies that it cannot be eliminated locally in any other coordinate system. Thus the associated forces, referred to as tidal forces, cannot be eliminated.

The covariant derivative $\nabla_{\mu_1}(\nabla_{\mu_2} V^\mu(x))$ may be worked out by the application of (1.1.23) to the tensor field components $T_{\mu_2}{}^\mu(x) = \nabla_{\mu_2} V^\mu(x)$. By interchanging the indices μ_1, μ_2, the following commutator, as applied to arbitrary vector field components $V^\mu(x)$, then readily arises[20]

$$[\nabla_{\mu_1}, \nabla_{\mu_2}] V^\mu(x) = R^\mu{}_{\nu\mu_1\mu_2} V^\nu(x). \tag{1.1.40}$$

Immediate consequences of the expression of the Riemann curvature in (1.1.39) are the identities

$$R^\mu{}_{\nu\rho\sigma} + R^\mu{}_{\rho\sigma\nu} + R^\mu{}_{\sigma\nu\rho} = 0, \qquad R^\mu{}_{\nu\rho\sigma} = -R^\mu{}_{\nu\sigma\rho}. \tag{1.1.41}$$

Some properties of the Riemann curvature tensor that also follow directly by using the explicit expression of the connection given in (1.1.49) below are

$$R_{\mu\nu\rho\sigma} = -R_{\nu\mu\rho\sigma} = -R_{\mu\nu\sigma\rho}, \; R_{\mu\nu\rho\sigma} = R_{\rho\sigma\mu\nu}, \; \left(g_{\mu\lambda} R^\lambda{}_{\nu\rho\sigma} = R_{\mu\nu\rho\sigma}\right). \tag{1.1.42}$$

In a local lorentz coordinate system, the connection vanishes *locally*, but not its derivative, otherwise the Riemann curvature will be zero. In such a coordinate system, at the point in question, it is readily verified that

$$\partial_\lambda R^\mu{}_{\nu\rho\sigma} + \partial_\rho R^\mu{}_{\nu\sigma\lambda} + \partial_\sigma R^\mu{}_{\nu\lambda\rho} = 0, \tag{1.1.43}$$

and by invoking the principle of equivalence, we may replace the ordinary derivatives by covariant ones in making a transition to curved spacetime. The following identity then emerges

$$R^\mu{}_{\nu\rho\sigma;\lambda} + R^\mu{}_{\nu\sigma\lambda;\rho} + R^\mu{}_{\nu\lambda\rho;\sigma} = 0, \tag{1.1.44}$$

holding in every coordinate system, and is referred to as the Bianchi identity.

The Riemann curvature was expressed in terms of the connection and its derivative. The latter, in turn, may be expressed in terms of the metric $g_{\mu\nu}(x)$ and its derivative as follows. Since the metric $g_{\mu\nu}(x)$ may be used to lower coordinate indices, this means, in particular, that for the covariant derivatives of vector components we may write

$$g_{\mu\nu}(x) V^\mu{}_{;\rho}(x) = V_{\nu\,;\rho}(x) = \left(g_{\mu\nu}(x) V^\mu(x)\right)_{;\rho}, \tag{1.1.45}$$

---

[20]See also Problem 1.2.

from which we conclude that the covariant derivative of the metric is zero

$$g_{\mu\nu\,;\rho}(x) = 0. \tag{1.1.46}$$

In a coordinate system in which all the components of the connection vanish at a point x, the above equation implies that the *first* derivative of the metric vanishes:

$$\frac{\partial g_{\mu\nu}(x)}{\partial x^\rho} = 0, \quad \text{if} \quad \Gamma_{\alpha\beta}{}^\gamma(x) = 0 \quad \text{at point } x. \tag{1.1.47}$$

Using the expression for the covariant derivative of a second rank tensor, as given in (1.1.21), in the equation

$$g_{\mu\sigma\,;\nu} + g_{\sigma\nu\,;\mu} - g_{\nu\mu\,;\sigma} = 0, \tag{1.1.48}$$

gives

$$\Gamma_{\mu\nu}{}^\rho = \frac{1}{2} g^{\rho\sigma} \left(\partial_\mu g_{\nu\sigma} + \partial_\nu g_{\mu\sigma} - \partial_\sigma g_{\mu\nu} \right). \tag{1.1.49}$$

An important tensor that we will encounter in the sequel, obtained from the Riemann tensor, is the Ricci tensor defined by

$$R_{\nu\sigma} = R^\mu{}_{\nu\mu\sigma} = \partial_\rho \Gamma_{\nu\sigma}{}^\rho - \partial_\sigma \Gamma_{\nu\mu}{}^\mu + \Gamma_{\mu\kappa}{}^\mu \Gamma_{\nu\sigma}{}^\kappa - \Gamma_{\sigma\kappa}{}^\mu \Gamma_{\nu\mu}{}^\kappa, \tag{1.1.50}$$

satisfying the symmetry condition[21]

$$R_{\nu\sigma} = R_{\sigma\nu}, \tag{1.1.51}$$

and the scalar curvature

$$R = g^{\nu\sigma} R_{\nu\sigma}. \tag{1.1.52}$$

From (1.1.42), we may be rewrite (1.1.44) as

$$R^{\mu\nu}{}_{\rho\sigma\,;\lambda} - R^{\nu\mu}{}_{\sigma\lambda\,;\rho} - R^{\mu\nu}{}_{\rho\lambda\,;\sigma} = 0.$$

Upon contraction over μ, ρ, this gives

$$R^\nu{}_{\sigma\,;\lambda} - R^{\nu\mu}{}_{\sigma\lambda\,;\mu} - R^\nu{}_{\lambda\,;\sigma} = 0.$$

Finally upon contraction over ν, σ, gives

$$R_{;\lambda} - R^\mu{}_{\lambda\,;\mu} - R^\nu{}_{\lambda\,;\nu} = 0,$$

---

[21]Note that $\partial_\sigma \Gamma_{\nu\mu}{}^\mu = \partial_\sigma \partial_\nu \ln[\sqrt{-g}\,]$, see Problem 1.4, (ii).

which may be rewritten as $\left(\nabla_\mu \, g^{\mu\nu} = 0, \; g^{\mu\nu} \, \nabla_\mu = \nabla^\nu \right)$

$$\nabla_\mu \left(R^{\mu\nu} - \frac{1}{2} g^{\mu\nu} R \right) = 0. \tag{1.1.53}$$

This suggests to introduce the tensor

$$G^{\mu\nu} \;=\; R^{\mu\nu} - \frac{1}{2} g^{\mu\nu} R, \qquad \text{with} \quad \nabla_\mu \, G^{\mu\nu} = 0, \tag{1.1.54}$$

which is referred to as the Einstein tensor, satisfying a covariant generalization of a conservation law, and has a unique role in the formulation of Einstein's theory of gravitation as we will see in the next section.

To make contact with conventional perturbative field theory, in which a field may be introduced for gravitation, one may use the fact that the metric $g_{\mu\nu}(x)$, used to measure intervals in curved spacetime, will deviate from the Minkowski metric due to gravity. Upon introducing a coupling parameter κ, which will be expressed in terms of the gravitational constant, one introduces a gravitational field $h_{\mu\nu}(x)$ defined through

$$g_{\mu\nu}(x) \;=\; \eta_{\mu\nu} + \kappa \, h_{\mu\nu}(x), \tag{1.1.55}$$

describing fluctuations about the Minkowski metric. In a weak coupling limit, the field $h_{\mu\nu}(x)$, which responds to gauge transformations, as will be investigated later, describes a massless particle of spin 2—the graviton, with a corresponding propagator which mediates the gravitational interaction between all particles. This and general quantum aspects of gravitation will be the subject of Sect. 1.3 and other sections.

Before closing this section and in view of application in curved spacetime, we recall the definition of the Levi-Civita *symbol*

$$\varepsilon^{\mu\nu\rho\sigma} = \begin{cases} +1, & \text{if } (\mu, \nu, \rho, \sigma) \text{ is an even permutation of } (0, 1, 2, 3), \\ -1, & \text{if } (\mu, \nu, \rho, \sigma) \text{ is an odd permutation of } (0, 1, 2, 3), \\ 0, & \text{otherwise.} \end{cases} \tag{1.1.56}$$

It is defined in this way in every coordinate system. Accordingly, $\varepsilon^{\mu\nu\lambda\sigma}$, is not a tensor for the simple reason that under a coordinate transformation with labels: $x \rightarrow \bar{x}$,

$$\frac{\partial \bar{x}^0}{\partial x^{\nu_1}} \frac{\partial \bar{x}^1}{\partial x^{\nu_2}} \frac{\partial \bar{x}^2}{\partial x^{\nu_3}} \frac{\partial \bar{x}^3}{\partial x^{\nu_4}} \, \varepsilon^{\nu_1 \nu_2 \nu_3 \nu_4} = \det \left[\frac{\partial \bar{x}^\mu}{\partial x^\nu} \right], \tag{1.1.57}$$

is not simply $\bar{\varepsilon}^{\,0123}$, where in writing the equality above we have used the definition of a determinant. On other hand, under such a general coordinate transformation

$$d\bar{x}^{\mu} = \frac{\partial \bar{x}^{\mu}}{\partial x^{\nu}} dx^{\nu}, \qquad (1.1.58)$$

and from (1.1.26) $\left| \det \left[\partial \bar{x}^{\mu} / \partial x^{\nu} \right] \right| = [-\bar{g}(\bar{x})]^{1/2} / [-g(x)]^{1/2}$. From this we learn that

$$\eta^{\mu_1 \mu_2 \mu_3 \mu_4} = \frac{1}{\sqrt{-g}} \varepsilon^{\mu_1 \mu_2 \mu_3 \mu_4}, \qquad (1.1.59)$$

up to a sign, due to the sign of the Jacobian of coordinate transformation $\det[\partial \bar{x}^{\mu} / \partial x^{\nu}]$, transforms as a tensor under a general coordinate transformation, and defines a (pseudo)-tensor.

Now consider the expression

$$g_{\mu_1 \nu_1} g_{\mu_2 \nu_2} g_{\mu_3 \nu_3} g_{\mu_4 \nu_4} \varepsilon^{\nu_1 \nu_2 \nu_3 \nu_4} = \mp(-g), \qquad (1.1.60)$$

where $\mp$ corresponds to the cases for which $(\mu_1, \mu_2, \mu_3, \mu_4)$ is an even/odd permutation of $(0, 1, 2, 3)$, respectively. Thus we may introduce a *symbol*

$$\varepsilon_{\mu\nu\rho\sigma} = \begin{cases} -1, & \text{if } (\mu, \nu, \rho, \sigma) \text{ is an even permutation of } (0, 1, 2, 3), \\ +1, & \text{if } (\mu, \nu, \rho, \sigma) \text{ is an odd permutation of } (0, 1, 2, 3), \\ 0, & \text{otherwise.} \end{cases} \qquad (1.1.61)$$

to rewrite (1.1.60) as

$$g_{\mu_1 \nu_1} g_{\mu_2 \nu_2} g_{\mu_3 \nu_3} g_{\mu_4 \nu_4} \varepsilon^{\nu_1 \nu_2 \nu_3 \nu_4} = (-g) \varepsilon_{\mu_1 \mu_2 \mu_3 \mu_4}, \qquad (1.1.62)$$

from which

$$g_{\mu_1 \nu_1} g_{\mu_2 \nu_2} g_{\mu_3 \nu_3} g_{\mu_4 \nu_4} \eta^{\nu_1 \nu_2 \nu_3 \nu_4} = \eta_{\mu_1 \mu_2 \mu_3 \mu_4}, \qquad (1.1.63)$$

thus introducing the covariant version of $\eta^{\nu_1 \nu_2 \nu_3 \nu_4}$ given by

$$\eta_{\mu_1 \mu_2 \mu_3 \mu_4} = \sqrt{-g}\, \varepsilon_{\mu_1 \mu_2 \mu_3 \mu_4}. \qquad (1.1.64)$$

We finally discuss an important geometrical aspect which shows how one may always define a coordinate system in such a way that locally, at a given point p, the connection vanishes. More precisely, consider a given point p in spacetime with coordinate label $x^{\mu}\big|_p \equiv x_p^{\mu}$, in a given coordinate system with coordinate labels $\{x^{\mu}\}$, and let $\Gamma_{\mu\sigma}^{\ \nu}(x_p)$ denote the connection determined at the point in question. Now let us see how one may introduce another coordinate system such that the connection in this new coordinate system vanishes *at* the point p, whose physical meaning will be spelled out below. To this end, we may introduce a new coordinate system with coordinate labels $\{\bar{x}^\mu\}$ by the following transformation

$$\bar{x}^{\,\mu} = (x^\mu - x_p^\mu) + \frac{1}{2}\,\Gamma_{v\sigma}{}^\mu(x_p)\,(x^v - x_p^v)(x^\sigma - x_p^\sigma), \tag{1.1.65}$$

where note that $\bar{x}^{\,\mu}\big|_p = 0$. We take the partial derivative of this equation with respect to $\bar{x}^\rho$ to obtain

$$\delta^\mu{}_\rho = \frac{\partial x^\mu}{\partial \bar{x}^\rho} + \Gamma_{v\sigma}{}^\mu(x_p)\frac{\partial x^v}{\partial \bar{x}^\rho}(x^\sigma - x_p^\sigma). \tag{1.1.66}$$

Note, in particular, that

$$\frac{\partial x^\mu}{\partial \bar{x}^\rho}\bigg|_p = \delta^\mu{}_\rho, \tag{1.1.67}$$

$$\frac{\partial \bar{x}^{\,\mu}}{\partial x^v}\bigg|_p = \delta^\mu{}_v, \qquad\qquad \frac{\partial^2 x^\mu}{\partial \bar{x}^\lambda \partial \bar{x}^\rho}\bigg|_p = -\Gamma_{\rho\lambda}{}^\mu(x_p). \tag{1.1.68}$$

The partial derivative of (1.1.66) with respect to $\bar{x}^\lambda$ reads

$$0 = \frac{\partial^2 x^\mu}{\partial \bar{x}^\lambda \partial \bar{x}^\rho} + \Gamma_{v\sigma}{}^\mu(x_p)\frac{\partial x^v}{\partial \bar{x}^\rho}\frac{\partial x^\sigma}{\partial \bar{x}^\lambda} + \Gamma_{v\sigma}{}^\mu(x_p)\frac{\partial^2 x^v}{\partial \bar{x}^\lambda \partial \bar{x}^\rho}(x^\sigma - x_p^\sigma). \tag{1.1.69}$$

Upon multiplying the latter by $\partial \bar{x}^\kappa / \partial x^\mu$ gives

$$\frac{\partial \bar{x}^\kappa}{\partial x^\mu}\left[\frac{\partial^2 x^\mu}{\partial \bar{x}^\lambda \partial \bar{x}^\rho} + \Gamma_{v\sigma}{}^\mu(x_p)\frac{\partial x^v}{\partial \bar{x}^\rho}\frac{\partial x^\sigma}{\partial \bar{x}^\lambda} + \Gamma_{v\sigma}{}^\mu(x_p)\frac{\partial^2 x^v}{\partial \bar{x}^\lambda \partial \bar{x}^\rho}(x^\sigma - x_p^\sigma)\right] = 0. \tag{1.1.70}$$

Now we recall the transformation rule of the connection in (1.1.14) with conveniently relabeled indices

$$\overline{\Gamma}_{\rho\lambda}{}^\kappa(\bar{x}) = \frac{\partial \bar{x}^\kappa}{\partial x^\mu}\frac{\partial x^v}{\partial \bar{x}^\rho}\frac{\partial x^\sigma}{\partial \bar{x}^\lambda}\,\Gamma_{v\sigma}{}^\mu(x) + \frac{\partial \bar{x}^\kappa}{\partial x^\mu}\frac{\partial^2 x^\mu}{\partial \bar{x}^\lambda \partial \bar{x}^\rho}. \tag{1.1.71}$$

Upon subtracting the equation in (1.1.70) from the one just above, the following exact equation emerges

$$\overline{\Gamma}_{\rho\lambda}{}^\kappa(\bar{x}) = \left[\Gamma_{v\sigma}{}^\mu(x) - \Gamma_{v\sigma}{}^\mu(x_p)\right]\frac{\partial \bar{x}^\kappa}{\partial x^\mu}\frac{\partial x^v}{\partial \bar{x}^\rho}\frac{\partial x^\sigma}{\partial \bar{x}^\lambda}$$

$$- \Gamma_{v\sigma}{}^\mu(x_p)\frac{\partial \bar{x}^\kappa}{\partial x^\mu}\frac{\partial^2 x^v}{\partial \bar{x}^\lambda \partial \bar{x}^\rho}(x^\sigma - x_p^\sigma). \tag{1.1.72}$$

At the point p in the new coordinate system, we learn from this equation and (1.1.67)/(1.1.68), that

$$\overline{\Gamma}_{\rho\lambda}{}^{\kappa}(\bar{x})\Big|_{p} = 0, \tag{1.1.73}$$

whose importance cannot be overemphasized. Such a coordinate system is referred to as a geodesic coordinate system.

The possibility of finding a coordinate system such that at a given point p, the connection vanishes, is physically quite interesting. By referring to (1.1.34), we learn that at the point p in question, in this coordinate system, the acceleration of the particle vanishes

$$\frac{d^2\bar{x}^{\mu}}{d\tau^2}\Big|_{p} = 0, \tag{1.1.74}$$

meaning that gravity is locally, i.e., *at* the point p, is wiped out. This is to be compared with the particle considered in the falling elevator in Fig. 1.1, and is intimately related to the principle of equivalence. Einstein's viewpoint in formulating a theory of gravitation based on the equivalence principle is certainly remarkable and goes to the heart of the mathematical foundation of general relativity.

1.2 Lagrangians for Gravitation: The Einstein-Hilbert Action, Einstein's Equation of GR, Energy-Momentum Tensor, Higher-Order Derivatives Lagrangians

A simple action that one may introduce to describe gravity involves the scalar curvature $R(x)$, given in (1.1.52), defined by

$$W = \frac{1}{\kappa^2}\int (dx)\sqrt{-g(x)}\, R(x), \quad R(x) = R_{\mu\nu}(x)\, g^{\mu\nu}(x), \quad g(x) = \det[g_{\mu\nu}(x)], \tag{1.2.1}$$

where we have necessarily introduced the invariant volume element $\sqrt{-g(x)}(dx)$ in (1.1.26) to ensure invariance under general coordinate transformations. The Ricci tensor $R_{\mu\nu}$ is given in (1.1.50). κ^2 is a parameter introduced for dimensional reasons. The action in (1.2.1) is referred to the Einstein-Hilbert action. The theory based on this action will be compared to that of Newton's theory of gravitation.

Under an infinitesimal transformation [22] $\delta g^{\mu\nu}$

$$\delta(\sqrt{-g}\, g^{\mu\nu} R_{\mu\nu}) = \sqrt{-g}\,(R_{\mu\nu} - \frac{1}{2}g_{\mu\nu}R)\,\delta g^{\mu\nu}, \tag{1.2.2}$$

---

[22]See Problem 1.9.

leading to the field equation

$$R_{\mu\nu} - \frac{1}{2} g_{\mu\nu} R = 0, \qquad \nabla^{\mu}\left(R_{\mu\nu} - \frac{1}{2} g_{\mu\nu} R\right) = 0, \qquad \nabla^{\nu}\left(R_{\mu\nu} - \frac{1}{2} g_{\mu\nu} R\right) = 0,$$

$$(1.2.3)$$

where $R_{\mu\nu} - \frac{1}{2} g_{\mu\nu} R \equiv G_{\mu\nu}$ is the Einstein tensor in (1.1.54). The second equality in the above equation was derived in (1.1.53), and the third one follows from the symmetry of $R_{\mu\nu}$, $g_{\mu\nu}$ in their indices.

Partial integration reads slightly in a different manner in curved spacetime. In view of applications of partial integrations we note, for example, that for a scalar field $\phi(x)$, and a vector field $S^{\mu}(x)$ one has

$$\sqrt{-g(x)}\big(\nabla_{\mu}\phi(x)\big)S^{\mu}(x) = -\sqrt{-g(x)}\,\phi(x)\,\nabla_{\mu}S^{\mu}(x), \qquad (1.2.4)$$

up to ordinary total derivatives, where ∇_{μ} denotes the covariant derivative introduced in the previous section. For a vector field $S^{\mu}(x)$, and a tensor field $F^{\mu\nu}(x)$,

$$\sqrt{-g(x)}F_{\mu\nu}(x)\,\nabla^{\mu}S^{\nu}(x) = -\sqrt{-g(x)}\,\big(\nabla^{\mu}F_{\mu\nu}(x)\big)\,S^{\nu}(x), \qquad (1.2.5)$$

up to ordinary total derivatives. The presence of the $\sqrt{-g(x)}$ factor should be noted in these two equations. For proofs of (1.2.4), (1.2.5), with further generalizations, see Problem 1.10.

From (1.2.2), (1.2.5), (1.1.53), in particular, we learn that the action W is invariant under the gauge transformations

$$\delta g^{\mu\nu}(x) = -\big(\nabla^{\mu}\Lambda^{\nu}(x) + \nabla^{\nu}\Lambda^{\mu}(x)\big), \qquad (1.2.6)$$

where the minus sign is introduced for convenience, for given vector fields $\Lambda^{\nu}(x)$, as a consequence of the fact that

$$\sqrt{-g}\left(R_{\mu\nu} - \frac{1}{2} g_{\mu\nu} R\right)\delta g^{\mu\nu}$$

$$= -\sqrt{-g}\left[\Lambda^{\nu}\nabla^{\mu}(R_{\mu\nu} - \frac{1}{2} g_{\mu\nu} R) + \Lambda^{\mu}\nabla^{\nu}(R_{\mu\nu} - \frac{1}{2} g_{\mu\nu} R)\right] = 0, \qquad (1.2.7)$$

up to ordinary total derivatives in its integrand.

Also the equality $g_{\sigma\mu}\delta g^{\mu\nu} = -g^{\mu\nu}\delta g_{\sigma\mu}$, gives the following equivalent infinitesimal gauge transformations for the metric $g_{\sigma\mu}$,

$$\delta g_{\sigma\mu}(x) = \big(\nabla_{\sigma}\Lambda_{\mu}(x) + \nabla_{\mu}\Lambda_{\sigma}(x)\big). \qquad (1.2.8)$$

with an overall plus sign for the variation of the metric relative to its inverse matrix in (1.2.6).

For future reference, the above transformation in (1.2.8) may be rewritten in an equivalent form as follows. To this end, we may write, in detail,

$$\delta g_{\sigma\mu}(x) = g_{\mu\lambda}\nabla_\sigma\Lambda^\lambda(x) + g_{\sigma\lambda}\nabla_\mu\Lambda^\lambda(x)$$

$$= g_{\mu\lambda}\partial_\sigma\Lambda^\lambda(x) + g_{\sigma\lambda}\partial_\mu\Lambda^\lambda + \Lambda^\lambda\partial_\lambda g_{\sigma\mu} + \Lambda^\gamma\left(g_{\mu\lambda}\Gamma_{\sigma\gamma}{}^\lambda + g_{\sigma\lambda}\Gamma_{\mu\gamma}{}^\lambda - \partial_\gamma g_{\sigma\mu}\right),$$

$$(1.2.9)$$

where in the writing the second line, we have used the definition of the covariant derivative of a vector field, and then added and subtracted the term $\Lambda^\gamma\partial_\gamma g_{\sigma\mu} \equiv \Lambda^\lambda\partial_\lambda g_{\sigma\mu}$. We recognize the last term on the right-hand side of (1.2.9) as $-\Lambda^\gamma\nabla_\gamma g_{\sigma\mu}$ which is zero. Hence (1.2.8) may be equivalently rewritten simply as

$$\delta g_{\sigma\mu}(x) = g_{\sigma\lambda}\partial_\mu\Lambda^\lambda + g_{\mu\lambda}\partial_\sigma\Lambda^\lambda(x) + \Lambda^\lambda\partial_\lambda g_{\sigma\mu}. \qquad (1.2.10)$$

At this stage, it is important to bring the above theory, based on the Einstein-Hilbert action, in contact with that of Newton's theory of gravitation. The potential energy $\varphi(\mathbf{X})$, in the latter, due to a mass distribution with mass density $\rho(\mathbf{X})$ is given by

$$\varphi(\mathbf{X}) = -G\int d^3\mathbf{X}' \frac{\rho(\mathbf{X}')}{|\mathbf{X}-\mathbf{X}'|}, \qquad (1.2.11)$$

where $G \equiv G_N$ is Newton's gravitational constant. The potential energy $\varphi(\mathbf{X})$ satisfies the equation[23]

$$\partial_j\partial^j\,\varphi(\mathbf{X}) = 4\pi\,G\,\rho(\mathbf{X}). \qquad (1.2.12)$$

Now we introduce a source $S_{\mu\nu}$ term to the field equation in (1.2.3) and write, more generally,

$$R_{\mu\nu} - \frac{1}{2}g_{\mu\nu}R = S_{\mu\nu}. \qquad (1.2.13)$$

We may rewrite this as

$$R_{\mu\nu} = \left(S_{\mu\nu} - \frac{1}{2}g_{\mu\nu}S\right), \qquad (1.2.14)$$

where $S = S^\mu{}_\mu$. In particular

$$R_{00} = \left(S_{00} - \frac{1}{2}g_{00}S\right). \qquad (1.2.15)$$

---

[23]We have used the familiar relation $\partial_j\partial^j\left(1/|\mathbf{X}-\mathbf{X}'|\right) = -4\pi\,\delta^3(\mathbf{X}-\mathbf{X}')$.

From (1.1.50), R_{00} may be written in terms of components of the Riemann tensor as

$$R_{00} = R^0{}_{000} + R^i{}_{0i0} = -R^i{}_{00i},\qquad (1.2.16)$$

where, in writing the last equality, we have, from (1.1.42), used the facts that $R^{00}{}_{\mu\nu} = 0$, $R^{i0}{}_{j0} = -R^{i0}{}_{0j}$.

We further consider the equation of geodesic deviation equation in (1.1.38) for $\mu = i$, which may be rewritten as

$$\frac{D^2 \delta x^i}{d\tau^2} = R^i{}_{00j}(x)\, \delta x^j \frac{dx^0}{d\tau} \frac{dx^0}{d\tau} + \cdots ,\qquad (1.2.17)$$

while the Newtonian one is given by

$$\frac{d^2 \delta X^i}{dt^2} = -\frac{1}{c^2}\left(\partial^i \partial_j \varphi(\mathbf{X})\right) \delta X^j \frac{dX^0}{d\tau} \frac{dX^0}{d\tau}.\qquad (1.2.18)$$

Working in units, in which $c = 1$, we may, upon comparison of these two equations and making use, in the process, of (1.2.12) and (1.2.16), that we may take $R_{00} \approx 4\pi G\, T_{00}$, where we have identified the mass density with T_{00} of the energy-momentum tensor associated with matter. We may identify S_{00} in (1.2.15) with $4\pi G\, T_{00}$ by noting that with $g_{00} \approx -1$, $T = -T_{00} + T_{ii} \approx -T_{00}$, with T_{ii} negligible in comparison to T_{00}, for low speeds of the underlying matter particles, we obtain for the right-hand side of (1.2.15) the expression $8\pi G\, (T_{00}/2)$. Hence as a tensor equation, the field equation (1.2.14) is taken as

$$R_{\mu\nu} = 8\pi G \left(T_{\mu\nu} - \frac{1}{2} g_{\mu\nu} T\right),\qquad (1.2.19)$$

or equivalently as

$$R_{\mu\nu} - \frac{1}{2} g_{\mu\nu} R = 8\pi G\, T_{\mu\nu},\qquad (1.2.20)$$

which is the celebrated Einstein's equation of general relativity.

The above doesn't yet tell us how to generate the energy-momentum tensor and, in turn, define the parameter κ in (1.2.1). To this end, we introduce a Lagrangian density for matter, and generalize the expression of the action in (1.2.1) to

$$W = \frac{1}{\kappa^2} \int (dx)\sqrt{-g(x)}\, R(x) + \int (dx)\sqrt{-g(x)}\, \mathscr{L}_{\text{matter}}(x).\qquad (1.2.21)$$

In order to obtain (1.2.20), we define the energy-momentum tensor of matter by

$$T_{\mu\nu}(x) = \frac{-2}{\sqrt{-g(x)}}\frac{\delta}{\delta g^{\mu\nu}(x)} \int (dx')\sqrt{-g(x')}\, \mathscr{L}_{\text{matter}}(x'),\qquad (1.2.22)$$

which leads from (1.2.1), (1.2.2), (1.2.21) to a parameter κ given by

$$\kappa^2 = 16\pi G. \tag{1.2.23}$$

in units $\hbar = 1$, $c = 1$. Re-inserting $\hbar$ and c, the fundamental Planck length is given by $\sqrt{\hbar G/c^3} \approx 1.616 \times 10^{-33}$ cm, with $G \equiv G_N = 6.709 \times 10^{-39}\, \hbar c^5/\mathrm{GeV}^2$.

For example, for a scalar field $\phi(x)$, we may take for the action for matter the integral

$$W_{\text{matter}} = -\frac{1}{2} \int (dx)\sqrt{-g}\,(g^{\mu\nu}\partial_\mu\phi\,\partial_\nu\phi + m^2\phi^2). \tag{1.2.24}$$

The corresponding energy-momentum tensor is derived in Problem 1.11.

Immediate generalizations of the Einstein-Hilbert Action in (1.2.1), are obtained by adding such terms to it as

$$\int \sqrt{-g}\,(dx)\,\lambda, \quad \int \sqrt{-g}\,(dx)\,R^2, \quad \int \sqrt{-g}\,(dx)\,R^{\mu\nu}R_{\mu\nu}, \quad \dots, \tag{1.2.25}$$

up to proportionality constants, where λ is referred to as a cosmological constant. The remaining actions are of higher orders in derivatives than the Einstein-Hilbert action and are more problematic in conventional field theory studies.[24] We will consider such terms in Sect. 1.8.

1.3   Quantum Particle Aspect of Gravitation: The Graviton and Polarization Aspects

The quantum particle associated with the gravitational field, the so-called graviton, emerges by considering the small fluctuation of the metric about the Minkowski one as the limit of the full metric where the gravitational field becomes weaker and the particle becomes identified.

We consider the action of general relativity,

$$W = \int (dx)\sqrt{-g}\, g^{\mu\nu}R_{\mu\nu}, \qquad g = \det[g_{\mu\nu}], \tag{1.3.1}$$

$$R_{\mu\nu} = \partial_\rho\Gamma_{\mu\nu}{}^\rho - \partial_\nu\Gamma_{\mu\rho}{}^\rho + \Gamma_{\rho\sigma}{}^\rho\Gamma_{\mu\nu}{}^\sigma - \Gamma_{\nu\sigma}{}^\rho\Gamma_{\mu\rho}{}^\sigma, \tag{1.3.2}$$

$$\Gamma_{\mu\nu}{}^\rho = \frac{1}{2}g^{\rho\sigma}\big(\partial_\mu g_{\nu\sigma} + \partial_\nu g_{\mu\sigma} - \partial_\sigma g_{\mu\nu}\big), \tag{1.3.3}$$

---

[24]Much work has been done with higher order derivatives and, in general, problems arise with unitarity (positivity) problems in conventional perturbative field theories methods, see, e.g., [59, 63].

where for the simplicity of the notation we have set the parameter $\kappa = 1$. One may restore it at any stage. The fluctuation about the Minkowski metric is obtained by writing

$$g_{\mu\nu} = \eta_{\mu\nu} + h_{\mu\nu}. \tag{1.3.4}$$

One eventually then carries out an expansion of the action in "powers" of $h_{\mu\nu}$ by using, in the process, the Minkowski metric to lower and raise indices. Details of such expansions are summarized in Box 1.1.

We may write $\det[g_{\mu\nu}]$ as $\det[\eta_{\alpha\beta}] \det[\delta^\mu{}_\nu + \eta^{\mu\sigma}h_{\sigma\nu}]$ and note that $\eta^{\mu\sigma}h_{\sigma\nu} = h^\mu{}_\nu$, where we have used the relation $\eta^{\mu\nu}\eta_{\nu\lambda} = \delta^\mu{}_\lambda$. The expansion of $\sqrt{-\det[g_{\mu\nu}]}$ is easily obtained upon expanding the logarithm in the following expression and taking the trace of the resulting matrix

$$\sqrt{-\det[g_{\mu\nu}]} = \exp\frac{1}{2}(\mathrm{Tr}\ln[\delta^\mu{}_\nu + h^\mu{}_\nu]) = \exp\frac{1}{2}\left(h^\mu{}_\mu - \frac{1}{2}h^\mu{}_\nu h^\nu{}_\mu + \mathcal{O}(h^3)\right). \tag{1.3.5}$$

These useful expansions are readily generated, where $h = h^\mu{}_\mu$, and we have used the fact that $g^{\mu\lambda}g_{\lambda\nu} = \delta^\mu{}_\nu$.

Box 1.1: Expansion of basic terms in "powers" of $h_{\mu\nu}$

$$\sqrt{-g} = 1 + \frac{1}{2}h + \frac{1}{8}h^2 - \frac{1}{4}h_{\mu\nu}h^{\mu\nu} + \mathcal{O}(h^3),$$

$$g^{\mu\nu} = \eta^{\mu\nu} - h^{\mu\nu} + h^{\mu\lambda}h_\lambda{}^\nu + \mathcal{O}(h^3),$$

$$\Gamma_{\mu\nu}{}^\rho = \frac{1}{2}(\partial_\mu h_\nu{}^\rho + \partial_\nu h_\mu{}^\rho - \partial^\rho h_{\mu\nu}) + \mathcal{O}(h^2),$$

$$R_{\mu\nu} = R^{(1)}_{\mu\nu} + R^{(2)}_{\mu\nu} + \mathcal{O}(h^3), \quad \partial_\mu\Gamma_{\nu\rho}{}^\rho = \frac{1}{2}\partial_\mu\partial_\nu h + \mathrm{O}(h^2),$$

$$R^{(1)}_{\mu\nu} = \frac{1}{2}\left(\partial^\rho\partial_\mu h_{\nu\rho} + \partial^\rho\partial_\nu h_{\mu\rho} - \Box h_{\mu\nu} - \partial_\mu\partial_\nu h\right),$$

$$\eta^{\mu\nu}R^{(2)}_{\mu\nu} = \frac{1}{2}\partial_\lambda h\,\partial_\rho h^{\rho\lambda} - \frac{1}{4}\partial_\rho h\partial^\rho h - \frac{1}{2}\partial_\rho h_{\sigma\lambda}\partial^\sigma h^{\rho\lambda} + \frac{1}{4}\partial^\rho h_{\sigma\lambda}\partial_\rho h^{\sigma\lambda},$$

up to a total derivative.

To second order, we have (see Box 1.1)

$$\sqrt{-g}\,g^{\mu\nu}R_{\mu\nu} = -\left(h^{\mu\nu} - \frac{1}{2}\eta^{\mu\nu}h\right)R^{(1)}_{\mu\nu} + \eta^{\mu\nu}R^{(2)}_{\mu\nu} + \mathrm{O}(h^3), \tag{1.3.6}$$

up to a total derivative. This leads to,

$$\sqrt{-g}\, g^{\mu\nu} R_{\mu\nu} = \frac{1}{2}\left(-\frac{1}{2}\partial^\sigma h_{\mu\nu}\,\partial_\sigma h^{\mu\nu} + \partial_\mu h^{\mu\sigma}\,\partial_\nu h^\nu{}_\sigma - \partial_\sigma h^{\sigma\mu}\,\partial_\mu h + \frac{1}{2}\partial^\mu h\,\partial_\mu h\right),$$

(1.3.7)

to second order, again up to a total derivative. From Appendix II, at the end of this volume, we recognize the expression between the round brackets in (1.3.7) as the Lagrangian density of a massless spin 2-particle.

Covariant descriptions of the graviton in second order and first order formalisms are dealt with in the following two subsections, respectively.

1.3.1   Second Order Covariant Formalism

The Lagrangian density in (1.3.7) is, up to a total derivative, invariant under the transformation

$$h^{\mu\nu} \rightarrow h^{\mu\nu} + \partial^\mu \xi^\nu + \partial^\nu \xi^\mu,$$

(1.3.8)

or equivalently under the transformation of the combination

$$\left(h^{\mu\nu} - \frac{1}{2}\eta^{\mu\nu}h\right) \rightarrow \left(h^{\mu\nu} - \frac{1}{2}\eta^{\mu\nu}h\right) + \partial^\mu \xi^\nu + \partial^\nu \xi^\mu - \eta^{\mu\nu}\partial\xi.$$

(1.3.9)

Accordingly, upon taking the ∂_μ derivative of the above equation, we may infer that one may always choose $\Box\,\xi^\nu = -\partial_\mu(h^{\mu\nu} - \frac{1}{2}\eta^{\mu\nu}h)$ so that the *new* transformed combination satisfies the gauge condition

$$\partial_\mu\left(h^{\mu\nu} - \frac{1}{2}\eta^{\mu\nu}h\right) = 0,$$

(1.3.10)

referred often as the harmonic gauge .

We provide a covariant description of the graviton and work with the following convenient constraint:

$$\partial_\mu\left(h^{\mu\nu} - \frac{1}{2}\eta^{\mu\nu}h\right) = \chi^\nu,$$

(1.3.11)

for some vector field χ^ν. This constraint may be derived directly from a Lagrangian density having the following structure

$$\mathscr{L} = \mathscr{L}_G + h^{\mu\nu}T_{\mu\nu} - 2\chi_\nu\,\partial_\mu\left(h^{\mu\nu} - \frac{1}{2}\eta^{\mu\nu}h\right) + \chi_\mu\,\chi^\mu,$$

(1.3.12)

where $\mathscr{L}_G$ is the Lagrangian density *within* the round brackets in (1.3.7) for a spin 2 massless particle. We have added an external source coupling to the field $h^{\mu\nu}$ which simplifies the work quite a bit.

Variation with respect to the field $h^{\mu\nu}$ gives

$$-\Box h_{\mu\nu} + \partial_\mu \partial^\lambda h_{\lambda\nu} + \partial_\nu \partial^\lambda h_{\lambda\mu} - \partial_\mu \partial_\nu h + \eta_{\mu\nu}(\Box h - \partial_\sigma \partial_\lambda h^{\sigma\lambda})$$
$$= T_{\mu\nu} + (\partial_\mu \chi_\nu + \partial_\nu \chi_\mu - \eta_{\mu\nu}\partial_\sigma \chi^\sigma). \tag{1.3.13}$$

Upon taking the ∂^μ-derivative of the later equation gives

$$\partial^\mu(\partial_\mu \chi_\nu + \partial_\nu \chi_\mu - \eta_{\mu\nu}\partial_\sigma \chi^\sigma) = \Box \chi_\nu = -\partial^\mu T_{\mu\nu}, \tag{1.3.14}$$

On the other hand, a variation of the Lagrangian density $\mathscr{L}$ with respect to χ_ν, gives the constraint in (1.3.11). Upon replacing χ^ν by $\partial_\mu(h^{\mu\nu} - \eta^{\mu\nu}h/2)$ everywhere in (1.3.13), leads to the basic equation

$$-\Box\left(h^{\mu\nu} - \frac{1}{2}\eta^{\mu\nu}h\right) = T^{\mu\nu}, \tag{1.3.15}$$

or equivalently to

$$-\Box h^{\mu\nu} = \left(T^{\mu\nu} - \frac{1}{2}\eta^{\mu\nu}T\right) = \frac{(\eta^{\mu\sigma}\eta^{\nu\rho} + \eta^{\mu\rho}\eta^{\nu\sigma} - \eta^{\mu\nu}\eta^{\sigma\rho})}{2}T_{\sigma\rho}. \tag{1.3.16}$$

The graviton propagator may be simply read from this equation to be

$$\Delta^{\mu\nu\sigma\rho}(x-x') = \int \frac{(dp)}{(2\pi)^4} e^{ip(x-x')} \Delta^{\mu\nu\sigma\rho}(p), \tag{1.3.17}$$

$$\Delta^{\mu\nu\sigma\rho}(p) = \frac{(\eta^{\mu\sigma}\eta^{\nu\rho} + \eta^{\mu\rho}\eta^{\nu\sigma} - \eta^{\mu\nu}\eta^{\sigma\rho})}{2(p^2 - i\epsilon)}. \tag{1.3.18}$$

The propagator $\Delta^{\mu\nu\sigma\rho}(x-x')$, in turn, satisfies

$$-\Box \Delta^{\mu\nu\sigma\rho}(x-x') = \frac{(\eta^{\mu\sigma}\eta^{\nu\rho} + \eta^{\mu\rho}\eta^{\nu\sigma} - \eta^{\mu\nu}\eta^{\sigma\rho})}{2}\delta^{(4)}(x-x'). \tag{1.3.19}$$

One may solve for $h^{\mu\nu}(x)$ in (1.3.16) to obtain

$$h^{\mu\nu}(x) = \int (dx')\Delta_+(x-x')\frac{(\eta^{\mu\sigma}\eta^{\nu\rho} + \eta^{\mu\rho}\eta^{\nu\sigma} - \eta^{\mu\nu}\eta^{\sigma\rho})}{2}T_{\sigma\rho}(x'). \tag{1.3.20}$$

We introduce a causally arranged conserved detector source $\tilde{T}_{\mu\nu}(x)$, with $x^0 > x'^0$, to obtain from (1.3.20), the expression

$$i\int (dx)\tilde{T}_{\mu\nu}(x)h^{\mu\nu}(x)$$

$$= \int \frac{d^3\mathbf{p}}{(2\pi)^3 2p^0}[i\,\tilde{T}^*_{\mu\nu}(p)]\frac{(\eta^{\mu\sigma}\eta^{\nu\rho} + \eta^{\mu\rho}\eta^{\nu\sigma} - \eta^{\mu\nu}\eta^{\sigma\rho})}{2}[i\,T_{\sigma\rho}(p)], \quad p^0 = |\mathbf{p}|.$$

$$\tag{1.3.21}$$

We may introduce a completeness relation in 4 dimensions as follows[25]

$$\eta^{\mu\nu} = \frac{(p+\bar{p})^{\mu}(p+\bar{p})^{\nu}}{(p+\bar{p})^2} + \frac{(p-\bar{p})^{\mu}(p-\bar{p})^{\nu}}{(p-\bar{p})^2} + \sum_{\lambda=1,2} e_{\lambda}^{\mu} e_{\lambda}^{\nu}, \quad p^2 = 0,$$

(1.3.22)

expanded in terms of the orthogonal system $\{(p+\bar{p})^{\mu}, (p-\bar{p})^{\mu}, e_1^{\mu}, e_2^{\mu}\}$, with $p = (p^0, \mathbf{p})$, $\bar{p} = (p^0, -\mathbf{p})$, and with e_1^{μ}, e_2^{μ} real orthonormal vectors. (1.3.22) may be simplified further to

$$\eta^{\mu\nu} = \frac{p^{\mu}\bar{p}^{\nu} + \bar{p}^{\mu}p^{\nu}}{p\bar{p}} + \sum_{\lambda=1,2} e_{\lambda}^{\mu} e_{\lambda}^{\nu},$$

(1.3.23)

For a conserved source, i.e., for $p^{\mu} T_{\mu\nu}(p) = 0$, $p^{\nu} T_{\mu\nu}(p) = 0$, we may replace expression (1.3.23) for the metric in (1.3.21), and effectively make the substitution

$$\frac{(\eta^{\mu\sigma}\eta^{\nu\rho} + \eta^{\mu\rho}\eta^{\nu\sigma} - \eta^{\mu\nu}\eta^{\sigma\rho})}{2} \to \sum_{\lambda,\lambda'=1,2} e_{\lambda\lambda'}^{\mu\nu} e_{\lambda\lambda'}^{\sigma\rho},$$

(1.3.24)

where

$$e_{\lambda\lambda'}^{\mu\nu} = \frac{1}{2}\left[e_{\lambda}^{\mu} e_{\lambda'}^{\nu} + e_{\lambda'}^{\mu} e_{\lambda}^{\nu} - \delta_{\lambda\lambda'} \sum_{\gamma=1,2} e_{\gamma}^{\mu} e_{\gamma}^{\nu} \right],$$

(1.3.25)

as obtained by using, in the process, (1.3.23). On the other hand, we may write

$$\sum_{\lambda,\lambda'=1,2} e_{\lambda\lambda'}^{\mu\nu} e_{\lambda\lambda'}^{\sigma\rho} = \sum_{\xi=1,2} \varepsilon_{\xi}^{\mu\nu} \varepsilon_{\xi}^{\sigma\rho},$$

(1.3.26)

$$\varepsilon_1^{\mu\nu} = \frac{1}{\sqrt{2}}(e_{11}^{\mu\nu} - e_{22}^{\mu\nu}) = \sqrt{2}\, e_{11}^{\mu\nu} = -\sqrt{2}\, e_{22}^{\mu\nu},$$

(1.3.27)

$$\varepsilon_2^{\mu\nu} = \frac{1}{2}(e_{12}^{\mu\nu} + e_{21}^{\mu\nu}) = \sqrt{2}\, e_{12}^{\mu\nu} = \sqrt{2}\, e_{21}^{\mu\nu}.$$

(1.3.28)

---

[25]Such a completeness relation is conveniently described by Schwinger [56, pp. 15–17]. See also Appendix II at the end of this volume.

Thus (1.3.21) takes the form

$$
i \int (dx)\tilde{T}_{\mu\nu}(x)h^{\mu\nu}(x) = \int \sum_{\xi=1,2} \frac{d^3\mathbf{p}}{(2\pi)^3 2p^0}[i\,\varepsilon_\xi^{\mu\nu}\tilde{T}_{\mu\nu}^*(p)][i\,\varepsilon_\xi^{\sigma\rho}T_{\sigma\rho}(p)]. \tag{1.3.29}
$$

Here $[i\,\varepsilon_\xi^{\sigma\rho}T_{\sigma\rho}(p)]$ is an amplitude that the source has emitted a graviton of momentum p and polarization specified by ξ, for different p, ξ in (1.3.27), with $h^{\mu\nu}(x)$ denoting the field generated at point x in the forward time direction of the source. For $\partial^\mu T_{\mu\nu} = 0$, we note that χ_ν in (1.3.14) satisfies the equation of a free field.

In a momentum description, with propagation vector $\mathbf{k} = |\mathbf{k}|(0,0,1)$, and $e_1^\mu = (0,1,0,0)$, $e_2^\mu = (0,0,1,0)$, (1.3.25)–(1.3.28) give, with $\xi = 1,2$,

$$
\varepsilon_1^{11} = \frac{1}{\sqrt{2}} = -\varepsilon_1^{22}, \qquad \varepsilon_2^{12} = \frac{1}{\sqrt{2}} = \varepsilon_2^{21}, \tag{1.3.30}
$$

with all the other components of $\varepsilon_\xi^{\mu\nu}$ equal to zero.

The components of the polarization tensors ε_2^{ab}, $a,b = 1,2$, may be considered in reference to a coordinate system rotated clockwise by an angle $45°$ of the original coordinate system defined by the $x = x^1 - x^2 = y$−plane. Remembering that a second rank tensor transforms as the product of two vectors, then the corresponding components are given by the following expressions in this new coordinate system ($\sin\theta = \cos\theta = 1/\sqrt{2}$)

$$
\varepsilon_2'^{11} = \varepsilon_2^{11}\cos^2\theta + \varepsilon_2^{22}\sin^2\theta - \varepsilon_2^{12}\sin\theta\cos\theta - \varepsilon_2^{21}\sin\theta\cos\theta = -1/\sqrt{2}, \tag{1.3.31}
$$

$$
\varepsilon_2'^{22} = \varepsilon_2^{11}\sin^2\theta + \varepsilon_2^{22}\cos^2\theta + \varepsilon_2^{12}\sin\theta\cos\theta + \varepsilon_2^{21}\sin\theta\cos\theta = +1/\sqrt{2}, \tag{1.3.32}
$$

$$
\varepsilon_2'^{12} = \varepsilon_2^{11}\sin\theta\cos\theta - \varepsilon_2^{22}\sin\theta\cos\theta + \varepsilon_2^{12}\cos^2\theta - \varepsilon_2^{21}\sin^2\theta = 0, \tag{1.3.33}
$$

$$
\varepsilon_2'^{21} = \varepsilon_2^{11}\sin\theta\cos\theta - \varepsilon_2^{22}\sin\theta\cos\theta - \varepsilon_2^{12}\sin^2\theta + \varepsilon_2^{21}\cos^2\theta = 0, \tag{1.3.34}
$$

Thus,

$$
\left(\varepsilon_1^{11} = \frac{1}{\sqrt{2}} = -\varepsilon_1^{22}\right), \quad \left(\varepsilon_2'^{11} = -\frac{1}{\sqrt{2}} = -\varepsilon_2'^{22}\right), \tag{1.3.35}
$$

define the two polarization states of gravitons. These are pictorially represented in Fig. 1.5. The deformation of a ring of test particles in such a gravitational field is formally depicted in Fig. 1.6.

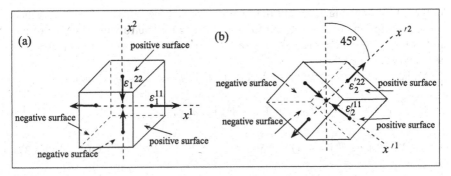

Fig. 1.5 $\left(\varepsilon_1^{11} = 1/\sqrt{2} = -\varepsilon_1^{22}\right)$, $\left(\varepsilon_2'^{11} = -1/\sqrt{2} = -\varepsilon_2'^{22}\right)$ define the two polarization states of a graviton propagating along the x^3—direction. The polarizations $\left(\varepsilon_2'^{11}, \varepsilon_2'^{22}\right)$ may be considered to be described in a coordinate system $x'^1 - x'^2$ in (**b**) obtained by a $45°$ c.w. rotation of the coordinate system $x^1 - x^2$ in (**a**). Note, for example, that since ε_1^{11} in (**a**) is positive, then at a negative surface the direction of the *arrow* must be in the negative x^1- direction to ensure that ε_1^{11} is positive, and so on for the directions of the other *arrows*

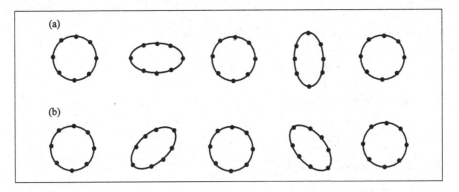

Fig. 1.6   The deformation of a ring of test particles in such a gravitational field is formally depicted at different cycles for the two polarizations in (1.3.35), respectively, corresponding to parts (**a**) and (**b**)

1.3.2   First Order Formalism

The purpose of this subsection is to introduce a first order formulation of linearized gravity.

To the above end, we introduce two fields $\xi^{\mu\nu}$, $\Gamma_{\mu\nu}{}^\lambda$ which are symmetric in the indices μ, ν. As before the Minkowski metric is given by $[\eta_{\mu\nu}] = \text{diag}[-1, 1, 1, 1]$. We define an action involving the two fields as follows

$$\mathscr{A} = \int (\mathrm{d}x)\, \xi^{\mu\nu} \left(\partial_\mu \Gamma_{\rho\nu}{}^\rho - \partial_\rho \Gamma_{\mu\nu}{}^\rho\right) + \eta^{\mu\nu}\left(\Gamma_{\lambda\rho}{}^\rho \Gamma_{\mu\nu}{}^\lambda - \Gamma_{\lambda\mu}{}^\rho \Gamma_{\rho\nu}{}^\lambda\right.$$

$$\left. + \partial_\rho \Gamma_{\mu\nu}{}^\rho - \partial_\mu \Gamma_{\rho\nu}{}^\rho\right), \tag{1.3.36}$$

proposed by Schwinger, except that we have added the last two total divergence terms, multiplying $\eta^{\mu\nu}$, at the end for convenience for later treatments. One of the tasks of this subsection is to show that this action leads to the equations of a massless spin two field.

Variation of the action with respect to the field $\Gamma_{\alpha\beta}{}^{\gamma}$ gives the equation

$$\frac{1}{2}(\partial_\mu \xi^{\mu\beta} \delta^{\alpha}{}_{\gamma} + \partial_\nu \xi^{\nu\alpha} \delta^{\beta}{}_{\gamma}) - \partial_\gamma \xi^{\alpha\beta}$$

$$+ (\Gamma_\gamma{}^{\beta\alpha} + \Gamma_\gamma{}^{\alpha\beta}) - \frac{1}{2}(\Gamma_\mu{}^{\mu\alpha} \delta^{\beta}{}_{\gamma} + \Gamma_\mu{}^{\mu\beta} \delta^{\alpha}{}_{\gamma}) - \Gamma_{\gamma\rho}{}^{\rho} \eta^{\alpha\beta} = 0. \qquad (1.3.37)$$

For $\alpha = \gamma$, this gives

$$\Gamma_\mu{}^{\mu\beta} = \partial_\mu \xi^{\mu\beta}, \qquad (1.3.38)$$

and for α contracted with β, gives

$$\Gamma_{\gamma\mu}{}^{\mu} = -\frac{1}{2} \partial_\gamma \xi, \qquad \xi \equiv \xi^{\mu}{}_{\mu}. \qquad (1.3.39)$$

Upon substituting the last two equations back in (1.3.37) leads to

$$\Gamma_{\alpha\beta\gamma} + \Gamma_{\alpha\gamma\beta} = \partial_\alpha(\xi_{\beta\gamma} - \frac{1}{2} \eta_{\beta\gamma} \xi). \qquad (1.3.40)$$

This suggests to define a field

$$h_{\alpha\beta} = \xi_{\alpha\beta} - \frac{1}{2} \eta_{\alpha\beta} \xi, \qquad (h^{\alpha}{}_{\alpha} \equiv h = -\xi). \qquad (1.3.41)$$

Using the fact that we may write

$$\Gamma_{\alpha\beta\gamma} = \frac{1}{2}(\Gamma_{\alpha\beta\gamma} + \Gamma_{\alpha\gamma\beta} + \Gamma_{\beta\gamma\alpha} + \Gamma_{\beta\alpha\gamma} - \Gamma_{\gamma\alpha\beta} - \Gamma_{\gamma\beta\alpha}), \qquad (1.3.42)$$

and (1.3.40) gives

$$\Gamma_{\alpha\beta\gamma} = \frac{1}{2}(\partial_\alpha h_{\beta\gamma} + \partial_\beta h_{\alpha\gamma} - \partial_\gamma h_{\alpha\beta}). \qquad (1.3.43)$$

Finally the variation of the action with respect to $\xi^{\sigma\lambda}$, gives

$$\frac{1}{2}(\partial_\sigma \Gamma_{\rho\lambda}{}^{\rho} + \partial_\lambda \Gamma_{\rho\sigma}{}^{\rho}) - \partial_\rho \Gamma_{\sigma\lambda}{}^{\rho} = 0, \qquad (1.3.44)$$

which upon using the expression for $\Gamma_{\alpha\beta\gamma}$ given in (1.3.43) in the above equation, leads to ($\Box = \eta^{\mu\nu} \partial_\mu \partial_\nu$)

$$- \Box h_{\sigma\lambda} + \partial_\sigma \partial_\rho h_\lambda{}^{\rho} + \partial_\lambda \partial_\rho h_\sigma{}^{\rho} - \partial_\sigma \partial_\lambda h = 0, \qquad (1.3.45)$$

and to its contracted version

$$\Box\, h - \partial_\alpha \partial_\beta h^{\alpha\beta} = 0. \tag{1.3.46}$$

Upon multiplying the latter equation by $\eta_{\sigma\lambda}$ and adding the resulting equation to the one in (1.3.45) leads to

$$-\Box\, \xi_{\sigma\lambda} + \partial_\sigma \partial^\mu \xi_{\mu\lambda} + \partial_\lambda \partial^\mu \xi_{\mu\sigma} - \eta_{\sigma\lambda} \partial_\mu \partial_\nu \xi^{\mu\nu} = 0, \quad \xi_{\sigma\lambda} = h_{\sigma\lambda} - \frac{1}{2}\eta_{\sigma\lambda} h, \tag{1.3.47}$$

and (1.3.46) becomes replaced by

$$\Box\, \xi + 2\,\partial_\mu \partial_\nu \xi^{\mu\nu} = 0. \tag{1.3.48}$$

We note that the action $\mathscr{A}$ in (1.3.36) is invariant under the gauge transformations[26]

$$\delta\, \xi^{\mu\nu} = \partial^\mu \Lambda^\nu + \partial^\nu \Lambda^\mu - \eta^{\mu\nu} \partial_\alpha \Lambda^\alpha, \quad \delta\, \Gamma_{\alpha\beta\gamma} = \partial_\alpha \partial_\beta \Lambda_\gamma \tag{1.3.49}$$

up to a total divergence in its integrand. We may rewrite

$$\Lambda^\mu = \left(\eta^{\mu\nu} - \frac{\partial^\mu \partial^\nu}{\Box}\right)\Lambda_\nu + \partial^\mu \frac{\partial \cdot \Lambda}{\Box} \equiv \Lambda_{\rm T}^\mu + \Lambda_{\rm L}^\mu,$$

$$\partial \cdot \Lambda_{\rm T} = 0, \quad \partial \cdot \Lambda = \partial \cdot \Lambda_{\rm L}. \tag{1.3.50}$$

We choose ($\xi = \xi^\mu{}_\mu$)

$$\partial \cdot \Lambda_{\rm L} = \frac{1}{2}\xi, \quad \text{i.e.,} \quad \Box\, \Lambda_{\rm L}^\nu = \partial^\nu \partial \cdot \Lambda_{\rm L} = \frac{1}{2}\partial^\nu \xi, \tag{1.3.51}$$

$$\Box\, \Lambda_{\rm T}^\nu = -\left(\partial_\mu \xi^{\mu\nu} + \frac{1}{2}\partial^\nu \xi\right), \tag{1.3.52}$$

where note that the latter equation is transversal and satisfies equation (1.3.48) upon taking the ∂_ν derivative, as it should. Thus the choice of gauge leads to a *new* field, which we also denote by $\xi^{\mu\nu}$ for simplicity of the notation, satisfying

$$\Box\, \xi_{\sigma\lambda} = 0, \quad (\partial_\mu \xi^{\mu\nu} = 0, \ \partial_\nu \xi^{\mu\nu} = 0, \ \xi = 0), \tag{1.3.53}$$

consistent with two polarization states of the graviton.

Note that the gauge transformation in (1.3.49), in general, provides an effective scaling factor for the combination $(\eta^{\mu\nu} - \xi^{\mu\nu})$ for a small fluctuation $\xi^{\mu\nu}$ about the

---

[26] See Problem 1.12.

Minkowski metric:

$$(\eta^{\mu\nu} - \xi^{\mu\nu}) + \delta(\eta^{\mu\nu} - \xi^{\mu\nu}) \simeq (1 + \partial_\alpha \Lambda^\alpha)(\eta^{\mu\nu} - \xi^{\mu\nu} - \partial^\mu \Lambda^\nu - \partial^\nu \Lambda^\mu), \quad (1.3.54)$$

unlike the transformation of $(\eta^{\mu\nu} - h^{\mu\nu})$,

$$(\eta^{\mu\nu} - h^{\mu\nu}) + \delta(\eta^{\mu\nu} - h^{\mu\nu}) \simeq (\eta^{\mu\nu} - h^{\mu\nu} - \partial^\mu \Lambda^\nu - \partial^\nu \Lambda^\mu). \quad (1.3.55)$$

The simplest application of the graviton propagator derived in (1.3.18) is of graviton emission by a given source. This is discussed next.

1.3.3   The Quanta of Gravitation in Evidence: Graviton Emission and Gravitational Radiation

As we have seen in the beginning of this section, the quantum particle associated with the gravitational field, the so-called graviton, emerges by considering the small fluctuation of the metric about the Minkowski one as the limit of the full metric where the gravitational field becomes weaker and the particle becomes identified.[27]

The vacuum-to-vacuum transition in the presence of a conserved external source $T^{\mu\nu} = T^{\nu\mu}$, $\partial_\mu T^{\mu\nu} = 0$ in a covariant gauge, as obtained from (1.3.20), reads

$$\langle\, 0_+ | 0_- \rangle$$

$$= \exp\left[\frac{i}{2}\int (dx)(dx')T^{\mu\nu}(x)\left[\eta_{\mu\sigma}\eta_{\nu\rho} - \frac{1}{2}\eta_{\mu\nu}\eta_{\sigma\rho}\right](8\pi G)D_+(x-x')T^{\sigma\rho}(x'),$$

$$(1.3.56)$$

$$D_+(x-x') = \int \frac{(dp)}{(2\pi)^4}\, e^{ip(x-x')}D_+(p) = \int \frac{(dp)}{(2\pi)^4}\frac{e^{ip(x-x')}}{p^2 - i\epsilon}. \quad (1.3.57)$$

The numerical factor $(8\pi G)$, multiplying $D_+(x-x')$, is introduced in conformity with Newton's law of gravitation and follows by considering $\approx$ static sources, $T^{\mu\nu} \approx \eta^{\mu 0}\eta^{\nu 0}T^{00}$, $T_1^{00}(\mathbf{x}) = M_1\delta^3(\mathbf{x} - \mathbf{R})$, $T_2^{00}(\mathbf{x}') = M_2\delta^3(\mathbf{x}' - \mathbf{R}')$, leading to[28]

$$\int d^3\mathbf{x}\, d^3\mathbf{x}'\, T^{00}(\mathbf{x})\left[g_{00}\,g_{00} - \frac{1}{2}g_{00}\,g_{00}\right]T^{00}(\mathbf{x}')$$

$$\times (8\pi G)\int \frac{d^3\mathbf{p}}{(2\pi)^3}\, e^{i\mathbf{p}\cdot(\mathbf{x}-\mathbf{x}')}\, D_+(p)\Big|_{p^0=0} = \frac{GM_1 M_2}{|\mathbf{R} - \mathbf{R}'|}, \quad G \equiv G_N. \quad (1.3.58)$$

---

[27]Gravitational waves have been detected from the merger of two black holes 1.3 billion light-years from the Earth via the Laser Interferometer Gravitational Wave Observatory (LIGO). See B. P. Abbott et al.: Phys. Rev. Lett. 116, 061102 (2016), Astrophys. J. Lett. 818, L22 (2016).

[28]See also (5.10.56) in Vol. I [43], as well as [57].

For a conserved source $T^{\mu\nu}$: $p_\mu T^{\mu\nu}(p) = 0$, in the momentum description, and in computing the vacuum-to vacuum transition probability, as obtained from (1.3.56), we may, from (1.3.24), (1.3.26), replace $\left[\eta^{\mu\sigma} \eta^{\nu\rho} - \frac{1}{2} \eta^{\mu\nu} \eta^{\sigma\rho} \right]$ by $\sum_{\xi=1,2} \varepsilon_\xi^{\mu\nu} \varepsilon_\xi^{\sigma\rho}$, where $\varepsilon_1^{\sigma\rho}, \varepsilon_2^{\sigma\rho}$ are graviton polarization tensors, to obtain

$$|\langle 0_+|0_-\rangle|^2 = \exp\left[-8\pi G \int \frac{d^3\mathbf{p}}{(2\pi)^3 2|\mathbf{p}|} \sum_{\xi=1,2} |\varepsilon_\xi^{\mu\nu} T_{\mu\nu}(p)|^2 \right], \tag{1.3.59}$$

satisfying the necessary condition $|\langle 0_+ | 0_-\rangle|^2 \le 1$.

Almost an identical analysis as the one carries out for photons emission my an external source,[29] the probability that a source $T^{\mu\nu}$ emits N gravitons of arbitrary momenta and polarizations is given by

$$\text{Prob}[N] = \frac{1}{N!} \left(8\pi G \int \frac{d^3\mathbf{p}}{(2\pi)^3 2|\mathbf{p}|} \sum_{\xi=1,2} |\varepsilon_\xi^{\mu\nu} T_{\mu\nu}(p)|^2 \right)^N |\langle 0_+ | 0_-\rangle|^2, \tag{1.3.60}$$

which represents a Poisson distribution, written in a convenient notation,

$$\text{Prob}[N] = \frac{(\langle N \rangle)^N}{N!} e^{-\langle N \rangle}, \tag{1.3.61}$$

with $\langle N \rangle$ denoting the average number of gravitons emitted given by

$$\langle N \rangle = 8\pi G \int \frac{d^3\mathbf{p}}{(2\pi)^3 2|\mathbf{p}|} \sum_{\xi=1,2} |\varepsilon_\xi^{\mu\nu} T_{\mu\nu}(p)|^2, \quad p^0 = |\mathbf{p}|, \tag{1.3.62}$$

and

$$|\langle 0_+ | 0_-\rangle|^2 = \exp[-\langle N \rangle]. \tag{1.3.63}$$

Accordingly, if $N(E)$ denotes the average number of gravitons emitted with energy in the interval $(E, E + dE)$, with $\int_0^\infty N(E)\, dE = \langle N \rangle$, and $E = |\mathbf{p}|$, we obtain

$$N(E) = 4\pi G \frac{|\mathbf{p}|}{(2\pi)^3} \int d\Omega \sum_{\xi=1,2} |\varepsilon_\xi^{\mu\nu} T_{\mu\nu}(p)|^2, \tag{1.3.64}$$

---

[29]See (5.5.7)–(5.5.11) in Vol. I [43]. See also Manoukian [41, 42].

and a total energy

$$\int_0^\infty EN(E)\,\mathrm{d}E \;=\; 4\pi\,\mathrm{G}\int_0^\infty \mathrm{d}|\mathbf{p}|\,\frac{|\mathbf{p}|^2}{(2\pi)^3}\int \mathrm{d}\Omega \sum_{\xi=1,2}\left|\varepsilon_\xi^{\mu\nu}T_{\mu\nu}(p)\right|^2$$

$$=\; \frac{\mathrm{G}}{\pi}\int_0^\infty \frac{\mathrm{d}p^0}{(2\pi)}\,(p^0)^2\int \mathrm{d}\Omega \sum_{\xi=1,2}\left|\varepsilon_\xi^{\mu\nu}T_{\mu\nu}(p^0,p^0\mathbf{n})\right|^2.$$

$$(1.3.65)$$

With $e_\lambda^\mu \;=\; (0,\mathbf{e}_\lambda)$ in (1.3.25), we have from (1.3.26), this volume, and from (4.7.231), (4.7.232) of Vol. I, now with e_λ^i real, $\varepsilon_\xi^{\mu\nu} \;=\; (1/2)\,(\eta^{\mu i}\eta^{\nu j}+\eta^{\mu j}\eta^{\nu i})\,\varepsilon_\xi^{ij}$

$$\sum_{\xi=1,2}\varepsilon_\xi^{ij}\varepsilon_\xi^{\ell m} \;=\; \pi^{ij,\ell m}, \qquad (1.3.66)$$

where from (4.7.128), (4.7.231)–(4.7.232) of Vol. I,

$$\pi^{ij,\ell m} \;=\; \frac{1}{2}\,[\,\pi^{i\ell}\pi^{jm}+\pi^{im}\pi^{j\ell}-\pi^{ij}\pi^{\ell m}\,], \qquad \pi^{ij} \;=\; \delta^{ij}-\frac{p^i p^j}{\mathbf{p}^2}, \qquad (1.3.67)$$

and (1.3.65) becomes

$$\int_0^\infty EN(E)\,\mathrm{d}E \;=\; \frac{\mathrm{G}}{\pi}\int_0^\infty \frac{\mathrm{d}p^0}{(2\pi)}\,(p^0)^2\int \mathrm{d}\Omega \sum_{\xi=1,2}\left|\varepsilon_\xi^{ij}T_{ij}(p^0,p^0\mathbf{n})\right|^2. \qquad (1.3.68)$$

On the other hand

$$T^{ij}(p^0,p^0\mathbf{n}) \;=\; \int \mathrm{d}x^0\,e^{ip^0 x^0}\int \mathrm{d}^3x\,e^{-ip^0\mathbf{n}\cdot\mathbf{x}}\,T^{ij}(x). \qquad (1.3.69)$$

The slow motion approximation of the constitutes of matter consists of taking[30]

$$\int \mathrm{d}^3x\,e^{-ip^0\mathbf{n}\cdot\mathbf{x}}\,T^{ij}(x) \;\simeq\; \int \mathrm{d}^3x\,T^{ij}(x), \qquad (1.3.70)$$

which allows us to write

$$p^0\,T^{ij}(p^0,p^0\mathbf{n}) \;\simeq\; i\int \mathrm{d}x^0\,e^{ip^0 x^0}\partial_0\int \mathrm{d}^3x\,T^{ij}(x). \qquad (1.3.71)$$

---

[30] See, e.g., [45, p. 989].

We subsequently use the relation

$$\int d^3\mathbf{x}\, T^{ij}(x) = \frac{1}{2}\frac{\partial^2}{\partial x^{0^2}}\int d^3\mathbf{x}\, x^i x^j\, T^{00}(x),\tag{1.3.72}$$

obtained from the conservation law of $T^{\mu\nu}$, and the integrals

$$\int d\Omega = 4\pi,\tag{1.3.73}$$

$$\int d\Omega\, \frac{p^i p^j}{\mathbf{p}^2} = \frac{4\pi}{3}\delta^{ij},\tag{1.3.74}$$

$$\int d\Omega\, \frac{p^i p^j p^\ell p^m}{\mathbf{p}^4} = \frac{4\pi}{15}\left[\delta^{i\ell}\delta^{jm} + \delta^{im}\delta^{j\ell} + \delta^{ij}\delta^{\ell m}\right],\tag{1.3.75}$$

which lead to

$$\int d\Omega \sum_{\xi=1,2}\varepsilon_\xi^{ij}\varepsilon_\xi^{\ell m} = \int d\Omega\, \pi^{ij,\ell m} = \frac{4\pi}{5}\left(\delta^{i\ell}\delta^{jm} + \delta^{im}\delta^{j\ell}\right) - \frac{8\pi}{15}\delta^{ij}\delta^{\ell m}$$

$$= \frac{8\pi}{5}\left(\frac{1}{2}\left[\delta^{i\ell}\delta^{jm} + \delta^{im}\delta^{j\ell}\right] - \frac{1}{3}\delta^{ij}\delta^{\ell m}\right).\tag{1.3.76}$$

From (1.3.66), (1.3.68), (1.3.71), (1.3.72), (1.3.76), we obtain

$$\int_0^\infty EN(E)\,dE = \frac{G}{5c^5}\int_{-\infty}^\infty dt\left[\left(\frac{d}{dt}\right)^3 Q^{ij}\right]^2,\tag{1.3.77}$$

where we have re-inserted the fundamental constant c - the speed of light, and

$$Q^{ij} = \int d^3\mathbf{x}\left[x^i x^j - \frac{\delta^{ij}}{3}|\mathbf{x}|^2\right]\frac{T^{00}}{c^2}.\tag{1.3.78}$$

The expression for the power (loss) then emerges as[31]

$$P = \frac{G}{5c^5}\left[\left(\frac{d}{dt}\right)^3 Q^{ij}\right]^2,\tag{1.3.79}$$

Note that Q^{ij}, may be expressed in terms of the mass density $\rho(x) = T^{00}(x)/c^2$. The expression in (1.3.79) is often referred to as the gravitational quadrupole radiation formula.

---

[31] This subsection is based on [38].

1.4   Quantum Fluctuation About a Background Metric

We consider quantum fluctuations about a given classical background metric $g_{\mu\nu}$ which satisfies Einstein's field equation.[32] To this end, we carry out the following replacement of the classical metric $g_{\mu\nu}$

$$g_{\mu\nu} \rightarrow g_{\mu\nu} + h_{\mu\nu}, \tag{1.4.1}$$

in the Einstein-Hilbert action in (1.2.1), suppressing the depence of the action on the parameter κ, with $h_{\mu\nu}$ now describing quantum fluctuations. This leads to the new action denoted by $\overline{W}$:

$$W = \int \sqrt{-g(x)} (\mathrm{d}x)\, R(x) \rightarrow \int \sqrt{-g(x)} (\mathrm{d}x)\, R(x) \Big|_{[g_{\mu\nu} \rightarrow (g_{\mu\nu} + h_{\mu\nu})]} \equiv \overline{W}. \tag{1.4.2}$$

From (1.2.10), we may infer that for a given background metric $g_{\mu\nu}$, the new action $\overline{W}$ is invariant under infinitesimal transformation of $h_{\mu\nu}$

$$\delta h_{\mu\nu} = (g_{\mu\lambda} + h_{\mu\lambda})\, \partial_\nu \Lambda^\lambda + (g_{\nu\lambda} + h_{\nu\lambda})\, \partial_\mu \Lambda^\lambda + \Lambda^\lambda \partial_\lambda (g_{\mu\nu} + h_{\mu\nu}). \tag{1.4.3}$$

Introducing the covariant derivative ∇_λ defined in terms of the background metric $g_{\mu\nu}$, we may rewrite the above transformation as[33]

$$\delta h_{\mu\nu} = (g_{\mu\lambda} + h_{\mu\lambda})\, \nabla_\nu \Lambda^\lambda + (g_{\nu\lambda} + h_{\nu\lambda})\, \nabla_\mu \Lambda^\lambda + \Lambda^\lambda \nabla_\lambda h_{\mu\nu}, \tag{1.4.4}$$

where in writing the last term we have used the fact that $\nabla_\lambda g_{\mu\nu} = 0$.

We now develop the theory of quantum fluctuations, about a given background metric $g_{\mu\nu}$, with the latter satisfying Einstein's field equation, as follows. We carry out the analysis up to second order in $h_{\mu\nu}$ starting from the action $\overline{W}$. In particular, we note that indices are lowered via the metric $g_{\mu\nu}$, and raised via its inverse $g^{\mu\nu}$. That is, covariance is defined in terms of the background metric. Also we note that the covariant derivative ∇_μ is defined in terms of the background metric $g_{\sigma\lambda}$ and is *independent* of $h_{\sigma\lambda}$. For simplicity of the notation, any coupling parameter has been absorbed in $h_{\mu\nu}$.

As in the entries of Box 1.1 (Sect. 1.3) we have, up to second order, in particular,

$$\sqrt{-g} \rightarrow \sqrt{-g}\,(1 + \frac{1}{2} h + \frac{1}{8} h^2 - \frac{1}{4}\, h_{\sigma\lambda}\, h^{\lambda\sigma}), \tag{1.4.5}$$

$$g^{\mu\nu} \rightarrow g^{\mu\nu} - h^{\mu\nu} + h^{\mu\lambda}\, h_\lambda{}^\nu, \tag{1.4.6}$$

---

[32] The background field method was introduced by DeWitt [21].
[33] See Problem 1.13

where $g^{\mu\lambda}g_{\lambda\nu} = \delta^\mu{}_\nu$, and from Problem 1.14, in particular,

$$\Gamma_{\mu\nu}{}^\sigma \rightarrow \Gamma_{\mu\nu}{}^\sigma + \frac{1}{2}\left(\nabla_\mu h_\nu{}^\sigma + \nabla_\nu h_\mu{}^\sigma - \nabla^\sigma h_{\mu\nu}\right) - \frac{1}{2}h^{\sigma\lambda}\left(\nabla_\mu h_{\nu\lambda} + \nabla_\nu h_{\mu\lambda} - \nabla_\lambda h_{\mu\nu}\right).$$

We will see, however, only the first correction in $\delta\Gamma_{\mu\nu}{}^\sigma$ will contribute as given in (1.4.10) below.

The investigation of the transformation rule of the Ricci tensor $R_{\mu\nu}$ is more involved. This has been considered in Problem 1.7, part (iii), to first order. Let us consider this afresh for additional clarity and for generalization to second order. Let $\delta_1\Gamma_{\mu\nu}{}^\rho + \delta_2\Gamma_{\mu\nu}{}^\rho$ denote the correction of the connection up to second order. From (1.1.49), we explicitly have up to second order, using the notation: $R_{\mu\nu} \rightarrow R_{\mu\nu} + \delta R_{\mu\nu}$,

$$\delta R_{\mu\nu} =$$

$$\left[\partial_\rho(\delta_1\Gamma_{\mu\nu}{}^\rho + \delta_2\Gamma_{\mu\nu}{}^\rho) + \Gamma_{\rho\kappa}{}^\rho(\delta_1\Gamma_{\mu\nu}{}^\kappa + \delta_2\Gamma_{\mu\nu}{}^\kappa) - \Gamma_{\rho\mu}{}^\kappa(\delta_1\Gamma_{\nu\kappa}{}^\rho + \delta_2\Gamma_{\nu\kappa}{}^\rho)\right.$$

$$\left. -\Gamma_{\rho\nu}{}^\kappa(\delta_1\Gamma_{\mu\kappa}{}^\rho + \delta_2\Gamma_{\mu\kappa}{}^\rho)\right] - \left[\partial_\nu(\delta_1\Gamma_{\mu\rho}{}^\rho + \delta_2\Gamma_{\mu\rho}{}^\rho) - \Gamma_{\nu\mu}{}^\kappa(\delta_1\Gamma_{\rho\kappa}{}^\rho + \delta_2\Gamma_{\rho\kappa}{}^\rho)\right]$$

$$+ \left[\delta_1\Gamma_{\rho\kappa}{}^\rho\delta_1\Gamma_{\mu\nu}{}^\kappa - \delta_1\Gamma_{\nu\kappa}{}^\rho\delta_1\Gamma_{\mu\rho}{}^\kappa\right], \qquad (1.4.7)$$

by conveniently re-arranging the order of the various terms, and by relabeling dummy indices. We recognize the expressions within the first two pairs of square brackets, as the difference between two covariant derivatives: $[\nabla_\rho(\delta_1\Gamma_{\mu\nu}{}^\rho + \delta_2\Gamma_{\mu\nu}{}^\rho) - \nabla_\nu(\delta_1\Gamma_{\mu\rho}{}^\rho + \delta_2\Gamma_{\mu\rho}{}^\rho)]$. Accordingly the explicit change in the Ricci tensor, up to second order, may be rewritten as

$$\delta R_{\mu\nu} = \nabla_\rho(\delta_1\Gamma_{\mu\nu}{}^\rho + \delta_2\Gamma_{\mu\nu}{}^\rho) - \nabla_\nu(\delta_1\Gamma_{\mu\rho}{}^\rho + \delta_2\Gamma_{\mu\rho}{}^\rho)$$

$$+ \delta_1\Gamma_{\rho\kappa}{}^\rho\delta_1\Gamma_{\mu\nu}{}^\kappa - \delta_1\Gamma_{\nu\kappa}{}^\rho\delta_1\Gamma_{\mu\rho}{}^\kappa. \qquad (1.4.8)$$

To first order, this reduces to the expression given in Problem 1.7 (iii) as expected. The first two terms give rise to the following contributions to the new action

$$\sqrt{-g}\, g^{\mu\nu}\, \nabla_\rho(\delta_1\Gamma_{\mu\nu}{}^\rho + \delta_2\Gamma_{\mu\nu}{}^\rho), \quad -\sqrt{-g}\, g^{\mu\nu}\, \nabla_\nu(\delta_1\Gamma_{\mu\rho}{}^\rho + \delta_2\Gamma_{\mu\rho}{}^\rho). \qquad (1.4.9)$$

From part (i) of Problem 1.5, both of these terms have the general structure

$$\sqrt{-g}\, g^{\mu\nu}\, \nabla_\mu\xi_\nu = \partial_\mu(\sqrt{-g}\, \xi^\mu),$$

and as total derivatives do not contribute to the new action. We may thus restrict the change in the connection in (1.4.8) to the first order

$$\delta_1 \Gamma_{\mu\nu}{}^{\kappa} = \frac{1}{2} \left[\nabla_{\mu} h_{\nu}{}^{\kappa} + \nabla_{\nu} h_{\mu}{}^{\kappa} - \nabla^{\kappa} h_{\mu\nu} \right]. \tag{1.4.10}$$

The expression which *replaces* the background term $\sqrt{-g}\, R$, taking into account of the quantum fluctuations, is now readily obtained from (1.4.5)–(1.4.8), (1.4.10), by omitting, in the process, the corrections terms in (1.4.9) coming from $\delta R_{\mu\nu}$, and is given by

$$\sqrt{-g}\, \mathscr{L} =$$

$$\sqrt{-g} \left[R - (R^{\mu\nu} - \frac{1}{2} g^{\mu\nu} R) h_{\mu\nu} - \frac{1}{4} R h_{\mu\nu} (h^{\mu\nu} - \frac{1}{2} g^{\mu\nu} h) + R_{\mu\nu} h^{\mu}{}_{\lambda} (h^{\lambda\nu} - \frac{1}{2} g^{\lambda\nu} h) \right]$$

$$+ \frac{1}{2} \sqrt{-g} \left[-\frac{1}{2} \nabla_{\rho} h^{\mu\nu} \nabla^{\rho} h_{\mu\nu} + \nabla_{\rho} h^{\mu\nu} \nabla_{\mu} h_{\nu}{}^{\rho} - \nabla_{\rho} h \nabla_{\mu} h^{\mu\rho} + \frac{1}{2} \nabla_{\rho} h \nabla^{\rho} h \right]. \tag{1.4.11}$$

As expected, the expression within the second pair of square brackets coincides, up to total derivatives, with the one within the round brackets in (1.3.7) when the covariant derivatives are replaced by partial ones, i.e., for a Minkowski metric.

We consider the covariant gauge generalization of the one in (1.3.11)

$$\nabla_{\mu} (h^{\mu\nu} - \frac{1}{2} g^{\mu\nu} h) = \chi^{\nu}, \tag{1.4.12}$$

referred to as the DeWitt gauge, which may be implemented by adding the term

$$\frac{1}{2} \sqrt{-g} \left[\chi_{\mu} \chi^{\mu} - 2 \chi_{\nu} \nabla_{\mu} (h^{\mu\nu} - \frac{1}{2} g^{\mu\nu} h) \right], \tag{1.4.13}$$

to $\sqrt{-g}\, \mathscr{L}$ in (1.4.11), leading to

$$\sqrt{-g}\, \mathscr{L} + \frac{1}{2} \sqrt{-g} \left[\chi_{\mu} \chi^{\mu} - 2 \chi_{\nu} \nabla_{\mu} (h^{\mu\nu} - \frac{1}{2} g^{\mu\nu} h) \right]. \tag{1.4.14}$$

We may use the gauge condition (1.4.12) to eliminate χ^{μ} in (1.4.14) in favor of the field $h^{\mu\nu}$. This gives for the gauge fixing term

$$\frac{1}{2} \sqrt{-g} \left[\chi_{\mu} \chi^{\mu} - 2 \chi_{\nu} \nabla_{\mu} (h^{\mu\nu} - \frac{1}{2} g^{\mu\nu} h) \right]$$

$$= -\frac{1}{2} \sqrt{-g} \left[g_{\nu\sigma} \nabla_{\mu} h^{\mu\nu} \nabla_{\lambda} h^{\lambda\sigma} - \nabla_{\mu} h^{\mu\nu} \nabla_{\nu} h + \frac{1}{4} \nabla_{\rho} h \nabla^{\rho} h \right]. \tag{1.4.15}$$

We apply the result of partial integration rules spelled out in (1.2.5), and more generally in Problem 1.10, to (1.4.14) to obtain

$$\sqrt{-g}\,\mathcal{L} + \frac{1}{2}\sqrt{-g}\left[\chi_\mu\,\chi^\mu - 2\,\chi_\nu\,\nabla_\mu(h^{\mu\nu} - \frac{1}{2}\,g^{\mu\nu}h)\right]$$

$$= \sqrt{-g}\left[R - (R^{\mu\nu} - \frac{1}{2}\,g^{\mu\nu}R)\,h_{\mu\nu}\right] + \sqrt{-g}\,\frac{1}{2}\,h^{\mu\nu}M_{\mu\nu,\lambda\sigma}\,h^{\lambda\sigma}, \qquad (1.4.16)$$

where $(\Box = g^{\mu\nu}\partial_\mu\partial_\nu)$,

$$M_{\mu\nu,\lambda\sigma} = \frac{1}{2}\,g_{\lambda\mu}\,g_{\sigma\nu}\,\Box - \frac{1}{4}g_{\mu\nu}\,g_{\lambda\sigma}\,\Box - g_{\nu\sigma}[\nabla_\lambda, \nabla_\mu]$$

$$+ 2\,R_{\lambda\mu}\,g_{\nu\sigma} - R_{\mu\nu}\,g_{\lambda\sigma} + \frac{1}{4}\,R\,g_{\mu\nu}\,g_{\lambda\sigma} - \frac{1}{2}\,R\,g_{\mu\lambda}\,g_{\nu\sigma}, \qquad (1.4.17)$$

where symmetrization of $M_{\mu\nu,\lambda\sigma}$ over $\mu \leftrightarrow \nu$, $\lambda \leftrightarrow \sigma$, $(\mu,\nu) \leftrightarrow (\lambda\sigma)$ is understood. From Problem 1.2 (ii),

$$[\nabla_\lambda, \nabla_\mu]\,h^{\lambda\sigma} = R_{\lambda\mu}\,h^{\lambda\sigma} + R^\sigma{}_{\lambda\rho\mu}\,h^{\lambda\rho} = R_{\lambda\mu}\,h^{\lambda\sigma} - R^\sigma{}_{\lambda\mu\rho}\,h^{\lambda\rho}. \qquad (1.4.18)$$

Accordingly, for the term quadratic in $h^{\mu\nu}$ in (1.4.16), we obtain

$$\sqrt{-g}\,\mathcal{L}_{h^2} = \sqrt{-g}\,\frac{1}{2}\,h^{\mu\nu}\,M_{\mu\nu,\lambda\sigma}\,h^{\lambda\sigma}, \qquad (1.4.19)$$

$$M_{\mu\nu,\lambda\sigma} = \frac{1}{2}\,g_{\lambda\mu}\,g_{\sigma\nu}\,\Box\ - \frac{1}{4}g_{\mu\nu}\,g_{\lambda\sigma}\,\Box\ + R_{\nu\lambda\mu\sigma}$$

$$+ R_{\lambda\mu}\,g_{\nu\sigma} - R_{\mu\nu}\,g_{\lambda\sigma} + \frac{1}{4}\,R\,g_{\mu\nu}g_{\lambda\sigma} - \frac{1}{2}\,R\,g_{\mu\lambda}\,g_{\nu\sigma}, \qquad (1.4.20)$$

and upon symmetrization

$$M_{\mu\nu,\lambda\sigma} = \frac{1}{4}\left(g_{\lambda\mu}g_{\sigma\nu} + g_{\sigma\mu}g_{\lambda\nu} - g_{\mu\nu}g_{\lambda\sigma}\right)\Box\ + \frac{1}{2}\left(R_{\nu\lambda\mu\sigma} + R_{\mu\lambda\nu\sigma}\right)$$

$$+ \frac{1}{4}\left(R_{\lambda\mu}g_{\nu\sigma} + R_{\lambda\nu}g_{\mu\sigma} + R_{\sigma\mu}g_{\nu\lambda} + R_{\sigma\nu}g_{\mu\lambda}\right) - \frac{1}{2}\left(R_{\mu\nu}g_{\lambda\sigma} + R_{\lambda\sigma}g_{\mu\nu}\right)$$

$$- \frac{1}{4}\,R\left(g_{\lambda\mu}g_{\sigma\nu} + g_{\sigma\mu}g_{\lambda\nu} - g_{\mu\nu}g_{\lambda\sigma}\right). \qquad (1.4.21)$$

Note that we may write

$$M^{\mu\nu,\gamma\kappa} = E^{\mu\nu\sigma\rho}\left[\delta^{\gamma\kappa}{}_{\sigma\rho}\Box\ - Q^{\gamma\kappa}{}_{\sigma\rho}\right], \qquad (1.4.22)$$

where

$$E^{\mu\nu,\sigma\rho} = \frac{1}{4}\left(g^{\mu\sigma}g^{\nu\rho} + g^{\nu\sigma}g^{\mu\rho} - g^{\mu\nu}g^{\sigma\rho}\right), \quad \left[\delta^{\gamma\kappa}{}_{\sigma\rho}\Box - Q^{\gamma\kappa}{}_{\sigma\rho}\right] \equiv -F^{\gamma\kappa}{}_{\sigma\rho},$$
(1.4.23)

$$\delta^{\gamma\kappa}{}_{\sigma\rho} = \frac{1}{2}\left(\delta^{\gamma}{}_{\sigma}\delta^{\kappa}{}_{\rho} + \delta^{\gamma}{}_{\rho}\delta^{\kappa}{}_{\sigma}\right),$$
(1.4.24)

$$Q^{\gamma\kappa}{}_{\sigma\rho} = -\left(R_{\sigma}{}^{\gamma}{}_{\rho}{}^{\kappa} + R_{\rho}{}^{\gamma}{}_{\sigma}{}^{\kappa}\right) + \frac{1}{2}R\left(\delta^{\gamma}{}_{\sigma}\delta^{\kappa}{}_{\rho} + \delta^{\kappa}{}_{\sigma}\delta^{\gamma}{}_{\rho} - g^{\gamma\kappa}g_{\sigma\rho}\right)$$

$$+ R_{\sigma\rho}g^{\gamma\kappa} + R^{\gamma\kappa}g_{\sigma\rho} - \frac{1}{2}\left(\delta^{\gamma}{}_{\sigma}R^{\kappa}{}_{\rho} + \delta^{\gamma}{}_{\rho}R^{\kappa}{}_{\sigma} + \delta^{\kappa}{}_{\sigma}R^{\gamma}{}_{\rho} + \delta^{\kappa}{}_{\rho}R^{\gamma}{}_{\sigma}\right),$$
(1.4.25)

and $\delta^{\gamma\kappa}{}_{\sigma\rho}$ is the symmetric Kronecker delta. We have also defined $F^{\gamma\kappa}{}_{\sigma\rho}$ in (1.4.23) which will play an important role in the sequel.

The first term $\sqrt{-g}\,R$ in (1.4.11) corresponds to the classical background. We may invoke the Einstein field equation satisfied by $g_{\mu\nu}$, to set the second term $-\sqrt{-g}\left(R_{\mu\nu} - (g_{\mu\nu}/2)R\right)$ equal to zero. Thus, in addition to the ghost field contribution to be given below, only the quadratic part $\sqrt{-g}\,\mathscr{L}_{h^2}$ in $h^{\mu\nu}$ will be considered in the sequel.

The Faddeev-Popov ghost field contribution may be inferred from the gauge function $\nabla_{\mu}\left(h^{\mu\nu} - (g^{\mu\nu}/2)h\right)$ and the gauge transformation of the field $h_{\mu\nu}$ in (1.4.4), as done for the non-abelian theory case in (6.3.32) in Volume I as follows. Restricting to the term giving only the "kinetic part" of the ghost contribution

$$\nabla_{\mu}\left(h^{\mu\nu} - \frac{1}{2}g^{\mu\nu}h\right) \rightarrow \frac{\delta}{\delta\Lambda^{\mu}}\left[g^{\nu}{}_{\lambda}\Box + \nabla_{\lambda}\nabla^{\nu} - \nabla^{\nu}\nabla_{\lambda}\right]\Lambda^{\lambda},$$
(1.4.26)

and using the identity $\left[\nabla_{\lambda}\nabla^{\nu} - \nabla^{\nu}\nabla_{\lambda}\right]\Lambda^{\lambda} = R_{\lambda}{}^{\nu}\Lambda^{\lambda}$ in Problem 1.2 (i), the "kinetic part" (KP) of the ghost field follows to be,

$$\sqrt{-g}\,\mathscr{L}_{\text{ghost}}\Big|_{\text{KP}} = \sqrt{-g}\,\bar{\zeta}^{\mu}\left(g_{\mu\nu}\Box + R_{\mu\nu}\right)\zeta^{\nu},$$
(1.4.27)

where the ghost vector field ζ^{μ} obeys the "wrong" statistics.

The action integral for consideration, including the ghost field, may be then written as

$$\mathscr{A} = -\int(dx)\sqrt{-g}\left[\frac{1}{2}h^{\mu\nu}\left[-M_{\mu\nu,\lambda\sigma}\right]h^{\lambda\sigma} + \bar{\zeta}^{\mu}\left[-g_{\mu\nu}\Box - R_{\mu\nu}\right]\zeta^{\nu}\right],$$
(1.4.28)

where $M_{\mu\nu,\lambda\sigma}$ is defined in (1.4.22).

To study the one loop contribution to the effective action of quantum general relativity based on the description of quantum fluctuation about a classic metric just developed, we consider first a useful technique, referred to as the Schwinger-DeWitt technique, to this end.

1.5   The Schwinger-DeWitt Technique

The Schwinger technique[34] of a parametric representation of a propagator and the generation of an effective action was introduced and applied for a particular case in Sect. 3.7 of Vol. I. In the present section, we consider further aspects of this technique for wider applications with additional generalizations due to Bryce DeWitt.[35] This powerful and elegant technique is referred to as the Schwinger-DeWitt technique. This will be used in the next section for studying the one loop contribution to the effective action of quantum general relativity based on the description of quantum fluctuation about a classic metric developed in the previous section.

For the simplest treatment of this technique, consider first the propagator associated with a free scalar field in n-dimensional Minkowski spacetime

$$\Delta_+(x - x') = \int \frac{(dp)}{(2\pi)^n} \frac{e^{ip(x-x')}}{(p^2 + m^2 - i\epsilon)}, \quad (dp) = dp^0 dp^1 \ldots dp^{n-1}. \quad (1.5.1)$$

By working in n-dimensional spacetime, we write[36]

$$\frac{1}{(p^2 + m^2 - i\epsilon)} = i \int_0^\infty ds \, \exp[-i s(p^2 + m^2 - i\epsilon)], \quad (1.5.2)$$

$$\int_{-\infty}^\infty dp^0 \exp[i s \, (p^0)^2] = \sqrt{\frac{\pi}{s}} \, e^{i\frac{\pi}{4}}, \quad (1.5.3)$$

$$\int d^{n-1}k \, \exp\left[-i\left((k^1)^2 + \cdots + (k^{n-1})^2\right)\right] = \left(\sqrt{\frac{\pi}{s}} \, e^{-i\frac{\pi}{4}}\right)^{n-1}. \quad (1.5.4)$$

Hence from (1.5.2), the propagator in (1.5.1) may be rewritten as

$$\Delta_+(x - x') = i \int_0^\infty ds \, e^{-is(m^2 - i\epsilon)} \int \frac{(dp)}{(2\pi)^n} \, e^{ip(x-x')} e^{-isp^2}. \quad (1.5.5)$$

Upon completing the squares in the expression $[sp^2 - p(x - x')]$, in the p-integral of the above equation,

$$sp^2 - p(x - x') = s\left(p - \frac{(x - x')}{2s}\right)^2 - \frac{(x - x')^2}{4s}, \quad (1.5.6)$$

---

[34]Schwinger [55].

[35]See the monumental work of DeWitt, e.g., in [21] and references therein.

[36]Such useful integrals are also considered in details in Appendix II of Vol. I [43].

and using the integrals in (1.5.3), (1.5.4), we obtain

$$\Delta_+(x-x') = \frac{(\mathrm{i})^2}{(4\pi\,\mathrm{i})^{n/2}} \int_0^\infty \frac{ds}{s^{n/2}} \exp\left[\mathrm{i}\,\frac{(x-x')^2}{4s}\right] e^{-\mathrm{i}s\,(m^2-\mathrm{i}\epsilon)}, \tag{1.5.7}$$

$(x-x')^2 = (\mathbf{x}-\mathbf{x}')^2 - (x^0 - x'^0)^2.$

An equally important parametric relation involving the propagator, and in view of applications to the effective action in quantum gravity, is the evaluation of the simple functional integral of the form in (II.13)–(II.18) in Appendix II at the end of this volume, involving a scalar field ϕ:

$$e^{\mathrm{i}\Gamma} = \int (\mathscr{D}\phi) \exp\left[-\mathrm{i}\,\frac{1}{2}\phi\left(-\Box + m^2 - \mathrm{i}\epsilon\right)\phi\right]$$

$$= \frac{1}{\sqrt{\det[-\Box + m^2 - \mathrm{i}\,\epsilon]}} = \exp\left(-\frac{1}{2}\,\mathrm{Tr}\ln[-\Box + m^2 - \mathrm{i}\epsilon]\right), \tag{1.5.8}$$

where the trace Tr is over the spacetime variable x, and here $\Box = \eta^{\mu\nu}\partial_\mu\partial_\nu$, with $\eta^{\mu\nu}$ denoting the (inverse of the) Minkowski metric. From the above expression, we have

$$\Gamma = \frac{\mathrm{i}}{2}\,\mathrm{Tr}\ln[-\Box + m^2 - \mathrm{i}\epsilon]. \tag{1.5.9}$$

From Appendix B of this chapter, dealing with a parametric representation of the logarithm of a matrix (operator), we may rewrite the latter as

$$\Gamma = -\frac{\mathrm{i}}{2} \int_0^\infty \frac{ds}{s} \,\mathrm{Tr}\left[\exp\left(-\mathrm{i}s(-\Box + m^2 - \mathrm{i}\epsilon)\right)\right], \tag{1.5.10}$$

up to an additive constant independent of the propagator. Upon introducing the spacetime matrix elements

$$\langle x \,|\, \exp\left(-\mathrm{i}s(-\Box + m^2 - \mathrm{i}\epsilon)\right) | \,x'\rangle = K(x,x';s), \tag{1.5.11}$$

$$K(x,x';s)\Big|_{s\to 0} = \delta^{(n)}(x-x'), \tag{1.5.12}$$

we may finally rewrite (1.5.10) as

$$\Gamma = \frac{1}{2} \int (dx) \int_0^\infty \frac{ds}{\mathrm{i}\,s}\, K(x,x;s). \tag{1.5.13}$$

A closed expression for $K(x,x';s)$ may be explicitly obtained. To this end by inserting the identity operator in p-representation in which $\Box$ becomes a multiplicative operator, we obtain

$$K(x, x'; s) = \int \langle x | p \rangle \frac{(dp)}{(2\pi)^n} \langle p | \exp\left(-is(-\Box + m^2 - i\epsilon)\right) | x' \rangle$$

$$= \int \frac{(dp)}{(2\pi)^n} e^{ip(x-x')} \exp\left[-is(p^2 + m^2 - i\epsilon)\right]. \tag{1.5.14}$$

From (1.5.5)–(1.5.7), we obtain

$$K(x, x'; s) = \frac{i}{(4\pi i)^{n/2}} \frac{1}{s^{n/2}} \exp\left[i\frac{(x - x')^2}{4s}\right] e^{-is(m^2 - i\epsilon)}. \tag{1.5.15}$$

The above equation allows us to re-express Γ in the following manner

$$\Gamma = \frac{1}{2} \int (dx) \int_0^\infty \frac{ds}{is} \frac{i}{(4\pi is)^{n/2}} \exp\left[i\frac{(x - x')^2}{4s}\right] e^{-is(m^2 - i\epsilon)}\bigg|_{x'=x}. \tag{1.5.16}$$

Note the damping produced by the $e^{-\epsilon s}$ factor for $s \to \infty$. On the other hand, a singularity arises for s in the neighborhood of the origin. We will see in Sect. 1.7 how such a singularity is handled in the quantum general relativity via dimensional regularization.

As a generalization, now consider the equation of a Green function $G(x, x')$ satisfying an equation of the type

$$\sqrt{-g(x)}\, \Omega_x G(x, x') = \delta^{(n)}(x, x'). \tag{1.5.17}$$

where Ω_x is some operator involving covariant derivatives. The above equation may be rewritten in a matrix notation as

$$\Omega G = \mathbb{1}, \quad \Omega_x \langle x | G | x' \rangle = \langle x | \mathbb{1} | x' \rangle = \langle x | x' \rangle = \frac{\delta^{(n)}(x, x')}{\sqrt{-g(x)}}, \tag{1.5.18}$$

$g(x) | x \rangle = | x \rangle g(x)$.

We may express the matrix G in the following convenient form

$$G = i \int_0^\infty ds\, e^{-is(\Omega - i\epsilon)}, \tag{1.5.19}$$

$$G(x, x') = \langle x | G | x' \rangle = i \int_0^\infty ds\, \langle x | e^{-is(\Omega - i\epsilon)} | x' \rangle. \tag{1.5.20}$$

Upon setting

$$K(x, x'; s) = \langle x | e^{-is(\Omega - i\epsilon)} | x' \rangle, \tag{1.5.21}$$

we have

$$i\frac{\partial}{\partial s}K(x,x';s) = \Omega_x K(x,x';s).$$ (1.5.22)

Because of the similarity with the heat equation, without the i factor, $K(x,x';s)$ is referred to as the "heat kernel".

From (1.5.20), (1.5.22)

$$\Omega_x G(x,x') = -\int_0^\infty ds\,\frac{\partial}{\partial s}K(x,x';s) = K(x,x';s)\Big|_{s\to 0} = \langle x|x'\rangle,$$ (1.5.23)

assuming that $K(x,x';s)\Big|_{s\to\infty} = 0$. This gives the normalization condition

$$K(x,x';s)\Big|_{s\to 0} = \frac{\delta^{(n)}(x,x')}{\sqrt{-g(x)}}.$$ (1.5.24)

In studying the one loop contribution, in a loop expansion carried out in the next section, we will be particularly interested in an expression as

$$\frac{i}{2}\,\mathrm{Tr}\,[\ln\Omega] = -\frac{i}{2}\int (dx)\sqrt{-g(x)}\int_0^\infty \frac{ds}{s}\langle x|e^{-is(\Omega-i\epsilon)}|x\rangle$$

$$= \frac{1}{2}\int (dx)\sqrt{-g(x)}\int_0^\infty \frac{ds}{is}K(x,x;s),$$ (1.5.25)

up to an additive constant independent of Ω, where we have used the parametric representation of the logarithm of a matrix (operator) derived in the Appendix B of this chapter. Tr involves a trace over the spacetime variable as well, and it is understood that the trace over the other variables pertinent to the problem in hand have been carried out.

Guided by the expression in (1.5.15), $K(x,x;s)$ is written as

$$K(x,x;s) = \frac{i}{(4\pi is)^{n/2}}F(x;s),$$ (1.5.26)

for some $F(x;s)$, leading to

$$\mathrm{Tr}\,[\ln\Omega] = \frac{1}{2}\int (dx)\sqrt{-g(x)}\int_0^\infty \frac{ds}{is}\frac{i}{(4\pi is)^{n/2}}F(x;s).$$ (1.5.27)

Assuming that the integrand vanishes sufficiently rapidly for $s\to\infty$, singularities may, nevertheless, arise from s in the neighborhood of the origin. To investigate the nature of such singularities near the origin, one carries out an expansion of $F(x;s)$

of the form[37]

$$F(x ; s) = \sum_{k \geq 0} (i s)^k a_k(x), \qquad (1.5.28)$$

in the neighborhood of the origin. The coefficients $a_k(x)$, referred to as DeWitt coefficients, are obtained from the coincident limit $x' \to x$ of recursion relations derived from the application in (1.5.22).[38] From power counting, for $n = 4$, it is easy to see that for $s \to 0$, the divergences occur for $k = 0, 1, 2$. Dimensional regularization, however, wipes out the singularities for $k = 0, 1$, and the divergence of the expression of $\mathrm{Tr}[\ln \Omega]$ in (1.5.27) comes solely from the DeWitt coefficient $a_2(x)$. The corresponding analysis will be carried out in Sect. 1.7.

In the next section, the one loop contribution to the effective action of quantum general relativity is obtained, and the Schwinger-DeWitt technique is applied to it to rewrite it in a manageable form for further analysis. Dimensional regularization is applied to the one loop contribution in Sect. 1.7. This is followed by Sect. 1.8 which deals, in general, with renormalization aspects of quantum gravity.

1.6 Loop Expansion and One-Loop Contribution to Quantum General Relativity

For simplicity of the notation, let χ denote the collection of fields, assuming Hermitian, in the theory, suppressing all of its relevant indices with the action of the theory in consideration denoted by $S[\chi]$. In the presence of an external source J coupled to the field χ, the path integral expression has the general structure

$$\int \mu[\chi] \mathscr{D}\chi \, \exp\left[\frac{i}{\hbar}(S[\chi] + J\chi)\right] \equiv e^{iW[J]}, \qquad (1.6.1)$$

where we re-inserted the constant $\hbar$ for bookkeeping purposes, and where the measure $\mu(\chi)$, for the problem at hand, is to be specified below. The functional derivatives

$$(-i)\hbar \frac{\delta}{\delta J(x)} W = \langle \chi(x) \rangle \equiv \chi_c(x), \qquad (1.6.2)$$

---

[37]Note that unlike (1.5.15), in the problem in hand to be applied to the effective action in quantum general relativity, there is no $\exp[-i m^2 s]$ term in the present case.

[38]DeWitt [21]. For a systematic analysis of the determination of the DeWitt coefficients see, e.g., [8, 12].

generate classical fields χ_c representing expectation values of corresponding quantum fields. One then defines an effective action $\Gamma[\chi_c]$ by the equation

$$e^{i\Gamma[\chi_c]/\hbar} = \int \mu[\chi]\mathscr{D}\chi \, \exp\left[\frac{i}{\hbar}\left(S[\chi]-[\chi-\chi_c]\Gamma'[\chi_c]\right)\right], \qquad \Gamma'[\chi_c] = \frac{\delta}{\delta\chi_c}\Gamma[\chi_c].$$
$$(1.6.3)$$

The loop expansion is defined by an expansion of the effective action in powers of $\hbar$:

$$\Gamma[\chi_c] = S[\chi_c] + \sum_{N\geq 1}(\hbar)^N \Gamma^{[N]}[\chi_c], \qquad \Gamma'[\chi_c] = S'[\chi_c] + \cdots , \qquad (1.6.4)$$

where $\Gamma^{[1]}[\chi_c]$, for example, is the one loop contribution to the effective action. To obtain the expression for the latter, we express

$$\chi - \chi_c = \sqrt{\hbar}\,\xi, \qquad\qquad\qquad\qquad (1.6.5)$$

$$S[\chi] = S[\chi_c] + \sqrt{\hbar}\,\xi\,S'[\chi_c] + \frac{1}{2}\hbar\,\xi\,S''[\chi_c]\xi + \cdots , \qquad\qquad (1.6.6)$$

where note that from the second expansion in (1.6.4), the term $\sqrt{\hbar}\xi\, S'[\chi_c]$ in (1.6.6) cancels out within the round brackets in (1.6.3). Accordingly,

$$e^{i\Gamma^{[1]}[\chi_c]} = \int \tilde{\mu}[\xi]\mathscr{D}\xi \, \exp\left[i\left(-\frac{1}{2}\xi M\xi\right)\right], \qquad M = -S''[\chi_c], \qquad (1.6.7)$$

where the measure $\tilde{\mu}(\xi)\mathscr{D}\xi$ will be determined for the problem at hand.

For the quantum general relativity, in the one loop approximation, the Lagrangian density, with quantum field $h^{\mu\nu}$ is given in (1.4.28). The ghost factor in the latter may instead be included in the measure of the path integral as a Faddeev-Popov determinant of the ghost operator. The action in question may be then replaced by

$$\mathscr{A} = \frac{1}{2}\int (dx)\sqrt{-g}\, h^{\mu\nu}\,[M_{\mu\nu,\lambda\sigma}]\,h^{\lambda\sigma}, \qquad\qquad (1.6.8)$$

where

$$M^{\mu\nu,\gamma\kappa} = E^{\mu\nu\sigma\rho}\left[\delta^{\gamma\kappa}{}_{\sigma\rho}\square - Q^{\gamma\kappa}{}_{\sigma\rho}\right], \qquad \square = g^{\mu\nu}\nabla_\mu\nabla_\nu, \qquad (1.6.9)$$

with $E^{\mu\nu\alpha\beta}$ and $Q^{\gamma\kappa}{}_{\alpha\beta}$ spelled out in (1.4.23), (1.4.25), respectively.

We may rewrite the action integral $\mathscr{A}$ in (1.6.8) conveniently as

$$\mathscr{A} = \frac{1}{2}\int (dx)\sqrt{-g(x)}\, h^{\mu\nu}(x)\,[M_{\mu\nu,\lambda\sigma}]\,h^{\lambda\sigma}(x) = \frac{1}{2}\int (dx)\sqrt{-g(x)}\,\langle h\,|\,x\rangle\langle x\,|M\,h\rangle$$

$$= \frac{1}{2} \int [(dx)\sqrt{-g(x)}] \int [(dx')\sqrt{-g(x')}] \langle h|x\rangle \langle x|M|x'\rangle \langle x'|h\rangle$$

$$\mapsto = -\frac{1}{2} \sum_{ij} h^i [E(-F)]_i{}^j h_j. \qquad (1.6.10)$$

with the last expression represented in a discrete-variable notation, where i, j will be understood to label not only components (as indices) but also spacetime points, consistent with the transition from the second to the third line in the above equation. In detail

$$[-F]_{\gamma\kappa}{}^{\sigma\rho} = \left[-\delta_{\gamma\kappa}{}^{\sigma\rho} \Box + Q_{\gamma\kappa}{}^{\sigma\rho} \right], \qquad \Box = g^{\mu\nu} \nabla_\mu \nabla_\nu, \qquad (1.6.11)$$

and $\delta_{\gamma\kappa}{}^{\sigma\rho}$ is the symmetric Kronecker delta in (1.4.24).

The matrix E is independent of x, accordingly the latter generates the expression $\langle x|e^{-isE}|x\rangle \propto \delta^{(n)}(0)$. This contribution may be removed by involving the factor $\sqrt{\det[E_i{}^j]}$ in the measure, to remove it, in turn, from the matrix $[(E(-F))_i{}^j]$ appearing in $h^i [E(-F)]_i{}^j h_j$ in (1.6.10). We also include the Faddeev-Popov factor, corresponding to the expression in (1.4.27), in the measure. The one loop contribution to the effective action is from (1.6.7), (2.6.31) in Vol. I given through[39]

$$e^{i\Gamma^{[1]}} = \frac{\sqrt{\det[E_i{}^j]}}{\sqrt{\det[(E(-F))_i{}^j]}} \det[(-N)_i{}^j], \qquad (1.6.12)$$

or,

$$\Gamma^{[1]} = -\frac{i}{2} \int_0^\infty \frac{ds}{s} \sum_j [e^{-is(-F-i\epsilon)}]_j{}^j + i \int_0^\infty \frac{ds}{s} \sum_j [e^{-is(-N-i\epsilon)}]_j{}^j$$

$$(1.6.13)$$

and from (1.6.10) in a more manageable form as[40]

$$\Gamma^{[1]} = \frac{1}{2} \int (dx)\sqrt{-g(x)} \int_0^\infty \frac{ds}{is} K(x,x;s) - \int (dx)\sqrt{-g(x)} \int_0^\infty \frac{ds}{is} K_{\text{ghost}}(x,x;s),$$

$$(1.6.14)$$

using the notation in (1.5.25), and note the minus sign between the two terms on the right-hand of the above equation, and where

$$-N_\mu{}^\nu = [-\delta_\mu{}^\nu \Box - R_\mu{}^\nu], \qquad \Box = g^{\mu\nu} \nabla_\mu \nabla_\nu. \qquad (1.6.15)$$

---

[39] See (II.13)–(II.18) in Appendix II at the end of this volume. For a reader who is not familiar with such integrals, may wish also to consult Sect. 2.6 of Vol. I [43], and in particular (2.6.19) in it.

[40] Note that $K(x,x;s)$ means setting $x' = x$ in $K(x,x';s)$ prior to considering limits on the s variable.

In the following section dimensional regularization is applied to the expression in (1.6.14) to extract its divergent part.

1.7   Dimensional Regularization of the One Loop Contribution to Quantum General Relativity

The one loop contribution to the effective action $\Gamma^{[1]}$ of quantum general relativity was derived in the previous section with its expression displayed in (1.6.14). The first term, for example, in the expression for $\Gamma^{[1]}$ is given by[41]:

$$\frac{1}{2}\int (\mathrm{d}x)\sqrt{-g(x)}\int_0^\infty \frac{\mathrm{d}s}{\mathrm{i}\,s} K(x,x;\,s) = \frac{1}{2}\int (\mathrm{d}x)\sqrt{-g(x)}\int_0^\infty \frac{\mathrm{d}s}{\mathrm{i}\,s}\frac{\mathrm{i}}{(4\pi\,\mathrm{i}\,s)^{n/2}} F(x;\,s),$$

(1.7.1)

where we have used the definition in (1.5.26). For $n = 4$, we note that in the neighborhood of the origin: $s \simeq 0$, that upon expanding $F(x\,;\,s)$ in powers of s, as given in (1.5.28) and below in (1.7.6), the above integral diverges at the lower end for $k = 0, 1, 2$ in the expansion of $F(x\,;\,s)$.

In dimensional regularization, the above integral is defined by analytic continuation from a domain in which $n < 0$, where it converges, to a value close to the actual dimension of spacetime. Assuming the integrand in the above integral vanishes rapidly for $s \to \infty$, and through integration by parts twice, for $n < 0$, using, in the process, the equality

$$\frac{1}{s^{(n/2)+1}} = \frac{4}{n(n-2)}\frac{\mathrm{d}^2}{\mathrm{d}s^2}\frac{1}{s^{(n/2)-1}},$$

(1.7.2)

we obtain for the integral in (1.7.1)

$$\frac{2}{n(n-2)}\frac{1}{(4\pi\,\mathrm{i})^{n/2}}\int (\mathrm{d}x)\sqrt{-g(x)}\int_0^\infty \frac{\mathrm{d}s}{s^{(n/2)-1}}\frac{\mathrm{d}^2}{\mathrm{d}s^2}F(x;\,s).$$

(1.7.3)

To investigate the behavior of this integral, as we analytically continue to the neighborhood of the physical dimension of spacetime, near $s \simeq 0$, we proceed as follows. Let $\ell \ll 1$ be a *fixed* number, so that we can expand $F(x\,;\,s)$ in powers of s, for $s \leq \ell$. We break the above integral into two parts:

$$\frac{2}{n(n-2)}\frac{1}{(4\pi\,\mathrm{i})^{(n/2)}}\int (\mathrm{d}x)\sqrt{-g(x)}\int_0^\ell \frac{\mathrm{d}s}{s^{(n/2)-1}}\frac{\mathrm{d}^2}{\mathrm{d}s^2}F(x;\,s)$$

$$+\frac{2}{n(n-2)}\frac{1}{(4\pi\,\mathrm{i})^{(n/2)}}\int (\mathrm{d}x)\sqrt{-g(x)}\int_\ell^\infty \frac{\mathrm{d}s}{s^{(n/2)-1}}\frac{\mathrm{d}^2}{\mathrm{d}s^2}F(x;\,s).$$

(1.7.4)

---

[41]Note that $K(x,x\,;\,s)$ means setting $x' = x$ in $K(x,x'\,;s)$ prior to considering limits on the s variable.

Now we analytically continue from the domain in which $n < 0$ to $n \to D = 4-\varepsilon$, where $0 \lesssim \varepsilon \ll \ell$. The second integral above, does not involve the origin on the s-axis.

To the above end, assuming then the finiteness of the second integral, we may, to extract the divergent part of the integral in (1.7.3), i.e., the ε-dependent part in a limiting sense, consider the first integral in (1.7.4), which may be rewritten as[42]

$$\frac{2}{(4-\varepsilon)(2-\varepsilon)} \frac{1}{(4\pi i)^{2-(\varepsilon/2)}} \int (dx)\sqrt{-g(x)} \int_0^\ell \frac{ds}{s^{1-(\varepsilon/2)}} \frac{d^2}{ds^2}[F(x\,;s) + s^2 a_2(x)]$$

$$-\frac{2}{(4-\varepsilon)(2-\varepsilon)} \frac{1}{(4\pi i)^{2-(\varepsilon/2)}} \int (dx)\sqrt{-g(x)} \int_0^\ell \frac{ds}{s^{1-(\varepsilon/2)}} \frac{d^2}{ds^2}[s^2 a_2(x)],$$

$$(1.7.5)$$

obtained by adding to and subtracting $s^2 a_2(x)$ from $F(x\,;s)$, where, in the neighborhood of the origin $s \simeq 0$, we carry out an expansion

$$F(x\,;s) = \sum_{k \geq 0} (is)^k a_k(x), \qquad (1.7.6)$$

as given in (1.5.28). Since

$$\frac{d^2}{ds^2}[F(x\,;s) + s^2 a_2(x)] \propto s, \quad \text{for} \quad s \simeq 0, \qquad (1.7.7)$$

the first integral exists in (1.7.5) for $\varepsilon \to 0$.

Thus the divergent part of the entire integral in (1.7.3), i.e., its ε-dependent part in a limiting sense, is given by

$$-\frac{2}{(4-\varepsilon)(2-\varepsilon)} \frac{1}{(4\pi i)^{2-(\varepsilon/2)}} \int (dx)\sqrt{-g(x)} \int_0^\ell \frac{ds}{s^{1-(\varepsilon/2)}} \frac{d^2}{ds^2}[s^2 a_2(x)]\bigg|_{\text{div}}$$

$$\to -\frac{1}{\varepsilon} \frac{1}{(4\pi i)^2} \int (dx)\sqrt{-g(x)}\, a_2(x) = \frac{1}{\varepsilon} \frac{1}{16\pi^2} \int (dx)\sqrt{-g(x)}\, a_2(x). \qquad (1.7.8)$$

From this, we may infer that the divergent part of the one loop contribution to the effective action $\Gamma^{[1]}$ in quantum general relativity in (1.6.14) is given by

$$\Gamma^{[1]}\bigg|_{\text{div}} = \frac{1}{\varepsilon} \frac{1}{16\pi^2} \int (dx)\sqrt{-g(x)}\,[\,a_2(x) - 2a_{2\text{ghost}}(x)\,]. \qquad (1.7.9)$$

---

[42]We note that due to dimensional reasons, we may introduce a mass parameter μ_D, in the process of considering the limit $\varepsilon \to 0$, when making the transition $(dx) \to (\mu_D)^{-\varepsilon} d^D x$ and finally back to (dx). There is no need to write this explicitly here. See Appendix III of Vol. I [43], in particular, (III.8).

The next section deals with renormalizability aspects of quantum general relativity and quantum gravity, in general.

1.8   Renormalization Aspects of Quantum Gravity: Explicit Structures of One- and Two-Loop Divergences of Quantum GR, The Full Theory of GR Versus Higher Derivatives Theories: The Low Energy Regime

Let us summarize how the divergent part of the one loop contribution $\Gamma^{[1]}$ to quantum general relativity, conveniently displayed here

$$\Gamma^{[1]} = \frac{1}{2} \int (dx) \sqrt{-g(x)} \int_0^\infty \frac{ds}{i\,s} K(x,x\,;\,s) - \int (dx) \sqrt{-g(x)} \int_0^\infty \frac{ds}{i\,s} K_{\text{ghost}}(x,x\,;\,s),$$

(1.8.1)

and given in (1.6.14), was derived. In reference to (1.5.22),

$$i\frac{\partial}{\partial s} K(x,x'\,;\,s) = -F_x\,K(x,x';\,s),$$

(1.8.2)

$$i\frac{\partial}{\partial s} K_{\text{ghost}}(x,x'\,;\,s) = -N_x\,K_{\text{ghost}}(x,x';\,s),$$

(1.8.3)

in a matrix notation in their indices, where in detail

$$[-F]_{\gamma\kappa}{}^{\sigma\rho} = \left[-\delta_{\gamma\kappa}{}^{\sigma\rho}\,\Box + Q_{\gamma\kappa}{}^{\sigma\rho} \right],$$

(1.8.4)

$$-N_\mu{}^\nu = [-\delta_\mu{}^\nu\,\Box - R_\mu{}^\nu], \qquad\qquad \Box = g^{\mu\nu}\nabla_\mu\nabla_\nu.$$

(1.8.5)

as defined in (1.6.11), (1.6.15), respectively, where $\delta_{\gamma\kappa}{}^{\sigma\rho}$ is the symmetric Kronecker delta given in (1.4.24), and $Q^{\gamma\kappa}{}_{\alpha\beta}$ is given in (1.4.25).

We first write

$$K(x,x\,;\,s) = \frac{i}{(4\pi\,i\,s)^{n/2}}\,F(x\,;\,s),$$

(1.8.6)

$$K_{\text{ghost}}(x,x\,;\,s) = \frac{i}{(4\pi\,i\,s)^{n/2}}\,F_{\text{ghost}}(x\,;\,s),$$

(1.8.7)

guided by the expression in (1.5.15). In the neighborhood of small but fixed s,

$$F(x\,;\,s) = \sum_{k\geq 0} (i\,s)^k a_k(x),$$

(1.8.8)

$$F_{\text{ghost}}(x\,;\,s) = \sum_{k\geq 0} (i\,s)^k a_{k\,\text{ghost}}(x).$$

(1.8.9)

The process of dimensional regularization carried out in the previous section, gives the following expression for the divergent part of the one loop contribution $\Gamma^{[1]}$

$$\Gamma^{[1]}\Big|_{\text{div}} = \frac{1}{\varepsilon}\frac{1}{16\pi^2}\int (dx)\sqrt{-g(x)}\left[\,a_2(x) - 2a_{2\text{ghost}}(x)\,\right], \qquad (1.8.10)$$

in $4 - \varepsilon$ dimensional spacetime with $\varepsilon \gtrsim 0$.

The DeWitt coefficients $a_k(x)$, $a_{k\text{ghost}}(x)$ are determined from recursion relations set up by (1.8.2), (1.8.3), at coincident points $x' = x$, prior to[43] considering a limit on the variable s, taken in the order just described.[44]

Quantum general relativity is one-loop finite on the mass shell, *i.e., for* $R_{\mu\nu} = 0$ and hence also $R = 0$. One reaches this conclusion without even carrying out the explicit computations involved in determining the coefficients $a_2(x)$, $a_{2\text{ghost}}(x)$ in (1.8.10). To establish this, note that the metric on the right-hand side of (1.4.1) involving the quantum fluctuation may be written as

$$\bar{g}_{\mu\nu} = g_{\mu\nu} + \kappa\,h_{\mu\nu}, \qquad (1.8.11)$$

where $g_{\mu\nu}$ satisfies the field equations $R_{\mu\nu} = 0$. The quadratic term in $h_{\mu\nu}$, for example, involves a κ^2 factor. From (1.6.6), we learn that a loop expansion in power of $\hbar$ is equivalent to an expansion in terms of the parameter[45] κ^2. The dimensionality of κ^2 is [Length]^2. The action involves a factor $1/\kappa^2$ and the dimensionality of (dx) is [Length]^4. Since the Riemann curvature tensor involves two derivatives, the integral in (1.8.10) must have the following general structure

$$\Gamma^{[1]}\Big|_{\text{div}} = \frac{1}{\varepsilon}\frac{1}{16\pi^2}\int (dx)\sqrt{-g(x)}\left[\,\tilde{a}\,R_{\mu\nu}\,R^{\mu\nu} + \tilde{b}\,R^2 + \tilde{c}\,\Box R + \tilde{d}\,R_{\mu\nu\lambda\sigma}\,R^{\mu\nu\lambda\sigma}\,\right], \qquad (1.8.12)$$

simply for dimensional reasons, with $\hbar$ set equal to one, for dimensionless numericals a, b, c, d, with the expression within the square brackets in the integrand of

---

[43]Note, for example, referring to (1.5.15), we have the following expression for its right-hand side for $x' = x$:

$$\frac{i}{(4\pi i)^{n/2}}\frac{1}{s^{n/2}}\,e^{-is(m^2 - i\epsilon)},$$

while $K(x, x'; s) \to \delta^{(n)}(x - x')$ for $s \to 0$, and the order in which limits are taken is important.
[44]DeWitt [21], See also [8, 12].
[45]In analogy to a one dimensional integral

$$\int_{-\infty}^{\infty} dz\, e^{-z^2}[1 + a_n(\kappa z)^n],$$

note that in addition to one in the square brackets, corresponding to one-loop, only even powers of n contribute to the above integral.

dimensionality $[\text{Length}]^{-4}$. As we have seen in the previous section, the $1/\varepsilon$ factor arises from dimensional regularization. Since $\sqrt{-g}\, \nabla_\mu \nabla^\mu R = \partial_\mu (\sqrt{-g}\, \nabla^\mu R)$, i.e., it is a total derivative, $\tilde{c}$ may be set equal to zero. On the other hand a key result proved in Appendix C of this chapter, shows that

$$\delta \left(\sqrt{-g} \left[R_{\mu\nu\sigma\lambda} R^{\mu\nu\sigma\lambda} - 4 R_{\mu\nu} R^{\mu\nu} + R^2 \right] \right), \tag{1.8.13}$$

is a total derivative. This is the content of the Euler-Poincaré characteristic and is also referred to as the Gauss-Bonnet Theorem in four dimensional spacetime. We may thus effectively replace $R_{\mu\nu\lambda\sigma} R^{\mu\nu\lambda\sigma}$ in (1.8.12) by $\left(4 R_{\mu\nu} R^{\mu\nu} - R^2 \right)$. The following explicit structure then emerges

$$\left. \Gamma^{[1]} \right|_{\text{div}} = \frac{1}{\varepsilon} \frac{1}{16\pi^2} \int (\mathrm{d}x) \sqrt{-g(x)} \left[a R_{\mu\nu} R^{\mu\nu} + b R^2 \right], \tag{1.8.14}$$

for some dimensionless numericals a, b. Thus irrespective of the values of the numerical constants a, b, quantum general relativity is one-loop finite[46] on the mass shell[47] thanks to the existence of the Euler-Poincaré characteristic which holds in four dimensional spacetime.

A fairly detailed analysis gives explicit expressions[48] for the DeWitt coefficients in $\int (\mathrm{d}x) \sqrt{-g(x)} \left[a_2(x) - 2 a_{2\,\text{ghost}}(x) \right]$ relevant to $\left. \Gamma^{[1]} \right|_{\text{div}}$, or equivalently to the coefficients a and b in (1.8.14) leading to

$$\left. \Gamma^{[1]} \right|_{\text{div}} = \frac{1}{\varepsilon} \frac{1}{16\pi^2} \int (\mathrm{d}x) \sqrt{-g(x)} \left(\frac{7}{20} R_{\mu\nu} R^{\mu\nu} + \frac{1}{120} R^2 \right), \tag{1.8.15}$$

a celebrated result first derived by 't Hooft and Veltman [61] by completely different methods than of the Schwinger-DeWitt technique.[49]

---

[46]In any case, one may modify the metric $g_{\mu\nu}$ to $g_{\mu\nu}$ plus a linear combination of $R_{\mu\nu}$ and $g_{\mu\nu} R$, as indicated in (1.8.18), denoting the new expression, say, by $\underline{g}_{\mu\nu}$ to write

$$\Gamma^{[0]}[g_{\mu\nu}] + \left. \Gamma^{[1]} \right|_{\text{div}} [g_{\mu\nu}] = \Gamma^{[0]}[\underline{g}_{\mu\nu}] + \cdots,$$

where $\Gamma^{[0]}[\underline{g}_{\mu\nu}]$ is the Einstein-Hilbert action with metric $\underline{g}_{\mu\nu}$, and "$\cdots$" is of the order of contributions of two loops.

[47]The finiteness of quantum general relativity in one-loop was also discovered independently of 't Hooft and Veltman [61] by Korepin, as an undergraduate student of L.D. Faddeev: Korepin [35], with English translation in [36]. See also [35] contribution in (Feynman et al. [28, p. 225]).

[48]See, e.g., the fairly detailed and systematic analysis given in [12], and especially Eq. (5.58) in it. See also [8].

[49]See also [34] for finiteness off the mass shell.

One may then carry out the modification

$$\frac{1}{\kappa^2} \sqrt{-g} R \rightarrow \sqrt{-g} \left[\frac{R}{\kappa^2} + a_{\text{bare}} R^2 + b_{\text{bare}} R_{\mu\nu} R^{\mu\nu} + \cdots \right], \qquad (1.8.16)$$

in the Einstein-Hilbert action, and set

$$a_{\text{bare}} = a - \frac{1}{16\pi^2 \varepsilon} \frac{1}{120}, \quad b_{\text{bare}} = b - \frac{1}{16\pi^2 \varepsilon} \frac{7}{20}, \qquad (1.8.17)$$

to cancel out the $1/\varepsilon$ singularities in (1.8.15).

A Lagrangian density having the sum of such quadratic terms R^2, $R_{\mu\nu} R^{\mu\nu}$, however, involves four derivatives. As a consequence of this the kinetic quadratic part in the field $h^{\mu\nu}$ in the Lagrangian density now contains the higher order term $\Box^2$, thus leading to a propagator which vanishes like $1/k^4$ instead of $1/k^2$, for $k^2 \rightarrow \infty$, and destroys the positivity required by a quantum theory in a perturbative setting.[50]

One may, instead carry out a wavefunction transformation to absorb the $1/\varepsilon$ singularities in (1.8.15)

$$g_{\mu\nu} \rightarrow g_{\mu\nu} + \frac{\kappa^2}{16\pi^2 \varepsilon} \left(\frac{11}{60} g_{\mu\nu} R - \frac{7}{20} R_{\mu\nu} \right), \qquad (1.8.18)$$

thus ensuring positivity.[51]

Before discussing in what sense quantum general relativity is non-renormalizable, let us first investigate the divergence problem of quantum general relativity in two and multi loops.

1.8.1 Two and Multi Loops

Consider the two-loop contribution. κ^2 is the expansion parameter in the loop expansion. With $\hbar$ set equal to one, the divergent part in two loops must be proportional to the dimensionless integral:

$$\kappa^2 \int (\mathrm{d}x) \sqrt{-g} \, F(x), \qquad (1.8.19)$$

---

[50][59]. Unitarity (positivity) of such a theory in a non-perturbative setting has been elaborated upon by Tomboulis [63].

[51]For the extension of the above alternative to higher order loops with a field redefinition, in the process of renormalization, such that the right behavior of a propagator for $p^2 \rightarrow \infty$ is maintained thus avoiding non-positivity conditions, by expanding the metric around the Minkowski one: $g_{\mu\nu} = \eta_{\mu\nu} + \kappa h_{\mu\nu}$, see [2]. See also [29].

where, because of dimensional reasons, $F(x)$ must be of dimensionality $[\text{Length}]^{-6}$. Clearly, any invariant of the form proportional to $R_{\mu\nu}(x)$, or $R(x)$, i.e., having the structures $R_{\mu\nu}(x)R^{\mu\nu}(x)$ $R(x)R(x)$ cannot contribute to $F(x)$ on the mass shell. On other hand, any term having a structure such as:

$$\sqrt{-g}\left(\nabla_\sigma R_{\mu\nu}\right)\left(\nabla^\sigma R^{\mu\nu}\right), \qquad \sqrt{-g}\left(\nabla_\sigma R\right)\left(\nabla^\sigma R\right),$$

$$\sqrt{-g}\left(\nabla^\mu R_{\mu\nu}\right)\left(\nabla^\nu R\right), \qquad \sqrt{-g}\left(\nabla^\mu R_{\mu\nu}\right)\left(\nabla_\sigma R^{\sigma\nu}\right), \qquad (1.8.20)$$

may, from Problem 1.10, be written as a total derivative $\partial_\mu(\sqrt{-g}\,\xi^\mu)$, *on the mass shell*, for some vector ξ^μ, and hence cannot contribute to the integral in (1.8.19). Accordingly, only invariant products of three Riemann tensors, with each tensor involving two derivatives, may contribute to $F(x)$ above. In Appendix D of this chapter, we prove the remarkable result that all such invariants, on the mass shell, are proportional to the invariant

$$R^{\mu\nu}{}_{\sigma\lambda}\,R^{\sigma\lambda}{}_{\kappa\rho}\,R^{\kappa\rho}{}_{\mu\nu}, \qquad (1.8.21)$$

from which, we may infer that the divergent part of the effective action in two loops must be proportional to

$$\kappa^2\int(\mathrm{d}x)\sqrt{-g(x)}\ R^{\mu\nu}{}_{\sigma\lambda}(x)\ R^{\sigma\lambda}{}_{\kappa\rho}(x)\ R^{\kappa\rho}{}_{\mu\nu}(x). \qquad (1.8.22)$$

On the other hand, since the one-loop contribution is finite, the divergence, in the two-loop contribution, must come from an overall integration as all the sub-integrations are finite by dimensional regularization, implying that the two-loop divergence must be proportional to a single power of $1/\varepsilon$ instead of a quadratic expression for the leading term.[52] Accordingly the general expression for the two-loop divergent part of quantum general relativity may be written to have the explicit form

$$\Gamma^{[2]}\Big|_{\mathrm{div}} = c_2\,\frac{\kappa^2}{\varepsilon}\,\frac{1}{(16\,\pi^2)^2}\int(\mathrm{d}x)\sqrt{-g(x)}\ R^{\mu\nu}{}_{\sigma\lambda}(x)\ R^{\sigma\lambda}{}_{\kappa\rho}(x)\ R^{\kappa\rho}{}_{\mu\nu}(x), \qquad (1.8.23)$$

involving only one numerical constant c_2. If the latter constant were zero the theory would be two-loop finite. Unfortunately, as has been confirmed by two groups,[53] the constant c_2 turns up to be not zero, implying that perturbative quantum general relativity is non-renormalizable as we now discuss.

We note that the Einstein-Hilbert action contain terms of second order in derivatives in the integrand, while $R_{\mu\nu}R^{\mu\nu}$, R^2, $R^{\mu\nu}{}_{\sigma\lambda}\,R^{\sigma\lambda}{}_{\kappa\rho}\,R^{\kappa\rho}{}_{\mu\nu}\ \ldots$ are of higher

---

[52]For the spelling out of this underlying technical detail see [17, 18, 60].

[53]Goroff and Sagnotti [30], van de Ven [65]. The computation of the coefficient c_2 is quite tedious and these two groups had to resort to the computer for its evaluation.

orders. If such terms are added to the original integrand, one may be then tempted to combine the coefficients $\bar{c}_1$, $\bar{d}_1$, $\bar{c}_2$ in

$$\int (dx) \sqrt{-g} \left[\bar{c}_1 R_{\mu\nu} R^{\mu\nu} + \bar{d}_1 R^2 + \bar{c}_2 R^{\mu\nu}{}_{\sigma\lambda} R^{\sigma\lambda}{}_{\kappa\rho} R^{\kappa\rho}{}_{\mu\nu} + \cdots \right] \qquad (1.8.24)$$

with such terms as in (1.8.15), (1.8.23), to define new coefficients which are to be determined experimentally. However, such terms, in turn, will, in general, generate divergent terms in the effective action of arbitrary high order loops, giving rise to an infinite number of parameters that are to be determined experimentally, and the theory would lose its predictive power. This is what is meant that the theory is perturbatively non-renormalizable. It is traced back to the fact that the expansion coupling κ^2 has a dimensionality of positive power of length: $[\text{Length}]^2$. Non-renormalizability still holds when scalars, Yang-Mills fields field as well as the photon, are included in the theory. Here it is important to note that $R_{\mu\nu} \neq 0$ in the presence of such fields and finiteness even in one-loop fails.[54]

Let us investigate how the divergences in the perturbative loop-expansion changes with the increase of the number of loops. To this end, we note by working in the momentum presentation, that the graviton propagator given in (1.3.18), (an internal line to a graph) carrying a momentum p goes like $1/p^2$, and a vertex involves two derivatives and goes like p^2. Two lines joining two vertices generate a loop with momentum integration goes like p^4. Three lines joining two vertices generate two loops. Moreover, a graph[55] involving ℓ^{int} lines, L loops, and V vertices satisfy the relation

$$L = \ell^{\text{int}} - V + 1. \qquad (1.8.25)$$

By power counting, we may define the superficial degree of divergence of a graph g by

$$d(g) = 4L + 2V - 2\ell^{\text{int}}, \qquad (1.8.26)$$

where the factor 4 multiplying the number of loops L corresponds to the dimension of spacetime. The factor 2 multiplying the number of vertices V corresponds to the two powers of momenta in a vertex part. The factor -2 multiplying the number of lines ℓ^{int} corresponds to the two powers of the inverse of the momenta associated with a propagator. From (1.8.25), (1.8.26), we may solve for $d(g)$ in terms of the number of loops L:

$$d(g) = 2 + 2L. \qquad (1.8.27)$$

---

[54]See, e.g., [20], also [61].

[55]Here one is considering a connected proper graph, where proper means that it cannot become disconnected by cutting a single line.

Thus we see that the superficial degree of divergence increases without bound with the number of loops.

The increase of the superficial degree of divergence $d(g)$ with the number of loops, naturally urges one to investigate the role of the renormalizabilty of a corresponding quantum gravity involving higher order derivatives terms added to the Einstein-Hilbert action. This is discussed next.

1.8.2 Higher Order Derivatives Corrections

In particular, a simple modification to the Einstein-Hilbert action by adding to it the term

$$\int (\mathrm{d}x)\sqrt{-g}\left[\bar{c}_1 R_{\mu\nu}R^{\mu\nu} + \bar{d}_1 R^2\right] \tag{1.8.28}$$

leads, as a consequence of the presence of the higher order term $\Box^2$ in the kinetic quadratic part in $h^{\mu\nu}$ in the Lagrangian density, to a propagator which falls of as $1/p^4$, instead of $1/p^2$ at high momenta, as discussed earlier. Now the total number of derivatives is 4. This simply implies to replace the -2 multiplying the number of lines ℓ^{int} in the superficial degree of divergence $d(g)$ in (1.8.26) by -4, and the 2 multiplying V by 4. From (1.8.25), (1.8.26), this tells that the superficial degree of divergence in this case becomes replaced by

$$d(g) = 4. \tag{1.8.29}$$

independently of the number of loops, and the renormalizability of the modified theory may be then established[56] with only a finite number of parameters subjected to an experimental determination. Unfortunately, such a theory involves ghosts in a perturbative treatment, due to the rapid damping of the propagator at high energies faster than $1/p^2$, and gives rise, in turn, to negative probabilities.[57]

We have seen that in a perturbative setting, the non-renormalizability of quantum GR implies the need of an infinite number of parameters are to be determined experimentally and hence is not of any practical value. These infinite number of parameters arise as one must add an infinite number of new terms to the Lagrangian density of GR, referred to as counter terms, to make the theory finite at the cost of having an infinite number of parameters to be determined in a perturbative setting. A non-perturbative approach, however, to the renormalization of such a theory may be

---

[56]See [59].

[57]A quantum viewpoint treatment of a higher derivative modification of the linearized Einstein-Hilbert action in (1.2.1) obtained by replacing R in it by $R + aR^2$ with $a > 0$, shows that the theory involves, in addition to the graviton, a massive scalar particle which cannot be gauged away and no ghosts arise in the theory—see [42].

developed, referred to as asymptotic safety, and is discussed next for completeness but rather briefly.

One may consider the above generalizations of GR involving all such infinite number of terms added to the Lagrangian density of GR each of which satisfies the same symmetry as in GR, implying the invariance of the generated action under general coordinate transformations, in a non-perturbative setting as follows. A renormalization group analysis, which relates the theory at different energy scales, is carried out with the main interest to investigate the theory at high energies at which ultraviolet divergences may appear. One envisages and subsequently consistently checks that effective couplings $g_i(k)$ corresponding to the (dimensionless) couplings defined in the theory,[58] approach *fixed points*: $g_i(k) \rightarrow g_i^*$ at high energy $k \rightarrow \infty$, i.e., the g_i^* are zeros of the beta functions[59] of the renormalization group. The tacit assumption in defining an asymptotically safe theory, which is then consistently verified via solutions of the renormalization group, is that one may restrict to those theories involving only a subclass of the couplings, referred to as relevant couplings, such that only a finite number of parameters are to be determined.[60] The corresponding theory is then referred to as asymptotically safe, and considered to be consistent and meaningful in a non-perturbative sense.[61] In this sense a non-perturbative treatment would establish a renormalizable quantum gravity if it is asymptotically safe. The great challenge is, of course, not only establish ultraviolet finiteness of the full theory, that no unphysical singularities in the theory would arise,[62] and most importantly, as just mentioned, that only a *finite* number of parameters need to be adjusted if the theory is to be considered reliable. Much work has been done on this[63] and further work is needed. Here one introduces an ultraviolet cut-off in the effective action as well as suppress energy modes less

---

[58]These are referred to dimensionless couplings in the sense that if the mass dimension of a dimensionful coupling $\tilde{g}$ is d, and $\tilde{g}(k)$ denotes the coupling defined at a renormalization point with energy scale k, then one may define its dimensionless counterpart coupling by $g(k) = (k)^{-d}\tilde{g}(k)$. The couplings are also restricted to those that do not change when a point-transformation of a field is carried since, in reference, to physical quantities the latter do not depend on how the fields are defined. Such inessential couplings g' give rise to partial derivatives of the Lagrangian density $\partial\mathscr{L}/\partial g'$ which are at most total derivatives (or zero) when the Euler-Lagrange equations are used. See [67].

[59]We have encountered beta functions in abelian and non-abelian gauge theories in much details in Vol. I [43].

[60]One assumes at least one of the relevant couplings $g_i^* \neq 0$. The reason is that when one considers the application of the renormalization group to a physical quantity Q, such as a total cross section of a process, and D is the dimensionality of Q, and E is some energy characterizing, then the high energy behavior of the quantity $(E)^{-D}R$ is governed by $g(E)$ defined in terms of the relevant couplings.

[61]This approach was introduced by Weinberg [67].

[62]Anselmi [2], Gomis and Weinberg [29].

[63]See, e.g., [46, 48, 49, 64].

than the energy k at which the theory is being considered.[64] We will not, however, consider asymptotic safety further, as it is beyond the scope of this book, and refer the reader to the references just mentioned.

One may generally improve the ultraviolet behavior of a theory when more symmetry is built into it. In supersymmetry, in particular, where there is a symmetry between fermionic and bosonic fields, and each bosonic degree of freedom has a supersymmetric fermionic degree of freedom counterpart and vice versa,[65] the ultraviolet behavior of the theory improves. This is because often divergent fermionic contributions from loops are canceled by the corresponding superpartner bosonic contributions and vice versa.[66] As a consequence of this, the divergence problem in a supersymmetric generalization of general relativity, referred to as supergravity, turns out to be better than the one in general relativity. Divergences, nevertheless, seem to appear in higher order loops, with variations depending on the number of supercharges (Sect. 2.1) involved in the theory and the dimensionality taken for spacetime.[67]

General relativity is expected to emerge as a low energy limit of a more fundamental theory, as the former has been quite successful in the low energy classical regime. Referring to (1.8.24), if one formally combines the coefficients $\bar{c}_1, \bar{d}_1, \ldots$ with a priori existing coefficients of such identical terms in a modified Einstein-Hilbert action version which includes such identical structures, to generate new finite parameters in the resulting theory, the number of parameters that have to be taken from experiments increase with no bound as we go to higher order loops. The modified terms beyond the Einstein-Hilbert action, however, involve higher order derivatives, and in a momentum description involve powers of momenta. In this sense, one may expect that such modifications are not important at low energies or equivalently in computations dealing with the long distance behavior of the theory, and one may expect, in turn, that the Einstein-Hilbert action to dominate in this regime. A brief account of this is given below.

1.8.3   The Low Energy Regime: Quantum GR as an Effective Field Theory and Modification of Newton's Gravitational Potential

One is led to believe that Einstein's general relativity is a low energy effective theory as the low energy limit of a more complicated theory, and as such it provides a reliable description of gravitation at low energies.

---

[64]This is in the spirit of Wilson's [68] (see also [69]) non-perturbative analysis of renormalizability of a theory which treats renormalizable and non-renormalizable theories on equal footings.

[65]This is treated in Chap. 2.

[66]See, e.g., Sect. 2.10.

[67]See, e.g., [19].

One may argue that the non-renormalizability of a quantum theory based on GR is due to the fact that one is trying to use it at energies which are far beyond its range of validity. As a matter of fact the derivatives occurring in the theory, in a momentum description, may be treated as small at sufficiently low energies treating it as an effective theory. Several applications on such an approach have been carried out as determining corrections to Newton's gravitational potential.

The modified terms beyond the Einstein-Hilbert action as counter terms needed to redefine coupling parameters in a perturbative setting, through the process of renormalization discussed above in this section, involve higher order derivatives, and in a momentum description involve powers of momenta. In this sense, one may expect that such modifications are not important at low energies or equivalently in computations dealing with the long distance behavior of the theory, and one may expect, in turn, that the Einstein-Hilbert action to dominate in this regime.

As ultraviolet divergences are associated with the small distance behavior of the theory, one tries to separate low energy effects from high energy ones, in view of applications in the low energy regime even if the theory has unfavorable ultraviolet behavior such as in quantum gravity. For interesting applications of such effective theories, in long distance applications, see, e.g., Bjerrum-Bohr et al. [15, 16]; Donoghue [25–27].

The Einstein-Hilbert action, in the presence of matter, may be written as

$$
W = \frac{1}{\kappa^2} \int \sqrt{-g(x)} \, (\mathrm{d}x) \, R(x) + \int \sqrt{-g(x)} \, (\mathrm{d}x) \, \mathscr{L}_{\text{matter}}(x), \tag{1.8.30}
$$

where for a scalar field, for example, the expression for $\mathscr{L}_{\text{matter}}(x)$ is given in Problem 1.11. To treat pure GR theory, as an effective action, one includes *all* possible higher order couplings involving the Riemann tensor $R_{\alpha\beta\gamma\,\rho}$ and its contractions, and make the replacement

$$
\frac{1}{\kappa^2} \sqrt{-g} \, R \rightarrow \sqrt{-g} \left[\frac{R}{\kappa^2} + a_1 R^2 + a_2 R_{\mu\nu} R^{\mu\nu} + \cdots \right], \tag{1.8.31}
$$

the coefficients of which are then adjusted after eliminating the divergences in each loop computation. With the divergent term of the Einstein-Hilbert action given in (1.8.15), the elimination of the divergence in the loop level is not difficult. One must, however include higher derivative contributions to the matter Lagrangian density as well.

For a scalar field contribution, one must also simultaneously carry out a similar generalization leading to such a modification:

$$
-\sqrt{-g} \left[\frac{1}{2} \left(g^{\mu\nu} \partial_\mu \phi \partial_\nu \phi + m^2 \phi^2 \right) \right] \rightarrow
$$

$$
\sqrt{-g} \left[-\frac{1}{2} \left(g^{\mu\nu} \partial_\mu \phi \, \partial_\nu \phi + m^2 \phi^2 \right) + b_1 \partial_\mu \phi \, \partial_\nu \phi R^{\mu\nu} + \right.
$$

$$
\left. + \left(c_1 g^{\mu\nu} \partial_\mu \phi \, \partial_\nu \phi + c_2 m^2 \phi^2 \right) R + \cdots \right], \tag{1.8.32}
$$

in discussing finiteness problems in loop expansions of the theory in the presence of a scalar field.

At this stage assuming the validity of a perturbation expansion to second order in Newton's coupling constant[68] G_N, we may use a dimensional analysis to write down corrections that may arise in the modification of Newton's static potential $U_N(r) = -G_N m_1 m_2 / r$ for the interaction between two massive spin-0 particles of masses m_1 and m_2 separated by a finite distance r. The two possible corrections to Newton's potential to second order in G_N, based on dimensional analysis alone, are then

$$\frac{G_N^2 \, m_1 m_2 (m_1 + m_2)}{c^2 \, r^2}, \quad \frac{G_N^2 \, m_1 m_2 \hbar}{c^3 \, r^3}, \tag{1.8.33}$$

conveniently parametrized. We note that a quantum correction naturally arises here proportional to $\hbar$. Accordingly, to second order in G_N, the modified static potential between the two particles becomes

$$U(r) = -\frac{G_N m_1 m_2}{r}\left[1 + \alpha \frac{G_N(m_1 + m_2)}{c^2 \, r} + \beta \frac{G_N \hbar}{c^3 \, r^2}\right], \tag{1.8.34}$$

where α and β are dimensionless constants.

As computations of amplitudes are done in the momentum description, and for a static potential computations are carried out in 3 dimensional space, we may perform a 3 dimensional Fourier transforms of the $1/r^2$ and $1/r^3$ terms in (1.8.33) to investigate the nature of contributions that arise in a long distance treatment of the above problem.

To the above end, we recall that

$$\int d^3\mathbf{r} \, \frac{e^{i\mathbf{k}\cdot\mathbf{r}}}{r} = \frac{4\pi}{\mathbf{k}^2}. \tag{1.8.35}$$

From dimensional analysis

$$\int d^3\mathbf{r} \, \frac{e^{i\mathbf{k}\cdot\mathbf{r}}}{r^2} = a\frac{1}{|\mathbf{k}|}, \tag{1.8.36}$$

where a is a constant to be determined. Using the well known relation

$$\frac{\partial}{\partial k^i}\frac{\partial}{\partial k^i}\frac{1}{|\mathbf{k}|} = -4\pi\delta^3(\mathbf{k}), \tag{1.8.37}$$

---

[68]That is, in particular, no corrections are assumed having the structures $G_N \ln(G_N)$, $G_N^2 \ln(G_N)$.

and applying the $(\partial/\partial k^i)(\partial/\partial k^i)$ to both sides of (1.8.36) gives $-(2\pi)^3\,\delta^3(\mathbf{k}) = -4\,\pi\,a\,\delta^3(\mathbf{k})$, from which $a = 2\,\pi^2$. That is

$$\int d^3\mathbf{r}\,\frac{e^{i\mathbf{k}\cdot\mathbf{r}}}{r^2} = \frac{2\,\pi^2}{|\mathbf{k}|} \equiv \frac{2\,\pi^2}{\sqrt{\mathbf{k}^2}}. \qquad (1.8.38)$$

The interesting case by far is the Fourier transform of $1/r^3$. Here we are dealing with a dimensionless integral and hence the answer must be logarithmic in $\mathbf{k}^2$:

$$\int d^3\mathbf{r}\,\frac{e^{i\mathbf{k}\cdot\mathbf{r}}}{r^3} = b\ln(\mathbf{k}^2). \qquad (1.8.39)$$

Upon applying $(\partial/\partial k^i)(\partial/\partial k^i)$ to both sides of this equation gives

$$-\int d^3\mathbf{r}\,\frac{e^{i\mathbf{k}\cdot\mathbf{r}}}{r} = b\,\frac{2}{\mathbf{k}^2}, \qquad (1.8.40)$$

which from (1.8.35) gives $b = -2\,\pi$. That is [69]

$$\int d^3\mathbf{r}\,\frac{e^{i\mathbf{k}\cdot\mathbf{r}}}{r^3} = -2\,\pi\ln(\mathbf{k}^2). \qquad (1.8.41)$$

The moral of the above Fourier transforms is that only non-analytic terms such as $\ln(\mathbf{k}^2)$, $1/\sqrt{\mathbf{k}^2}$, contribute to the corrections to the potential at low energy (long distance). As a matter of fact analytic terms as in

$$\int \frac{d^3\mathbf{k}}{(2\pi)^3}\,e^{i\mathbf{k}\cdot\mathbf{r}}\,[\,1, \mathbf{k}^2, \ldots] = [\delta^3(\mathbf{r}), -\nabla^2\delta^3(\mathbf{r}), \ldots], \qquad (1.8.42)$$

which vanish far away from the origin, and originate from the ultraviolet (short distance) behavior of the theory.

Thus by trying to separate low energy effects from high energy ones, in view of applications in the low energy regime even if the theory has unfavorable ultraviolet behavior, one has been able to evaluate expressions like in (1.8.34) at long distances.[70]

---

[69] We note that the integral may not be of a higher power in $\ln(\mathbf{k}^2)$.

[70] For further details, see the above mentioned references. Recent recorded values of α and β in (1.8.34) are $\alpha = 3$, and $\beta = 41/10\pi$ [15].

1.9   Introduction to Loop Quantum Gravity

"Loop Quantum Gravity" (LQG) is a non-perturbative background independent formulation of quantum gravity. It is very interesting in the sense that space itself emerges from the theory. The theory stems from general relativity and ends up to be quite remote from it, but it retains the basic desirable ingredient of background independence. In "Loop quantum Gravity", we will encounter a quantum field theory in three dimensional space.

Our interest is measuring areas of surfaces in such a 3 dimensional space. We will see that the latter is written in terms of a field $E_i{}^a$, obtained directly from the definition of area of a surface, and is referred to, generally, as a gravitational "electric" field, which, in turn, is expressed in terms of triads $e^j{}_b$ in this space. Here $a, b = 1, 2, 3$, pertain to this space, while i, j denote Euclidean space indices. The metric on the 3 dimensional space is simply given by $q_{ab} = e^i_a \delta_{ij} e^j_b$. In a technical language, $E_i{}^a$ is called a densitized triad which determines the geometry of the 3 dimensional space. We then define, a canonical conjugate variable to $E_i{}^a$, referred to as a connection, and finally impose equal commutation relations on this pair of canonical conjugate field variables promoting the formalism to the quantum world.

LQG is described by an SU(2) gauge theory involving this connection giving rise to the parallel transport of a spinor along a curve and leads naturally to the concept of what is called a holonomy, as a propagator of parallel transport of a spinor. It also involves the densitized triad, mentioned above, as the canonical conjugate of the connection which determines the geometry of space instead of the metric or of its departure from the Minkowski metric. These two basic fields: the connection and the densitized triad, are referred to Ashtekar variables.[71] The theory is formulated in terms of these variables, where the connection dependence arises through the holonomy, and of quantum loops [72] expressed in terms of the holonomies. It may be stated that in this formalism, a priori, one is dealing with a quantum field theory in three dimensional space formulated in terms of the Ashtekar variables.

The states of space are described by the superposition of so-called spin-network states.[73] The basic idea goes to Penrose [47] whose interest was to construct the concept of space from combining angular momenta. Spin-network states are functionals of the connection and are defined in terms of a graph, its lines and vertices and are represented in a particular way which will be spelled out as we go along. For the time being, it suffices to say that they diagonalize the so-called area operator, as a geometrical operator, providing a discrete spectrum for the latter with the smallest possible non-zero value given by the order of the Planck length squared: $\hbar G/c^3 \sim 10^{-66}\,\text{cm}^2$. The emergence of space in terms of "quanta of geometry",

---

[71] Ashtekar [4, 5], Ashtekar and Isham [6], see also Sen [58].
[72] Rovelli and Smolin [52], Rovelli and Smolin [53].
[73] Rovelli and Smolin [52], [54], Baez [9], [10], Sen [58].

providing a granular structure of space, is a major and beautiful prediction of the theory.

In Sect. 1.9.1, a time-slicing approach of spacetime is formalized which eventually leads to develop loop quantum gravity within a three-dimensional geometry, where a time coordinate becomes an irrelevant variable. The concept of a gravitational field in the language of loop quantum gravity, as one of the Ashtekar variables, is introduced in Sect. 1.9.2, and the concept of a holonomy and some of its basic properties are established in Sect. 1.9.3, which involves the second Ashtekar variable—the connection. Via the quantization of the gravitational field, the concept of an area in the 3D space, as introduced in Sect. 1.9.2, is promoted to a functional differential operator acting on functionals of the connection. Spin-networks and spin-network states are introduced, and their intricacies are spelled out, in Sect. 1.9.4. The discrete spectrum of the area operator, by developing in the process an eigenvalue equation for it, is investigated in Sect. 1.9.5 establishing the quantized nature of area measurements in space at small distances of the order of the Plank length squared.

I will develop the subject matter in the simplest possible way, hoping that more physics students who are finding it difficult and beyond reach, would be exposed to this beautiful and unique theory which I feel is essential. We concentrate on a "thin slice" of the theory and refer the interested reader in more details, in the underlying constraints of the theory, of detailed references, and of further developments to Rovelli [50], Thiemann [62], Rovelli and Vidotto [51].

1.9.1 The ADM Formalism and Intricacies of the Underlying Geometry

The ADM [74] formalism, also known as a $3+1$ decomposition of spacetime, consists of time slicing of spacetime, by introducing, in the process, a sequence of 3D spaces, Σ labeled by a continuous coordinate t. Let Σ denote such a space, to be introduced in this formalism, labeled by a coordinate label t. We may define a time evolution (time-like) vector t^{μ} to specify the direction in which a time derivative is taken. The temporal distance between such hypersurfaces, labeled by coordinates t and $t + dt$, is denoted by $N\,dt$, and N is referred to as the "lapse". Thus by introducing a unit timelike vector $n^{\mu} : n^{\mu}n_{\mu} = -1$, perpendicular to the hypersurface Σ labeled by t, we may generate the following figure showing the same point P at different times. The vector N^{μ} in Σ, referred to as the "shift vector", was introduced to move from the point in question in Σ to the point where the vector Nn^{μ} is erected. By a

---

[74]ADM refers to Arnowitt, Deser and Misner [3].

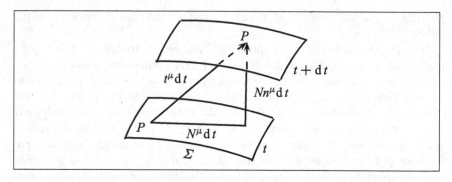

Fig. 1.7 The ADM 3+1 decomposition of spacetime, where N is referred to as the lapse, N^μ as the shift vector, and n^μ is a unit timelike vector: $n^\mu n_\mu = -1$

convenient parametrization

$$n^\mu = \frac{1}{N}(1, -N^a), \quad n_\mu = (-N, 0), \quad N^\mu = (0, N^a), \quad \text{we obtain,}$$

$$\text{with} \quad N^\mu n_\mu = 0, \quad t^\mu = N^\mu + Nn^\mu, \quad t^\mu = (1, 0, 0, 0).$$

These give

$$g_{00} = g_{\mu\nu} t^\mu t^\nu = \left[-(N)^2 + g_{ab} N^a N^b\right], \qquad g_{0a} = t^\mu g_{\mu a} = N_a, \qquad (1.9.1)$$

$$-ds^2 = g_{\mu\nu}\, dX^\mu\, dX^\nu = \left[-(N)^2 + N_a N^a\right] dt^2 + 2N_a\, dx^a\, dt + q_{ab}\, dx^a\, dx^b. \tag{1.9.2}$$

Thus the interval is specified completely in terms of the metric[75] q_{ab} of Σ, the lapse N, and the shift vector N^a (Fig. 1.7).

The ADM observation was that one may carry out tensor analysis of spatial tensor fields on Σ_t using the spatial metric q_{ab}, and this is unaffected by coordinate choices at later times. The time coordinate variable t label then becomes irrelevant in the underlying analysis. The lapse and the shift vector would tell us on how a coordinate system may be continued as one moves away from the surface $t = $ const..

Now we restrict the analysis to the three dimensional geometry in Σ with metric q_{ab}. We may introduce triads e^i_a, where the indices a run over $1, 2, 3$, with the indices $i, j, \ldots$ denoting Euclidean space labelings, and write the metric q_{ab}, as

$$q_{ab} = e^i_a \delta_{ij} e^j_b, \qquad \det[q_{ab}] = (\det[e^i_a])^2. \tag{1.9.3}$$

---

[75]This means, in particular, that $N_a = q_{ab}N^b$. Note that q_{ab} is rather a standard notation for this metric.

In the next section, we will see how the concept of the gravitation field may be introduced, as one of the key variables in the theory, and eventually introduce the concept of holonomy.[76] Thanks to the Ashtekar variables, introduced from the subsequent analysis, the non-abelian gravitational field character, in particular, is reduced to a non-abelian Yang-Mills gauge field one with an underlying SU(2) gauge group in the "Loop Quantum Gravity" formulation.[77]

1.9.2   Gravitational "Electric" Flux Across a Surface in 3D Space

Consider a surface S in Σ parametrized by a pair of variables (λ^1, λ^2). When λ^2 is kept fixed, and λ^1 is made to vary, one traces a curve with tangent having components $e_a^i(\partial x^a/\partial \lambda^1) \equiv v_1^i$ locally in a tangent plane. Similarly, when λ^1 is kept fixed, and λ^2 is made to vary, one traces a curve with tangent having components $e_a^i(\partial x^a/\partial \lambda^2) \equiv v_2^i$ locally in the tangent plane. An oriented surface element, in the tangent plane, may be then defined locally by $d\sigma_i = \varepsilon_{ijk} v_1^j v_2^k d\lambda^1 d\lambda^2$. The area of the given surface emerges as

$$\mathscr{A}[S] = \int_S \sqrt{d\sigma_i \, d\sigma_i} = \int_S d\lambda^1 d\lambda^2 \sqrt{v_1 \cdot v_1 \, v_2 \cdot v_2 - (v_1 \cdot v_2)^2}, \qquad (1.9.4)$$

where we have used the identity $\varepsilon_{ijk} \varepsilon_{ilm} = \delta_{jl}\delta_{km} - \delta_{jm}\delta_{kl}$. We may introduce a vector n_a perpendicular to both $\partial x^a/\partial \lambda^1$, $\partial x^a/\partial \lambda^2$, by

$$n_a = \varepsilon_{abc} \frac{\partial x^b}{\partial \lambda^1} \frac{\partial x^c}{\partial \lambda^2},$$

and note that with $u_i^a \equiv \partial x^a/\partial \lambda^i$,

$$\frac{1}{2} \varepsilon_{ijk} \varepsilon^{abc} e_b^j e_c^k n_a = \frac{1}{2} \varepsilon_{ijk} \varepsilon^{abc} e_b^j e_c^k \varepsilon_{ade} u_1^d u_2^e = \varepsilon_{ijk} v_1^j v_2^k \equiv E_i,$$

$$(1.9.5)$$

which allows us to introduce a field with components

$$E_i^a = \frac{1}{2} \varepsilon_{ijk} \varepsilon^{abc} e_b^j e_c^k, \qquad E_i^a n_a = E_i, \qquad (1.9.6)$$

---

[76]For holonomies in quantum mechanics and applications, see Manoukian [40, Sect. 8.13, p. 524].

[77]We recall an elementary aspect of group theory, showing that an SU(2) transformation induces a 3 dimensional rotation in Euclidean space with metric δ_{ij} as follows. For (x^1, x^2, x^3) as components of a three-vector in Euclidean space, write $X = \sigma_i x^i$, where the σ_i are the Pauli matrices. Then $\det X = -x^i x^i$. For an SU(2) matrix M which induces a transformation on X via $X \to X' = M X M^{-1}$, one has $\det X' = \det X$, i.e., $x'^i x'^i = x^i x^i$.

and rewrite the expression for the area in (1.9.4) simply as

$$\mathscr{A}[S] = \int_S d\lambda^1 d\lambda^2 \sqrt{E_i E_i}. \tag{1.9.7}$$

The area $\mathscr{A}$ may be written as a Riemann sum, by partitioning the surface S into N infinitesimal surface elements $S_1, \ldots S_N$, for $N \to \infty$, as follows

$$\mathscr{A}[S] = \lim_{N \to \infty} \sum_{k=1}^{N} \sqrt{E_i[S_k] E_i[S_k]}, \tag{1.9.8}$$

an expression that will be useful later.

In the loop quantum gravity, the field $E_i{}^a$, is termed as the gravitational field, rather than the metric or its deviation from the Minkowski one. In a technical language is called a densitized triad. It is also referred to as the gravitational "electric" field, and one may consider the flux of the field across a surface S given by

$$E_i[S] = \int_S d\lambda^1 d\lambda^2 E_i{}^a n_a. \tag{1.9.9}$$

In order to define a line of force of gravitation piercing such a surface, we introduce next the concept of a holonomy which connects and provides a link between two points along a curve. The canonical conjugate field variable to $E_i{}^a$ is denoted by $A^i{}_a$. In terms of this field, we define, for the subsequent analysis, the field variable Γ_a, referred to as an SU(2) connection,

$$\Gamma_a = \left(\frac{\sigma_i}{2}\right) \Gamma^i{}_a = \omega \left(\frac{\sigma_i}{2}\right) A^i{}_a.$$

Here ω is a free parameter whose nature will be discussed later, and the σ_i denote the Pauli matrices.

1.9.3   Concept of a Holonomy and Some of its Properties

The connection Γ_a, tells us how a spinor $\chi = (\chi^1 \ \chi^2)^\top$ is parallel transported along a curve, parametrized by variable s. As we have done in (6.1.5) in Chap. 6 of Vol. I, we have the similar equation $(dx^a/ds = \dot{\gamma}^a(s))$

$$\frac{d}{ds}\chi(\gamma(s)) = i\dot{\gamma}^a(s) \Gamma_a(\gamma(s)) \chi(\gamma(s)), \tag{1.9.10}$$

in matrix form, which is easily integrated[78] from $\underline{s}$ to $\bar{s}$, corresponding to two points on the curve, to give

$$\chi(\gamma(\bar{s})) = \mathscr{P}\left(\exp\left[\,i\int_{\underline{s}}^{\bar{s}} ds\,\dot{\gamma}^{\,a}(s)\,\Gamma_a(\gamma(s))\,\right]\right)\chi(\gamma(\underline{s})), \tag{1.9.11}$$

where $\mathscr{P}$ means path-ordering, exactly as is done for time-ordering. That is, in particular,

$$\mathscr{P}[\,\Gamma_a(\gamma(s_1))\,\Gamma_b(\gamma(s_2))\,] = \Gamma_b(\gamma(s_2))\,\Gamma_a(\gamma(s_1)) \quad \text{if} \quad s_2 > s_1,$$

along the curve, and so on. The object

$$h_\gamma(\bar{s},\underline{s}) = \mathscr{P}\left(\exp\left[\,i\int_{\underline{s}}^{\bar{s}} ds\,\dot{\gamma}^{\,a}(s)\,\Gamma_a(\gamma(s))\,\right]\right), \tag{1.9.12}$$

is referred to as a holonomy.[79] The holonomy is the basic variable used in the "Loop Quantum Gravity" formulation.

For two curves introduced via the equation

$$\gamma(s) = \begin{cases} \gamma_1(s), & \underline{s} < s, \\[2mm] \gamma_2(s), & s < \bar{s}, \end{cases} \tag{1.9.13}$$

joining at the point s, we obviously have

$$h_\gamma(\bar{s},\underline{s}) = h_{\gamma_2}(\bar{s},s)\,h_{\gamma_1}(s,\underline{s}). \tag{1.9.14}$$

Under an SU(2) gauge transformation, via a unitary matrix $V(x)$ locally, we have the well known gauge transformation[80]

$$\Gamma_a(x) \rightarrow \Gamma_a^V(x) = V(x)\left(\Gamma_a(x) - \frac{\partial_a}{i}\right)V^{-1}(x). \tag{1.9.15}$$

In view of the above transformation law, we investigate the gauge transformation of $h_\gamma(\bar{s},\underline{s})$. To this end, we consider a limiting cases of a holonomy which is a trivial one involving zero integration, where one doesn't move from the starting point and

---

[78] See Problem 1.19.
[79] Some authors define a holonomy as the trace of the expression in (1.9.12).
[80] See, e.g., Chap. 6, (6.1.10) of Vol. I [43].

is given by

$$h_\gamma(\bar{s}, \underline{s})\big|_{\bar{s}=\underline{s}} = 1, \tag{1.9.16}$$

Given an SU(2) - transformation via $V(x(s))$, define

$$Q(s) = V(x(s)) h_\gamma(s, \underline{s}). \tag{1.9.17}$$

Upon taking the derivative d/ds of the latter equation, using (1.9.10), and the facts that $(d/ds)V(x(s)) = \dot{\gamma}^{\,a}(s)\,\partial_a V(x(s))$, $(\partial_a V)\,V^{-1} = -V\,\partial_a V^{-1}$, we obtain

$$\begin{aligned}
\frac{d}{ds} Q(s) &= \dot{\gamma}^{\,a}(s)\left[\partial_a V(x(s)) + i\, V(x(s))\, \Gamma_a(x(s))\right] h_\gamma(s, \underline{s}) \\
&= \dot{\gamma}^{\,a}(s)\left[\partial_a V(x(s)) + i\, V(x(s))\, \Gamma_a(x(s))\right] V^{-1}(x(s))\, V(x(s))\, h_\gamma(s, \underline{s}) \\
&= i\, \dot{\gamma}^{\,a}(s)\left[V(x(s))\left(\Gamma_a(x) - \frac{1}{i}\,\partial_a\right) V^{-1}(x(s))\right] Q(s), \\
&= i\, \dot{\gamma}^{\,a}(s)\, \Gamma^V(x(s))\, Q(s), \tag{1.9.18}
\end{aligned}$$

where in writing the last line, we have used the gauge transformation rule for $\Gamma_a(x)$ in (1.9.15), and have also used, in the process, the definition of $Q(s)$ in (1.9.17) all over again. The solution of (1.9.18) is given by

$$Q(\bar{s}) = h_\gamma^V(\bar{s}, \underline{s})\, Q(\underline{s}), \tag{1.9.19}$$

where

$$h_\gamma^V(\bar{s}, \underline{s}) = \mathscr{P}\left(\exp\left[\,i\int_{\underline{s}}^{\bar{s}} ds\, \dot{\gamma}^{\,a}(s)\, \Gamma_a^V(\gamma(s))\,\right]\right). \tag{1.9.20}$$

The last two equations give from (1.9.16), (1.9.17), the explicit gauge transformation of $h_\gamma(\bar{s}, \underline{s})$,

$$h_\gamma^V(\bar{s}, \underline{s}) = V(x(\bar{s}))\, h_\gamma(\bar{s}, \underline{s})\, V^{-1}(x(\underline{s})). \tag{1.9.21}$$

Hence $\mathrm{Tr}\,[\,h_\gamma(\underline{s}, \underline{s})\,]$, referred to as Wilson loop,[81] corresponding to a *closed loop*, is *invariant under an* SU(2) *transformation* given in (1.9.15).

The two fields $A^j{}_b$ and $E_i{}^a$ are canonical conjugate variables, and the Hamiltonian of general relativity may be expressed in terms of the fields $E_i{}^a$, $\Gamma^j{}_b$.[82] We will not

---

[81] See also Sect. 6.12. of Vol. I [43].
[82] See, e.g., [50, 62].

need the corresponding explicit expression of the Hamiltonian here, however. In particular, the Poisson brackets are given by ($\Gamma^i{}_a \equiv \omega A^i{}_a$)

$$\{A^i_a, E_j{}^b\}_{\text{PB}} = \delta^i_j \, \delta^b_a \, \delta^3(x, x'), \qquad \{\Gamma^i_a, E_j{}^b\}_{\text{PB}} = \omega \, \delta^i{}_j \, \delta^b_a \, \delta^3(x, x'), \qquad (1.9.22)$$

where x, x' are in Σ, and $\hbar$, c, have been set equal to one,[83] and ω is taken as

$$\omega = 8\pi \beta G_N, \qquad (1.9.23)$$

where G_N is Newton's constant, and β, a numerical, is a free parameter of loop quantum gravity, referred to as a Barbero-Immirzi constant.[84] Its value is usually chosen by making a comparison with physically established results ($0 < \beta < 1$) such as in studies of the entropy of black holes, via the Bekestein-Hawking Entropy Formula. The latter expression is given Appendix E of this chapter.

A transition to a quantum formulation may be achieved by a functional representation of the densitized triad $E_i{}^a$ as follows

$$\hat{E}_i{}^a(x) = (-i)\frac{\delta}{\delta A^i{}_a(x)}, \qquad (1.9.24)$$

consistent with (1.9.22), now as a commutator, with $A^i{}_a(x)$ acting as a multiplicative operator.

Consider the application of the flux in (1.9.9), now as an operator, to a holonomy with the underlying curve γ crossing the surface S, at a given point, where on one side of the surface the curve is denoted by γ_1, and on the other side it is denoted by γ_2, as shown in Fig. 1.8a: going, say, in the direction from γ_2 to γ_1. In view of application to the Riemann sum in (1.9.8), we may consider the surface S to be arbitrary small so that only one line crosses it. To the above end, we consider variation of a holonomy $h_\gamma(\bar{s}, \underline{s})$ about some point $s : \underline{s} < s < \bar{s}$. From (1.9.12)–(1.9.14), we may first write ($\tau = \sigma/2$)

$$\delta h_\gamma(\bar{s}, \underline{s}) = h_\gamma\left(\bar{s}, s + \frac{\Delta s}{2}\right) \{\,.\,\} \, h_\gamma\left(s - \frac{\Delta s}{2}, \underline{s}\right), \qquad (1.9.25)$$

$$\{\,.\,\} = 1 + i \int_{s-\frac{\Delta s}{2}}^{s+\frac{\Delta s}{2}} ds_1 \dot{\gamma}^a(s_1) \, \tau_i \, \Gamma^i{}_a(x(s_1)) + \frac{(i)^2}{2} \int_{s-\frac{\Delta s}{2}}^{s+\frac{\Delta s}{2}} ds_2 \int_{s-\frac{\Delta s}{2}}^{s+\frac{\Delta s}{2}} ds_1 \, \dot{\gamma}^a(s_1)\dot{\gamma}^b(s_2)$$

$$\times \left[\left(\tau_i \tau_j \, \Theta(s_1 - s_2) + \tau_j \tau_i \, \Theta(s_2 - s_1) \right) \Gamma^i{}_a(x(s_1)) \Gamma^j{}_b(x(s_2)) \right] + \cdots. \qquad (1.9.26)$$

---

[83]In this section, the variables $x, x', \ldots$ are understood to correspond to three components.

[84]Barbero [11], Immirzi [33].

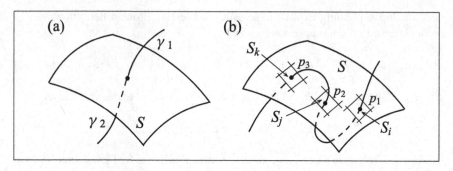

Fig. 1.8 (a) A *curve*, with parts denoted by γ_1, γ_2, as it pierces an area element—see (1.9.28). (b) A *curve* piercing an area several times—see (1.9.30).

From the above two equations, we obtain by functional differentiation

$$(-\mathrm{i})\frac{\delta}{\delta A^i{}_a(x)}A^j{}_b(x(s)) = -\mathrm{i}\,\delta^j{}_i\,\delta^a{}_b\,\delta^3(x-x(s)),$$

in the process, with $\Delta s \to 0$,

$$\widehat{E}_i[S]\,h_\gamma(\bar{s},\underline{s}) = 8\pi\beta G_\mathrm{N}\int_S \mathrm{d}\lambda^1\mathrm{d}\lambda^2 \int_{s-\frac{\Delta s}{2}}^{s+\frac{\Delta s}{2}} \mathrm{d}s_1\,\varepsilon_{abc}$$

$$\times \frac{\partial x^a}{\partial\lambda^1}\frac{\partial x^b}{\partial\lambda^2}\frac{\partial x^c}{\partial s_1}\,\delta^3\big(x(\lambda^1,\lambda^2)-x(s_1)\big)\,h_{\gamma_1}(\bar{s},s)\,\tau_i\,h_{\gamma_2}(s,\underline{s}). \tag{1.9.27}$$

We recognize $\varepsilon_{abc}(\partial x^a/\partial\lambda^1)(\partial x^b/\partial\lambda^2)(\partial x^c/\partial s_1)$ as the Jacobian of transformation of variables of integration leading to integrations over the variables x^1,x^2,x^3, thus obtaining the simple expression

$$\widehat{E}_i[S]\,h_\gamma(\bar{s},\underline{s}) = 8\pi\beta G_\mathrm{N}\,h_{\gamma_1}(\bar{s},s)\,\tau_i\,h_{\gamma_2}(s,\underline{s}), \tag{1.9.28}$$

since we have considered the intersection of the curve at a single point as shown in Fig. 1.8a.

Similarly, referring to the second term in (1.9.26), we have by taking the limit $\Delta s \to 0$

$$\widehat{E}_i[S]\,\widehat{E}_i[S]\,h_\gamma(\bar{s},\underline{s}) = (8\pi\beta G_\mathrm{N})^2\,\tau_i\,\tau_i\,h_\gamma(\bar{s},\underline{s}) = (8\pi\beta G_\mathrm{N})^2\,\frac{1}{2}\left(\frac{1}{2}+1\right)h_\gamma(\bar{s},\underline{s}), \tag{1.9.29}$$

where we have, in the process, used the elementary property: $\Theta(s_1-s_2)+\Theta(s_2-s_1)=1$, $\tau_i\tau_i=3I/4$, and (1.9.14).

As a further generalization, consider a curve γ with no parts of it touching, which crosses a surface S at several points as shown in Fig. 1.8b, then clearly by partitioning S into very fine cells so that at most one intersection occurs per cell, by using (1.9.28) for each cell involving an intersection, and summing over these cells, we obtain

$$\widehat{E}_i[S]\,h_\gamma \;=\; 8\pi\beta G_N \sum_p \delta_p\, h_{\gamma_1}^p\, \tau_i\, h_{\gamma_2}^p, \tag{1.9.30}$$

as a sum over all points of intersection, with the δ_p taking the values ± 1.

Before closing this subsection, we consider higher spin representations of SU(2) as well. To this end, irreducible representations of this group are labeled by the spin j taking values in $\{1/2, 1, 3/2, 2, \ldots\}$. The generators of the group will be denoted by $T_i^{(j)}, i = 1, 2, 3$, satisfying the quantization condition $T_i^{(j)} T_i^{(j)} = j(j+1)I$, with $[T_i^{(j)}, T_k^{(j)}] = i\varepsilon_{ikl}\, T_l^{(j)}$, $T_i^{1/2} = \sigma_i/2 \equiv \tau_i$.

Let $R^j[A, h_\gamma] = \langle\, h_\gamma \,|\, R^j \,\rangle$, denote the spin j irreducible matrix representation of the holonomy of the connection Γ^i or A^i, along a curve γ. For spin $1/2$, the so-called fundamental representation, $R^{1/2}[A, h_\gamma] = h_\gamma$. The action of the operator $\widehat{E}[S]$ on $R^j[A, h_\gamma]$ is the same as the one given in (1.9.28), for the curve γ intersecting the surface at a given point, obtained by simply replacing τ_i by $T_i^{(j)}$, i.e.,

$$\widehat{E}_i[S]\, R^j\left[A, h_\gamma\right] \;=\; 8\pi\beta G_N\, R^j\left[A, h_{\gamma_1}\right]\, T_i^{(j)} R^j\left[A, h_{\gamma_2}\right], \tag{1.9.31}$$

and similarly for the case involving several intersections.[85]

Now we are ready to define the concept of spin networks, construct space itself and derive formally a major result of loop quantum gravity.

1.9.4 Definition of Spin Networks, Spin Network States, States of Geometry

To introduce the concept of "Spin Networks" and "Spin Network States", we introduce first the terminology of a graph in this formalism. A graph $\mathscr{G}$ is a collection of oriented curves, say, $\gamma_1, \gamma_2, \ldots, \gamma_L$, referred to as links, that may join only at their end points. These end points are called nodes, or vertices and will be denoted by $\mathscr{N}_1, \ldots, \mathscr{N}_N$. The number of lines meeting a given node is referred to as the valence of the node and is necessarily non-zero. We associate spins $j_1, j_2, \ldots, j_L$ with the L links which may take values in $\{1/2, 1, 3/2, \ldots\}$. In Fig. 1.9, a simple graph is shown involving three links with two three-valent nodes, and an n-valent

---

[85]We will not need actual explicit representations of the so-called Wigner type matrices $R^j[A, h_\gamma]$. For details of Wigner matrices $D^{(j)}$ in the theory of angular-momentum & spin, in general, see Manoukian [40].

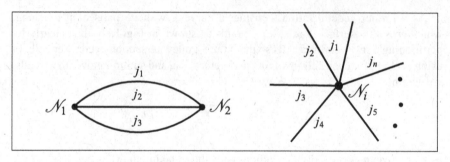

Fig. 1.9 A simple graph is shown involving three links with two three-valent nodes, as well as an n-valent node, with n lines impinging on it, with spins denoted generally by $j_1, j_2, \ldots, j_n$

node, say $\mathcal{N}_i$ involving, by definition, n lines impinging on it, with spins denoted generally by $j_1, j_2, \ldots, j_n$.

In order to define spin network states, we have to combine these spins, and most importantly, generate invariant, i.e., gauge invariant, states under SU(2) transformations. To this end, a spin j may be described by a vector $v_m^{(j)}$ with $2j + 1$ indices. Accordingly, when we combine n spins, we generate a vector space with elements $v_{m_1 \ldots m_n}^{(j_1 \ldots j_n)}$ which we know from elementary quantum mechanics of addition of (angular momenta) spins, may involve several total spins. Gauge invariance implies to restrict to the subspace which is invariant under SU(2) transformations, referred to as the singlet space. For a trivalent node $n = 3$, one would then obviously have the elementary restriction for the three spins: $|j_1 - j_2| \leq j_3 \leq j_1 + j_2$. In this case, we identify the single element $v_{m_1 m_2 m_3}^{(j_1\ j_2\ j_3)}$, simply with the Clebsch-Gordan coefficient $\langle j_1, m_1 : j_2, m_2 \, | \, j_3, -m_3 \rangle$, up to a normalization factor.[86] Such vector elements (tensors) are referred to as intertwiners. For a 4-valent node, and a 5-valent node, for example, one may represent the nodes with their lines impinging on them as shown in Fig. 1.10 where one introduces so-called virtual links and consider couplings of the spins at the resulting nodes 1 and 2 in case (a), and at nodes 1, 2, 3 in case (b). That is, for these nodes, in such general cases, as well as for n - valent nodes, they may be described as contractions of three-valent nodes. These decompositions also show that the dimensions of the invariant vector spaces may be greater than one.

The collection of all the links, with associated spins j_ℓ, the nodes, with associated intertwiners v_i of a given graph $\mathcal{G}$, is referred to as a spin-network, represented by $(\mathcal{G}, j_\ell, v_i)$. A spin-network state, with connection Γ, or A, may be symbolically represented by

$$\Psi_S^{\mathcal{G}}[A] \, = \, \otimes_\ell R^{(j_\ell)}[A, h_{\gamma_\ell}] \otimes_i v_i, \tag{1.9.32}$$

[86]Clebsch-Gordan coefficients are proportional to so-called $3j$ symbols, see, e.g, Manoukian [40, p. 289].

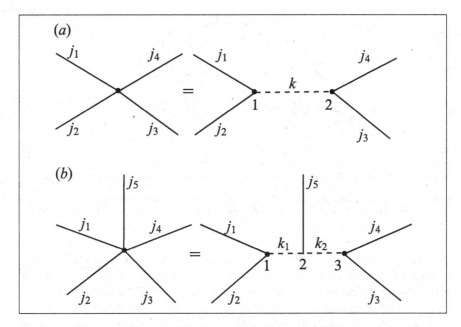

Fig. 1.10 A 4-valent node and a 5-valent node are shown, respectively, in parts (**a**) and (**b**). They are expressed in terms of 3-valent nodes. The *dashed lines* denote virtual links and are introduced so that contractions between nodes may, in general, be described as contractions between three-valent nodes

defined with appropriate contractions. The main thing to note here is that the dependence of a spin-network state on the connection, arises from the dependence of the spin j_ℓ irreducible matrix representations $R^{(j_\ell)}[A, h_{\gamma_\ell}]$ of the holonomy of the connection along the curves γ_ℓ as explicitly displayed above. The action of the operator $\widehat{E}_i[S]$ on these functionals have been spelled out in (1.9.31). Functionals which depend on the connection through its holonomy are referred to as cylindrical functionals.

Summarizing then, in reference to a graph $\mathscr{G}$, a spin-network is defined with SU(2) representations of spins associated with its links, and SU(2) invariant tensors attached to its nodes, referred to as intertwiners. We will see that the spin-network states diagonalize the area operator. The quantum states of the geometry of space are defined as linear superpositions of spin-network states.

1.9.5   Quanta of Geometry

Here we set an eigenvalue equation for the area operator $\widehat{\mathscr{A}}[S]$ as obtained from the expression in (1.9.8), with densitized triad $E_i{}^a(x)$ promoted to a quantum operator $\widehat{E}_i{}^a(x)$ represented as a functional differential operator as given in (1.9.24) [see also (1.9.32)].

A spin-network is shown below, with links intersecting some surfaces (represented by thinner lines in side views).

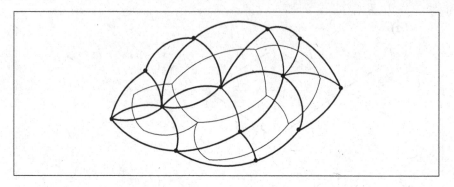

Fig. 1.11 A spin-network with links intersecting some surfaces drawn (represented by *thinner lines* in side views). A spin network represents a quantum state of the geometry of space. Classically, it looks continuous

Given a spin-network, let S be any surface such that no nodes of the spin-network lie on it. Surfaces of this type are shown above. Since the links may join only at their end points (at nodes), we may partition, the surface S into fine cells S_k such that at most one link intersects each cell. This allows us to use the limit of a Riemann sum, as given in (1.9.8), and the explicit expression of the action of the operator $\widehat{E}_i[S]\widehat{E}_i[S]$, for a given surface in (1.9.29) for one intersection, corresponding to spin 1/2, to conclude that for a spin-network state $\Psi_S^{\mathscr{G}}[A]$,

$$\widehat{\mathscr{A}}\,[S]\,\Psi_S^{\mathscr{G}}[A] = \frac{8\pi\beta\hbar G_N}{c^3} \sum_p \sqrt{j_p(j_p+1)}\,\Psi_S^{\mathscr{G}}[A], \qquad (1.9.33)$$

where we have used the fact the so-called Casimir operator $T_i^{(j)}T_i^{(j)} = j(j+1)I$, and it commutes with all the generators of the algebra. Here we have restored the fundamental constants $\hbar$, c. The sum is over all intersection points p, with the spin of the corresponding link intersecting the surface being j_p. This remarkable result shows the quantization of area in units of the Planck length squared $\hbar G_N/c^3 \sim 10^{-66}\,\mathrm{cm}^2$ as its smallest non-zero value. It is defined by ultimate boundaries of space itself at such microscopic distances. Obviously when there are no intersections of the links with the surface,[87] one gets zero—a hole in the quantum space. This gives an interesting granular structure to space. It is important to emphasize that a spin network is not embedded in space but defines space itself. Loops are generated from links attached to nodes. Loop quantum gravity, unlike other approaches, offers an intriguing description of space generated by such a set of loops with nothing else in between (Fig. 1.11).

---

[87]For the situation when some of the nodes and/or nodes and links lie on the surface, the analysis becomes quite involved, and we refer the reader for the underlying details to Rovelli [50, p. 254].

Summary

Let us recapitulate the basic points of this beautiful formalism of quantum gravity. Although one has started from general relativity, and applied a time-slicing of spacetime, one ended up with a theory quite far from its origin, but still retained the very basic and desirable ingredient of background independence. The resulting non-perturbative theory is an SU(2) gauge theory. It involves a densitized triad $E_j{}^b$ obtained directly from the definition of area of a surface, and a connection $\Gamma^i{}_a$ (or $A^i{}_a$), which, through the expression of a holonomy, tells us how a spinor is parallel transported along a curve, where $A^i{}_a$ and $E_j{}^b$ are canonical conjugate variables. Here space is generated by the so-called spin networks, described by spin-network states. A quantum state of space is given by a superposition of spin-network states. The latter are also referred to as s-knots. A spin-network is a functional of the connection. It is defined in terms of a graph, its vertices, referred to as nodes, and a collection of curves, referred to as its links. The latter are labeled by spins j, and may join only at their end points at the nodes. Along, every link one introduces the spin-j irreducible matrix representation of the holonomy of the SU(2) connection along the corresponding curve. On the other hand, the nodes are represented by tensors, referred to as intertwiners, which are essentially contractions of Clebsch-Gordan coefficients. The spin-network states emerge by contracting the matrix elements of the matrices of the various j-spins representations and the intertwiners resulting in gauge invariant expressions. The spin-networks diagonalize the area operator providing a discrete spectrum for the latter.

I have provided the simplest possible approach to this very interesting theory and, as mentioned in the Introduction to this section on "Loop Quantum Gravity", we hope, as a result, more physics students will be able to benefit from it. For more intricate details, consistency and further developments, defining other geometrical objects, including the treatment of constraints in the theory, as well as studies of the world-history of spin-networks and other applications, we refer the reader to [50, 51, 62].

Appendix A: Variation of a Determinant

In varying $\sqrt{-g}$ with respect to $g_{\mu\nu}$ where $g(x) = \det[g_{\mu\nu}(x)]$, we need to know how to vary a determinant. This is explicitly used in Problem 1.3, and in obtaining the field equation in (1.2.2). This appendix shows you how to do that.

Given two matrices N, M, integrate the following from $\lambda = 0$ to $\lambda = 1$:

$$\frac{d}{d\lambda} e^{\lambda N} e^{-\lambda M} = e^{\lambda N}(N - M) e^{-\lambda M}, \tag{A-1.1}$$

to obtain

$$e^N e^{-M} = 1 + \int_0^1 d\lambda \, e^{\lambda N}(N - M) e^{-\lambda M}, \tag{A-1.2}$$

or for $N = M + \delta M$

$$e^N - e^M = \int_0^1 d\lambda\, e^{\lambda N}(N - M)\, e^{-\lambda M}\, e^M. \tag{A-1.3}$$

$$\delta e^M = \int_0^1 d\lambda\, e^{\lambda M}\, (\delta M)\, e^{-\lambda M}\, e^M. \tag{A-1.4}$$

We multiply the latter equation from the right by e^{-M}, and take the trace to obtain

$$\mathrm{Tr}\left[(\delta e^M)\, e^{-M}\right] = \mathrm{Tr}\left[\delta M\right] = \mathrm{Tr}\left[e^{-M}\delta e^M\right], \tag{A-1.5}$$

where the second equality follows from the fact that

$$\mathrm{Tr}\left[e^{\lambda M}\, (\delta M)\, e^{-\lambda M}\right] = \mathrm{Tr}\left[(\delta M)\, e^{-\lambda M}e^{\lambda M}\right] = \mathrm{Tr}\left[\delta M\right]. \tag{A-1.6}$$

For a given matrix A, let $M = \ln[A]$. (A-1.5) then gives

$$\mathrm{Tr}\left[A^{-1}\delta A\right] = \delta\left[\mathrm{Tr}\,(\ln[A])\right]. \tag{A-1.7}$$

On the other hand for a c-number a: $\delta e^a = e^a\,\delta a$, and for $a = \mathrm{Tr}\,(\ln[A])$, $e^{\mathrm{Tr}\ln[A]} = \det[A]$. This c-number equation then gives

$$\delta \det[A] = \det[A]\,\delta\left[(\mathrm{Tr}\ln[A])\right], \tag{A-1.8}$$

which upon comparison with (A-1.7) leads to

$$\delta \det[A] = \det[A]\,\mathrm{Tr}\left[A^{-1}\delta A\right]. \tag{A-1.9}$$

Appendix B: Parametric Integral Representation of the Logarithm of a Matrix

For a given matrix M into consideration, one may conveniently introduce another matrix N commuting with M, such as the unit matrix, and express

$$\ln M - \ln N = \int_0^1 \frac{(M - N)\, d\lambda}{N + (M - N)\lambda} = i(M - N)\int_0^\infty ds\, e^{-is(N - i\epsilon)}\int_0^1 d\lambda\, e^{-is\lambda(M - N)}. \tag{B-1.1}$$

An integration of the right-hand side over λ gives the following parametric integral representation

$$\ln M = -\int_0^\infty \frac{ds}{s} e^{-is(M-i\epsilon)} + \left[\ln N + \int_0^\infty \frac{ds}{s} e^{-is(N-i\epsilon)}\right], \qquad (B\text{-}1.2)$$

where we note that the second term, within the square brackets, is independent of the matrix M in consideration. Of particular interest is the trace of $\ln M$ given by

$$\mathrm{Tr}[\ln M] = -\int_0^\infty \frac{ds}{s} \mathrm{Tr}\left[e^{-is(M-i\epsilon)}\right] + \mathrm{Tr}\left[\ln N + \int_0^\infty \frac{ds}{s} e^{-is(N-i\epsilon)}\right], \qquad (B\text{-}1.3)$$

where, again the second term on the right-hand side is independent of the matrix M. In a path integral representation, through exponentiation of the expression in (B-1.3), such a second term just gives rise to an overall multiplicative constant factor to the generating functional $\langle 0_+|0_-\rangle$ independent of the matrix M in consideration.

Appendix C: Content of the Euler-Poincaré Characteristic

The purpose of this Appendix is to establish the following fundamental result in four dimensional spacetime that

$$\delta\left(\sqrt{-g}\left[R_{\mu\nu\sigma\lambda} R^{\mu\nu\sigma\lambda} - 4 R_{\mu\nu}R^{\mu\nu} + R^2\right]\right), \qquad (C\text{-}1.1)$$

is a total derivative.[88] This is the content of the so-called Euler-Poincaré characteristic which holds true in four dimensional spacetime. It is also known as the Gauss-Bonnet Theorem in four dimensions. According to Problem 1.16, this is established by considering equivalently the far simpler expression

$$\delta\left(\sqrt{-g}\, \eta^{\mu\nu\alpha\beta} \eta^{\sigma\lambda\gamma\kappa} R_{\mu\nu\sigma\lambda} R_{\alpha\beta\gamma\kappa}\right).$$

To establish that this is a total derivative, we make use of the following results:[89]

$$\delta\sqrt{-g} = \frac{1}{2}\sqrt{-g}\, g^{\rho\rho'}\delta g_{\rho\rho'}, \qquad (C\text{-}1.2)$$

$$\delta\eta^{\mu\nu\sigma\rho} = -\frac{1}{2}\eta^{\mu\nu\sigma\rho}\, g^{\rho\rho'}\delta g_{\rho\rho'}, \qquad (C\text{-}1.3)$$

---

[88]DeWitt [22, 23], 't Hooft and Veltman [61].

[89]Note all the indices here, including α, β, are tensorial indices in curved spacetime with metric $g_{\mu\nu}$, and not Lorentz ones.

$$\delta R_{\mu\nu\sigma\lambda} = \delta g_{\mu\rho} R^{\rho}{}_{\nu\sigma\lambda} + g_{\mu\rho}(\nabla_{\sigma}\,\delta\Gamma_{\nu\lambda}{}^{\rho} - \nabla_{\lambda}\,\delta\Gamma_{\nu\sigma}{}^{\rho}), \qquad \text{(C-1.4)}$$

$$\eta^{\sigma\lambda\gamma\kappa}\,\nabla_{\sigma}\,R_{\alpha\beta\gamma\kappa} = 0, \qquad \text{(C-1.5)}$$

$$\nabla_{\rho}\eta^{\mu\nu\alpha\beta} = 0, \qquad \text{(C-1.6)}$$

established, respectively, in Problems 1.3, (i), Problem 1.15, Problem 1.7, (ii), Problem 1.17, Problem 1.18.

From (C-1.2)–(C-1.4), the variation in question is given by

$$\delta\left(\sqrt{-g}\,\eta^{\mu\nu\alpha\beta}\,\eta^{\sigma\lambda\gamma\kappa}R_{\mu\nu\sigma\lambda}R_{\alpha\beta\gamma\kappa}\right) = \frac{1}{2}g^{\rho\rho'}\delta g_{\rho\rho'}\left(\sqrt{-g}\,\eta^{\mu\nu\alpha\beta}\eta^{\sigma\lambda\gamma\kappa}R_{\mu\nu\sigma\lambda}R_{\alpha\beta\gamma\kappa}\right)$$

$$+ 2\left(-\frac{1}{2}\right)g^{\rho\rho'}\delta g_{\rho\rho'}\left(\sqrt{-g}\,\eta^{\mu\nu\alpha\beta}\eta^{\sigma\lambda\gamma\kappa}R_{\mu\nu\sigma\lambda}R_{\alpha\beta\gamma\kappa}\right)$$

$$+ 2\,\delta g_{\mu\rho}\sqrt{-g}\,\eta^{\mu\nu\alpha\beta}\,\eta^{\sigma\lambda\gamma\kappa}R^{\rho}{}_{\nu\sigma\lambda}R_{\alpha\beta\gamma\kappa}$$

$$+ 4\sqrt{-g}\,\eta^{\mu\nu\alpha\beta}\eta^{\sigma\lambda\gamma\kappa}\nabla_{\sigma}\left(\delta\Gamma_{\nu\lambda}{}^{\rho}\right)g_{\mu\rho}R_{\alpha\beta\gamma\kappa}\,. \qquad \text{(C-1.7)}$$

The first two expressions on the right-hand side of the above equation may be combined and give

$$-\frac{1}{2}g^{\rho\rho'}\delta g_{\rho\rho'}\left(\sqrt{-g}\,\eta^{\mu\nu\alpha\beta}\eta^{\sigma\lambda\gamma\kappa}R_{\mu\nu\sigma\lambda}R_{\alpha\beta\gamma\kappa}\right).$$

From (C-1.5), (C-1.6) and the fact that $\nabla_{\sigma}g_{\mu\rho} = 0$, the last term in (C-1.7) may be equivalently rewritten as

$$4\sqrt{-g}\,\nabla_{\sigma}\left(\eta^{\mu\nu\alpha\beta}\,\eta^{\sigma\lambda\gamma\kappa}\,\delta\Gamma_{\nu\lambda}{}^{\rho}g_{\mu\rho}\,R_{\alpha\beta\gamma\kappa}\right),$$

which from Problem 1.5 (i) is a total derivative.

Therefore as an intermediary step we have

$$\delta\left(\sqrt{-g}\,\eta^{\mu\nu\alpha\beta}\eta^{\sigma\lambda\gamma\kappa}R_{\mu\nu\sigma\lambda}R_{\alpha\beta\gamma\kappa}\right)$$

$$= -\frac{1}{2}g^{\rho\rho'}\delta g_{\rho\rho'}\left(\sqrt{-g}\,\eta^{\mu\nu\alpha\beta}\eta^{\sigma\lambda\gamma\kappa}R_{\mu\nu\sigma\lambda}R_{\alpha\beta\gamma\kappa}\right)$$

$$+ 2\,\delta g_{\mu\rho}\sqrt{-g}\,\eta^{\mu\nu\alpha\beta}\eta^{\sigma\lambda\gamma\kappa}R^{\rho}{}_{\nu\sigma\lambda}\,R_{\alpha\beta\gamma\kappa}$$

$$+ 4\sqrt{-g}\,\nabla_{\sigma}\left(\eta^{\mu\nu\alpha\beta}\eta^{\sigma\lambda\gamma\kappa}\,\delta\Gamma_{\nu\lambda}{}^{\rho}\,g_{\mu\rho}\,R_{\alpha\beta\gamma\kappa}\right). \qquad \text{(C-1.8)}$$

We rewrite the second term on the right-hand side of the above equation as

$$2\,\delta g_{\mu\rho}\sqrt{-g}\,\eta^{\mu\nu\alpha\beta}\eta^{\sigma\lambda\gamma\kappa}R_{\rho'\nu\sigma\lambda}\,g^{\rho\rho'}R_{\alpha\beta\gamma\kappa}\,. \qquad \text{(C-1.9)}$$

To investigate the nature of this second term on the right-hand side of (C-1.8), we proceed as follows.

A totally anti-symmetric tensor of rank 5, i.e., having five indices, in four dimensional spacetime, is necessarily zero due to the simple fact that at least two of its indices must be equal. Now consider the part

$$\delta g_{\mu\rho} R_{\rho'\nu\sigma\lambda} R_{\alpha\beta\gamma\kappa},$$

in this expression, concentrating on the following five indices $(\mu, \rho', \nu, \alpha, \beta)$. Suppressing the other indices in the just given part, we may, in turn, consider the expression:

$$\left(\delta g_{\mu} . R_{\rho'\nu} .. R_{\alpha\beta} .. \right) \eta^{\mu\nu\alpha\beta}. \tag{C-1.10}$$

By interchanging in turn the index ρ' with the other indices within the round brackets, we may generate the expression

$$\left(\delta g_{\mu} . R_{\rho'\nu} .. R_{\alpha\beta} .. - \delta g_{\rho'} . R_{\mu\nu} .. R_{\alpha\beta} .. - \delta g_{\mu} . R_{\nu\rho'} .. R_{\alpha\beta} .. \right.$$
$$\left. - \delta g_{\mu} . R_{\alpha\nu} .. R_{\rho'\beta} .. - \delta g_{\mu} . R_{\beta\nu} .. R_{\alpha\rho'} .. \right) \eta^{\mu\nu\alpha\beta} = 0, \tag{C-1.11}$$

where due to the anti-symmetry property of the multiplicative factor $\eta^{\mu\nu\alpha\beta}$, the above must be equal to zero as indicated, since an antisymmetric tensor of rank 5 is zero as mentioned above. By re-inserting the suppressed fixed indices, the above equation leads, upon multiplying by $g^{\rho\rho'} \eta^{\sigma\lambda\gamma\kappa}$, to

$$\delta g_{\mu\rho} R^{\rho}{}_{\nu\sigma\lambda} R_{\alpha\beta\gamma\kappa} \eta^{\mu\nu\alpha\beta} \eta^{\sigma\lambda\gamma\kappa} = g^{\rho\rho'} \delta g_{\rho'\rho} R_{\mu\nu\sigma\lambda} R_{\alpha\beta\gamma\kappa} \eta^{\mu\nu\alpha\beta} \eta^{\sigma\lambda\gamma\kappa}$$
$$+ \quad \delta g_{\mu\rho} R_{\nu}{}^{\rho}{}_{\sigma\lambda} R_{\alpha\beta\gamma\kappa} \eta^{\mu\nu\alpha\beta} \eta^{\sigma\lambda\gamma\kappa}$$
$$+ \quad \delta g_{\mu\rho} R_{\alpha\nu\sigma\lambda} R^{\rho}{}_{\beta\gamma\kappa} \eta^{\mu\nu\alpha\beta} \eta^{\sigma\lambda\gamma\kappa}$$
$$+ \quad \delta g_{\mu\rho} R_{\beta\nu\sigma\lambda} R_{\alpha}{}^{\rho}{}_{\gamma\kappa} \eta^{\mu\nu\alpha\beta} \eta^{\sigma\lambda\gamma\kappa}. \tag{C-1.12}$$

In the second line simply write $R_{\nu}{}^{\rho}{}_{\sigma\lambda} = -R^{\rho}{}_{\nu\sigma\lambda}$. In the third line exchange $(\sigma\lambda) \leftrightarrow (\gamma\kappa)$, and $\alpha \leftrightarrow \beta$. In the fourth line first write $R_{\alpha}{}^{\rho}{}_{\gamma\kappa} = -R^{\rho}{}_{\alpha\gamma\kappa}$ then exchange $(\sigma\lambda) \leftrightarrow (\gamma\kappa)$, and $\alpha \leftrightarrow \nu$, and use, in the process, the anti-symmetry of $\eta^{\sigma\lambda\gamma\kappa}$ and, in particular, the anti-symmetry of $R_{\beta\alpha\sigma\lambda}$ in its first two indices. These allow us to rewrite (C-1.12) as

$$\delta g_{\mu\rho} R^{\rho}{}_{\nu\sigma\lambda} R_{\alpha\beta\gamma\kappa} \eta^{\mu\nu\alpha\beta} \eta^{\sigma\lambda\gamma\kappa} = g^{\rho\rho'} \delta g_{\rho'\rho} R_{\mu\nu\sigma\lambda} R_{\alpha\beta\gamma\kappa} \eta^{\mu\nu\alpha\beta} \eta^{\sigma\lambda\gamma\kappa}$$
$$- 3 \, \delta g_{\mu\rho} R^{\rho}{}_{\nu\sigma\lambda} R_{\alpha\beta\gamma\kappa} \eta^{\mu\nu\alpha\beta} \eta^{\sigma\lambda\gamma\kappa}. \tag{C-1.13}$$

We may now solve for $\delta g_{\mu\rho} R^{\rho}{}_{\nu\sigma\lambda} R_{\alpha\beta\gamma\kappa} \eta^{\mu\nu\alpha\beta} \eta^{\sigma\lambda\gamma\kappa}$, to obtain

$$\delta g_{\mu\rho} R^{\rho}{}_{\nu\sigma\lambda} R_{\alpha\beta\gamma\kappa} \eta^{\mu\nu\alpha\beta} \eta^{\sigma\lambda\gamma\kappa} = \frac{1}{4} g^{\rho\rho'} \delta g_{\rho'\rho} R_{\mu\nu\sigma\lambda} R_{\alpha\beta\gamma\kappa} \eta^{\mu\nu\alpha\beta} \eta^{\sigma\lambda\gamma\kappa}.$$

(C-1.14)

This equation allows us to rewrite the term displayed in (C-1.9) as

$$2 \delta g_{\mu\rho} \sqrt{-g} \, \eta^{\mu\nu\alpha\beta} \eta^{\sigma\lambda\gamma\kappa} R_{\rho'\nu\sigma\lambda} g^{\rho\rho'} R_{\alpha\beta\gamma\kappa}$$

$$= \frac{1}{2} g^{\rho\rho'} \delta g_{\rho\rho'} \left(\sqrt{-g} \, \eta^{\mu\nu\alpha\beta} \eta^{\sigma\lambda\gamma\kappa} R_{\mu\nu\sigma\lambda} R_{\alpha\beta\gamma\kappa} \right).$$

(C-1.15)

Thus the first and second term on the right-hand side of the intermediary step in (C-1.8) cancel out. All told, we finally obtain

$$\delta \left(\sqrt{-g} \, \eta^{\mu\nu\alpha\beta} \eta^{\sigma\lambda\gamma\kappa} R_{\mu\nu\sigma\lambda} R_{\alpha\beta\gamma\kappa} \right) = 4 \sqrt{-g} \, \nabla_{\sigma} \left(\eta^{\mu\nu\alpha\beta} \eta^{\sigma\lambda\gamma\kappa} \delta \Gamma_{\nu\lambda}{}^{\rho} g_{\mu\rho} R_{\alpha\beta\gamma\kappa} \right),$$

(C-1.16)

which from Problem 1.5 (i), may be rewritten in the form $4 \partial_{\sigma} (\sqrt{-g} \, \xi^{\sigma})$, and is a total derivative.

Appendix D: Invariant Products of Three Riemann Tensors

In this appendix we show that the invariant products of three Riemans tensors on the mass shell, i.e., when $R_{\mu\nu} = 0$ and hence also $R = 0$, are necessarily proportional to the invariant

$$R^{\mu\nu}{}_{\sigma\lambda} R^{\sigma\lambda}{}_{\kappa\rho} R^{\kappa\rho}{}_{\mu\nu}.$$

(D-1.1)

At the very outset, we note the following:

(1) The condition $R_{\mu\nu} = 0$ $(R = 0)$, and the anti-symmetry conditions of $R_{\mu\nu\sigma\lambda}$ in the interchange $\mu \leftrightarrow \nu$ or of $\sigma \leftrightarrow \lambda$, imply that in order to generate a non-trivial, i.e., non-vanishing, invariant out a product of three Riemann, no two indices within each of the Riemann tensors may be set equal.

(2) Exactly two indices must be common between any two of Riemann tensors in the product. If four indices are common between two of the Riemann tensors are common, no non-trivial invariant may be formed out of this scalar and the third Riemann fourth rank tensor. If three indices are common between two Riemann tensors, no non-trivial invariant may be constructed out of this second rank tensor and the third Riemann fourth rank tensor. Similarly if only one index is common between two of the Riemann tensors, no non-trivial invariant may be constructed of this sixth rank tensor and the third Riemann fourth rank tensor. Again we use the mass shell condition $R_{\mu\nu} = 0$ $(R = 0)$.

A direct application of the cyclic relation of the permutation of the indices of the Riemann tensor in the first equation in (1.1.41), which holds irrespective of the mass shell condition, gives

$$R^{\lambda\sigma}{}_{\kappa\rho} = -R^{\lambda}{}_{\kappa\rho}{}^{\sigma} - R^{\lambda}{}_{\rho}{}^{\sigma}{}_{\kappa},$$
(D-1.2)

from which

$$R^{\mu\nu}{}_{\sigma\lambda} R^{\lambda\sigma}{}_{\kappa\rho} = -R^{\mu\nu}{}_{\sigma\lambda} R^{\lambda}{}_{\kappa\rho}{}^{\sigma} - R^{\mu\nu}{}_{\sigma\lambda} R^{\lambda}{}_{\rho}{}^{\sigma}{}_{\kappa}.$$
(D-1.3)

Upon relabeling the indices on the right-hand side of the above equation, and using the elementary properties in (1.1.42), the above equation leads to

$$R^{\mu\nu}{}_{\sigma\lambda}R^{\lambda\sigma}{}_{\kappa\rho} = 2 R^{\mu\nu}{}_{\sigma\lambda} R^{\lambda}{}_{\kappa}{}^{\sigma}{}_{\rho}.$$
(D-1.4)

Upon multiplying the latter by $R^{\kappa\rho}{}_{\mu\nu}$ gives the identity

$$R^{\mu\nu}{}_{\sigma\lambda} R^{\lambda}{}_{\kappa}{}^{\sigma}{}_{\rho} R^{\kappa\rho}{}_{\mu\nu} = -\frac{1}{2} R^{\mu\nu}{}_{\sigma\lambda} R^{\sigma\lambda}{}_{\kappa\rho} R^{\kappa\rho}{}_{\mu\nu},$$
(D-1.5)

where we have used the relation $R^{\lambda\sigma}{}_{\kappa\rho} = -R^{\sigma\lambda}{}_{\kappa\rho}$ on the right-hand side of the above equation. With relabeled indices, we may apply the identity in (D-1.3) to $R^{\kappa\rho}{}_{\mu\nu} R^{\mu\nu}{}_{\sigma\lambda}$ in the above equation to obtain

$$R^{\kappa\rho}{}_{\mu\nu} R^{\nu}{}_{\sigma}{}^{\mu}{}_{\lambda} R^{\lambda}{}_{\kappa}{}^{\sigma}{}_{\rho} = \frac{1}{4} R^{\mu\nu}{}_{\sigma\lambda} R^{\sigma\lambda}{}_{\kappa\rho} R^{\kappa\rho}{}_{\mu\nu}.$$
(D-1.6)

More appropriately, by relabeling indices, this may be rewritten as

$$R^{\mu\nu}{}_{\sigma\lambda} R^{\lambda}{}_{\rho}{}^{\sigma}{}_{\kappa} R^{\kappa}{}_{\mu}{}^{\rho}{}_{\nu} = \frac{1}{4} R^{\mu\nu}{}_{\sigma\lambda} R^{\sigma\lambda}{}_{\kappa\rho} R^{\kappa\rho}{}_{\mu\nu}.$$
(D-1.7)

We denote the left-hand side of the above equation by

$$R^{\mu\nu}{}_{\sigma\lambda} R^{\lambda}{}_{\rho}{}^{\sigma}{}_{\kappa} R^{\kappa}{}_{\mu}{}^{\rho}{}_{\nu} \equiv (\mu\nu\sigma\lambda)(\lambda\rho\sigma\kappa)(\kappa\mu\rho\nu).$$
(D-1.8)

Due to the identity in (D-1.4), and the fact that $R_{\mu\nu\sigma\lambda}$ is odd in the interchange $\mu \leftrightarrow \nu$, or of $\sigma \leftrightarrow \lambda$, and even in the interchange $(\mu, \nu) \leftrightarrow (\sigma, \lambda)$, it is sufficient to establish that the following two invariants

$$(\mu\nu\sigma\lambda)(\sigma\rho\mu\kappa)(\lambda\kappa\nu\rho), \quad (\mu\nu\sigma\lambda)(\mu\rho\sigma\kappa)(\lambda\kappa\nu\rho),$$
(D-1.9)

are, in turn, also proportional to the invariant in (D-1.1). [Note that in the products of the Riemann tensors corresponding to the ones in (D-1.9) and the one in (D-1.8) any two of the Riemann tensors, in a given product, have exactly two indices in common.]

To the above end, consider the invariant $R^{\mu\nu}{}_{\sigma\lambda} R^{\sigma\lambda}{}_{\kappa\rho} R^{\kappa\rho}{}_{\mu\nu}$, and concentrate on the indices in $R^{\mu\nu}{}_{..} R^{\sigma\lambda}{}_{..} R^{\kappa\cdot}{}_{..}$, suppressing the other indices. Using the fact that a totally anti-symmetric tensor with five indices must be zero in four dimensional spacetime,[90] we can construct the following totally anti-symmetric tensor in four dimensions:

$$
\begin{aligned}
0 =&\left(R^{\mu\nu}{}_{..} R^{\sigma\lambda}{}_{..} - R^{\sigma\nu}{}_{..} R^{\mu\lambda}{}_{..} - R^{\lambda\nu}{}_{..} R^{\sigma\mu}{}_{..} - R^{\mu\sigma}{}_{..} R^{\nu\lambda}{}_{..} - R^{\mu\lambda}{}_{..} R^{\sigma\nu}{}_{..} \right) R^{\kappa\cdot}{}_{..} \\
&- \left(R^{\kappa\nu}{}_{..} R^{\sigma\lambda}{}_{..} - R^{\sigma\nu}{}_{..} R^{\kappa\lambda}{}_{..} - R^{\lambda\nu}{}_{..} R^{\sigma\kappa}{}_{..} - R^{\kappa\sigma}{}_{..} R^{\nu\lambda}{}_{..} - R^{\kappa\lambda}{}_{..} R^{\sigma\nu}{}_{..} \right) R^{\mu\cdot}{}_{..} \\
&- \left(R^{\mu\kappa}{}_{..} R^{\sigma\lambda}{}_{..} - R^{\sigma\kappa}{}_{..} R^{\mu\lambda}{}_{..} - R^{\lambda\kappa}{}_{..} R^{\sigma\mu}{}_{..} - R^{\mu\sigma}{}_{..} R^{\kappa\lambda}{}_{..} - R^{\mu\lambda}{}_{..} R^{\sigma\kappa}{}_{..} \right) R^{\nu\cdot}{}_{..} \\
&- \left(R^{\mu\nu}{}_{..} R^{\kappa\lambda}{}_{..} - R^{\kappa\nu}{}_{..} R^{\mu\lambda}{}_{..} - R^{\lambda\nu}{}_{..} R^{\kappa\mu}{}_{..} - R^{\mu\kappa}{}_{..} R^{\nu\lambda}{}_{..} - R^{\mu\lambda}{}_{..} R^{\kappa\nu}{}_{..} \right) R^{\sigma\cdot}{}_{..} \\
&- \left(R^{\mu\nu}{}_{..} R^{\sigma\kappa}{}_{..} - R^{\sigma\nu}{}_{..} R^{\mu\kappa}{}_{..} - R^{\kappa\nu}{}_{..} R^{\sigma\mu}{}_{..} - R^{\mu\sigma}{}_{..} R^{\nu\kappa}{}_{..} - R^{\mu\kappa}{}_{..} R^{\sigma\nu}{}_{..} \right) R^{\lambda\cdot}{}_{..} \, .
\end{aligned}
$$

$$\tag{D-1.10}$$

Inserting the remaining fixed indices, for example, in $R^{\mu\cdot}{}_{..}$, $R^{\nu\cdot}{}_{..}$, we obtain from the original expression $R^{\mu\rho}{}_{\mu\nu}$, $R^{\nu\rho}{}_{\mu\nu}$, which vanish on the mass shell. This means that the second and third lines in the above equation do not contribute on the mass shell. Similarly, within the first pair of round brackets only the first term gives a non-zero contribution. Within the fourth pair of round brackets, only the second and fourth terms are non-vanishing, and in the fifth pair of round brackets, only the third and the fifth terms give non-zero contributions. All told, (D-1.10) implies that

$$
\begin{aligned}
R^{\mu\nu}{}_{\sigma\lambda} R^{\sigma\lambda}{}_{\kappa\rho} R^{\kappa\rho}{}_{\mu\nu} =& - R^{\kappa\nu}{}_{\sigma\lambda} R^{\mu\lambda}{}_{\kappa\rho} R^{\sigma\rho}{}_{\mu\nu} - R^{\mu\kappa}{}_{\sigma\lambda} R^{\nu\lambda}{}_{\kappa\rho} R^{\sigma\rho}{}_{\mu\nu} \\
& - R^{\kappa\nu}{}_{\sigma\lambda} R^{\sigma\mu}{}_{\kappa\rho} R^{\lambda\rho}{}_{\mu\nu} - R^{\mu\kappa}{}_{\sigma\lambda} R^{\sigma\nu}{}_{\kappa\rho} R^{\lambda\rho}{}_{\mu\nu},
\end{aligned}
\tag{D-1.11}
$$

which upon relabeling the indices on the right-hand side of the equation, and using the elementary properties of the Riemann curvature tensor in (1.1.42), we obtain the following identity

$$
R^{\mu\nu}{}_{\sigma\lambda} R^{\sigma\lambda}{}_{\kappa\rho} R^{\kappa\rho}{}_{\mu\nu} = 4\, R^{\mu\nu}{}_{\sigma\lambda} R^{\sigma\rho}{}_{\mu\kappa} R^{\lambda\kappa}{}_{\nu\rho}.
\tag{D-1.12}
$$

Or more appropriately,

$$
R^{\mu\nu}{}_{\sigma\lambda} R^{\sigma\rho}{}_{\mu\kappa} R^{\lambda\kappa}{}_{\nu\rho} = \frac{1}{4}\, R^{\mu\nu}{}_{\sigma\lambda} R^{\sigma\lambda}{}_{\kappa\rho} R^{\kappa\rho}{}_{\mu\nu},
\tag{D-1.13}
$$

---

[90]As mentioned in Appendix C, this is due to the fact that two of its indices must be equal implying the vanishing of a totally anti-symmetric tensor with five (or more) indices in four dimensional spacetime.

establishing that the first invariant in (D-1.9) is proportional to the invariant in question on the right-hand of the above equation.

On the other hand, we may rewrite the basic identity in (D-1.13) as

$$
\frac{1}{4} R^{\mu\nu}{}_{\sigma\lambda} R^{\sigma\lambda}{}_{\kappa\rho} R^{\kappa\rho}{}_{\mu\nu} = R^{\mu\nu}{}_{\sigma\lambda} R^{\sigma\rho}{}_{\mu\kappa} R^{\lambda\kappa}{}_{\nu\rho} = R^{\mu\nu}{}_{\sigma\lambda} \left(-R_{\mu}{}^{\sigma\rho}{}_{\kappa} - R_{\mu}{}^{\rho}{}_{\kappa}{}^{\sigma} \right) R^{\lambda\kappa}{}_{\nu\rho}
$$

$$
= -R_{\mu}{}^{\sigma\rho}{}_{\kappa} R^{\lambda\kappa}{}_{\nu\rho} R^{\mu\nu}{}_{\sigma\lambda} + R^{\mu\nu}{}_{\sigma\lambda} R_{\mu}{}^{\rho\sigma}{}_{\kappa} R^{\lambda\kappa}{}_{\nu\rho} = -R^{\mu\nu}{}_{\sigma\lambda} R^{\lambda}{}_{\rho}{}^{\sigma}{}_{\kappa} R^{\kappa}{}_{\mu}{}^{\rho}{}_{\nu}
$$

$$
+ R^{\mu\nu}{}_{\sigma\lambda} R_{\mu}{}^{\rho\sigma}{}_{\kappa} R^{\lambda\kappa}{}_{\nu\rho} = -\frac{1}{4} R^{\mu\nu}{}_{\sigma\lambda} R^{\sigma\lambda}{}_{\kappa\rho} R^{\kappa\rho}{}_{\mu\nu} + R^{\mu\nu}{}_{\sigma\lambda} R_{\mu}{}^{\rho\sigma}{}_{\kappa} R^{\lambda\kappa}{}_{\nu\rho},
$$

$$
\text{(D-1.14)}
$$

where in writing the fourth equality, we have relabeled the indices of the first term, and in writing the fifth equality we have used the identity in (D-1.7) for the term just mentioned. The above equation gives for the very last on its extreme right-hand side the identity

$$
R^{\mu\nu}{}_{\sigma\lambda} R_{\mu}{}^{\rho\sigma}{}_{\kappa} R^{\lambda\kappa}{}_{\nu\rho} = \frac{1}{2} R^{\mu\nu}{}_{\sigma\lambda} R^{\sigma\lambda}{}_{\kappa\rho} R^{\kappa\rho}{}_{\mu\nu}, \tag{D-1.15}
$$

establishing that the second invariant in (D-1.9) is proportional to the invariant in question as well.

The remarkable property of invariant products of three Riemann curvature tensors, on the mass shell, being proportional to the invariant product in (D-1.1) is a key identity in investigating the divergent part of quantum general relativity in the two-loop contribution as discussed above within the text.[91]

Appendix E: Bekenstein-Hawking Entropy Formula of a Black Hole

The Bekenstein-Hawking Entropy formula[92] was briefly discussed in the introduction to this chapter. It relates the entropy of a Black Hole (BH) to the surface area of the event horizon. It arises as a consequence of investigations by Hawking[93] that a BH is a thermodynamic object, it radiates and has a temperature associated with it.[94] To find its temperature of a BH, note that the so-called thermodynamic partition function $e^{-\beta H}$, in performing averages (by taking a trace), where H is the

---

[91]Identities similar to the one in (D-1.12), as follows from (D-1.10) on the mass shell, were derived by different methods in [66].

[92]Bekenstein [14].

[93]Hawking [31, 32].

[94]Particle emission from a BH is formally explained through virtual pairs of particles created near the horizon with one particle falling into the BH while the other becoming free outside the horizon.

Hamiltonian of a system,[95] may be obtained from the time development unitary operator $e^{-itH/\hbar}$, by the substitution: $t \rightarrow -i\hbar\beta$. The trace operation gives rise to a periodicity condition on β.

The infinitesimal distance squared in Minkowski space, i.e., in flat spacetime is given by $ds^2 = -\eta_{\mu\nu} dx^\mu dx^\nu$, with metric $\eta_{\mu\nu} = \text{diag}[-1, 1, 1, 1]$. For a spherically symmetric (uncharged, non-rotating) body of mass M, for example, the gravitational effect of the body causes a curvature of spacetime around it, which amounts in replacing the line element squared by $ds^2 = -g_{\mu\nu} dx^\mu dx^\nu$, where $g_{\mu\nu}$ is the corresponding metric which takes this distortion into account. The line element squared outside the body now takes the form

$$ds^2 = \left(\left[1 - \frac{2\,G_N M}{c^2\,r}\right]^{1/2} c\,dt\right)^2 - \left(\left[1 - \frac{2\,G_N M}{c^2\,r}\right]^{-1/2} dr\right)^2 - r^2\left[(d\theta)^2 + (\sin\theta\,d\phi)^2\right],$$

$$(E\text{-}1.1)$$

where the point $r = R_{BH} = 2\,G_N M/c^2$ at which the coefficient of $(dr)^2$ blows out defines the radius of the BH, specifying the location of the horizon.[96]

The above metric is referred to as the Schwarzschild metric, and R_{BH} is also referred to as the Schwarzschild radius.

We note that near the horizon $r \gtrsim 2\,G_N M/c^2$

$$d\left[1 - \frac{2\,G_N M}{c^2\,r}\right]^{1/2} = \frac{G_N M}{c^2\,r^2}\left[1 - \frac{2\,G_N M}{c^2\,r}\right]^{-1/2} dr \simeq \frac{c^2}{4\,G_N M}\left[1 - \frac{2\,G_N M}{c^2\,r}\right]^{-1/2} dr.$$

$$(E\text{-}1.2)$$

This suggests that near the horizon, we may define the variable

$$\rho = \frac{4\,G_N M}{c^2}\left[1 - \frac{2\,G_N M}{c^2\,r}\right]^{1/2},$$

$$(E\text{-}1.3)$$

carry out the transformation $t \rightarrow -i\hbar\beta$, as discussed earlier, to obtain

$$-\,ds^2 \rightarrow \left(\rho\,d\left[\hbar\frac{c^3}{4\,G_N M}\beta\right]\right)^2 + (d\rho)^2 + \frac{4\,G_N^2 M^2}{c^4}\left[(d\theta)^2 + (\sin\theta\,d\phi)^2\right].$$

$$(E\text{-}1.4)$$

The first two terms represent the line element squared in a two dimensional space (a disc), with $[\hbar c^3/4G_N M]\beta$ representing an angle. For $[\hbar c^3/4G_N M]\beta = 2\pi$ and $\beta = 1/k_B T$, where k_B is the Boltzmann constant, the temperature of the black hole

---

[95]See, e.g., [39].

[96]For a pedestrian approach, this may be roughly inferred from Newton's theory of gravitation that the escape speed of a particle in the gravitational field of a spherically symmetric massive body of mass M, at a distance r, is obtained from the inequality $v^2/2 - G_N M/r < 0$, and by formally replacing v by the ultimate speed c gives for the critical radius $R_{CRITICAL} = 2\,G_N M/c^2$ such that for $r < R_{CRITICAL}$ a particle cannot escape.

is thus given by[97]

$$T_{BH} = \frac{\hbar c^3}{8\pi G_N M k_B},$$
(E-1.5)

where note that a very massive black hole is cold.

Recall that entropy S represents a measure of the amount of disorder with associate encoded information, and invoking the thermodynamic interpretation of a BH, we may write

$$\frac{\partial S}{\partial(M c^2)} = \frac{1}{T},$$
(E-1.6)

which upon integration with boundary condition that for $M \to 0$, $S \to 0$, gives the celebrated result

$$S_{BH} = \frac{c^3 k_B}{4 G_N \hbar} A, \qquad A = 4\pi \left(\frac{2 G_N M}{c^2}\right)^2,$$
(E-1.7)

referred to as the Bekenstein-Hawking Entropy formula of a BH, where A is the (surface) area of the horizon. Using the fact that $G_N \hbar / c^3 = \ell_P^2$ denotes the Plank length squared, we may appropriately rewrite (E-1.7) as

$$S_{BH} = k_B \frac{A}{4 \ell_P^2}.$$
(E-1.8)

As mentioned earlier, the proportionality of entropy to the area rather than to the volume of a BH horizon should be noted.

Problems

1.1. Establish the anti-symmetry property:
$e_{\beta\mu}(\partial_\nu e_\alpha{}^\mu + \Gamma_{\nu\sigma}{}^\mu e_\alpha{}^\sigma) = -e_{\alpha\mu}(\partial_\nu e_\beta{}^\mu + \Gamma_{\nu\sigma}{}^\mu e_\beta{}^\sigma),$
obtained by simply interchanging the Lorentz indices α, β.

---

[97]A pedestrian approach in determining the temperature is the following. By comparing the expression of energy expressed in terms of the wavelength of radiation λ: $E = h c / \lambda$, with the expression $E = k_B T$, gives $T = h c / k_B \lambda$. On dimensional grounds $\lambda \sim 2 G_N M / c^2$, which gives $T \sim \pi \hbar c^3 / G_N M k_B$. This leads the expression in (E-1.5) up to a proportionality constant.

1.2. Show that:

(i) $[\nabla_\alpha, \nabla_\beta]\,\xi^\alpha = R^\alpha{}_{\sigma\alpha\beta}\,\xi^\sigma = R_{\sigma\beta}\,\xi^\alpha$ for a vector field ξ^α.

(ii) $[\nabla_\alpha, \nabla_\nu]\,h^{\alpha\beta} = R_{\sigma\nu}\,h^{\sigma\beta} + R^\beta{}_{\sigma\alpha\nu}\,h^{\sigma\alpha}$ for a symmetric tensor field $h^{\alpha\beta}$.

1.3. Show that:

(i) $\delta\sqrt{-g} = \dfrac{1}{2}\sqrt{-g}\,g^{\alpha\beta}\,\delta g_{\alpha\beta} = -\dfrac{1}{2}\sqrt{-g}\,g_{\alpha\beta}\,\delta g^{\alpha\beta}$, where $g = \det[g_{\mu\nu}]$.

(ii) $\partial_\mu\sqrt{-g} = \dfrac{1}{2}\sqrt{-g}\,g^{\alpha\beta}\partial_\mu g_{\alpha\beta} = \sqrt{-g}\,\Gamma_{\mu\sigma}{}^\sigma$.

1.4. Establish the equalities:

$$\text{(i)}\; g^{\mu\sigma}\,\Gamma_{\mu\sigma}{}^\alpha = -\frac{1}{\sqrt{-g}}\,\partial_\mu(\sqrt{-g}\,g^{\alpha\mu}).$$

$$\text{(ii)}\; \delta^\mu{}_\nu\,\Gamma^\nu_{\mu\sigma} = \frac{\partial_\sigma\sqrt{-g}}{\sqrt{-g}} = \partial_\sigma \ln[\sqrt{-g}\,].$$

1.5. (i) For a vector field ξ^μ, show that $\sqrt{-g}\,\nabla_\mu\xi^\mu = \partial_\mu(\sqrt{-g}\,\xi^\mu)$.

(ii) For a scalar field ϕ, show that $\sqrt{-g}\,g^{\mu\nu}\partial_\mu\phi\,\partial_\nu\phi = -\sqrt{-g}\,\phi(g^{\mu\nu}\nabla_\mu\nabla_\nu\phi)$, up to a total derivative.

(iii) For a *fixed* real scalar field ϕ, show that $\delta[\sqrt{-g}\,(g^{\mu\nu}\partial_\mu\phi\,\partial_\nu\phi + m^2\phi^2)]$ is given by

$$\sqrt{-g}\,[\partial_\mu\phi\,\partial_\nu\phi - \frac{1}{2}g_{\mu\nu}(\partial^\rho\phi\,\partial_\rho\phi + m^2\phi^2)]\delta g^{\mu\nu}.$$

1.6. Use (1.1.14) to infer that although $\Gamma_{\rho\nu}{}^\gamma$ is not a tensor, the variation $\delta\Gamma_{\rho\nu}{}^\gamma$ is a tensor.

1.7. Show that the infinitesimal variations of the Riemann tensor, the Ricci tensor and the connection may be written as:

$$\text{(i)}\; \delta R^\mu{}_{\nu\rho\sigma} = \nabla_\rho\,\delta\Gamma_{\nu\sigma}{}^\mu - \nabla_\sigma\,\delta\Gamma_{\nu\rho}{}^\mu.$$

$$\text{(ii)}\; \delta R_{\mu\nu\rho\sigma} = \delta g_{\mu\lambda}\,R^\lambda{}_{\nu\rho\sigma} + g_{\mu\lambda}\big(\nabla_\rho\,\delta\Gamma_{\nu\sigma}{}^\lambda - \nabla_\sigma\,\delta\Gamma_{\nu\rho}{}^\lambda\big).$$

$$\text{(iii)}\;\; \delta R_{\nu\sigma} = \nabla_\rho\,\delta\Gamma_{\nu\sigma}{}^\rho - \nabla_\sigma\,\delta\Gamma_{\nu\mu}{}^\mu.$$

$$\text{(iv)}\; \delta\Gamma_{\mu\nu}{}^\rho = \frac{1}{2}g^{\rho\sigma}\big(\nabla_\mu\delta g_{\nu\sigma} + \nabla_\nu\delta g_{\mu\sigma} - \nabla_\sigma\delta g_{\mu\nu}\big).$$

1.8. Establish that $\sqrt{-g}\,g^{\mu\nu}\,\delta R_{\mu\nu} = 0$, up to a total derivative.

1.9. Verify that $\delta(\sqrt{-g}\,g^{\mu\nu}R_{\mu\nu}) = \sqrt{-g}(R_{\mu\nu} - \frac{1}{2}g_{\mu\nu}R)\delta g^{\mu\nu}$.

1.10. Derive the following fundamental equalities which hold true *up to* total derivatives. These are generalization of results to be used in partial integrations in curved spacetime.

(i) $\sqrt{-g}\,(\nabla_\mu T^{\mu_1\cdots\mu_k\mu_{k+1}\cdots\mu_n})\,S_{\mu_1\cdots\mu_k}{}^\mu{}_{\mu_{k+1}\cdots\mu_n}$

$\qquad = -\sqrt{-g}\,T^{\mu_1\cdots\mu_k\mu_{k+1}\cdots\mu_n}\,\nabla_\mu S_{\mu_1\cdots\mu_k}{}^\mu{}_{\mu_{k+1}\cdots\mu_n},$

for n rank and $n+1$ rank tensors $T^{\mu_1\cdots\mu_k\mu_{k+1}\cdots\mu_n}$, $S_{\mu_1\cdots\mu_k}{}^\mu{}_{\mu_{k+1}\cdots\mu_n}$, respectively,

(ii) $\sqrt{-g}\,(\nabla_\mu T^{\mu_1\cdots\mu_k\mu\mu_{k+1}\cdots\mu_n})\,S_{\mu_1\cdots\mu_k\mu_{k+1}\cdots\mu_n}$

$\qquad = -\sqrt{-g}\,T^{\mu_1\cdots\mu_k\mu\mu_{k+1}\cdots\mu_n}\,\nabla_\mu S_{\mu_1\cdots\mu_k\mu_{k+1}\cdots\mu_n},$

for $n+1$ rank and n rank tensors $T^{\mu_1\cdots\mu_k\mu\mu_{k+1}\cdots\mu_n}$, $S_{\mu_1\cdots\mu_k\mu_{k+1}\cdots\mu_n}$, respectively. In particular, for $n = 0$, i.e., for a scalar field ϕ, (i) reads

(iii) $\qquad\qquad \sqrt{-g}\,(\nabla_\mu\phi)\,S^\mu = -\sqrt{-g}\,\phi\,\nabla_\mu S^\mu.$

1.11. The action of a massive real scalar field is defined by:

$$W_{\text{matter}} = -\frac{1}{2}\int (dx)\sqrt{-g}(g^{\mu\nu}\partial_\mu\phi\,\partial_\nu\phi + m^2\phi^2).$$

(i) Derive the field equation for the field ϕ.
(ii) The energy-momentum tensor is defined in (1.2.22) by

$$T_{\mu\nu} = -\frac{2}{\sqrt{-g}}\frac{8}{8g^{\mu\nu}}W_{\text{matter}}. \qquad \text{Find } T_{\mu\nu}.$$

(iii) Show that $\nabla_\mu T^{\mu\nu} = 0$, $\nabla_\nu T^{\mu\nu} = 0$.

1.12. Show that the action in (1.3.36) of first order formulation of linearized gravity is invariant under the infinitesimal gauge transformations:

$$\delta\,\xi^{\mu\nu} = \partial^\mu\Lambda^\nu + \partial^\nu\Lambda^\mu - \eta^{\mu\nu}\,\partial_\alpha\Lambda^\alpha, \quad \delta\,\Gamma_{\alpha\beta\gamma} = \partial_\alpha\partial_\beta\Lambda_\gamma.$$

1.13. Show that the expressions in (1.4.3) and (1.4.4) are the same.

1.14. Find $\delta\Gamma_{\mu\nu}{}^\sigma$ up to second order in $h_{\mu\nu}$, with the latter describing a fluctuation about a background metric $g_{\mu\nu}$.

1.15. Given the tensor $\eta^{\mu\nu\lambda\sigma} = \varepsilon^{\mu\nu\lambda\sigma}/\sqrt{-g}$, with $\varepsilon^{\mu\nu\lambda\sigma}$ totally anti-symmetric, and $\varepsilon^{0123} = +1$. Show that, under an infinitesimal variation $\delta g^{\rho\kappa}$, $\delta\eta^{\mu\nu\lambda\sigma} = (1/2)\eta^{\mu\nu\lambda\sigma}g_{\rho\kappa}\delta g^{\rho\kappa} = -(1/2)\eta^{\mu\nu\lambda\sigma}g^{\rho\kappa}\delta g_{\rho\kappa}$.

1.16. Show that $R_{\mu\nu\rho\sigma}R_{\lambda\kappa\gamma\,\epsilon}\,\eta^{\mu\nu\lambda\kappa}\,\eta^{\rho\sigma\gamma\,\epsilon} = 4\big(4R_{\mu\nu}R^{\mu\nu} - R_{\mu\nu\rho\sigma}R^{\mu\nu\rho\sigma} - R^2\big)$.

1.17. Show that $\eta^{\rho\sigma\alpha\lambda}\nabla_\lambda R_{\mu\nu\rho\sigma} = 0$.

1.18. Show that $\nabla_\lambda\eta^{\mu\nu\alpha\beta} = 0$.

1.19. Show that the solution of the equation in (1.9.10) is given by the one in (1.9.11), thus leading to the explicit expression for the holonomy in (1.9.12).

Recommended Reading

DeWitt, B. (2014). *The global approach to quantum field theory*. Oxford: Oxford University Press.

Kiefer, C. (2012). *Quantum gravity* (3rd ed.) Oxford: Oxford University Press.

Manoukian, E. B. (2015). Quantum viewpoint of quadratic $f(R)$ gravity, constraints, vacuum-to-vacuum transition amplitude and particle content. *General Relativity and Gravitation, 47:78,* 1–11.

Manoukian, E. B. (2016). *Quantum field theory I: Foundations and abelian and non-abelian gauge theories*. Dordrecht: Springer.

Manoukian, E. B. (1983). *Renormalization*. New York: Academic.

Oriti, D. (Ed.). (2009). *Approaches to quantum gravity*. Cambridge: Cambridge University Press.

Rovelli, C. (2008). *Quantum gravity*. Cambridge: Cambridge University Press.

Rovelli, C., & Vidotto, F. (2015). *Covariant loop quantum gravity*. Cambridge: Cambridge University Press.

Thiemann, T. (2007). *Modern canonical quantum gravity*. Cambridge: Cambridge University Press.

References

1. Ansari, M. H. (2008). Generic degeneracy and entropy in loop quantum gravity. *Nuclear Physics B, 795,* 635–644.
2. Anselmi, D. (2003). Absence of higher derivatives in the renormalization of propagators in quantum field theory with infinitely many couplings. *Classical Quantum Gravity, 20,* 2344–2378.
3. Arnowitt, R. S., Deser, S., & Misner, W. (2008). The dynamics of general relativity. *General Relativity and Gravitation, 40,* 1997–2027. (Reprinted from *Gravitation: an introduction to current research*, (Chap. 7) In L. Witten (Ed.) New York: Wiley (1962)).
4. Ashtekar, A. (1986). New variables for classical and quantum gravity. *Physics Review Letter, 57,* 2244–2247.
5. Ashtekar, A. (1987). New Hamiltonian formulation of general gravity. *Physics Review D, 36,* 1587–1602.
6. Ashtekar, A., & Isham, C. J. (1992). Representations of the Holonomy algebras of gravity and non-Abelian gauge theories. *Classical and Quantum Gravity, 9,* 1433–1467. hep–th/9202053.
7. Ashtekar, A., & Lewandoski, J. (1997). Quantum theory of gravity I: Area operators. *Classical and Quantum Gravity, 14,* A55–A81.
8. Avramidi, I. G. (2000). *Heat kernel and quantum gravity*. Heidelberg: Springer.
9. Baez, J. C. (1994). *Knots and quantum gravity*. Oxford: Oxford University Press.
10. Baez, J. C. (1996). Spin networks in nonperturbative quantum gravity. In L. Kauffman (Ed.), *Interface of knot theory and physics*. Providence, RI: American Mathematical Society.

11. Barbero, F. (1995). Real Ashtekar variables for Lorentzian signature space times. *Physical Review D, 51*, 5507–5510.
12. Barvinsky, A. O., & Vilkovisky, G. A. (1985). The generalized Schwinger-DeWitt technique in gauge theories and quantum gravity. *Physics Reports, 119*, 1–74.
13. Barvinsky, A. O. & Vilkovisky, G. A. (1987). The effective action in quantum field theory: Two-loop approximation. In I. Batalin, C. J. Isham, & G. A. Vilkovisky (Eds.), *Quantum field theory and quantum statistics* (Vol. I, pp. 245–275). Bristol: Adam Hilger.
14. Bekenstein, J. D. (1973). Black holes and entropy. *Physics Review D, 7*, 2333–2346.
15. Bjerrum-Bohr, N. E. J., Donoghue, J. F., & Holstein, B. R. (2003). Quantum gravitational corrections to the non-relativistic scattering potential of two masses. *Physical Review D, 67*, 084033. (Erratum: *Physical Review D, 71*, 069903 (2005)).
16. Bjerrum-Bohr, N. E. J., Donoghue, J. F. & Holstein, B. R. (2003b). Quantum corrections to the Schwarzchild and Kerr metrics. *Physical Review, 68*, 084005–084021.
17. Caswell, W. E., & Kennedy, A. D. (1982). Simple approach to renormalization theory. *Physical Review D, 25*, 392–408.
18. Chase, M. K. (1982). Absence of leading divergences in two-loop quantum gravity. *Nuclear Physics B, 203*, 434–444.
19. Deser, S. (2000). Infinities in quantum gravities. *Annalen der Physik, 9*, 299–306.
20. Deser, S., van Nieuwenhuizen, P., & Boulware, D. (1975). Uniqueness and nonrenormalizability of quantum gravitation. In G. Shaviv & J. Rosen (Eds.), *General relativity and gravitation*. New York: Wiley.
21. DeWitt, B. (1965). *Dynamical theory of groups and fields*. New York: Gordon and Breach.
22. DeWitt, B. (1967). Quantum theory of gravity. II. the manifestly covariant theory. *Physical Review, 162*, 1195–1239.
23. DeWitt, B. (1967). Quantum theory of gravity. III. Applications of the covariant theory. *Physical Review, 162*, 1239–1256.
24. DeWitt, B. (2014). *The global approach to quantum field theory*. Oxford: Oxford University Press.
25. Donoghue, J. F. (1994). Leading correction to the newtonian potential. *Physics Review Letter, 72*, 2996.
26. Donoghue, J. F. (1994). General relativity as an effective field theory, the leading corrections. *Physical Review D, 50*, 3874–3888. (gr-qg/9405057).
27. Donoghue, J. F. (1997). In C. Fernando & M. -J. Herrero (Eds.), Advanced school on effective theories. Almunecar: World Scientific. UMHEP-424, gr-qc/9512024.
28. Feynman, R. P., Morinigo, F. B., & Wagner, W. G. (1995). In B. Hatfield (Ed.) with a forword by J. Preskill & K. S. Thorne. *Feynman lectures on gravitation*. Addison-Wesley: Reading, MA.
29. Gomis, J., & Weinberg, S. (1996). Are nonrenormalizable gauge theories renormalizable? *Nuclear Physics B, 469*, 473–487. (hep-th/9510087).
30. Goroff, M. H., & Sagnotti, A. (1986). The ultraviolet behaviour of Einstein gravity. *Nuclear Physics B, 266*, 709–736.
31. Hawking, S. W. (1974). Black hole explosions? *Nature, 248*, 230–231.
32. Hawking, S. W. (1975). Particle creation by black holes. *Communications in Mathematical Physics, 43*, 199–220.
33. Immirzi, G. (1997). Quantum gravity and Regge calculus. *Nuclear Physics B Proceedings Supplements, 57*, 65–72.
34. Kallosh, R. E., Tarasov, O. V., & Tyutin, I. V. (1978). One-loop finiteness of quantum gravity off mass shell. *Nuclear Physics B, 137*, 145–163.
35. Korepin, V. (1974). Diploma Thesis, Leningrad State University.
36. Korepin, V. (2009). Cancellation of ultra-violet infinities in one loop gravity http://insti.physics.sunysb.edu/~Korepin/uvg.pdf. English Translation of 1974 Diploma.

37. Manoukian, E. B. (1983). *Renormalization*. New York: Academic.
38. Manoukian, E. B. (1990). A quantum viewpoint of gravitational radiation. *General Relativity and Gravitation, 22*, 501–505.
39. Manoukian, E. B. (1991). Derivation of the closed-time path of quantum field theory at finite temperatures. *Journal of Physics, G17*, L173–L175.
40. Manoukian, E. B. (2006). *Quantum theory: A wide spectrum*. AA Dordrecht: Springer.
41. Manoukian, E. B. (2015). Vacuum-to-vacuum transition amplitude and the classic radiation theory. *Radiation Physics and Chemistry, 106*, 268–270.
42. Manoukian, E. B. (2015). Quantum viewpoint of quadratic $f(R)$ gravity, constraints, vacuum-to-vacuum transition amplitude and particle content. *General Relativity and Gravitation, 47:78*, 1–11.
43. Manoukian, E. B. (2016). *Quantum field theory I: Foundations and abelian and non-abelian gauge theories*. Dordrecht: Springer.
44. Meissner, K. (2004). Black-hole entropy in loop quantum gravity. *Classical and Quantum Gravity, 21*, 5245–5251.
45. Misner, C. W., Thorne, K. S., & Wheeler, J. A. (1973). *Gravitation*. New York: W. H. Freeman and Company.
46. Niedermaier, M., & Reuter, M. (2006). The asymptotic safety scenario in quantum gravity. *Living Reviews in Relativity, 9*, 5.
47. Penrose, R. (1971). Angular momentum: An approach to combinatorial spacetime. In T. Bastin (Ed.), *Quantum theory and beyond* (pp. 151–180). Cambridge: Cambridge University Press.
48. Percacci, R. (2010). A short introduction to asymptotic safety. In *Proceedings of the "Time and Matter" Conference*, Budva, Montenegro.
49. Reuter, M. (1998). Nonperturbative evolution equation for quantum gravity. *Physical Review D, 57*, 971–985.
50. Rovelli, C. (2008). *Quantum gravity*. Cambridge: Cambridge University Press.
51. Rovelli, C., & Vidotto, F. (2015). *Covariant loop quantum gravity*. Cambridge: Cambridge University Press.
52. Rovelli, C., & Smolin, L. (1988). Knot theory and quantum gravity. *Physical Review Letters, 61*, 1155–1158.
53. Rovelli, C., & Smolin, L. (1990). Loop space representation of quantum general relativity. *Nuclear Physics B, 331*, 80–152.
54. Rovelli, C., & Smolin, L. (1995). Spin networks and quantum gravity. *Physical Review D, 52*, 5743–5759. gr–qc/9505006.
55. Schwinger, J. (1951). On gauge invariance and vacuum polarization. *Physical Review, 82*, 664–679.
56. Schwinger, J. (1969). *Particles and sources*. New York: Gordon and Breach.
57. Schwinger, J. (1976). Gravitons and photons: The methodological unification of source theory. *General Relativity and Gravitation, 7*, 251–256.
58. Sen, A. (1982). Gravity as a spin system. *Physics Letters B, 119*, 89–91.
59. Stelle, K. S. (1977). Renormalization of higher-derivative quantum gravity. *Physics Review D, 16*, 953–969.
60. 't Hooft, G., & Veltman, M. (1973). Diagrammar. CERN Report, 73-9.
61. 't Hooft, G., & Veltman, M. (1974). One-loop divergencies in the theory of gravitation. *Annales de l'Institut Henri Poincaré, 20*, 69–94.
62. Thiemann, T. (2007). *Modern canonical quantum gravity*. Cambridge: Cambridge University Press.
63. Tomboulis, E. T. (1984). Unitarity in higher-derivative quantum gravity. *Physics Review Letters, 52*, 1173–1176.
64. Vacca, G. P., & Zanusso, O. (2019). Asyptotic safety in Einstein gravity and scalar-fermion matter. *Physics Review Letters, 105*(1–4), 231601.
65. van de Ven, A. E. M. (1992). Two-loop quantum gravity. *Nuclear Physics B, 378*, 309–366.

66. van Nieuwenhuizen, P., & Wu, C. C. (1977). On integral relations for invariants constructed from three Riemann tensors and their applications in quantum gravity. *Journal of Mathematical Physics, 18*, 182–186.
67. Weinberg, S. (1979). Ultraviolet divergences in quantum theories of gravitation. In S. W. Hawking & W. Israel (Eds.), *General relativity: An Einstein centenary survey* (pp. 790–831). Cambridge: Cambridge University Press.
68. Wilson, K. G. (1975). The renormalization group: Critical phenomena and the Kondo problem. *Reviews of Modern Physics, 47*, 773.
69. Wilson, K. G., & Kogut, J. B. (1974). The Renormalization group and the ϵ expansion. *Physics Report, 12*, 75–199.

Chapter 2
Introduction to Supersymmetry

It is often stated that "Supersymmetry" is a theory with mathematical beauty.[1] Well, it is.[2] Imposing, a priori, a symmetry in developing quantum field theory interactions is quite important in the sense that, together with the criterion of renormalizability, it narrows down the type of interactions one was, a priori, set up to develop. New physics may also arise from the requirement of the invariance of a theory under some given symmetry. The discovery of the positron in Dirac's work by insisting the invariance of a quantum theory of the electron under Lorentz transformations, as dictated by special relativity, is the classic example of one such a physical consequence.[3]

Supersymmetry gives rise to symmetry transformations between fermions and bosons. As such a unifying principle, it necessarily predicts the existence of new particles, and gives rise to a complete symmetry between the known particles and particles of opposite statistics, their superpartners, referred to in short as sparticles, having the same masses, and unites them in symmetry multiplets, referred to as supermultiplets, such that the number of degrees of freedom of the bosonic and fermionic ones being equal.[4] This will be spelled out in Sect. 2.5. Since no bosons and fermions with equal masses are known at the present, this symmetry must be broken spontaneously or may be made broken by the addition of supersymmetry breaking terms to the initial Lagrangian density by making sure, in the process,

---

[1]It is worth remembering the famous statement made by Dirac in [12], in a different context, that "A theory with mathematical beauty is more likely to be correct than an ugly one that fits some experimental data".

[2]For an overall view of quantum field theory since its birth in 1926 see Chap. 1 of [26].

[3]This introduction is partly based on the general introductory chapter of Vol. I. to QFT [26].

[4]Here a degree of freedom is defined as follows. A real scalar field has one degree of freedom, while a complex one has two, and a massive real field of spin $j > 0$, has $(2j + 1)$ degrees of freedom. A Dirac spinor has $2(2 \times (1/2) + 1)$ degrees of freedom while a Majorana spinor has $(2 \times (1/2) + 1)$.

© Springer International Publishing Switzerland 2016
E.B. Manoukian, *Quantum Field Theory II*, Graduate Texts in Physics,
DOI 10.1007/978-3-319-33852-1_2

to preserve its renormalizabilty. The superpartners are expected to be heavier than their particles counterparts, otherwise they would have been observed so far. Supersymmetry is a theory with remarkable consequences and is the subject matter of this chapter.

Although Nature is not necessarily supersymmetric, the imposition of such a symmetry on a field theory has turned out to have several advantages over its non-supersymmetric counterpart many of which will be discussed later. For one thing, it puts restrictions on the relation between the various couplings in a theory and also improves, in general, the situation encountered in the basic divergence problem of quantum field theory. The unification of the gauge couplings of a supersymmetric version at high energies, also elaborated on it further below, is also more precise. As we will also see, supersymmetric theories have the very desirable property of having Hamiltonians with non-negative spectra.

By extending Minkowski space, supersymmetry may be elegantly formulated in superspace (Sect. 2.1), where one introduces, in the process, a Majorana spinor[5] θ, that is, a spinor which satisfies the relation $\theta = \mathscr{C}\overline{\theta}^{\mathrm{T}}$, where $\mathscr{C}$ is the charge conjugation matrix,[6] and $\overline{\theta} = \theta^{\dagger}\gamma^{0}$. The spinor $\theta = (\theta_a)$, with a denoting a spinor index, gives rise to fermionic degrees of freedom which are included together with those provided by the coordinates x^{μ}, thus defining generalized coordinate labels (x^{μ}, θ_a) in a mathematical description of superspace. The latter variables satisfy the basic relations $[x^{\mu}, \theta_a] = 0$, $\{\theta_a, \theta_b\} = 0$. Supersymmetry transformations may be introduced by defining a unitary operator which for infinitesimal transformations takes the form

$$U = 1 + \mathrm{i}\,\delta\overline{\epsilon}_a Q_a, \qquad \overline{Q} = -Q^{\mathrm{T}}\mathscr{C}^{-1}, \quad \delta\overline{\epsilon}\,Q = \overline{Q}\,\delta\epsilon, \qquad (2.1)$$

with $\mathscr{C}$ denoting the charge conjugation matrix, where the parameter ϵ is a Majorana spinor, and Q is referred to as the supercharge operator. Here we note how the need of a generator of fermionic type naturally arises, carrying a spinorial index, in order to describe transformations between fields of particles of half-odd integer spins, the fermions, and fields of particles of integer spins, the bosons, and vice versa.[7]

The operators $Q, Q^{\dagger}$ allow one to move within a supermultiplet involving bosonic and fermionic particles (Sect. 2.5) implementing, in the process, transformations between states differing by half-odd integer spins, relating states corresponding to different statistics. The fact that supersymmetry transformations lead to changes of

---

[5]The components of which are anti-commuting referred to as Grassmann variables. The nature of this additional variable (parameter) and relevant details to it will be exploited in detail in Sects. 2.1–2.5.

[6]See Appendix II at the end of the book to see how the charge conjugation matrix arises.

[7]An earlier analysis by Coleman and Mandula [7] implied that, under some underlying conditions, an extension of the Poincaré algebra was not possible if one restricts the new generators to be of bosonic types, in the sense that these additional bosonic generators simply commute with (i.e., decouple from) all the familiar generators of the Poincaré ones in Sect. 4.2 in Chap. 4 of Vol. I. A fairly detailed and lucid treatment of this is given in [41, pp. 12–22].

spin, changing in turn the statistics of a state, means quantum mechanically that a supercharge operator does not commute with spin: $[Q_a, J^{\mu\nu}] \neq 0$. Accordingly, the explicit derivation (Sect. 2.5) of this commutator is important as it embodies a key result in supersymmetry. Since the masses in a given supermultiplet are equal, this means quantum mechanically that $[Q_a, \text{Mass}^2] = 0$.

The super-Poincaré algebra involves anti-commutators, in addition to commutators, and the underlying algebra is referred to as a graded algebra. In particular, it is shown in (2.5.28) for $b = a$, with a sum over a, that

$$\frac{1}{4} \sum_a \{Q_a, Q_a^\dagger\} = P^0 = H, \qquad Q_a^\dagger = -(\gamma^0 \mathscr{C})_{ab} Q_b. \qquad (2.2)$$

Hence for a given state Ω,

$$\frac{1}{4} \sum_a \left(\|Q_a^\dagger \Omega\|^2 + \|Q_a \Omega\|^2 \right) = \langle \Omega | H | \Omega \rangle \geq 0. \qquad (2.3)$$

That is, the spectrum of a supersymmetric theory is non-negative. In particular, we note from this equation, that *if* the vacuum |vac⟩ is invariant under a supergauge transformation, as it should be in a supersymmetric theory, i.e., $Q_a|\text{vac}\rangle = 0$ for all a, then the vacuum energy is zero: $\langle \text{vac}|H|\text{vac}\rangle = 0$.

On the other hand, if, say there is a field χ with infinitesimal supersymmetry transformation $\delta\chi = [\chi, \delta\bar{\epsilon}Q]/i$, such that the vacuum expectation value of the latter does not vanish, i.e.,

$$0 \neq i\,\langle \text{vac}|\delta\chi|\text{vac}\rangle = \langle \text{vac}|\chi\delta\bar{\epsilon}Q|\text{vac}\rangle - \langle \text{vac}|(Q^\dagger\gamma^0)\delta\epsilon\,\chi|\text{vac}\rangle, \qquad (2.4)$$

where we have used the relations $\delta\bar{\epsilon}Q = \bar{Q}\delta\epsilon$, $\bar{Q} = Q^\dagger\gamma^0$, we must obviously have $Q_a|\text{vac}\rangle \neq 0$, for at least an a, otherwise an equality to zero for all a will be inconsistent with the non-vanishing of the vacuum expectation value of $\delta\chi$ above. The non-vanishing property $Q_a|\text{vac}\rangle \neq 0$, for an a, implies that the vacuum is not invariant under supersymmetry transformations and symmetry is said to be spontaneously broken. This in turn implies that for spontaneously broken supersymmetry, the vacuum energy satisfies the inequality $\langle \text{vac}|H|\text{vac}\rangle > 0$, i.e., it is strictly positive, since $\langle \text{vac}|Q_a^\dagger Q_a|\text{vac}\rangle > 0$, at least for an a. The situation is summarized in Fig. 2.1.

Unlike other discoveries, supersymmetry was not, a priori, invented under pressure set by experiments and was a highly intellectual achievement. Theoretically, however, it quickly turned out to be quite important in further developments of quantum field theory. Elaborating on what was mentioned earlier, in a supersymmetric extension of the standard model, the electroweak and strong effective couplings do merge at energies about 10^{16} GeV, signalling the possibility that these interactions are different manifestations of a single force in support of a grand unified theory of the fundamental interactions. Also gravitational effects are expected to be important

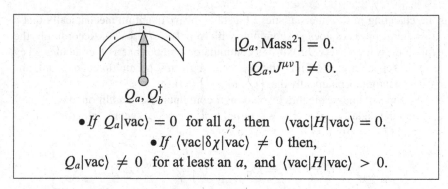

$$[Q_a, \text{Mass}^2] = 0.$$

$$[Q_a, J^{\mu\nu}] \neq 0.$$

- *If* $Q_a|\text{vac}\rangle = 0$ for all a, then $\quad \langle\text{vac}|H|\text{vac}\rangle = 0.$
  - *If* $\langle\text{vac}|\delta\chi|\text{vac}\rangle \neq 0$ then,
  $Q_a|\text{vac}\rangle \neq 0$ for at least an a, and $\langle\text{vac}|H|\text{vac}\rangle > 0.$

Fig. 2.1 $Q_a, Q_b^\dagger$ allow one to move within a supermuliplet, and if they do not annihilate a state, they generate states with spins differing by $\pm 1/2$. The latter implies that the commutator of the supercharge operator and spin does not vanish. The non-vanishing of the vacuum expectation value of a (scalar) field leads to spontaneous supersymmetry breaking. Note that if $\langle\text{vac}|\delta\chi|\text{vac}\rangle \propto \langle\text{vac}|\mathscr{F}|\text{vac}\rangle\delta\epsilon$, where $\mathscr{F}$ is a scalar field, then it is the non-vanishing of the vacuum expectation value of the *scalar* field is in question here

at the quantum level at the Planck energy of the order 10^{19} GeV, or possibly at even lower energies, giving the hope of a unified theory of the four fundamental interactions at high energies.

The enormous mass difference between the electroweak scale $\sim 10^2$ GeV and all the way up to a grand unification scale $\sim 10^{15}$ GeV, referred to as the hierarchy problem, should be of concern, as it is such an enormous shift in energy scale providing no hint as to what happens to the physics in between. In the standard model,[8] the vacuum expectation value of the Higgs boson sets the scale for the masses in the theory, such as for the masses of the vector bosons. The masses imparted to the initially massless vector bosons in the theory, for example, via the Higgs mechanism, using the parameters in the Lagrangian density, i.e., at the tree level, are in very good agreement with experimental results. On the other hand, if one introduces a large energy scale cut-off $\kappa \sim 10^{15}$ GeV, of the order of a grand unified energy scale, or the Planck energy scale $\kappa \sim 10^{19}$ GeV, at which gravitation may play a significant role, to compute the shift of the squared-mass of the Higgs boson (a scalar particle) due to radiative corrections, it turns out to be quadratic[9] in κ, which is quite large for such a large cut-off. This requires that the bare mass squared of the Higgs boson to be correspondingly large to cancel such a quadratic dependence on κ and obtain a physical mass of the Higgs boson of the order of magnitude of the minute energy[10] ($\sim 10^2$ GeV), in comparison, characterizing the standard model, and this seems quite unnatural for the cancelation of such huge quantities. The question, in turn, arises as what amounts for the enormous difference

---

[8] See Chap. 6, Vol. I [26] for details.

[9] See, e.g., [39].

[10] Aad et al. [1], Chatrchyan et al. [6].

between the energy scale of grand unification and the energy scale that characterizes the standard model. This unnatural cancelation of enormously large numbers has been termed a facet of the hierarchy problem. Supersymmetric theories have, in general, the tendency to cancel out such quadratic divergences, up to possibly of divergences of logarithmic type which are tolerable, thus protecting a scalar particle from acquiring such a large bare mass. So supersymmetry may have an important role to play here.

The name "Supersymmetry" for this symmetry is attributed to Salam and Strathdee as it first appears in the title of one of their papers [30]. The extension of the algebra of the Poincaré group to a superalgebra was first carried out by Gol'fand and Likhtman in [19] to construct supersymmetric field theory models, and with the implementation of spontaneous symmetry breaking by Volkov and Akulov in [40]. In 1974,[11] Wess and Zumino independently,[12] also developed supersymmetric models in 4 dimensions, and this work has led to an avalanche of papers on the subject and to a rapid development of the theory. In particular, supersymmetric extensions of the standard model were developed,[13] supergravity, as a supersymmetric extension of gravitational theory, was also developed.[14] Color confinement, within supersymmetric settings, were also analysed.[15] Of particular interest was also the development of the superspace concept as an extension of the Minkowski one, with generalized coordinates (x^μ, θ_a) as mentioned earlier. From the superspace concept, one then conveniently sets up SUSY invariant integrals as in defining supersymmetric actions (Sect. 2.4) needed in developing supersymmetric field theories.[16]

As generalizations of ordinary fields, one may introduce instead so-called superfields to develop supersymmetric theories. These fields are functions of superspace variables $(t, \mathbf{x}, \theta)$ instead of just the spacetime ones $(t, \mathbf{x})$, and, as ordinary fields, they may carry vector, spinor, and other indices. The main task in gauge theories, is to develop supersymmetric extension of Yang-Mills field theories, where the Yang-Mills field is a non-abelian *vector* gauge field. Accordingly, it is interesting to introduce a *pure* vector superfield. That is, a superfield for which its θ-independent part is an ordinary (Lorentz) vector field and hence the adjective "pure".[17] Surprisingly, an *explicit* expression of the *pure* vector superfield has been only recently derived [25]. One usually constructs supersymmetric extensions

---

[11]Wess and Zumino [42].

[12]These basic papers, together with other key ones, are conveniently collected in Ferrara [16].

[13]See Fayet [15], Dimopoulos and Georgi [11].

[14]See Freedman, van Nieuwenhuizen and Ferrara [17], Deser and Zumino [10].

[15]See Seiberg and Witten [35].

[16]See also Salam and Strathdee [31, 32].

[17]Pure vector superfield is not to be confused with so-called "vector" superfields, given the unfortunate name vector superfields, which are real scalar superfields depending on vector fields: see, e.g. [2, p. 56], [41, p. 117], [3, p. 81]. They are sometimes referred to as scalar (vector) superfields.

of Yang-Mills *vector* field theories, however, by introducing and using a *spinor* superfield to do so.[18]

Things are not much better for supergravity, as far as its renormalizability is concerned. Unlike GR, however, non-renormalizability of supergravity is revealed by necessarily going even to higher order loops in the theory.[19]

Superspace and the super-Poincaré transformations, and their group properties, generalizing the Poincaré ones, is the subject matter of Sect. 2.1. Basic properties of the Grassmann variable θ_a are spelled out in Sect. 2.2. Superdifferentiation and superintegrations are treated, respectively, in Sects. 2.3 and 2.4. Derivations of the fundamental anti-commutators/commutators of the super-Poincaré algebra as well as the spelling out the nature of the supermultiplets are given in Sect. 2.5. Superfields are generated in Sect. 2.6. Various supersymmetric field theories are developed in Sects. 2.7–2.9, with the supersymmetric gauge theories ones in Sect. 2.12. The removal of the so-called quadratic divergence associated with the self-energy of a scalar particle in a supersymmetric setting is exploited in Sect. 2.10 as it is critical in supersymmetry. Spontaneous symmetry breaking is treated in Sect. 2.11. Incorporation of supersymmetry in the standard model and the study of the fascinating subject of the unification of the gauge coupling constants as a much improvement over the one obtained for the non-supersymmetric one is given in Sect. 2.13. The local version of supersymmetry, where transformations between bosonic and fermion fields depend on the spacetime dependent parameter $\epsilon(x)$, that is at which point x of spacetime we are, is called supergravity and is an immediate generalization of general relativity with an underlying *structure* of spacetime. This occupies us in the last three sections of this chapter, where the full pure supergravity, including its underlying geometrical aspects, are derived culminating in Sect. 2.16.

We consider theories with only one spinor supercharge (Q_a), with four components, corresponding to ones referred to as $\mathcal{N} = 1$ (simple) supersymmetric field theories. With the very basics knowledge acquired by the reader of supersymmetry in this chapter, he or she may consult more advanced treatments for additional details and further developments of this fascinating subject.

2.1 Superspace: Arena of Sparticles—Particles

As discussed in the introduction to the chapter, to describe transformations between particles of half-odd integer spins, the fermions, and particles of integer spins, the bosons, one introduces a generator (generators) which necessarily carry a spinorial

---

[18]See [2, p. 74]), [41, p. 124]), [3, p. 92], [41], Weinberg in his monumental work on supersymmetry, on p. 124, line 13 from below, states that it is rather surprising that the need of a spinor field arises in such a case.

[19]See, e.g., [8, 9, 21, 36, 37]. Here we are referring to the case of $D = 4$ dimensions with one ($\mathcal{N} = 1$) supersymmetry charge operator in the theory. For supersymmetric extensions (i.e., for $1 \leq \mathcal{N} \leq 8$) and arbitrary dimensions ($4 \leq D \leq 11$) of the theory, see these references on the renormalizability problems of such extensions.

index. Accordingly, to accommodate such a symmetry, in a spacetime description as an extension of the Minkowski spacetime, one may introduce a four component spinor (θ_a), which commutes with the spacetime variable x^μ, as an additional coordinate tied together with x^μ in a way similar the time variable x^0 is tied together with those of the space ones $\mathbf{x}$ in special relativity. The resulting space with coordinates labels (x^μ, θ_a) is referred to as *superspace*. Supersymmetry is a spacetime symmetry. By the extension of Minkowski spacetime to superspace, we will, in later sections, develop supersymmetric field theories.

Before discussing on how this spinor θ is tied together with the spacetime coordinate x^μ, we note that the spinor θ is chosen as a Majorana spinor with anti-commuting components and, as mentioned above, that commute with x^μ:

$$\{\theta_a, \theta_b\} = 0, \qquad \theta_a \theta_a = 0, \qquad [\theta_a, x^\mu] = 0. \qquad (2.1.1)$$

Here we recall the definition of a Majorana spinor

$$\theta = \mathscr{C}\overline{\theta}^{\mathsf{T}}, \ \overline{\theta} = -\theta^{\mathsf{T}}\mathscr{C}^{-1}, \ \overline{\theta} = \theta^\dagger \gamma^0,$$

where the charge conjugation matrix $\mathscr{C}$ is defined by $\mathscr{C} = i\gamma^2\gamma^0$ and satisfies:[20]

$$\mathscr{C}^{-1}\gamma^\mu \mathscr{C} = -(\gamma^\mu)^{\mathsf{T}}, \ \mathscr{C}^\dagger = \mathscr{C}^{\mathsf{T}} = -\mathscr{C}.$$

Also for two anti-commuting Majorana spinors θ, χ, it is easily verified that[21]

$$\overline{\theta}\chi = \overline{\chi}\theta \ \text{(Hermiticity condition)}, \qquad \overline{\theta}\gamma^\mu \chi = -\overline{\chi}\gamma^\mu\theta. \qquad (2.1.2)$$

We introduce a supersymmetry transformation as a shift of θ by a Majorana spinor ϵ, with anti-commuting components, and anti-commuting with the components of θ: $\{\theta_a, \epsilon_b\} = 0$ as well. In order to tie θ_a with the spacetime coordinates x^μ, we also have to shift x^μ by a vector depending on θ, ϵ which must be Hermitian. As in special relativity the transformations are taken to be linear, now in x and θ, given by the simple supersymmetry transformation on the coordinates (x^μ, θ_a) in an 8-dimensional superspace:

$$x'^\mu = x^\mu + \frac{i}{2}\overline{\epsilon}\gamma^\mu\theta - b^\mu, \qquad (2.1.3)$$

$$\theta'_a = \theta_a + \epsilon_a. \qquad (2.1.4)$$

Here we have also introduced an ordinary spacetime translation of x^μ specified by the vector components b^μ. The i factor in (2.1.3) is necessary to ensure Hermiticity

---

[20] For details on the gamma matrices see Appendix I at the end of this volume. See also Appendix II to see how the charge conjugation matrix $\mathscr{C}$ arises.

[21] For additional such identities see (2.2.8)–(2.2.10).

according to the rule (second equality) in (2.1.2). Below we will generalize the transformations in (2.1.3), (2.1.4) to include Lorentz boosts and spacial rotations.

For a subsequent transformation $x' \to x''$, $\theta' \to \theta''$,

$$x''^{\mu} = x'^{\mu} + \frac{i}{2}\bar{\epsilon}'\gamma^{\mu}\theta' - b'^{\mu}, \tag{2.1.5}$$

$$\theta''_a = \theta'_a + \epsilon'_a, \tag{2.1.6}$$

from which we may infer that

$$x''^{\mu} = x^{\mu} + \frac{i}{2}(\bar{\epsilon} + \bar{\epsilon}')\gamma^{\mu}\theta - \left(b^{\mu} + b'^{\mu} - \frac{i}{2}\bar{\epsilon}'\gamma^{\mu}\epsilon\right), \tag{2.1.7}$$

$$\theta''_a = \theta_a + (\epsilon_a + \epsilon'_a). \tag{2.1.8}$$

That is, the supersymmetry transformations may be specified by the pair (ϵ, b) satisfying the group properties:

1. Group multiplication: $(\epsilon', b')(\epsilon, b) = (\epsilon + \epsilon', b + b' - \frac{i}{2}\bar{\epsilon}'\gamma\epsilon)$.
2. Identity $(0, 0) : (0, 0)(\epsilon, b) = (\epsilon, b)$.
3. Inverse $(\epsilon, b)^{-1} = (-\epsilon, -b) : (-\epsilon, -b)(\epsilon, b) = (0, 0)$.
4. Associativity Rule: $(\epsilon_3, b_3)[(\epsilon_2, b_2)(\epsilon_1, b_1)] = [(\epsilon_3, b_3)(\epsilon_2, b_2)](\epsilon_1, b_1)$.

In writing the last equality in property 3, we have used the fact that, according to the second equality in (2.1.2), $\bar{\epsilon}\gamma\epsilon = 0$.

We now extend the transformation in (2.1.3), (2.1.4) to include Lorentz boosts and spacial rotations. We recall that under a homogeneous Lorentz transformation, x^{μ} is transformed via the matrix $\Lambda^{\mu}{}_{\nu}$, of special relativity, while the four component spinor transforms via a matrix $[K_{ab}]$, $\theta \to K\theta$, under a Lorentz transformation, whose explicit structure is not needed here, satisfying the properties:[22]

$$\Lambda^{\mu}{}_{\nu}\gamma^{\nu} = K^{-1}\gamma^{\mu}K, \qquad K^{\dagger}\gamma^0 = \gamma^0 K^{-1}. \tag{2.1.9}$$

We define the supersymmetric transformations in superspace, known as Super-Poincaré Transformations, or Super-Inhomogeneous Lorentz Transformations, by

$$x' = \Lambda x + \frac{i}{2}\bar{\epsilon}\gamma K\theta - b, \tag{2.1.10}$$

$$\theta' = K\theta + \epsilon, \tag{2.1.11}$$

in a convenient matrix notation. These include supersymmetry transformations, translations as well as Lorentz boosts and spatial rotations. Consistency of such a transformation requires that its structure remains the same under subsequent

---

[22]See Appendix II at the end of this volume for such transformations.

transformations leading to group properties spelled out below as done for the pure supersymmetry ones above.

Under a subsequent transformation $(x', \theta') \rightarrow (x'', \theta'')$ one has

$$x'' = \Lambda' x' + \frac{i}{2} \bar{\epsilon}' \gamma K' \theta' - b', \qquad (2.1.12)$$

$$\theta'' = K' \theta' + \epsilon', \qquad (2.1.13)$$

from which we may infer that

$$x'' = [\Lambda' \Lambda] x + \frac{i}{2} [\bar{\epsilon}' + \bar{\epsilon} K'^{-1}] \gamma [K'K] \theta - [\Lambda' b + b' - \frac{i}{2} \bar{\epsilon}' \gamma K' \epsilon], \qquad (2.1.14)$$

$$\theta'' = [K'K] \theta + [\epsilon' + K' \epsilon]. \qquad (2.1.15)$$

where in writing the second term on the right-hand side of (2.1.14), we have used the first identity in (2.1.9) to write

$$\Lambda'^{\mu}{}_{\nu} \bar{\epsilon} \gamma^{\nu} K \theta = \bar{\epsilon} K'^{-1} \gamma^{\mu} K' K \theta. \qquad (2.1.16)$$

That is, a Super-Poincaré Transformation, or a Super-Inhomogeneous Lorentz Transformation, may be specified by the quadruplet $(\Lambda, K, \epsilon, b)$, and upon comparison of (2.1.14)/(2.1.15) with (2.1.10)/(2.1.11) we learn that upon a subsequent transformation:

$$\Lambda \rightarrow [\Lambda' \Lambda], \quad K \rightarrow [K'K], \quad \epsilon \rightarrow [\epsilon' + K\epsilon], \quad b \rightarrow [b + b' - \frac{i}{2} \bar{\epsilon}' \gamma K' \epsilon].$$

That is we have the following group properties of the transformations:

1. Group Multiplication:
   $(\Lambda', K', \epsilon', b')(\Lambda, K, \epsilon, b) = (\Lambda' \Lambda, K'K, \epsilon' + K' \epsilon, \Lambda' b + b' - \frac{i}{2} \bar{\epsilon}' \gamma K' \epsilon).$
2. Identity $(I, I, 0, 0)$:
   $(I, I, 0, 0)(\Lambda, K, \epsilon, b) = (\Lambda, K, \epsilon, b).$
3. Inverse $(\Lambda, K, \epsilon, b)^{-1} = (\Lambda^{-1}, K^{-1}, -K^{-1} \epsilon, -\Lambda^{-1} b)$:
   $(\Lambda^{-1}, K^{-1}, -K^{-1} \epsilon, -\Lambda^{-1} b)(\Lambda, K, \epsilon, b) = (I, I, 0, 0).$
4. Associativity Rule:
   $(\Lambda_3, K_3, \epsilon_3, b_3)[(\Lambda_2, K_2, \epsilon_2, b_2)(\Lambda_1, K_1, \epsilon_1, b_1)]$
   $= [(\Lambda_3, K_3, \epsilon_3, b_3)(\Lambda_2, K_2, \epsilon_2, b_2)](\Lambda_1, K_1, \epsilon_1, b_1).$

Here we note that

$$(\epsilon' + K' \epsilon)^{\dagger} \gamma^0 = \bar{\epsilon}' + \bar{\epsilon} K'^{-1},$$

as a result of the second identity in (2.1.9), relevant to the third entry on the right-hand side of the equality in property 1. Also that in writing the last equality in property 3, we have used the fact that according to (2.1.2) and the second identity in (2.1.9),

$$\bar{\epsilon} K \gamma^\mu K^{-1} \epsilon = \overline{(K^{-1}\epsilon)} \gamma^\mu (K^{-1}\epsilon) = 0.$$

In Sect. 2.6, we will use these group properties to develop the Super-Poincaré Algebra satisfied by the generators of the corresponding unitary operators acting on the so-called superfields and on particle supermultiplets. Next we derive basic properties involving products of components of the spinor θ and related summation formulae needed in developing supersymmetric field theories.

2.2 Basic Properties of Products of Components of the Spinor θ and Summation Formulae

In view of developing expressions for products of components of the spinor θ, and derive related key summation formulae needed in developing supersymmetric field theories, we first recall the definition of the charge conjugation matrix $\mathscr{C}$

$$\mathscr{C} = i\gamma^2 \gamma^0, \tag{2.2.1}$$

which satisfies the following basic identitites ($\sharp$ stands for any of the operations $(.)^\dagger, (.)^{-1}, (.)^\top$)

$$\mathscr{C}^\sharp = -\mathscr{C}, \quad \mathscr{C}^{-1}\gamma^\mu \mathscr{C} = -(\gamma^\mu)^\top, \quad [\mathscr{C}, \gamma^5] = 0, \quad \mathscr{C}^{-1}\gamma^5\gamma^\mu \mathscr{C} = (\gamma^5\gamma^\mu)^\top, \tag{2.2.2}$$

$$\mathscr{C}^{-1}[\gamma^\mu, \gamma^\nu]\mathscr{C} = -([\gamma^\mu, \gamma^\nu])^\top, \quad \mathscr{C}^{-1}\gamma^5[\gamma^\mu, \gamma^\nu]\mathscr{C} = -(\gamma^5[\gamma^\mu, \gamma^\nu])^\top. \tag{2.2.3}$$

In the Dirac and chiral representations, it is given, respectively, by[23]

$$(\mathscr{C})_{\text{Dirac}} = \begin{pmatrix} 0 & -i\sigma^2 \\ -i\sigma^2 & 0 \end{pmatrix}, \quad (\mathscr{C})_{\text{chiral}} = \begin{pmatrix} -i\sigma^2 & 0 \\ 0 & i\sigma^2 \end{pmatrix}. \tag{2.2.4}$$

We note that $\gamma^5\mathscr{C}$, $\gamma^5\gamma^\mu\mathscr{C}$ together with $\mathscr{C}$ are anti-symmetric matrices.

Once a charge conjugation matrix $\mathscr{C}$ has been defined, one may define a Majorana spinor θ, not to be confused with the Majorana representation of the gamma matrices, by the condition, which we record here for convenience by

---

[23]For details on the gamma matrices in various representations see Appendix I at the end of the book.

$$\theta = \theta_{\mathscr{C}},$$

$$\theta = \mathscr{C}\bar{\theta}^{\mathsf{T}}, \qquad \bar{\theta} = -\theta^{\mathsf{T}}\mathscr{C}^{-1}. \tag{2.2.5}$$

From these definitions, one may infer the general structure of a Majorana spinor, for example, in the *chiral* representation, to be of the form ($\bar{\theta} = \theta^{\dagger}\gamma^{0}$):

$$\theta = \begin{pmatrix} \theta_1 \\ \theta_2 \\ \theta_3 \\ \theta_4 \end{pmatrix}, \qquad \theta_1 = \theta_4^*, \qquad \theta_2 = -\theta_3^*. \tag{2.2.6}$$

Consider a Majorana spinor ξ with its components anti-commuting with those of θ. That is, for $a, b = 1, 2, 3, 4$,

$$\{\theta_a, \xi_b\} = 0. \tag{2.2.7}$$

Then for a given matrix $A = [A_{ab}]$,

$$\bar{\theta} A \xi = -\theta_a (\mathscr{C}^{-1})_{ab} A_{bc} \mathscr{C}_{cd} \bar{\xi}_d = +\bar{\xi}_d (\mathscr{C}^{-1})_{ab} A_{bc} \mathscr{C}_{cd} \theta_a.$$

Therefore (2.2.2), (2.2.3), give

$$\bar{\theta}\xi = \bar{\xi}\theta, \quad \bar{\theta}\gamma^5\gamma^\mu\xi = \bar{\xi}\gamma^5\gamma^\mu\theta, \quad \bar{\theta}\gamma^\mu\xi = -\bar{\xi}\gamma^\mu\theta, \quad \bar{\theta}\gamma^5\xi = \bar{\xi}\gamma^5\theta,$$
$$\tag{2.2.8}$$
$$\bar{\theta}\gamma^\mu\gamma^\nu\xi = \bar{\xi}\gamma^\nu\gamma^\mu\theta, \quad \bar{\theta}[\gamma^\mu,\gamma^\nu]\xi = -\bar{\xi}[\gamma^\mu,\gamma^\nu]\theta, \tag{2.2.9}$$

$$\bar{\theta}\gamma^5[\gamma^\mu,\gamma^\nu]\xi = -\bar{\xi}\gamma^5[\gamma^\mu,\gamma^\nu]\theta, \quad \bar{\theta}\gamma^5\gamma^\mu\gamma^\nu\xi = -\eta^{\mu\nu}\bar{\theta}\gamma^5\xi - \frac{1}{2}\bar{\xi}\gamma^5[\gamma^\mu,\gamma^\nu]\theta. \tag{2.2.10}$$

These, in particular, show that $\bar{\theta}\xi$ is Hermitian, and for θ, with *anti*-commuting components,

$$\bar{\theta}\gamma^\mu\theta = 0, \quad \bar{\theta}[\gamma^\mu,\gamma^\nu]\theta = 0, \quad \bar{\theta}\gamma^5\gamma^\mu\gamma^\nu\theta = -\eta^{\mu\nu}\bar{\theta}\gamma^5\theta. \tag{2.2.11}$$

The latter properties are interesting in the sense that if, say, we want to expand products of (anti-commuting) components of a Majorana spinor, we may do this solely in terms of $\bar{\theta}\theta$, $\bar{\theta}\gamma^5\theta$, $\bar{\theta}\gamma^5\gamma^\mu\theta$ as the first two other expressions given in the above last equation vanish. Here we recall that the matrices

$$I, \quad \gamma^5, \quad \gamma^\mu, \quad \gamma^5\gamma^\mu, \quad [\gamma^\mu,\gamma^\nu],$$

give rise to 16 independent 4×4 matrices. For example, for θ, with anti-commuting components, we may write

$$\overline{\theta}_b \theta_a = c_1 \, \delta_{ab} \, \overline{\theta}\theta + c_2 \, \gamma^5_{ab} \, \overline{\theta}\gamma^5\theta + c_3 \, (\gamma^5\gamma_\mu)_{ab} \, \overline{\theta}\gamma^5\gamma^\mu\theta. \qquad (2.2.12)$$

The coefficients c_1, c_2, c_3 are easily obtained by multiplying, in turn, by δ_{ab}, by γ^5_{ba}, and finally by $(\gamma^5\gamma^\sigma)_{ba}$, to obtain $c_1 = c_2 = c_3 = 1/4$, where recall, in particular, that $\{\gamma^5, \gamma^\mu\} = 0$, $\{\gamma^\mu, \gamma^\nu\} = -2\,\eta^{\mu\nu}$, $(\gamma^5)^2 = I$.

For further applications, we use the notation B for the matrices I, γ^5, $\gamma^5\gamma^\mu$, and rewrite (2.2.12) in the more elegant form, after multiplying it by $-\mathscr{C}_{bc}$, using in the process (2.2.5), and doing an obvious relabeling of the components of θ, as

$$\theta_a \theta_b = -\frac{1}{4}\, \mathscr{C}_{ca} \sum_B B_{bc}\, \overline{\theta}B\theta = \frac{1}{4}\sum_B (B\mathscr{C})_{ab}\, \overline{\theta}B\theta, \qquad (2.2.13)$$

where note from the properties in (2.2.2), (2.2.3) that

$$\mathscr{C}^{-1}B\,\mathscr{C} = B^\mathsf{T}, \quad \mathscr{C}^\mathsf{T} = \mathscr{C}^{-1} = -\mathscr{C}, \quad B\mathscr{C} = -(B\mathscr{C})^\mathsf{T}. \qquad (2.2.14)$$

Let A denote any of the matrices I, γ^5, $\gamma^5\gamma^\mu$. Then upon multiplying (2.2.13) by $(\overline{\theta}A)_a$, and using (2.2.14), we obtain the useful formula

$$\overline{\theta}A\theta\,\theta_b = -\frac{1}{4}\sum_B (BA\theta)_b\, \overline{\theta}B\theta, \qquad B \in \{I, \gamma^5, \gamma^5\gamma^\mu\}. \qquad (2.2.15)$$

Note the minus sign multiplying $1/4$ on the right-hand of the equation. For any two matrices $A_1\,A_2$, as the B matrices above, multiplying (2.2.15) by $(A_2)_{cb}$, and replacing A by A_1, give the following more general formula

$$\overline{\theta}A_1\theta\,(A_2\theta)_c = -\frac{1}{4}\sum_B \left(A_2 B A_1\theta\right)_c \overline{\theta}B\theta. \qquad (2.2.16)$$

As an application of this formula, consider the explicit evaluation of the sum of the following two terms, by choosing in turn $A_1 = A_2 = I$, $A_1 = A_2 = \gamma^5$,

$$\overline{\theta}\theta\,\theta_c + \overline{\theta}\gamma^5\theta\,(\gamma^5\theta)_c = -\frac{1}{4}\sum_B \left([B + \gamma^5 B\gamma^5]\theta\right)_c \overline{\theta}B\theta. \qquad (2.2.17)$$

Summation over B, representing the matrices I, γ^5, $\gamma^5\gamma^\mu$, shows that the right-hand side of the above equation is $-1/2$ of the left-hand side giving

$$\overline{\theta}\theta\,\theta_c = -\overline{\theta}\gamma^5\theta\,(\gamma^5\theta)_c, \qquad (2.2.18)$$

where again recall that $\{\gamma^5, \gamma^\mu\} = 0$, $(\gamma^5)^2 = 1$, to infer that

$$B + \gamma^5 B \gamma^5 = 0 \quad \text{for} \quad B = \gamma^5 \gamma^\mu. \tag{2.2.19}$$

Some basic properties of expansions of products of components of θ are given below. The first two equations as well as the first entry in (2.2.22) were established above in (2.2.12), (2.2.13), (2.2.18), respectively. The other identities are worked out in Problems 2.3, 2.4. These will be important later in dealing with superfields and in developing supersymmetric field theories.

$$\bar{\theta}_b \theta_a = \frac{1}{4}\left[\delta_{ab} \,\bar{\theta}\theta + \gamma^5_{ab} \,\bar{\theta}\gamma^5\theta + (\gamma^5\gamma_\mu)_{ab} \,\bar{\theta}\gamma^5\gamma^\mu\theta \right], \tag{2.2.20}$$

$$\theta_a \theta_b = \frac{1}{4}\left[\mathscr{C}_{ab} \,\bar{\theta}\theta + (\gamma^5\mathscr{C})_{ab}\bar{\theta}\gamma^5\theta + (\gamma^5\gamma_\mu\mathscr{C})_{ab}\bar{\theta}\gamma^5\gamma^\mu\theta \right]. \tag{2.2.21}$$

$$\bar{\theta}\theta \,\theta_a = -\bar{\theta}\gamma^5\theta \,(\gamma^5\theta)_a, \qquad \bar{\theta}\theta \,\bar{\theta}_a \quad = -\bar{\theta}\gamma^5\theta \,(\bar{\theta}\gamma^5)_a$$

$$\bar{\theta}\gamma^5\gamma^\mu\theta \,\theta_a = -\bar{\theta}\gamma^5\theta \,(\gamma^\mu\theta)_a, \qquad \bar{\theta}\gamma^5\gamma^\mu\theta \,\bar{\theta}_a = \quad \bar{\theta}\gamma^5\theta \,(\bar{\theta}\gamma^\mu)_a \tag{2.2.22}$$

$$(\bar{\theta}\theta)(\bar{\theta}\gamma^5\theta) = 0, \quad (\bar{\theta}\theta)(\bar{\theta}\gamma^5\gamma^\mu\theta) = 0, \quad (\bar{\theta}\gamma^5\gamma^\mu\theta)(\bar{\theta}\gamma^5\theta) = 0, \tag{2.2.23}$$

$$(\bar{\theta}\gamma^5\gamma^\mu\theta)(\bar{\theta}\gamma^5\gamma^\sigma\theta) = \eta^{\mu\sigma}\,(\bar{\theta}\gamma^5\theta)^2, \quad (\bar{\theta}\theta)^2 = -(\bar{\theta}\gamma^5\theta)^2, \tag{2.2.24}$$

$$\bar{\theta}\theta \,\theta_a\theta_b = -\frac{1}{4}\mathscr{C}_{ab}(\bar{\theta}\gamma^5\theta)^2, \quad \bar{\theta}\gamma^5\theta \,\theta_a\theta_b = \frac{1}{4}(\gamma^5\mathscr{C})_{ab}(\bar{\theta}\gamma^5\theta)^2, \tag{2.2.25}$$

$$\bar{\theta}\gamma^5\gamma^\sigma\theta \,\theta_a\theta_b = \frac{1}{4}(\gamma^5\gamma^\sigma\mathscr{C})_{ab}\,(\bar{\theta}\gamma^5\theta)^2. \tag{2.2.26}$$

$$\theta_a\theta_b\theta_c = \frac{1}{4}\bar{\theta}\gamma^5\theta\left[(\mathscr{C}\gamma^5)_{ab}\theta_c - (\mathscr{C}\gamma^5)_{ac}\theta_b + (\mathscr{C}\gamma^5)_{bc}\theta_a \right.$$
$$\left. - \mathscr{C}_{ab}(\gamma^5\theta)_c + \mathscr{C}_{ac}(\gamma^5\theta)_b - \mathscr{C}_{bc}(\gamma^5\theta)_a \right], \tag{2.2.27}$$

$$\theta_a\theta_b\theta_c\theta_d = \frac{1}{16}(\bar{\theta}\gamma^5\theta)^2\left[(\mathscr{C}\gamma^5)_{ab}(\mathscr{C}\gamma^5)_{cd} - (\mathscr{C}\gamma^5)_{ac}(\mathscr{C}\gamma^5)_{bd} \right.$$
$$\left. + (\mathscr{C}\gamma^5)_{bc}(\mathscr{C}\gamma^5)_{ad} - \mathscr{C}_{ab}\mathscr{C}_{cd} + \mathscr{C}_{ac}\mathscr{C}_{bd} - \mathscr{C}_{bc}\mathscr{C}_{ad} \right]. \tag{2.2.28}$$

Fierz identities involving the charge conjugation matrix as well the classic Fierz identity are given in Appendix A of this chapter.

2.3　Superderivatives and Products of Superderivatives

The next step needed in working in superspace is to see how the concept of a derivative is generalized in this case. In Sect. 2.1, we have seen that a Majorana spinor θ, with anti-commuting four components, is tied together with the spacetime coordinate x and commutes with it, and the resulting space with coordinate labels (x^μ, θ_a) is referred to as superspace. Clearly a covariant derivative and the consistent transformation rule for it in superspace must involve both x^μ and θ or $\overline{\theta}$, which we now consider.

We first recall with the rules developed for differentiation with respect to Grassmann variables in Sect. 2.4 in Chap. 1 of Vol. I, that for any matrix A involving product of gamma matrices we have $(\theta = \mathscr{C}\overline{\theta}^{\mathsf{T}}, \; \overline{\theta} = -\theta^{\mathsf{T}}\mathscr{C}^{-1}, \; \mathscr{C}^\dagger = \mathscr{C}^{-1} = \mathscr{C}^{\mathsf{T}} = -\mathscr{C})$

$$\frac{\partial}{\partial\overline{\theta}_a} \overline{\theta} A \theta = (A\theta)_a - (\overline{\theta}A\mathscr{C})_a = (A\theta)_a + (\theta^{\mathsf{T}}\mathscr{C}^{-1}A\mathscr{C})_a, \tag{2.3.1}$$

giving, in particular, from (2.2.2), (2.2.3)

$$\frac{\partial}{\partial\overline{\theta}_a} \overline{\theta}\gamma^5\theta = 2(\gamma^5\theta)_a, \quad \frac{\partial}{\partial\overline{\theta}_a} \overline{\theta}\gamma^5\gamma^\mu\theta = 2(\gamma^5\gamma^\mu\theta)_a,$$

$$\frac{\partial}{\partial\overline{\theta}_a} (\overline{\theta}\gamma^5\theta)^2 = 4(\overline{\theta}\gamma^5\theta)(\gamma^5\theta)_a. \tag{2.3.2}$$

Due to the Majorana characters of θ, ϵ, we may write

$$\overline{\epsilon}\gamma^\mu K\theta = -(K\theta)^\dagger\gamma^0\gamma^\mu\epsilon.$$

On the other hand, from the second identity in (2.1.9), $K^\dagger\gamma^0 = \gamma^0 K^{-1}$. Accordingly, the general Super-Poincaré Transformation including Lorentz boosts and spatial rotations given in (2.1.10), (2.1.11), may be rewritten as

$$x'^\mu = \Lambda^\mu{}_\nu x^\nu - \frac{i}{2}\overline{\theta}K^{-1}\gamma^\mu\epsilon - b^\mu, \tag{2.3.3}$$

$$\theta' = K\theta + \epsilon, \qquad \overline{\theta}' = \overline{\theta}K^{-1} + \overline{\epsilon}, \tag{2.3.4}$$

where a spinor ψ in Minkowski space transforms as $\psi \to K\psi$ under a Lorentz transformation,[24] with K satisfying the identities $\big((\Lambda^{-1})^{\mu\nu} = \Lambda^{\nu\mu}\big)$

$$\Lambda^\mu{}_\nu\gamma^\nu = K^{-1}\gamma^\mu K, \qquad K\gamma^\nu = \gamma^\mu\Lambda_\mu{}^\nu K. \tag{2.3.5}$$

---

[24]See Appendix II at the end of this volume.

The chain rule implies that

$$\frac{\partial}{\partial\overline{\theta}} = \frac{\partial}{\partial\overline{\theta}}\,\overline{\theta}\,'\,\frac{\partial}{\partial\overline{\theta}'} + \frac{\partial x\,'^{\mu}}{\partial\overline{\theta}}\,\partial\,'_{\mu}. \tag{2.3.6}$$

Hence upon carrying out the above differentiations with respect to $\overline{\theta}$ and multiplying the resulting equation from the left by K give

$$K\frac{\partial}{\partial\overline{\theta}} = \frac{\partial}{\partial\overline{\theta}'} \cdot -\frac{i}{2}\,(\gamma^{\mu}\epsilon)\partial\,'_{\mu}. \tag{2.3.7}$$

Finally solving for ϵ from (2.3.4), using the fact that $\partial\,'_{\mu} = \Lambda_{\mu}{}^{\nu}\partial_{\nu}$ together with the second identity in (2.3.5), give from (2.3.7) the transformation rule

$$\left(\frac{\partial}{\partial\overline{\theta}'} - \frac{i}{2}\,(\gamma^{\mu}\theta\,')\partial\,'_{\mu}\right) = K\left(\frac{\partial}{\partial\overline{\theta}} - \frac{i}{2}\,(\gamma^{\mu}\theta)\partial_{\mu}\right), \tag{2.3.8}$$

and, in turn, define the superderivative or super-covariant derivative

$$D = \frac{\partial}{\partial\overline{\theta}} - \frac{i}{2}\,(\gamma^{\mu}\theta)\partial_{\mu}. \tag{2.3.9}$$

From this expression of the super-covariant derivative in superspace, the following anti-commutation relations of the components of D_a emerge (see Problem 2.7)

$$\{D_a,\,D_b\} = -i\,(\gamma^{\mu}\mathscr{C})_{ab}\partial_{\mu}. \tag{2.3.10}$$

In analogy to $\overline{\theta}$, as obtained from θ, we may define the superderivative $\overline{D}$, which has the following anti-commutation relation with D: (see Problem 2.7)

$$\overline{D}_a = -D_b\,\mathscr{C}_{ba}^{-1}, \quad \{D_a,\,\overline{D}_b\} = i\,(\gamma^{\mu})_{ab}\,\partial_{\mu}. \tag{2.3.11}$$

For our subsequent applications in developing supersymmetric field theories, we also need some properties of products of superderivatives and related summations formulae. Some of the basic ones are given below.

Using, in the process, the basic properties of the charge conjugation matrix $\mathscr{C}$ in (2.2.2), (2.2.3), the following two equalities are easily derived in Problem 2.8:

$$\overline{D}\,\gamma^{\mu}D = -2i\,\partial^{\mu}, \quad \overline{D}\,[\gamma^{\mu},\gamma^{\nu}]D = 0. \tag{2.3.12}$$

Also the properties (2.2.2), (2.2.3), involving the charge conjugation matrix $\mathscr{C}$, and the anti-commutation relations of the superderivatives D, $\overline{D}$ in (2.3.10), (2.3.11), lead to the equalities (see Problem 2.9)

$$\overline{D}BD\,D_a = D_a\,\overline{D}BD - 2i\,(\gamma^{\mu}BD)_a\partial_{\mu}, \quad B = I,\,\gamma^5,\,\gamma^5\gamma^{\mu}. \tag{2.3.13}$$

Recalling the 16 independent 4×4 matrices I, γ^{μ}, γ^5, $\gamma^5\gamma^{\mu}$, $[\gamma^{\mu}, \gamma^{\nu}]$, and taking into account of the second equality in (2.3.12), one may expand the product $D_a \overline{D}_b$ in the following manner

$$D_a\overline{D}_b = a_1\delta_{ab}\overline{D}D + a_2(\gamma^{\mu})_{ab}\overline{D}\gamma_{\mu}D + a_3(\gamma^5)_{ab}\overline{D}\gamma^5D + a_4(\gamma^5\gamma^{\mu})_{ab}\overline{D}\gamma^5\gamma_{\mu}D. \tag{2.3.14}$$

By making use of the anti-commutation relation (2.3.11), and taking the trace (i.e., set $a = b$) of the above equation gives $a_1 = -1/4$. Using the first equality in (2.3.12), and, in turn, multiplying the above equation by $(\gamma^{\sigma})_{ba}$, $(\gamma^5)_{ba}$, $(\gamma^5\gamma^{\sigma})_{ba}$, the other coefficients are readily determined, giving

$$D_a\overline{D}_b = \frac{i}{2}(\gamma^{\mu})_{ab}\partial_{\mu} - \frac{1}{4}\sum_B (B)_{ab}\overline{D}BD, \tag{2.3.15}$$

with a summation over the matrices B defined in (2.3.13). From this, one may also write

$$D_aD_b = -\frac{i}{2}(\gamma^{\mu}\mathscr{C})_{ab}\partial_{\mu} + \frac{1}{4}\sum_B (B\mathscr{C})_{ab}\overline{D}BD. \tag{2.3.16}$$

Upon multiplying the latter by $(\overline{D}A)_a$, from the left, where $A = I$ or γ^5 or $\gamma^5\gamma^{\sigma}$, we also obtain

$$\overline{D}AD\,D_b = -\frac{i}{2}(\gamma^{\mu}AD)_b\partial_{\mu} - \frac{1}{4}\sum_B (BAD)_b\overline{D}BD. \tag{2.3.17}$$

The following operator identities[25] then follow from the applications of (2.3.13), (2.3.16), (2.3.17) and are worked out in Problem 2.10:

$$\overline{D}D\,D_a = D_a\overline{D}D - 2\,i\,(\gamma^{\mu}D)_a\partial_{\mu}, \tag{2.3.18}$$

$$\overline{D}\gamma^5D\,D_a = -\overline{D}D\,(\gamma^5D)_a, \tag{2.3.19}$$

$$D_a\overline{D}\gamma^5D = -\overline{D}D\,(\gamma^5D)_a - 2\,i\,(\gamma^5\gamma^{\mu}D)_a\partial_{\mu}, \tag{2.3.20}$$

$$(\overline{D}\gamma^5D)^2 = -(\overline{D}D)^2, \tag{2.3.21}$$

$$\overline{D}\gamma^5\gamma^{\sigma}D\,D_a = -\overline{D}D\,(\gamma^5\gamma^{\sigma}D)_a - i\,(\gamma^5[\gamma^{\sigma}, \gamma^{\mu}]D)_a\partial_{\mu}, \tag{2.3.22}$$

$$D_a\overline{D}\gamma^5\gamma^{\sigma}D = -\overline{D}D\,(\gamma^5\gamma^{\sigma}D)_a + 2\,i\,(\gamma^5D)_a\partial^{\sigma}. \tag{2.3.23}$$

---

[25]For an extensive treatment of the products of the superderivatives see [33].

2.4   Invariant Integration in Superspace

The first step in developing integration theory in superspace, one has to investigate the nature of the volume element

$$(\mathrm{d}x)\,(\mathrm{d}\theta) \;\equiv\; \mathrm{d}x^0 \mathrm{d}x^1 \mathrm{d}x^2 \mathrm{d}x^3 \mathrm{d}\theta_1 \mathrm{d}\theta_2 \mathrm{d}\theta_3 \mathrm{d}\theta_4, \tag{2.4.1}$$

under the Super-Poincaré Transformation in (2.1.10), (2.1.11) in superspace, in analogy to the volume element $(\mathrm{d}x)$ under the Lorentz transformation in Minkowski spacetime. That is, given a scalar (super)field $\Phi(x,\theta)$ defined by the condition

$$\Phi'(x',\theta') = \Phi(x,\theta), \tag{2.4.2}$$

we have to investigate the meaning of an integral such as

$$\int (\mathrm{d}x)(\mathrm{d}\theta)\, \Phi(x,\theta). \tag{2.4.3}$$

as an *invariant* integral in superspace. This is the essence of defining invariant actions in superspace as the first step in developing supersymmetric field theories.

Before doing this, let us see how one carries integrations with respect to Grassmann variables. Assuming the translational invariance of an integral, say with respect to the component θ_1: $\int \mathrm{d}\theta_1\,\theta_1$, under the substitution

$$\theta_1 \rightarrow \theta_1 + \theta_1^{(o)} = \theta_1', \tag{2.4.4}$$

with $\mathrm{d}\theta_1' = \mathrm{d}\theta_1$, i.e.,

$$\int \mathrm{d}\theta_1'\, \theta_1' = \int \mathrm{d}\theta_1\, \theta_1 \tag{2.4.5}$$

we have

$$\int \mathrm{d}\theta_1'\, \theta_1' = \int \mathrm{d}\theta_1\, (\theta_1 + \theta_1^{(o)}) = \int \mathrm{d}\theta_1\, \theta_1 + \left(\int \mathrm{d}\theta_1 \right)\theta_1^{(o)}, \tag{2.4.6}$$

we learn that

$$\int \mathrm{d}\theta_a = 0, \qquad \int \mathrm{d}\theta_a\, \theta_b = \delta_{ab}, \tag{2.4.7}$$

where we have normalized the numerical value of the last integral to be one for $a = b$.

We also note that the Dirac delta is simply given by

$$\delta(\theta_a - \theta'_a) = (\theta_a - \theta'_a), \tag{2.4.8}$$

and gives rise to the following integrals,

$$\int d\theta_a \, (\theta_a - \theta'_a) = 1, \qquad \int d\theta_a \, (\theta_a - \theta'_a) \, \theta_a = \theta'_a, \tag{2.4.9}$$

where we have used the fact that $\theta'_a \theta_a = -\theta_a \theta'_a$.

Upon setting

$$\delta^{(2)}(\theta_R) = \theta_4 \theta_3, \qquad \delta^{(2)}(\theta_L) = \theta_2 \theta_1, \tag{2.4.10}$$

with $(d\theta) = d\theta_1 d\theta_2 d\theta_3 d\theta_4$, the following useful integrals should be noted

$$\int (d\theta) \, \delta^{(2)}(\theta_R) \, \delta^{(2)}(\theta_L) = 1, \tag{2.4.11}$$

$$\int (d\theta) \, \frac{1}{8} \, (\overline{\theta} \gamma^5 \theta)^2 = 1, \tag{2.4.12}$$

$$\int (d\theta) \, \delta^{(2)}(\theta_R) \, \overline{\theta}\theta = 2, \qquad \int (d\theta) \, \delta^{(2)}(\theta_L) \, \overline{\theta}\theta = -2, \tag{2.4.13}$$

$$\int (d\theta) \, \delta^{(2)}(\theta_R) \, \overline{\theta}\theta_R = 0, \qquad \int (d\theta) \, \delta^{(2)}(\theta_L) \, \overline{\theta}\theta_L = 0, \tag{2.4.14}$$

$$\int (d\theta) \, \delta^{(2)}(\theta_R) \, \overline{\theta}\theta_L = 2, \qquad \int (d\theta) \, \delta^{(2)}(\theta_L) \, \overline{\theta}\theta_R = -2, \tag{2.4.15}$$

$$\int (d\theta) \, \delta^{(2)}(\theta_{R/L}) \, \overline{\theta}\gamma^5\theta = -2, \quad \int (d\theta) \, \delta^{(2)}(\theta_{R/L}) \, \overline{\theta}\gamma^5\gamma^\mu\theta = 0. \tag{2.4.16}$$

The 8-dimensional matrix associated with a Jacobian relating the variables (x', θ'), (x, θ) is easily worked out to be (see Problem 2.11)

$$M = \left[\frac{\partial(x'^\mu, \theta'_a)}{\partial(x^\nu, \theta_b)} \right] = \begin{pmatrix} \Lambda & T \\ 0 & K \end{pmatrix}, \qquad T = [-\frac{i}{2}(\overline{\epsilon}\gamma^\mu K)]. \tag{2.4.17}$$

The elements of the matrices Λ, K are c-numbers, while those in T are Grassmann ones, that is anti-commuting but commuting with those in Λ, K.

Remark Since the variables x^μ, θ_a are tied down together in a super-Lorentz transformation, it would be not only naïve but incorrect to say that under such a transformation, $dx'^\mu = \Lambda^\mu{}_\nu \, dx^\nu$, (in superspace, dx^μ has a more complicated transformation than that), and for a spinor $d\theta'_a = K_{ab} \, d\theta_b$, independently, and hence $(dx')(d\theta') = (\det\Lambda/\det K)(dx)(d\theta)$ according to a change of variables of commuting variables (x^μ) and anti-commuting[26] ones (θ_a). We will see that due to the presence of anti-commuting elements in the matrix in (2.4.17), one has to re-define the concept of a determinant and the trace of such matrices and we end up, nevertheless, obtain exactly the just mentioned expression for the Jacobian of transformation.

When a matrix such as the one in (2.4.17) involves Grassmann variables, the basic property $\det M_1 M_2 = \det M_1 \det M_2$, indeed breaks down for two such matrices M_1, M_2. The presence of the Grassmann variables in (2.4.17) will, self consistently allow us to extend the definition of a determinant as well as of the trace for such supermatrices and the corresponding operations will be denoted, respectively, by Sdet and STr.

The matrix M in (2.4.17), may be rewritten as (see Problem 2.12)

$$M = \exp\left(\ln\left[M\right]\right) = \exp\left(\ln\left[I - (I - M)\right]\right) = \exp\begin{pmatrix} A & \eta \\ 0 & B \end{pmatrix}, \qquad (2.4.18)$$

with

$$A = -\sum_{n \geq 1} \frac{(I - \Lambda)^n}{n}, \qquad B = -\sum_{n \geq 1} \frac{(I - K)^n}{n}, \qquad (2.4.19)$$

$$\eta = \sum_{n \geq 1} [(I - \Lambda)^{n-1} T + (I - \Lambda)^{n-2} T(I - K) + \cdots + T(I - K)^{n-1}] \frac{1}{n}. \qquad (2.4.20)$$

The elements of the matrices A and B are c-numbers, while those in η are anti-commuting ones.

The matrices in the exponential in (2.4.18) are special cases of matrices of the form

$$V = \begin{pmatrix} A & \eta \\ \xi & B \end{pmatrix} \qquad (2.4.21)$$

with A, B having c-number elements, while all the elements in η, ξ anti-commute.

For a matrix $X = \exp A$, with c-number elements, one has the useful identity $\det X = \exp \operatorname{Tr} A$. We now define supermatrices $M_1 = \exp V_1$, $M_2 = \exp V_2$, and for the superdeterminant of the product of two supermatrices, we must have

$$\operatorname{Sdet} M_i = \exp \operatorname{STr} V_i, \qquad \operatorname{Sdet}(M_1 M_2) = \exp \operatorname{STr}(V_1 + V_2). \qquad (2.4.22)$$

---

[26]See Sect. 2.1, (2.1.1).

By using the Baker-Campbell-Hausdorff expansion, we may write for the left-hand side of the second expression in the above equation

$$\text{Sdet}\,(M_1\,M_2) = \exp \text{STr}\left(V_1 + V_2 + \frac{1}{2}[V_1, V_2] + \cdots\right), \qquad (2.4.23)$$

and every term beyond $V_1 + V_2$ has the structure of the commutator of two matrices such as V_1 and V_2. According to the equality in the second equation in (2.4.22), the STr of any two such matrices V_1 and V_2 must vanish.

From (2.4.8), one explicitly has

$$[V_1, V_2] = \begin{pmatrix} [A_1, A_2] + \eta_1\xi_2 - \eta_2\xi_1 & (A_1\eta_2 - A_2\eta_1 + \eta_1 B_2 - \eta_2 B_1) \\ \xi_1 A_2 - \xi_2 A_1 + B_1\xi_2 - B_2\xi_1 & [B_1, B_2] + \xi_1\eta_2 - \xi_2\eta_1 \end{pmatrix}.$$
$$(2.4.24)$$

We note that in reference to the *matrices* on the diagonal of the commutator $[V_1, V_2]$, we always have $\text{Tr}\,[A_1, A_2] = 0 = \text{Tr}\,[B_1, B_2]$. On the other hand for the matrices η_1, η_2, ξ_1, ξ_2 with anti-commuting elements, the following equality for the traces hold

$$\text{Tr}\,(\eta_1\xi_2 - \eta_2\xi_1) = \text{Tr}\,(\xi_1\eta_2 - \xi_2\eta_1). \qquad (2.4.25)$$

Thus we learn that the *commutator* of any two matrices of the type given in (2.4.21), not only has the *same* structure as the type just mentioned, but also the ordinary traces of the two matrices on the diagonal are *equal*. This applies to *every* single term appearing in the exponential in (2.4.23) beyond the $(V_1 + V_2)$ term, being the commutator of two such matrices. Accordingly, for a general matrix V as given in (2.4.21), in order to satisfy the condition (2.4.22), one defines the supertrace

$$\text{STr}\begin{pmatrix} A & \eta \\ \xi & B \end{pmatrix} = \text{Tr}\,A - \text{Tr}\,B, \qquad (2.4.26)$$

with *opposite* relative signs, where Tr denotes the ordinary trace, and all the terms in the exponential in (2.4.23), beyond that $(V_1 + V_2)$, will automatically vanish, due to equalities as the one in (2.4.25).

The basic property embodied in this consistent definition of a determinant, now extended to supermatrices M_1, M_2 in (2.4.22), will be then satisfied as follows

$$\text{Sdet}\,M_1 M_2 = \text{Sdet}\,M_1 \; \text{Sdet}\,M_2. \qquad (2.4.27)$$

In the absence of Grassmann variables

$$V_1 \rightarrow [A_1], \quad V_2 \rightarrow [A_2], \qquad (2.4.28)$$

and (2.4.22) holds true on account that the trace of the commutator of any two matrices is zero.

Upon using the result in (2.4.26) for the matrix M in question in (2.4.17), (2.4.18), by noting, in the process, that

$$\det \Lambda = \exp\left(-\operatorname{Tr}\sum_{n \geq 1}(I - \Lambda)^n/n\right), \qquad (2.4.29)$$

and similarly for the matrix K, we obtain

$$\operatorname{Sdet} M = \frac{\det \Lambda}{\det K}. \qquad (2.4.30)$$

For Lorentz transformations, including 3D spatial rotations, $\det \Lambda = 1$. This also true for the matrix K implementing Lorentz transformations of spinors, that $\det K = 1$, thus establishing the invariance of the volume element in (2.4.1) in superspace.

Consider two matrices ξ, η, with elements of Grassmann type, and a matrix C, with c-number elements, whose inverse exists, which give rise to a matrix $\xi\,C^{-1}\eta$ with c-number elements. Quite generally, it is shown in Problem 2.13 that the superdeterminant of an arbitrary matrix of the type Y, in the following equation, is consistently given by

$$\operatorname{Sdet} Y = \operatorname{Sdet}\begin{pmatrix} C & \eta \\ \xi & D \end{pmatrix} = \frac{\det C}{\det(D - \xi C^{-1}\eta)}, \qquad (2.4.31)$$

provided the expression on the right-hand side of the equation exists,[27] where D is a matrix with c-number elements.

The above analysis establishes the invariance of the integral (2.4.3) in superspace.

Since the components of $\bar{\theta}$ may be written as a linear combination of the components of θ, and due to the nature of the Grassmann character of these components, we may spell out, in detail, the general structure of a scalar superfield $\Phi(x, \theta)$ to be

$$\Phi(x, \theta) = \phi(x) + [\phi^1(x)\,\theta_1 + \cdots + \phi^4(x)\,\theta_4]$$
$$+ [\phi^{43}(x)\,\theta_4\theta_3 + \phi^{42}(x)\,\theta_4\theta_2 + \cdots + \phi^{21}(x)\,\theta_2\theta_1]$$
$$+ [\phi^{432}(x)\,\theta_4\theta_3\theta_2 + \phi^{431}(x)\,\theta_4\theta_3\theta_1 + \phi^{421}(x)\,\theta_4\theta_2\theta_1 + \phi^{321}(x)\,\theta_3\theta_2\theta_1]$$
$$+ [\phi^{4321}(x)\,\theta_4\theta_3\theta_2\theta_1], \qquad (2.4.32)$$

---

[27]Much of the work on superanalysis is attributed to Berezin [4].

whose integral in superspace is simply given by

$$\int (\mathrm{d}x)\, \mathrm{d}\theta_1 \mathrm{d}\theta_2 \mathrm{d}\theta_3 \mathrm{d}\theta_4 \, \Phi(x,\theta) = \int (\mathrm{d}x)\, \phi^{4321}(x). \tag{2.4.33}$$

This established basic property will be quite useful later on.

2.5 Super-Poincaré Algebra and Supermultiplets

Supersymmetry transformations are carried out via generators, associated with unitary operators, acting on particle states, and carry spinor indices and hence are of fermionic type. These generators together with the ones of the Poncaré transformations satisfy a new algebra, referred to as the Super-Poincaré Algebra, involving commutators as well as anti-commutators. We consider theories with only one supersymmetry generator as a spinor with four components corresponding to ones referred to as $\mathcal{N} = 1$ (simple) supersymmetric field theories.

The mere fact that the supercharge operator, as will be derived below, *does not commute with the angular momentum operator* necessarily implies that the particle supermultiplets, associated with this algebra, involve particles with *different spins* and hence the derivation of the commutator in question is of importance. The fermionic character of the supersymmetric generator, as witnessed by working, in particular, with Grassmann variables, shows also that the number of particles in a supermultiplet are *finite* in number. Below we derive the details of this algebra and investigate the nature of supermultiplets associated with these symmetry transformations. The reader is advised to review first the content of Sect. 2.1 on superspace as well the section on the Poincaré Algebra in Sect. 4.2 in Vol. I [26].

In superspace, coordinates are labeled by (x^μ, θ_a), where θ is a Majorana spinor with anti-commuting components which commute with x^μ. For a supersymmetry transformation, not involving Lorentz transformations and spatial rotations, the transformation rule $x, \theta \to x', \theta'$, including translations, is defined by [see (2.1.3), (2.1.4)]

$$x' = x + \frac{\mathrm{i}}{2}\bar{\epsilon}\gamma\theta - b, \tag{2.5.1}$$

$$\theta' = \theta + \epsilon, \tag{2.5.2}$$

and a Super-Poincaré Transformation, which involves Lorentz transformations and spatial rotations, is defined by [see (2.1.10), (2.1.11)]

$$x' = \Lambda x + \frac{\mathrm{i}}{2}\bar{\epsilon}\gamma K\theta - b, \tag{2.5.3}$$

$$\theta' = K\theta + \epsilon, \tag{2.5.4}$$

with both sets written in convenient matrix notations. K denotes the spinor-Dirac representation (Sect. 2.4 in Vol. I, Sect. 2.1) of a Lorentz group. In the transformation rules (2.5.1)–(2.1.4), the spinor (θ_a) is tied together with the spacetime coordinate (x^μ) in a way similar that x^0 is tied together with x^i in special relativity. The transformation rules are also linear in (x^μ, θ_a). We consider infinitesimal transformations via a generator G spelled out in (2.5.5), (2.5.15) below, which results from considering infinitesimal transformations around a closed path represented pictorially by

emphasizing the reversal of the transformations in the third and the fourth segments of the path, described by successive unitary transformations $U_2^{-1} U_1^{-1} U_2 U_1$ with $U_j = 1 + i\,G_j$ for infinitesimal transformations.

For additional clarity we consider the simple supersymmetry transformations in (2.5.1), (2.5.2) first before tackling the ones in (2.5.3), (2.5.4). To this end, the generator of the corresponding transformation on operators and particle states is defined by

$$G = \delta\,b_\mu P^\mu + \delta\,\bar{\epsilon}_a\,Q_a, \tag{2.5.5}$$

where P^μ is the energy-momentum operator and the Q_a denote the components of the spinor generator of supersymmetry, and is defined to satisfy $Q = \mathscr{C}\overline{Q}^{\mathsf{T}}$, $\overline{Q} = (Q)^\dagger \gamma^0$. The latter as well as $\delta\bar{\epsilon}$ are taken as Majorana spinors which guaranty, in particular, the Hermiticity of the generator G on account of the easily derived property

$$\delta\,\bar{\epsilon}_a\,Q_a = \overline{Q}_a\,\delta\,\epsilon_a, \tag{2.5.6}$$

where the parameters ϵ_a and their variations anti-commute with the operators Q_a.

As mentioned above, we consider infinitesimal transformations forming a closed path described by

$$(\epsilon_2, b_2)^{-1}(\epsilon_1, b_1)^{-1}(\epsilon_2, b_2)(\epsilon_1, b_1) = (\epsilon, b), \tag{2.5.7}$$

where the group properties are given in Sect. 2.1 below (2.1.8), and use the general commutation rule involving the generators G_1, G_2, G

$$G = \frac{1}{i}[G_1, G_2], \qquad (2.5.8)$$

as follows from the structure $U_2^{-1} U_1^{-1} U_2 U_1 = U$, $U_j = 1 + iG_j$, $U = 1 + iG$, with corresponding infinitesimal parameters $(\delta\epsilon_1, \delta b_1)$, $(\delta\epsilon_2, \delta b_2)$, $(\delta\epsilon, \delta b)$, respectively.

The group property of the parameters are spelled out below (2.1.8), leads to

$$\delta b_\mu = -i\,\delta\bar{\epsilon}_2\,\gamma_\mu\,\delta\epsilon_1 = i\,\delta\bar{\epsilon}_{1a}\,\delta\bar{\epsilon}_{2b}\,(\gamma_\mu\,\mathscr{C})_{ab}, \qquad (2.5.9)$$

$$\delta\bar{\epsilon} = 0, \qquad (2.5.10)$$

where $\mathscr{C}$ is the charge conjugation matrix (cf. Sect. 2.2).

For the subsequent analysis, we replace the expressions for the generators, G_1, G_2, G, by their corresponding expressions given in (2.5.5) into (2.5.8), with G, for example, defined in (2.5.5), use the parametric relations in (2.5.9), (2.5.10), and note that

$$[\delta\bar{\epsilon}_{1a}\,Q_a, \delta\bar{\epsilon}_{2b}\,Q_b] = -\,\delta\bar{\epsilon}_{1a}\delta\bar{\epsilon}_{2b}\,\{Q_a, Q_b\}, \qquad (2.5.11)$$

converting, in the process, a commutator to an anti-commutator. Upon comparing the coefficients of identical products of the parameters $\delta\bar{\epsilon}_1$, $\delta\bar{\epsilon}_2$, δb_1, δb_2 on both sides of the resulting expression for (2.5.8), we obtain

$$[P^\mu, P^\nu] = 0, \quad [Q_a, P^\mu] = 0, \quad \{Q_a, Q_b\} = (\gamma_\mu\,\mathscr{C})_{ab}\,P^\mu. \qquad (2.5.12)$$

To generalize the above algebra to the Super-Poincaré one corresponding to the transformations in (2.5.3), (2.5.4), we first recall the explicit structures of the following infinitesimal deviations from the identity elements

$$\Lambda^\mu{}_\nu = \delta^\mu{}_\nu + \delta\omega^\mu{}_\nu, \qquad\qquad \delta\omega^{\mu\nu} = -\delta\omega^{\nu\mu}, \qquad (2.5.13)$$

$$K_{ab} = \delta_{ab} + \frac{i}{2}\,\delta\omega^{\mu\nu}(S_{\mu\nu})_{ab}, \qquad S_{\mu\nu} = \frac{i}{4}[\gamma_\mu, \gamma_\nu]. \qquad (2.5.14)$$

The generator for infinitesimal transformations on operators and particle states takes the form

$$G = \delta b_\mu\,P^\mu + \frac{1}{2}\,\delta\omega_{\mu\nu}\,J^{\mu\nu} + \delta\bar{\epsilon}_a\,Q_a. \qquad (2.5.15)$$

The group property in Sect. 2.1 below (2.1.16), corresponding to a closed path, as before, now with group multiplications of elements $(\Lambda, K, \epsilon, b)$, leads to the parametric relations

$$\delta\omega^{\mu\nu} = \delta\omega_2{}^{\mu\alpha}\,\delta\omega_{1\alpha}{}^{\nu} - \delta\omega_2{}^{\nu\alpha}\,\delta\omega_{1\alpha}{}^{\mu}, \tag{2.5.16}$$

$$\delta b^{\mu} = \delta\omega_2{}^{\mu}{}_{\alpha}\,\delta b_1^{\alpha} - \delta\omega_1{}^{\mu}{}_{\alpha}\,\delta b_2^{\alpha} + i\,\delta\bar{\epsilon}_{1a}\,\delta\bar{\epsilon}_{2b}\,(\gamma^{\mu}\mathscr{C})_{ab}, \tag{2.5.17}$$

$$\delta\epsilon_a = \frac{i}{2}\left(\delta\omega_2{}^{\mu\nu}(S_{\mu\nu}\,\delta\epsilon_1)_a - \delta\omega_1{}^{\mu\nu}(S_{\mu\nu}\,\delta\epsilon_2)_a\right), \tag{2.5.18}$$

where we have used, in the process,[28] the identities

$$\Lambda\gamma = K^{-1}\gamma\,K, \qquad \Lambda^{-1}\gamma = K\gamma\,K^{-1}, \tag{2.5.19}$$

in matrix notations in spinor indices. The expression for $\delta\epsilon_a$ also leads to

$$\delta\bar{\epsilon}_a = -\frac{i}{2}\left(\delta\omega_2{}^{\mu\nu}\,\delta\bar{\epsilon}_{1b} - \delta\omega_1{}^{\mu\nu}\,\delta\bar{\epsilon}_{2b}\right)(S_{\mu\nu})_{ba}, \tag{2.5.20}$$

where note that $(S_{\mu\nu})^{\dagger}\gamma^0 = \gamma^0(S_{\mu\nu})$, and $S_{\mu\nu}$ is defined (2.5.14).

Upon comparing the coefficients of identical products of the parameters $\delta\omega_1$, $\delta\omega_2$, $\delta\bar{\epsilon}_1$, $\delta\bar{\epsilon}_2$, δb_1, δb_2 on both sides of the resulting expression for (2.5.8), the Super-Poincaré Algebra now readily emerges:

$$\{Q_a, Q_b\} = (\gamma_{\mu}\mathscr{C})_{ab}\,P^{\mu}, \qquad \{Q_a, Q_a\} = 0, \tag{2.5.21}$$

$$[Q_a, J_{\mu\nu}] = (S_{\mu\nu})_{ab}\,Q_b, \tag{2.5.22}$$

$$[Q_a, P^{\mu}] = 0, \qquad [Q_a, P^2] = 0, \tag{2.5.23}$$

$$[P^{\mu}, P^{\nu}] = 0, \tag{2.5.24}$$

$$[P^{\mu}, J^{\sigma\lambda}] = i\,(\eta^{\mu\lambda}P^{\sigma} - \eta^{\mu\sigma}P^{\lambda}), \tag{2.5.25}$$

$$[J^{\mu\nu}, J^{\sigma\lambda}] = i\,(\eta^{\mu\sigma}J^{\nu\lambda} - \eta^{\nu\sigma}J^{\mu\lambda} + \eta^{\nu\lambda}J^{\mu\sigma} - \eta^{\mu\lambda}J^{\nu\sigma}). \tag{2.5.26}$$

where the second anti-commutation relation in (2.5.21) holds with or without a sum over a. The derived non-commutativity property of the generator Q with the angular momentum should be noted and is responsible for the fact that supermultiplets involve particles with different spins. This we consider next.

---

[28]See (2.1.9).

We recall the basic property of a Majorana spinor

$$\overline{Q} = -Q^\top \mathscr{C}^{-1}. \tag{2.5.27}$$

Therefore upon multiplying (2.5.21) by $\mathscr{C}_{bc}^{-1}$ we obtain

$$\{Q_a, \overline{Q}_c\} = -(\gamma_\mu)_{ac} P^\mu.$$

Upon multiplying the latter by $(\gamma^0)_{cb}$ gives

$$\{Q_a, Q_b^\dagger\} = -(\gamma_\mu \gamma^0)_{ab} P^\mu. \tag{2.5.28}$$

The commutation relation of Q_a with the angular momentum operator $J^i = \epsilon^{ijk} J^{jk}/2$ follows from (2.5.22) to be in matrix notation

$$[Q, \mathbf{J}] = \frac{1}{2} \Sigma\, Q, \qquad \Sigma = \begin{pmatrix} \sigma & 0 \\ 0 & \sigma \end{pmatrix}. \tag{2.5.29}$$

Its adjoint satisfies the equation

$$[\mathbf{J}, Q^\dagger] = \frac{1}{2} Q^\dagger \Sigma. \tag{2.5.30}$$

Also note, by now familiar, the commutation relations [see (2.5.25), (2.5.26)]

$$[P^0, \mathbf{J}] = 0, \qquad [J^3, \mathbf{J}^2] = 0. \tag{2.5.31}$$

We first treat massive supermultiplets. Consider a massive particle of mass m. By going to the rest frame of the particle, states may be labeled by (m, j, σ), with $P^\mu = (m, \mathbf{0})$. In the sequel, we suppress the dependence of the states on m. Upon setting

$$A_a = \frac{1}{\sqrt{m}} Q_a, \qquad A_a^\dagger = \frac{1}{\sqrt{m}} Q_a^\dagger, \tag{2.5.32}$$

in the process, and observing the simple structure of Σ in (2.5.29), (2.5.30), and using the fact that $\gamma^0 \gamma_0 = -I$, we obtain from (2.5.28)–(2.5.30), (2.5.32)

$$\{A_a, A_b^\dagger\} = \delta_{ab}, \quad \mathbf{J} A_a = A_b\left(\mathbf{J}\,\delta_{ba} - \frac{1}{2}\sigma_{ab}\right), \ a, b = 1, 2,$$

$$\mathbf{J} A_a^\dagger = A_b^\dagger\left(\mathbf{J}\,\delta_{ba} + \frac{1}{2}\sigma_{ba}\right), \qquad a, b = 1, 2,$$

$$J^3 A_1 = A_1\left(J^3 - \frac{1}{2}\right), \qquad J^3 A_2 = A_2\left(J^3 + \frac{1}{2}\right),$$

$$J^3 A_1^\dagger = A_1^\dagger \left(J^3 + \frac{1}{2} \right), \qquad J^3 A_2^\dagger = A_2^\dagger \left(J^3 - \frac{1}{2} \right), \qquad (2.5.33)$$

together with the elementary identities that follow from (2.5.21)

$$(A_1^\#)^2 = 0, \qquad (A_2^\#)^2 = 0, \qquad (2.5.34)$$

where $A^\#$ refers to the operator or its adjoint.

One can always define a normalized state $|\varphi\rangle$ such that $A_a |\varphi\rangle = 0$, $a = 1, 2$. For example, if $A_1 |\varphi\rangle \neq 0$, then you may consider the normalized state $A_1 |\varphi\rangle$ which is annihilated by A_1 and so on. Accordingly, given $A_a |\varphi\rangle = 0$, $a = 1, 2$, consider the following. For a given unit vector $\mathbf{n}$, set

$$e^{i\vartheta \mathbf{n} \cdot \mathbf{J}} A_a \, e^{-i\vartheta \mathbf{n} \cdot \mathbf{J}} = A_a[\vartheta, \mathbf{n}]. \qquad (2.5.35)$$

Then, from the second equation in the first line of (2.5.33), with $a, b = 1, 2$, its derivative with respect to ϑ is given by

$$\frac{\partial}{\partial \vartheta} A_a[\vartheta, \mathbf{n}] = -\frac{i\mathbf{n}}{2} \cdot \left(\boldsymbol{\sigma} A[\vartheta, \mathbf{n}] \right)_a, \qquad (2.5.36)$$

which leads from (2.5.35) to

$$e^{i\vartheta \mathbf{n} \cdot \mathbf{J}} A_a \, e^{-i\vartheta \mathbf{n} \cdot \mathbf{J}} = \left(e^{-i\vartheta \frac{\mathbf{n} \cdot \boldsymbol{\sigma}}{2}} \right)_{ab} A_b. \qquad (2.5.37)$$

That is $A_a[\vartheta, \mathbf{n}]$ annihilates $|\varphi\rangle$ as well. Hence one may carry out the following expansion with the property

$$0 = A_a \, e^{-i\vartheta \mathbf{n} \cdot \mathbf{J}} |\varphi\rangle = \sum_{j,\sigma} A_a \, |j, \sigma\rangle \langle j, \sigma| \, e^{-i\vartheta \mathbf{n} \cdot \mathbf{J}} |\varphi\rangle, \qquad (2.5.38)$$

which is true for *all* ϑ, $\mathbf{n}$. That is, the latter is true for *all* arbitrary coefficients $a_{j,\sigma}(\vartheta, \mathbf{n}) = \langle j, \sigma| e^{-i\vartheta \mathbf{n} \cdot \mathbf{J}} |\varphi\rangle$, as we may vary over the infinite possible values that may be taken by $(\mathbf{n}, \vartheta)$. Thus we may infer that one may set-up normalized spin states $|j, \sigma\rangle$ such that $A_a |j, \sigma\rangle = 0$, for $a = 1, 2$.

Given the above normalized spin states $|j, \sigma\rangle$, one may use the third equation in (2.5.33), together with the identities in (2.5.34), to obtain

$$[\mathbf{J}, A_1^\dagger A_2^\dagger] = \frac{1}{2} A_1^\dagger A_2^\dagger (\sigma_{11} + \sigma_{22}) = 0, \qquad (2.5.39)$$

where we have used the fact that $(\sigma_{11} + \sigma_{22}) \equiv \mathrm{Tr}\, \boldsymbol{\sigma} = \mathbf{0}$. That is,

$$\mathbf{J} A_1^\dagger A_2^\dagger = A_1^\dagger A_2^\dagger \mathbf{J}. \qquad (2.5.40)$$

We may, therefore, introduce new normalized spin states

$$|j,\sigma\rangle' = A_1^\dagger A_2^\dagger |j,\sigma\rangle, \tag{2.5.41}$$

orthogonal to the states $|j,\sigma\rangle$ as a consequence of the fact that the latter is annihilated by A_a.

Clearly, the non-vanishing states $A_1^\dagger |j,\sigma\rangle$, $A_2^\dagger |j,\sigma\rangle$, $|j,\sigma\rangle$, $|j,\sigma\rangle'$, are mutually orthogonal. On the other hand, by referring to the third equation in (2.5.33), involving the operators $A_1^\dagger$, $A_2^\dagger$, we note that $\mathbf{J} + \boldsymbol{\sigma}/2$ represents the addition of two angular momenta one of which corresponds to spin 1/2. Hence we may conclude that for $j > 0$, a massive supermultiplet contains also states of spin $j \pm 1/2$ with all of the corresponding particles having the *same* mass. Due to the Grassmann property of the spinor generators $(A_a^\sharp)^2 = 0$, and the anti-commutation relations in (2.5.34), *no* other states may be constructed by the applications of these generators and their products on the already obtained states above as they would either be zero or proportional to the pre-existing ones.

In particular, for a given $j > 0$ we have learnt that there are two sates with spin j. These states will correspond to bosonic or fermionic degrees of freedom, depending whether j is an integer or a half-odd integer, respectively. An elegant way of determining the number of fermionic and bosonic degrees of freedom in a supermultiplet is obtained in the following manner.

Since from (2.5.23), $[\mathbf{P}, Q_a^\dagger] = 0$, every particle in a supermultiplet has the same momentum, say, $\mathbf{p}$, and same mass, say, m. Suppose that $\mathbf{p} = |\mathbf{p}|(0,0,1)$. Also from (2.5.25), (2.5.28)

$$\{Q_1, Q_1^\dagger\} = P^0, \qquad [J^3, P^3] = 0,$$
$$[J^2, J^3] = 0, \qquad [J^2, P^3] = 2\,\mathrm{i}\,(J^2 P^1 - J^1 P^2), \tag{2.5.42}$$

and we may define states $|\mathbf{p}, j, \sigma\rangle$, satisfying

$$J^3 |\mathbf{p}, j, \sigma\rangle = \sigma |\mathbf{p}, j, \sigma\rangle, \quad \mathbf{p} = |\mathbf{p}|(0,0,1),$$
$$\mathrm{e}^{\mathrm{i}2\pi J^3} |\mathbf{p}, j, \sigma\rangle = (-1)^{2\sigma} |\mathbf{p}, j, \sigma\rangle \equiv (-1)^{2j} |\mathbf{p}, j, \sigma\rangle. \tag{2.5.43}$$

On the other hand, from (2.5.37), we may infer that

$$\mathrm{e}^{\mathrm{i}2\pi J^3} Q_1 \mathrm{e}^{-\mathrm{i}2\pi J^3} = (\mathrm{e}^{-\mathrm{i}\pi\sigma^3})_{1b} Q_b = -Q_1. \tag{2.5.44}$$

Now consider the following trace multiplied by $E = \sqrt{\mathbf{p}^2 + m^2}$,

$$E \sum_{j,\sigma} \langle \mathbf{p}, j, \sigma \,|\, \mathrm{e}^{\mathrm{i}\,2\pi\, J^3} |\, \mathbf{p}, j, \sigma \rangle = \sum_{j,\sigma} \langle \mathbf{p}, j, \sigma \,|\, \mathrm{e}^{\mathrm{i}\,2\pi\, J^3} P^0 |\, \mathbf{p}, j, \sigma \rangle$$

$$= \sum_{j,\sigma} \langle \mathbf{p}, j, \sigma \, | \, e^{i 2\pi J^3} \left(Q_1 Q_1^\dagger + Q_1^\dagger Q_1 \right) | \, \mathbf{p}, j, \sigma \rangle$$

$$= \sum_{j,\sigma} \langle \mathbf{p}, j, \sigma \, | \left(- Q_1 e^{i 2\pi J^3} Q_1^\dagger + e^{i 2\pi J^3} Q_1^\dagger Q_1 \right) | \, \mathbf{p}, j, \sigma \rangle = 0, \qquad (2.5.45)$$

where in writing the second equality we have used the first equation in (2.5.42), and in writing the third equality we have used (2.5.44), and in writing the last one we have used the fact that $\mathrm{Tr}\,[AB] = \mathrm{Tr}\,[BA]$. That is

$$0 = \mathrm{Tr}\left[e^{i 2\pi J^3} \right] = \sum_{j \geq 0} (-1)^{2j} N_j (2j + 1) = n_{\mathrm{b}} - n_{\mathrm{f}}, \qquad (2.5.46)$$

where N_j denotes number of particles with spin j, each with $(2j+1)$ spin degrees of freedom, and $n_{\mathrm{b/f}}$ denote the number of bosonic/fermionic degrees of freedom, respectively. Hence we learn that the number bosonic degrees of freedom is equal to the fermionic ones. For example for a given j, say $j = 1/2$, the above analysis shows that we have a supermultiplet with two $1/2$ spins with $2(2\times(1/2) + 1) = 4$ fermionic degrees of freedom, one spin 1 of 3 degrees of freedom and one spin 0 of 1 degree of freedom. All in all we have, 4 fermionic degrees of freedom and 4 bosonic ones.

For $j = 0$, we clearly have a massive supermuliplet involving two spin 0 states, and one of spin 1/2 with two degrees of freedom.

For massless supermultiplets, consider a particle with energy-momentum $P^\mu = (E, 0, 0, E)$, $E > 0$, i.e., which is moving along the 3-axis. For massless particles, working in the chiral representation of the Dirac matrices, (2.5.28) leads to

$$\{B_1, B_1^\dagger\} = 1, \qquad B_a = Q_a / \sqrt{2E}, \quad B_a^\dagger = Q_a^\dagger / \sqrt{2E}, \qquad (2.5.47)$$

and

$$\{B_2, B_2^\dagger\} = 0, \quad \{B_1, B_2^\dagger\} = 0, \quad \{B_2, B_1^\dagger\} = 0, \quad (B_1)^2 = 0, \quad (B_1^\dagger)^2 = 0. \qquad (2.5.48)$$

As in (2.5.33), we also have

$$J^3 B_1^\dagger = B_1^\dagger (J^3 + 1/2), \quad J^3 B_1 = B_1 (J^3 - 1/2). \qquad (2.5.49)$$

That is, if $\bar{\lambda}$ denotes the largest helicity in a supermultiplet, associated with some particle, then

$$B_1^\dagger \, |E, \bar{\lambda}\rangle = 0. \qquad (2.5.50)$$

Hence another state in the supermultiplet would be given by $B_1|E, \bar{\lambda}\rangle = |E, \bar{\lambda} - 1/2\rangle$. Also, $(B_1)^2 = 0$, implies that $B_1|E, \bar{\lambda} - 1/2\rangle = 0$. That is, a massless supermuliplet contains just the two states $|E, \bar{\lambda}\rangle$ and $|E, \bar{\lambda} - 1/2\rangle$. On the other hand, CPT invariance implies the existence also of its anti-supermultiplet consisting just of the states $|E, -\bar{\lambda}\rangle$ and $|E, -\bar{\lambda} + 1/2\rangle$ of opposite helicities. For example the superpartner of the photon, called the photino, is a massless fermion, of helicities $\pm 1/2$, and that of the graviton, the superpartner, called the gravitino, is a fermion, a Rarita-Schwinger massless particle, of helicities $\pm 3/2$. The photon and the graviton being so-called gauge fields, their superpartners are also each referred to as gauginos. In general supermultiplets with $\bar{\lambda} = 1/2, 1$ are referred to as chiral and vector supermultiplets, respectively.

It is important to reiterate, in closing, that the second commutation rule in (2.5.23) implies that all the particles within a supermultiplet have the same masses, for both cases with massive or massless particles.

2.6 A Panorama of Superfields

To develop supersymmetric field theories, we introduce so-called superfields defined on superspace, that is, functions of (x^μ, θ_a) with the latter specifying points in superspace (Sects. 2.1 and 2.4). One would then write a Lagrangian in terms of these fields and eventually obtain supersymmetric actions to describe the dynamics of the underlying particles and sparticles. We work exclusively in the *chiral* representation of the Dirac matrices[29] in which γ^5 is diagonal.

We recall that θ is a Majorana spinor with anti-commuting components, which, in turn, commute with x^μ. That is,

$$\theta = \mathscr{C}\bar{\theta}^\mathsf{T}, \quad \bar{\theta} = -\theta^\mathsf{T}\mathscr{C}^{-1}, \quad \{\theta_a, \theta_b\} = 0, \quad [\theta_a, x^\mu] = 0, \qquad (2.6.1)$$

where $\mathscr{C}$ is the charge conjugation matrix. Some pertinent equations, as given in Sect. 2.2, that will be needed and used repeatedly in this section, are, conveniently, summarized in Box 2.1.

---

[29]See Appendix I at the end of the book.

Box 2.1: Some basic properties involving the spinor θ and the charge conjugation matrix $\mathscr{C}$

$$\bar{\theta}_b\,\theta_a = \tfrac{1}{4}[\delta_{ab}\,\bar{\theta}\theta + \gamma^5_{ab}\,\bar{\theta}\gamma^5\theta + (\gamma^5\gamma_\mu)_{ab}\,\bar{\theta}\gamma^5\gamma^\mu\theta\,],$$

$$\bar{\theta}\theta\,\theta_c = -\bar{\theta}\gamma^5\theta\,(\gamma^5\theta)_c, \quad \bar{\theta}\theta\,\bar{\theta}_c = -\bar{\theta}\gamma^5\theta\,(\bar{\theta}\gamma^5)_c,$$

$$\bar{\theta}\gamma^5\gamma^\mu\theta\,\theta_c = -\bar{\theta}\gamma^5\theta\,(\gamma^\mu\theta)_c, \quad \bar{\theta}\gamma^5\gamma^\mu\theta\,\bar{\theta}_c = \bar{\theta}\gamma^5\theta\,(\bar{\theta}\gamma^\mu)_c,$$

$$(\bar{\theta}\theta)(\bar{\theta}\gamma^5\theta) = 0, \quad (\bar{\theta}\theta)(\bar{\theta}\gamma^5\gamma^\mu\theta) = 0, \quad (\bar{\theta}\gamma^5\gamma^\mu\theta)(\bar{\theta}\gamma^5\theta) = 0,$$

$$(\bar{\theta}\gamma^5\gamma^\mu\theta)(\bar{\theta}\gamma^5\gamma^\sigma\theta) = \eta^{\mu\sigma}(\bar{\theta}\gamma^5\theta)^2, \quad (\bar{\theta}\theta)^2 = -(\bar{\theta}\gamma^5\theta)^2.$$

$$\bar{\theta}\xi = \bar{\xi}\theta, \quad \bar{\theta}\gamma^5\gamma^\mu\xi = \bar{\xi}\gamma^5\gamma^\mu\theta, \quad \bar{\theta}\gamma^\mu\xi = -\bar{\xi}\gamma^\mu\theta, \quad \bar{\theta}\gamma^5\xi = \bar{\xi}\gamma^5\theta,$$

$$\bar{\theta}\gamma^\mu\gamma^\nu\xi = \bar{\xi}\gamma^\nu\gamma^\mu\theta, \quad \bar{\theta}[\gamma^\mu,\gamma^\nu]\xi = -\bar{\xi}[\gamma^\mu,\gamma^\nu]\theta,$$

$$\bar{\theta}\gamma^5[\gamma^\mu,\gamma^\nu]\xi = -\bar{\xi}\gamma^5[\gamma^\mu,\gamma^\nu]\theta, \quad \bar{\theta}\gamma^5\gamma^\mu\gamma^\nu\xi = -\eta^{\mu\nu}\bar{\theta}\gamma^5\xi - \tfrac{1}{2}\bar{\xi}\gamma^5[\gamma^\mu,\gamma^\nu]\theta,$$

$$\bar{\theta}\xi = \bar{\xi}\theta, \quad \bar{\theta}\gamma^5\gamma^\mu\xi = \bar{\xi}\gamma^5\gamma^\mu\theta, \quad \bar{\theta}\gamma^\mu\xi = -\bar{\xi}\gamma^\mu\theta, \quad \bar{\theta}\gamma^5\xi = \bar{\xi}\gamma^5\theta,$$

$$\bar{\theta}\gamma^\mu\theta = 0, \quad \bar{\theta}[\gamma^\mu,\gamma^\nu]\theta = 0, \quad \bar{\theta}\gamma^5\gamma^\mu\gamma^\nu\theta = -\eta^{\mu\nu}\bar{\theta}\gamma^5\theta.$$

$$\partial/\partial\bar{\theta}\,(\bar{\theta}\gamma^5\theta) = 2\gamma^5\theta, \quad \partial/\partial\bar{\theta}\,(\bar{\theta}\gamma^5\gamma^\mu\theta) = 2\gamma^5\gamma^\mu\theta,$$

$$\partial/\partial\bar{\theta}\,(\bar{\theta}\gamma^5\theta)^2 = 4(\bar{\theta}\gamma^5\theta)\gamma^5\theta.$$

$$\mathscr{C}^{-1}\gamma^\mu\mathscr{C} = -(\gamma^\mu)^{\mathsf{T}}, \quad [\mathscr{C},\gamma^5] = 0, \quad \mathscr{C}^{-1}\gamma^5\gamma^\mu\mathscr{C} = (\gamma^5\gamma^\mu)^{\mathsf{T}},$$

$$\mathscr{C}^\dagger = \mathscr{C}^{-1} = \mathscr{C}^{\mathsf{T}} = -\mathscr{C}.$$

$$D = \partial/\partial\bar{\theta} - \tfrac{1}{2}(\gamma^\mu\theta)\partial_\mu, \quad \bar{D} = -D^{\mathsf{T}}\mathscr{C}^{-1}.$$

$$\mathscr{C} = \begin{pmatrix} -i\sigma^2 & 0 \\ 0 & i\sigma^2 \end{pmatrix} = \begin{pmatrix} 0 & -1 & 0 & 0 \\ 1 & 0 & 0 & 0 \\ 0 & 0 & 0 & 1 \\ 0 & 0 & -1 & 0 \end{pmatrix}, \quad \mathscr{C}^{-1} = \begin{pmatrix} 0 & 1 & 0 & 0 \\ -1 & 0 & 0 & 0 \\ 0 & 0 & 0 & -1 \\ 0 & 0 & 1 & 0 \end{pmatrix}.$$

We here introduce several superfields together with some of their basic properties and relegate the relevant technical details of investigations to their respective subsections that follow. This makes the reading on this necessarily detailed study much easier. Needless to say these subsections form an integral part of this section and are to be read simultaneously with this material.

Scalar Superfield

The scalar superfield is defined by

$$\Phi(x,\theta) = A(x) - i\,\bar{\theta}\gamma^5\psi(x) + \frac{i}{4}\bar{\theta}\gamma^5\theta\,B(x) - \frac{1}{4}\bar{\theta}\theta\,G(x) - \frac{1}{4}\bar{\theta}\gamma^5\gamma^\mu\theta\,V_\mu(x)$$

$$+ \frac{i}{2\sqrt{2}}\bar{\theta}\gamma^5\theta\,\bar{\theta}[\chi(x) + \frac{i}{\sqrt{2}}\gamma\partial\psi(x)] - \frac{1}{16}(\bar{\theta}\gamma^5\theta)^2[\mathscr{D}(x) + \frac{1}{2}\Box A(x)].$$

$$(2.6.2)$$

Here $A(x)$, $B(x)$, $G(x)$, $\mathscr{D}(x)$ are Lorentz scalars, $V^\mu(x)$ is a Lorentz vector, and $\psi(x)$, $\chi(x)$ are spinors, $\Box = \partial_\mu \partial^\mu$. The numerical coefficients in this expansion, as well as the coefficients of $\overline{\theta}\gamma^5\theta\,\overline{\theta}$ and $(\overline{\theta}\gamma^5\theta)^2$, expressed in the above forms, are written for convenience. One may, of course, absorb the numerical coefficients in the fields and introduce other symbols for the coefficients just mentioned at the cost of complicating the algebra and the interpretation that follows from the analysis. In particular, it is important to emphasize that the θ independent part $A(x)$ of the above expression is a Lorentz scalar.

The supersymmetry transformations between spinor and boson field components arise in the following manner. For a supersymmetric transformation (see (2.1.10)/(2.1.11), expressed in terms of $\overline{\epsilon}$ with $b^\mu = 0$)

$$x'^\mu = x^\mu + \frac{i}{2}\overline{\epsilon}\gamma^\mu\theta,$$

$$\theta'_a = \theta_a + \epsilon_a, \tag{2.6.3}$$

for an infinitesimal $\overline{\epsilon}$, the scalar field is shown in Sect. 2.6.1 below in (2.6.74), to respond in the following manner:

$$\delta\Phi(x,\theta) = \overline{\epsilon}\Big(\frac{\partial}{\partial\overline{\theta}} + \frac{i}{2}\gamma^\mu\theta\partial_\mu\Big)\Phi(x,\theta). \tag{2.6.4}$$

In terms of the supersymmetry generator Q, the variation of a superfield

$$\delta\Phi = \frac{1}{i}[\Phi,\overline{\epsilon}Q],$$

gives the rules of transformations between fermions and bosons. Upon comparison of the expressions obtained from both sides of (2.6.4), as applied to the explicit expression of $\Phi(x,\theta)$ in (2.6.2), and using the basic rules in the above table (see also (2.2.20)–(2.2.28)), the following supersymmetry transformations between the fields arise:

$$\delta A = -i\overline{\epsilon}\gamma^5\psi, \tag{2.6.5}$$

$$\delta\psi = -\frac{1}{2}(B + i\gamma^5 G)\epsilon - \frac{1}{2}\gamma^\mu(\gamma^5\partial_\mu A + iV_\mu)\epsilon, \tag{2.6.6}$$

$$\delta B = \frac{1}{\sqrt{2}}\overline{\epsilon}(\chi + i\sqrt{2}\gamma^\mu\partial_\mu\psi), \tag{2.6.7}$$

$$\delta G = \frac{i}{\sqrt{2}}\overline{\epsilon}\gamma^5(\chi + i\sqrt{2}\gamma^\mu\partial_\mu\psi), \tag{2.6.8}$$

$$\delta V_\mu = \frac{i}{\sqrt{2}}\overline{\epsilon}(\gamma_\mu\chi - i\sqrt{2}\partial_\mu\psi), \tag{2.6.9}$$

$$\delta \chi = -\frac{1}{\sqrt{2}} \left(\frac{1}{2} [\gamma^\sigma, \gamma^\mu] \partial_\sigma V_\mu - i\gamma^5 \mathscr{D} \right), \tag{2.6.10}$$

$$\delta \mathscr{D} = \frac{1}{\sqrt{2}} \bar{\epsilon} \gamma^5 \gamma^\mu \partial_\mu \chi. \tag{2.6.11}$$

It is interesting to note that $\delta \mathscr{D}$, as given in (2.6.11), is a total derivative. This point is of great significance and we will come back to it shortly.

The product of scalar superfields is also a scalar superfields. That is, in the above notation,

$$\Phi_1(x, \theta) \, \Phi_2(x, \theta) = \Phi(x, \theta), \tag{2.6.12}$$

where the components fields of $\Phi(x, \theta)$ are expressed in terms of the component fields of $\Phi_1(x, \theta)$, $\Phi_2(x, \theta)$, and are spelled out in Sect. 2.6.1 below.

The importance of the scalar superfield stems from the following. A scalar superfield $\Phi(x, \theta)$ implies that the following integral is invariant under supersymmetry transformations (see Sect. 2.4)

$$\int (dx)(d\theta) \, \Phi(x, \theta) = \text{invariant}, \quad (dx) = dx^0 dx^1 dx^2 dx^3, \quad (d\theta) = d\theta_1 d\theta_2 d\theta_3 d\theta_4. \tag{2.6.13}$$

From the theory of integrations over Grassmann variables (Sect. 2.4) the only contribution to the above integral comes from the last term in (2.6.2) involving the factor $(\bar{\theta}\gamma^5\theta)^2$ having four θs. As a matter of fact from (2.2.28), we have explicitly

$$\theta_4\theta_3\theta_2\theta_1 = \frac{1}{8}(\bar{\theta}\gamma^5\theta)^2, \tag{2.6.14}$$

and from (2.6.2), the above integral becomes

$$\int (dx)(d\theta) \, \Phi(x, \theta) = -\frac{1}{2} \int (dx) \, [\mathscr{D}(x) + \frac{1}{2}\Box A(x)]. \tag{2.6.15}$$

This provides the starting point to generate supersymmetric actions, with $\mathscr{D}(x)$ real (Hermitian). In this *respect*, since $\Box A(x) = \partial_\mu(\partial^\mu A(x))$ is a total differential, the last term in (2.6.15), may be then omitted under the integral sign, leading from (2.6.15) to

$$\int (dx)(d\theta) \, \Phi(x, \theta) = -\frac{1}{2} \int (dx) \mathscr{D}(x). \tag{2.6.16}$$

Also from (2.6.11)

$$\delta \int (dx)(d\theta)\, \Phi(x,\theta) = -\frac{1}{2}\int (dx)\, \delta \mathscr{D}(x) = -\frac{1}{2\sqrt{2}}\int (dx)\, \partial_\mu(\bar{\epsilon}\gamma\,^5\gamma^\mu\chi(x)) = 0,$$

$$(2.6.17)$$

up to a surface integral, providing an explicit demonstration that *the integral in (2.6.16), corresponding to the $\mathscr{D}$ term of the scalar superfield is invariant under a supersymmetric transformation*, and hence suitable candidate for a supersymmetric action. For additional details concerning the scalar superfield see Sect. 2.6.1 below. In particular, it is shown in (2.6.80)–(2.6.86), that the product of scalar superfields is also a scalar superfield.

Obviously, one cannot arbitrarily set some of the fields of the superfield $\Phi(x,\theta)$ equal to zero since, according to (2.6.5)–(2.6.11), these field components transform between themselves. The field $\Phi(x,\theta)$ in (2.6.2) is, however, reducible, in the sense that a subset of the field components may be selected which transform among themselves. This leads us next to discuss another set of superfields.

Chiral Superfields

We have just mentioned that the scalar superfield Φ is reducible. That is, one may select a subset of its fields which transform among themselves. To this end, we first define the left-chiral and right-chiral fields as follows

$$\psi = \psi_L + \psi_R, \qquad \psi_L \equiv \left(\frac{1-\gamma\,^5}{2}\right)\psi, \quad \psi_R \equiv \left(\frac{1+\gamma\,^5}{2}\right)\psi, \qquad (2.6.18)$$

where ψ_L, ψ_R are referred to as left-chiral and right-chiral spinors, which may have, respectively, only two upper and only two lower non-vanishing components. Recall that in the chiral representation $\gamma\,^5$ is diagonal. Note that $\gamma\,^5\psi_{L/R} = \mp\psi_{L/R}$.

In reference to (2.6.5)–(2.6.11), we set

$$\chi = 0, \quad \mathscr{D} = 0, \quad iV_\mu = \partial_\mu A, \quad -iG = B \equiv i\mathscr{F}, \quad \psi_R = 0, \qquad (2.6.19)$$

and obtain a closed system of transformation of fields among themselves:

$$\delta\psi_L = -i\mathscr{F}\epsilon_L - \gamma\partial_\epsilon_R A, \qquad (2.6.20)$$

$$\delta\mathscr{F} = \bar{\epsilon}\gamma\partial\,\psi_L, \qquad (2.6.21)$$

$$\delta A = i\bar{\epsilon}\,\psi_L, \qquad (2.6.22)$$

thus generating a left-chiral superfield defined by

$$\Phi_L(x,\theta) = A(x) + i\bar{\theta}\psi_L(x) + \frac{1}{2}\bar{\theta}\theta_L\,\mathscr{F}(x) + \frac{i}{4}\bar{\theta}\gamma\,^5\gamma^\mu\theta\,\partial_\mu A(x)$$

$$- \frac{1}{4}\bar{\theta}\gamma\,^5\theta\,\bar{\theta}\gamma\partial\psi_L(x) - \frac{1}{32}(\bar{\theta}\gamma\,^5\theta)^2\,\Box A(x). \qquad (2.6.23)$$

Here ψ is a Majorana spinor, and $\psi_L \equiv (I - \gamma^5)\psi/2$ is its left-handed part, which is the reason why this scalar superfield is called a left-chiral superfield.

Here we may pose to note that the last term in the expression in (2.6.23) is a total derivative and hence its integral over superspace is zero (up to the total differential contribution). The $\mathscr{F}$ term in it, however, is quite important in generating action integrals for the following reason. Since $\theta = \theta_L + \theta_R$, accordingly, we have the following integral [see also (2.4.25)–(2.4.31)]

$$
\begin{aligned}
I\Big|_{\mathscr{F}} &= \int (dx)(d\theta)\, \delta^{(2)}(\theta_R)\, \Phi(x, \theta) \\
&= \frac{1}{2} \int (dx)(d\theta)\, \delta^{(2)}(\theta_R)\, \overline{\theta}\theta_L\, \mathscr{F}(x) = \int (dx)\, \mathscr{F}(x),
\end{aligned}
\tag{2.6.24}
$$

where recall (see Sect. 2.4) that $\int d\theta_3 d\theta_4\, \theta_4\theta_3 = 1$, and we have denoted the components of θ_R by θ_3, θ_4, $\delta^{(2)}(\theta_R) = \theta_4\theta_3$. We have also used the fact that

$$
\overline{\theta}\theta_L = -\theta^\top \mathscr{C}^{-1} \frac{1 - \gamma^5}{2} \theta = -\theta_1\theta_2 + \theta_2\theta_1 = 2\,\theta_2\theta_1.
\tag{2.6.25}
$$

Since under a supersymmetry transformation, $\delta\mathscr{F}$ is a total differential, as seen in (2.6.21), we learn that *the integral in (2.6.24) corresponding to the $\mathscr{F}$ term of the left-chiral superfield in (2.6.23) is invariant under a supersymmetry transformation* up to a total differential in the integrand. On the other hand the $\mathscr{D}$ part of the left-chiral field (2.6.23), involving $\Box A = \partial_\mu(\partial^\mu A)$, is a total derivative and is thus not interesting in setting up Lagrangian densities.

The expression of this field in (2.6.23) may be simplified by introducing, in the process, the variable

$$
\hat{x}^\mu = x^\mu + \frac{i}{4} \overline{\theta}\gamma^5\gamma^\mu\theta,
\tag{2.6.26}
$$

leading to the following simple expression for the left-chiral superfield

$$
\Phi_L(x, \theta) = A(\hat{x}) + i\overline{\theta}\psi_L(\hat{x}) + \frac{1}{2}\overline{\theta}\theta_L\, \mathscr{F}(\hat{x}),
\tag{2.6.27}
$$

as shown in Sect. 2.6.2 below. It is interesting to note that if one defines right-hand and left-hand superderivatives by

$$
D^{R/L} \equiv \frac{1 \pm \gamma^5}{2} D, \qquad \text{then} \quad D^R\Phi_L(x, \theta) = 0.
\tag{2.6.28}
$$

That is, the right-hand superderivative annihilates a left-chiral superfield. The latter is often taken as the definition of a left-chiral superfield.

We may, in turn, generate a right-chiral superfield as follows. In reference to (2.6.5)–(2.6.11), we set

$$\chi = 0, \quad \mathscr{D} = 0, \quad i V_\mu = -\partial_\mu A, \quad i G = B \equiv -i \mathscr{F}, \quad \psi_L = 0, \qquad (2.6.29)$$

where, needless to say, $\mathscr{F}(x)$, $A(x)$ may be chosen as new functions. We obtain a closed system of transformation of fields among themselves:

$$\delta \psi_R = i \mathscr{F} \epsilon_R + \gamma \partial \epsilon_L A, \qquad (2.6.30)$$

$$\delta \mathscr{F} = -\bar{\epsilon} \gamma \partial \psi_R, \qquad (2.6.31)$$

$$\delta A = -i \bar{\epsilon} \psi_R, \qquad (2.6.32)$$

thus generating a right-chiral superfield defined by

$$\Phi_R(x,\theta) = A(x) - i \bar{\theta} \psi_R(x) + \frac{1}{2} \bar{\theta} \theta_R \mathscr{F}(x) - \frac{i}{4} \bar{\theta} \gamma^5 \gamma^\mu \theta \, \partial_\mu A(x)$$

$$-\frac{1}{4} \bar{\theta} \gamma^5 \theta \, \bar{\theta} \gamma \partial \psi_R(x) - \frac{1}{32} (\bar{\theta} \gamma^5 \theta)^2 \Box A(x). \qquad (2.6.33)$$

The above expression simplifies, in turn, to

$$\Phi_R(x,\theta) = A(\hat{x}^\dagger) - i \bar{\theta} \psi_R(\hat{x}^\dagger) + \frac{1}{2} \bar{\theta} \theta_R \mathscr{F}(\hat{x}^\dagger), \qquad (2.6.34)$$

and

$$D^L \Phi_R(x,\theta) = 0, \qquad (2.6.35)$$

as is easily checked, which is often taken as the definition of a right-chiral superfield.

Using the fact, in the process of demonstration, that $(\bar{\theta} \psi_L)^\dagger = \bar{\theta} \psi_R$, as a consequence of the anti-commutativity of γ^0 and γ^5, we note that the Hermitian conjugate of a left-chiral superfield is given by

$$\Phi_L^\dagger = A^\dagger - i \bar{\theta} \psi_R + \frac{1}{2} \bar{\theta} \theta_R \mathscr{F}^\dagger - \frac{i}{4} \bar{\theta} \gamma^5 \gamma^\mu \theta \, \partial_\mu A^\dagger$$

$$-\frac{1}{4} \bar{\theta} \gamma^5 \theta \, \bar{\theta} \gamma \partial \psi_R - \frac{1}{32} (\bar{\theta} \gamma^5 \theta)^2 \Box A^\dagger, \qquad (2.6.36)$$

where we have used the facts that $(\bar{\theta} \gamma^5 \theta)^\dagger = -\bar{\theta} \gamma^5 \theta$, and the Majorana character of the field ψ: $(\bar{\theta} \psi_L)^\dagger = \bar{\theta} \psi_R$, $(\bar{\theta} \gamma^\mu \psi_L)^\dagger = -\bar{\theta} \gamma^\mu \psi_R$. That is, the Hermitian conjugate of a left-chiral superfield is a right-chiral one (and vice versa). For details leading to the properties of chiral fields above, see Sect. 2.6.2 below.

To generalize abelian and non-abelian gauge theories to their supersymmetric counterparts, we need to introduce supersymmetric vector fields. This is done next.

(Scalar-) Vector Superfields

A (scalar-) vector superfield is defined by imposing a reality condition on the general scalar superfield $\Phi(x, \theta) = \Phi^\dagger(x, \theta)$ in (2.6.2), which, for convenience, will be written as

$$\mathcal{V}(x, \theta) = S(x) - i\,\overline{\theta}\gamma^5\tilde{\psi}(x) + \frac{i}{4}\,\overline{\theta}\gamma^5\theta\,B(x) - \frac{1}{4}\,\overline{\theta}\theta\,G(x) - \frac{1}{4}\,\overline{\theta}\gamma^5\gamma^\mu\theta\,V_\mu(x)$$

$$+ \frac{i}{2\sqrt{2}}\,\overline{\theta}\gamma^5\theta\,\overline{\theta}[\,\chi(x) + \frac{i}{\sqrt{2}}\gamma\partial\tilde{\psi}(x)\,] - \frac{1}{16}\,(\overline{\theta}\gamma^5\theta)^2\,[\,\mathscr{D}(x) + \frac{1}{2}\,\Box S(x)\,].$$

$$(2.6.37)$$

Although this is a scalar superfield, it is referred to as a (scalar-) vector superfield, and quite often, unfortunately, it is also referred to as a vector superfield because it includes the Lorentz vector V^μ as a component field. We first consider the abelian case.

(Scalar-) Vector Superfields: Abelian Case

We note that in (2.6.37), S, B, G, $\mathscr{D}$ are (new) real Lorentz scalars, V^μ is a real Lorentz vector, and $\tilde{\psi}$, χ are Majorana spinors.

In analogy to ordinary gauge transformations, we define a supergauge transformation of, say, a left-chiral superfield as follows:

$$\Phi_L(x, \theta) \rightarrow e^{i e\,\Lambda(x, \theta)}\Phi_L(x, \theta), \qquad (2.6.38)$$

where $\Lambda(x, \theta)$ is a classical left-chiral superfield, with general structure that will be given in (2.6.97) below, which may be referred to as a gauge-parameter superfield, and e is a parameter. We are interested in defining supergauge invariant combinations of bilinear products $\Phi_L^\dagger(x, \theta) \ldots \Phi_L(x, \theta)$. To this end we impose the supergauge invariance of the combination

$$\Phi^\dagger_L(x, \theta)\,e^{-2 e\mathcal{V}(x, \theta)}\,\Phi_L(x, \theta) \rightarrow \Phi^\dagger_L(x, \theta)\,e^{-2 e\mathcal{V}(x, \theta)}\,\Phi_L(x, \theta), \qquad (2.6.39)$$

where the factor of 2 in the exponential is introduced for convenience, $e^{-2 e\mathcal{V}(x, \theta)}$ is referred to as the gauge connection. Thus from (2.6.38), one consistently obtains the supergauge transformation of $\mathcal{V}(x, \theta)$ to be

$$\mathcal{V}(x, \theta) \rightarrow \mathcal{V}(x, \theta) - \frac{i}{2}\,\left(\Lambda^\dagger(x, \theta) - \Lambda(x, \theta)\right) = \mathcal{V}'(x, \theta), \qquad (2.6.40)$$

where we note that

$$(i/2)\,\left(\Lambda^\dagger(x, \theta) - \Lambda(x, \theta)\right),$$

is real. Its explicit structure is given in Problem 2.16.

We work in a specific supergauge, referred to as the Wess-Zumino gauge, such that the gauge-parameter superfield Λ is chosen in such a manner as to gauge away the field components $(S, \tilde{\psi}, B, G)$ of $\mathscr{V}$, as shown in (2.6.98)–(2.6.104) below. In this supergauge, the vector superfield $\mathscr{V}$ then reduces to

$$\mathscr{V}(x, \theta) = -\frac{1}{4} \overline{\theta} \gamma^5 \gamma^\mu \theta \, V_\mu(x) + \frac{i}{2\sqrt{2}} \overline{\theta} \gamma^5 \theta \, \overline{\theta} \chi(x) - \frac{1}{16} (\overline{\theta} \gamma^5 \theta)^2 \mathscr{D}(x),$$

$$(2.6.41)$$

as given in Sect. 2.6.3 below, where as will be shown in (2.6.103), χ and $\mathscr{D}$ are gauge invariant, while the vector field components V_μ have the conventional gauge transformations of an abelian gauge theory

$$V_\mu(x) \rightarrow V_\mu(x) + \partial_\mu a(x).$$

$$(2.6.42)$$

The structure of the gauge-parameter Λ to be used now in this supergauge, is given by the simpler expression

$$\Lambda(x, \theta) = a(x) + \frac{i}{4} \overline{\theta} \gamma^5 \gamma^\mu \theta \, \partial_\mu a(x) - \frac{1}{32} (\overline{\theta} \gamma^5 \theta)^2 \, \Box \, a(x), \quad a(x) \equiv \mathrm{Re}\, a(x).$$

$$(2.6.43)$$

The advantage in working in the Wess-Zumino gauge is that since now $\mathscr{V}$ has no θ-independent part, $\exp(-2\,\mathrm{e}\,\mathscr{V})$ in (2.6.39) reduces from an infinite series to a *polynomial*. Note that the reality condition of $\mathscr{V}$ in (2.6.41) implies that $(\overline{\theta} \chi)^\dagger = \overline{\theta} \chi$, and χ is a Majorana spinor.

(Scalar-) Vector Superfields: Non-Abelian Case

Introducing Hermitian generators t_C of the underlying non-abelian group, we write a vector superfield, and, say, a left-chiral superfield, and a gauge-parameter left-chiral one as

$$\mathscr{V} = t_C \, \mathscr{V}_C, \quad \Phi_{\mathrm{L}} = t_C \, \Phi_{\mathrm{CL}}, \quad \Lambda = t_C \, \Lambda_C,$$

$$(2.6.44)$$

where $\mathscr{V}_C$, Φ_{CL}, Λ_C, are given, respectively, in (2.6.23), (2.6.37), (2.6.97) below, with every field component in them carrying now an index C. Implicit in this definition here, is that one will be using the adjoint representation of the underlying group. *To avoid confusion with other indices occurring in the formalism, we have used a rather standard notation of labeling the indices of the generators by capital letters, e.g., t_C, in this chapter.*

Due to the matrix structure of the underlying superfields, the supergauge transformation rule in (2.6.38) is, from (2.6.39), replaced by the more complicated structure applied to the gauge connection $\exp(-2\,g\,t_C\,\mathscr{V}_C)$, where g is a parameter, as follows:

$$\exp(-2g\mathcal{V}) \rightarrow \exp(ig\Lambda^\dagger)\exp(-2g\mathcal{V})\exp(-ig\Lambda) \equiv \exp(-2g\mathcal{V}'). \tag{2.6.45}$$

We work in the Wess-Zumino supergauge for which the gauge-parameter superfield Λ is chosen in such a manner as to gauge away the field components $(S, \tilde{\psi}, B, G)$ of $\mathcal{V}$. In this supergauge, the vector superfield $\mathcal{V}$ then reduces again to the structure given in (2.6.41) (see Sect. 2.6.3), but now for infinitesimal $a(x)$, (2.6.112)–(2.6.114) lead to

$$\mathcal{V}'_C = \mathcal{V}_C - \frac{1}{4}\bar{\theta}\gamma^5\gamma^\mu\theta\,\partial_\mu a_C + gf_{DEC}\,\mathcal{V}_D\,a_E, \qquad a(x) = \mathrm{Re}\,a(x), \tag{2.6.46}$$

where now $\mathcal{V}$ is the Wess-Zumino gauged field having the structure in (2.6.41), and the gauge parameter Λ to be used now in this supergauge is given in (2.6.43) with a real. Note that the structure constants f_{DEC} are real and totally antisymmetric. The above equation then leads to

$$V'_{C\mu} = V_{C\mu} + \partial_\mu a_C + g\,f_{DEC}\,V_{D\mu}\,a_E, \tag{2.6.47}$$

$$\chi'_C = \chi_C + gf_{DEC}\,\chi_D\,a_E, \tag{2.6.48}$$

$$\mathscr{D}'_C = \mathscr{D}_C + gf_{DEC}\,\mathscr{D}_D\,a_E. \tag{2.6.49}$$

Here we recognize (2.6.47) as the standard infinitesimal gauge transformation of a non-abelian gauge vector field. The so-called superpartner described by the spinor χ is referred to as the gaugino.

Pure Vector Superfields

The pure vector superfield is defined by[30]

$$\mathscr{V}^\mu = -\frac{1}{2g}\,(\mathscr{C}\gamma^\mu)_{ab}\,D^R_a\,e^{2g\mathscr{V}}\,D^L_b\,e^{-2g\mathscr{V}}, \tag{2.6.50}$$

Under the supergauge transformation (2.6.45),

$$\mathscr{V}^\mu \rightarrow -\frac{1}{2g}\,(\mathscr{C}\gamma^\mu)_{ab}\,D^R_a\,e^{ig\Lambda}\,e^{2g\mathscr{V}}\,e^{-ig\Lambda^\dagger}\,D^L_b\,e^{ig\Lambda^\dagger}\,e^{-2g\mathscr{V}}\,e^{-ig\Lambda}, \tag{2.6.51}$$

where we recall that Λ is left-chiral and hence $\Lambda^\dagger$ is right-chiral.[31] According to (2.6.28)/(2.6.35), they are, respectively, annihilated by $D^{R/L}$. That is,

$$D^R e^{ig\Lambda} = e^{ig\Lambda}\,D^R, \qquad D^L e^{ig\Lambda^\dagger} = e^{ig\Lambda^\dagger}\,D^L, \tag{2.6.52}$$

---

[30] Salam and Strathdee [30].

[31] Here we have also used the trivial identity $e^{2g\mathscr{V}}e^{-2g\mathscr{V}} = I$.

and we may rewrite (2.6.51) as

$$
\mathscr{V}^\mu \rightarrow -\frac{1}{2g} e^{ig\Lambda} (\mathscr{C}\gamma^\mu)_{ab} D_a^{\mathrm{R}} e^{2g\mathscr{V}} \left(D_b^{\mathrm{L}} e^{-2g\mathscr{V}} e^{-ig\Lambda} \right)
$$

$$
= -\frac{1}{2g} e^{ig\Lambda} (\mathscr{C}\gamma^\mu)_{ab} D_a^{\mathrm{R}} e^{2g\mathscr{V}} \left(D_b^{\mathrm{L}} e^{-2g\mathscr{V}} \right) e^{-ig\Lambda}
$$

$$
-\frac{1}{2g} e^{ig\Lambda} (\mathscr{C}\gamma^\mu)_{ab} \left(D_a^{\mathrm{R}} D_b^{\mathrm{L}} e^{-ig\Lambda} \right). \tag{2.6.53}
$$

The first equality in (2.6.52), allows us to replace the product $D_a^{\mathrm{R}} D_b^{\mathrm{L}}$ in the second term on the extreme right-hand side of (2.6.53) by their anti-commutator. This anti-commutator may be obtained from $\{D_{a'}, D_{b'}\}$ in (2.3.10) by multiplying it by $(1 + \gamma^5)_{aa'}(1 - \gamma^5)_{bb'}/4$ giving

$$
\{D_a^{\mathrm{R}}, D_b^{\mathrm{L}}\} = -i \left(\gamma^\sigma \mathscr{C} \frac{1-\gamma^5}{2} \right)_{ab} \partial_\sigma. \tag{2.6.54}
$$

This leads to

$$
\mathscr{V}^\mu \rightarrow e^{ig\Lambda} \mathscr{V}^\mu e^{-ig\Lambda} - \frac{1}{ig} e^{ig\Lambda} \partial_\mu e^{-ig\Lambda}, \tag{2.6.55}
$$

and showing that it transforms as a non-abelian gauge field.[32]
   Using the relations

$$
\{\gamma^5, \gamma^\mu\} = 0, \quad [\gamma^5, \mathscr{C}] = 0, \quad ((1-\gamma^5)/2)^2 = (1-\gamma^5)/2 \tag{2.6.56}
$$

(2.6.50) may be equivalently re-expressed as

$$
\mathscr{V}^\mu = -\frac{1}{2g} \left(\mathscr{C}\gamma^\mu \frac{1-\gamma^5}{2} \right)_{ab} D_a e^{2g\mathscr{V}} D_b e^{-2g\mathscr{V}}. \tag{2.6.57}
$$

We work exclusively in the Wess-Zumino supergauge in which $\mathscr{V}$ is reduced to the expression in (2.6.41), and we work directly in the general, non-abelian case, for which $\mathscr{V} = t_\alpha \mathscr{V}_\alpha$, with the (Hermitian) matrices t_α as the generators of the underlying group. The abelian gauge counterpart is then easily extracted.
   In the Wess-Zumino supergauge, since $\mathscr{V}$ has no θ independent part, $\exp(-2g\mathscr{V})$ reduces to the following polynomial

$$
e^{-2g\mathscr{V}} = 1 + \frac{g}{2}\bar{\theta}\gamma^5\gamma^\mu\theta V_\mu - \frac{ig}{\sqrt{2}}\bar{\theta}\gamma^5\theta\bar{\theta}\chi + \frac{g}{8}(\bar{\theta}\gamma^5\theta)^2[\mathscr{D} + g V^\nu V_\nu]. \tag{2.6.58}
$$

---

[32]See, e.g., [26], (6.1.30) in Chap. 6 of Vol. I.

As the explicit expression of the pure vector superfield $\mathcal{V}_\mu$ is perhaps not very well known,[33] a detailed derivation of it is given in Sect. 2.6.4 below. We here summarize the main results.

Pure Abelian Vector Superfield

The pure abelian vector superfield, in the Wess-Zumino supergauge, is given by

$$\mathcal{V}^\mu = V^\mu + \frac{i}{\sqrt{2}}\overline{\theta}\gamma^\mu\chi + \overline{\theta}\gamma^5\gamma_\nu\theta\,A^{\nu\mu} + \overline{\theta}\gamma^5\theta\,\overline{\theta}B^\mu + \frac{i}{4}(\overline{\theta}\gamma^5\theta)^2\partial_\nu[A^{\nu\mu} - \frac{i}{8}\partial^\nu V^\mu],$$

(2.6.59)

and with $F^{\mu\nu} = \partial^\mu V^\nu - \partial^\nu V^\mu$,

$$A^{\nu\mu} = \frac{i}{4}\partial^\mu V^\nu + \frac{1}{8}\varepsilon^{\alpha\beta\nu\mu}F_{\alpha\beta} + \frac{1}{4}\eta^{\nu\mu}\mathscr{D}, \quad B^\mu = \frac{1}{2\sqrt{2}}\left[\eta^{\mu\nu} - \frac{1}{2}\gamma^5\gamma^\mu\gamma^\nu\right]\partial_\nu\chi,$$

(2.6.60)

and note that the θ independent part of $\mathcal{V}_\mu$ is a Lorentz vector. Also note the order of the indices in $\partial^\mu V^\nu$ in the first term on the right-hand of the above equation.

Pure Non-Abelian Vector Superfield

The pure vector superfield $\mathcal{V}_\mu$, in the Wess-Zumino supergauge, is again given by the expression in (2.6.59) but now $\mathcal{V}_\mu = t_E\mathcal{V}_{E\mu}$,[34]

$$\mathcal{V}^\mu = V^\mu + \frac{i}{\sqrt{2}}\overline{\theta}\gamma^\mu\chi + \overline{\theta}\gamma^5\gamma_\nu\theta\,A^{\nu\mu} + \overline{\theta}\gamma^5\theta\,\overline{\theta}B^\mu + \frac{i}{4}(\overline{\theta}\gamma^5\theta)^2\partial_\nu[A^{\nu\mu} - \frac{i}{8}\partial^\nu V^\mu],$$

(2.6.61)

$$A^{\lambda\rho}(x) = \frac{i}{4}\partial^\lambda V^\rho(x) - \frac{i}{4}G^{\lambda\rho}(x) + \frac{1}{8}\varepsilon^{\lambda\rho\sigma\mu}G_{\sigma\mu}(x) + \frac{1}{4}\eta^{\lambda\rho}\mathscr{D}(x), \quad (2.6.62)$$

$$G_{\sigma\mu}(x) = \partial_\sigma V_\mu(x) - \partial_\mu V_\sigma(x) - ig\,[V_\sigma(x), V_\mu(x)], \quad (2.6.63)$$

$$B^\rho(x) = \frac{1}{2\sqrt{2}}(\eta^{\rho\sigma} - \frac{1}{2}\gamma^5\gamma^\rho\gamma^\sigma)\partial_\sigma\chi(x) + \frac{ig}{2\sqrt{2}}\gamma^\rho\gamma^\sigma\frac{1+\gamma^5}{2}[V_\sigma(x), \chi(x)].$$

(2.6.64)

$A^{\lambda\rho} = t_E A_E^{\lambda\rho}, B_a^\mu = t_E B_{Ea}^\mu$, where a is a spinor index,

$$[V_\sigma, \chi] = it_E f_{ECD}\,V_{C\sigma}\,\chi_D, \quad (2.6.65)$$

---

[33] The *explicit* expression of $\mathcal{V}_\mu$ has been derived recently in [25]. Note the γ^5 taken there is the minus of the present one.

[34] Manoukian [25].

as usual. Note the order of the indices $\partial^\lambda V^\rho$ now, in the first term on the right-hand side of (2.6.62). In Problem 2.17, $\mathcal{V}_\mu(x)$, in (2.6.61), is re-expressed as a function of $\hat{x}^\mu = x^\mu + (i/4)\bar{\theta}\gamma\,^5\gamma^\mu\theta$, with the latter introduced in (2.6.26), and is given by

$$\mathcal{V}^\rho(x,\theta) = \widetilde{V^\rho}(\hat{x}) + \frac{i}{\sqrt{2}}\,\bar{\theta}\gamma\,^\rho\chi(\hat{x})$$

$$-\bar{\theta}\gamma\,^5\gamma_\lambda\theta\left[\frac{i}{4}\,G^{\lambda\rho}(\hat{x}) + \frac{1}{8}\,\varepsilon^{\rho\sigma\mu\lambda}G_{\sigma\mu}(\hat{x}) - \frac{1}{4}\,\eta^{\lambda\rho}\mathscr{D}(\hat{x})\right]$$

$$-\ \bar{\theta}\gamma\,^5\theta\left[\frac{1}{2\sqrt{2}}\,\bar{\theta}\gamma\,^\rho\gamma^\sigma\frac{1+\gamma\,^5}{2}\Big((\partial_\sigma\chi)(\hat{x}) - i\,g[\,V_\sigma(\hat{x}),\chi(\hat{x})]\Big)\right], \qquad (2.6.66)$$

where, in the process, the expressions for $A^{\lambda\rho}$, given in (2.6.62), and B^ρ, given in (2.6.64), have been used, as well as the basic identity $\{\gamma\,^\rho,\gamma\,^\sigma\} = -2\eta^{\rho\sigma}$.

Spinor Superfields

We define a spinor superfield by the expression

$$\mathscr{W}_a = \frac{1}{2\sqrt{2}\,ig}\,\mathscr{C}_{bc}D\,^R_b D\,^R_c\,e^{2g\mathcal{V}}D\,^L_a\,e^{-2g\mathcal{V}}, \qquad (2.6.67)$$

where a is a spinor index. This expression should be compared with the expression of the pure vector superfield in (2.6.50). The present superfield, however, transforms as a field strength

$$\mathscr{W}_a \to e^{ig\,\Lambda}\,\mathscr{W}_a\,e^{-ig\,\Lambda} \qquad (2.6.68)$$

as shown in Sect. 2.6.5 below.

In the following, we work exclusively in the Wess-Zumino supergauge. In the latter supergauge, the spinor superfield is given by

$$\mathscr{W}_{Aa} = \exp\left[\frac{i}{4}\bar{\theta}\gamma\,^5\gamma^\mu\theta\partial_\mu\right]\times$$

$$\times\left(\chi_{Aa} + \frac{1}{\sqrt{2}}(\gamma^\mu\gamma^\nu\theta_L)_a\,G_{A\mu\nu} - i\,(\gamma\,\nabla\chi)_{Aa}\,\bar{\theta}\theta_L - i\,\sqrt{2}\,\mathscr{D}_A(\theta_L)_a\right), \qquad (2.6.69)$$

$$G_{A\mu\nu} = \partial_\mu V_{A\nu} - \partial_\nu V_{A\mu} + g f_{ABC}\,V_{B\mu}V_{C\nu}\,,$$

$$(\gamma\,\nabla\chi)_{Aa} = \Big(\gamma^\mu(\delta_{AC}\partial_\mu + g f_{ABC}\,V_{B\mu})\chi_C\Big)_a\,, \qquad (2.6.70)$$

where, as in (2.6.95), we have factored out the translational operator which simply changes the variable x in the fields above to $\hat{x}$, with the latter defined in (2.6.26). The presence of a spinor χ_{Aa} as a θ independent part of $\mathscr{W}_{Aa}$ should be noted. For the abelian counterpart simply set $f_{ABC} = 0$, and delete the corresponding indices.

Now we go into the details of the superfields encountered above.

2.6.1 The Scalar Superfield

A scalar superfield $\Phi(x, \theta)$, as a function of (x^μ, θ_a), may be expanded in powers of the components θ_a as follows

$$\Phi(x, \theta) = A(x) + \psi^a(x)\theta_a + \psi^{ab}\theta_a\theta_b + \psi^{abc}\theta_a\theta_b\theta_c + \psi^{abcd}\theta_a\theta_b\theta_c\theta_d, \quad (2.6.71)$$

terminating, of course, in fourth order due to the anti-commutativity of the components of θ, with the theta independent part $A(x)$ being a scalar.

By examining equations (2.2.21), (2.2.27), and (2.2.28), for the expansions of products of the components of θ, we note that the product of four of θs is proportional to $(\overline{\theta}\gamma^5\theta)^2$, the product of three is proportional to $(\overline{\theta}\gamma^5\theta)\theta$, (or equivalently $(\overline{\theta}\gamma^5\theta)\overline{\theta}$), and finally the product of two is a linear combination of $\overline{\theta}\theta$, $\overline{\theta}\gamma^5\theta$, and $\overline{\theta}\gamma^5\gamma^\mu\theta$. Hence one may introduce a scalar superfield as given in (2.6.2).

By definition, a scalar superfield satisfies

$$\Phi'(x', \theta') = \Phi(x, \theta), \quad (2.6.72)$$

and, in particular, for a supersymmetric transformation as given in (2.6.3),

$$\Phi'(x, \theta) = \Phi(x - \frac{i}{2}\overline{\epsilon}\gamma^\mu\theta, \theta - \epsilon) = \Phi(x, \theta) + \left(-\frac{i}{2}\overline{\epsilon}\gamma^\mu\theta\,\partial_\mu - \overline{\epsilon}\frac{\partial}{\partial\theta}\right)\Phi(x, \theta), \quad (2.6.73)$$

for infinitesimal $\overline{\epsilon}$, from which we define

$$\delta\Phi(x, \theta) = \Phi(x, \theta) - \Phi'(x, \theta) = \overline{\epsilon}\left(\frac{\partial}{\partial\theta} + \frac{i}{2}\gamma^\mu\theta\partial_\mu\right)\Phi(x, \theta). \quad (2.6.74)$$

We note that the above operation is almost the same as $\overline{\epsilon}D$ with the sign in the middle of D changed, however, to a plus sign.

We may relate the above response of the superfield to the action of the unitary operator $U = 1 + i\overline{\epsilon}Q$ [see (2.5.5)], for infinitesimal $\overline{\epsilon}$, with $\overline{\epsilon}Q = (\overline{\epsilon}Q)^\dagger$ as the generator of the supersymmetry transformation, via $\Phi' = U^\dagger\Phi U$ and

$$\delta\Phi = \frac{1}{i}[\Phi, \overline{\epsilon}Q]. \quad (2.6.75)$$

The transformation rules for the component fields in (2.6.2) may be readily obtained from (2.6.74), and, in particular, by the differentiation rules with respect to $\overline{\theta}$ given in Box 2.1 in the beginning of the section, after $\overline{\epsilon}(\partial/\partial\overline{\theta} + (i/2)\gamma^\mu\theta\partial_\mu)\Phi(x, \theta)$ in (2.6.74) is expressed in the original form as $\Phi(x, \theta)$ in (2.6.2).

Expressing $\overline{\epsilon}(\partial/\partial\overline{\theta} + (i/2)\gamma^\mu\theta\partial_\mu)\Phi(x, \theta)$ in the same form as $\Phi(x, \theta)$ is best illustrated by the application of the operation $\overline{\epsilon}\partial/\partial\overline{\theta}$ to the sixth term on the right-hand side of (2.6.2), by using in the process the differentiation rule with respect to the derivative $\partial/\partial\bar\theta$ in Box 2.1, and the use of the key equation for the expansion of $\theta_a\bar\theta_b = -\bar\theta_b\theta_a$ (the first entry in Box 2.1). This gives

$$\bar\epsilon\frac{\partial}{\partial\bar\theta}\left(\frac{i}{2\sqrt{2}}\bar\theta\gamma{}^5\theta\,\bar\theta[.]\right) = \frac{i}{2\sqrt{2}}\left(2\bar\epsilon\gamma{}^5\theta\,(\bar\theta[.]) + \bar\theta\gamma{}^5\theta\,(\bar\epsilon[.])\right)$$

$$= \frac{i}{4\sqrt{2}}\left(-\bar\theta\theta(\bar\epsilon\gamma{}^5[.]) + \bar\theta\gamma{}^5\theta(\bar\epsilon[.]) - \bar\theta\gamma{}^5\gamma^\mu\theta(\bar\epsilon\gamma_\mu[.])\right). \qquad (2.6.76)$$

expressed in terms of $\bar\theta\theta$, $\bar\theta\gamma{}^5\theta$, and $\bar\theta\gamma{}^5\gamma^\mu\theta$ using (2.2.20). [Note the plus sign in the second term in the extreme right-hand side of the equation.]

Another illustration of the same type is obtained by the application of the operation $(i/2)\,\bar\epsilon\gamma^\mu\theta\partial_\mu$ to the second term on the right-hand side of (2.6.2). The summation formula for $\theta_a\bar\theta_b$, in Box 2.1 gives

$$\frac{i}{2}\bar\epsilon\gamma^\mu\theta\partial_\mu(-i\bar\theta\gamma{}^5\psi) = \frac{1}{2}\bar\epsilon\gamma^\mu\theta\,\bar\theta\gamma{}^5\partial_\mu\psi$$

$$= -\frac{1}{8}\left(-\bar\theta\theta\,(\bar\epsilon\gamma{}^5\gamma\partial\psi) + \bar\theta\gamma{}^5\theta\,(\bar\epsilon\gamma\partial\psi) - \bar\theta\gamma{}^5\gamma^\sigma\theta\,(\bar\epsilon\gamma\partial\gamma_\sigma\psi)\right), \qquad (2.6.77)$$

where we have used $\{\gamma{}^5,\gamma^\mu\} = 0$, $(\gamma{}^5)^2 = 1$, and the usual notation $\gamma^\mu\partial_\mu \equiv \gamma\partial$.

The following equation should be noted which makes use of the vanishing property of the product of any two of $\bar\theta\theta$, $\bar\theta\gamma{}^5\theta$, $\bar\theta\gamma^\mu\theta$ (see also Box 2.1). This equation is obtained by the application of $(i/2)\,\bar\epsilon\gamma^\mu\theta\partial_\mu$ also to the sixth term on the right-hand side of (2.6.2) and gives:

$$\frac{i}{2}\bar\epsilon\gamma^\mu\theta\partial_\mu\left(\frac{i}{2\sqrt{2}}\bar\theta\gamma{}^5\theta\,\bar\theta[.]\right)$$

$$= -\frac{1}{4\sqrt{2}}\bar\theta\gamma{}^5\theta\,\bar\epsilon\gamma^\mu\theta\,\bar\theta\partial_\mu[.]$$

$$= \frac{1}{16\sqrt{2}}\bar\theta\gamma{}^5\theta\left(\bar\theta\theta\,\bar\epsilon\gamma\partial[.] - \bar\theta\gamma{}^5\theta\,\bar\epsilon\gamma{}^5\gamma\partial[.] - \bar\theta\gamma{}^5\gamma^\sigma\theta\,\bar\epsilon\gamma{}^5\gamma\partial\gamma_\sigma[.]\right)$$

$$= -\frac{1}{16\sqrt{2}}(\bar\theta\gamma{}^5\theta)^2\bar\epsilon\gamma{}^5\gamma\partial[.]. \qquad (2.6.78)$$

Finally we use the properties of the bilinear products $\bar\epsilon\ldots\theta$, given in Box 2.1, to rewrite them as $\bar\theta\ldots\epsilon$, and express the products $\bar\theta\theta\,\bar\theta$, $\bar\theta\gamma{}^5\gamma^\mu\theta\,\bar\theta$, respectively, as $-\bar\theta\gamma{}^5\theta\,\bar\theta\gamma{}^5$, and $\bar\theta\gamma{}^5\theta\,\bar\theta\gamma^\mu$, as shown on the second and third lines in Box 2.1. The transformation rules, given in (2.6.5)–(2.6.11), then readily emerge for the components fields of the scalar superfield $\Phi(x,\theta)$ in (2.6.2) by the application of the operation in (2.6.74).

The relation in (2.6.5), follows by comparing $\delta\Phi$ and

$$\bar{\epsilon}\left(\frac{\partial}{\partial\bar{\theta}} + \frac{i}{2}\gamma^\mu\theta\partial_\mu\right)\Phi,$$

for the terms involving no θs. The remaining relations (2.6.6) to (2.6.11) follow upon the comparisons, respectively, of the coefficients of $\bar{\theta}$, $\bar{\theta}\gamma\,^5\theta$, $\bar{\theta}\theta$, $\bar{\theta}\gamma\,^5\gamma^\mu\theta$, $\bar{\theta}\gamma\,^5\theta\,\bar{\theta}$ and $(\bar{\theta}\gamma\,^5\theta)^2$. In obtaining the expression for relation in (2.6.10), the transformation rule for ψ in (2.6.6), in the process, has been used. Similarly, in obtaining the expression for the relation in (2.6.11), the transformation rule for A in (2.6.5), in the process, has been used.

The transformation law

$$\delta\Phi = \frac{1}{i}[\Phi, \bar{\epsilon}Q],$$

via the generator $\bar{\epsilon}Q = \bar{Q}\epsilon$, leads to transformations between bosons and fermions as obtained through (2.6.5)–(2.6.11).

The product of scalar superfields is also a scalar superfields. That is, in the above notation,

$$\Phi_1(x,\theta)\,\Phi_2(x,\theta) = \Phi(x,\theta), \tag{2.6.79}$$

where the components fields of $\Phi(x,\theta)$ are expressed in terms of the component fields of $\Phi_1(x,\theta)$, $\Phi_2(x,\theta)$. By using, in the process, the vanishing property of the product of any two of $\bar{\theta}\theta$, $\bar{\theta}\gamma\,^5\theta$, $\bar{\theta}\gamma\,^5\gamma^\mu\theta$, and the expression for the expansion of $\theta_a\bar{\theta}_b$, as given in Box 2.1 in the beginning of this section, the components of the scalar superfield $\Phi(x,\theta)$ are readily expressed in terms of the components of the scalar superfields $\Phi_1(x,\theta)$, $\Phi_2(x,\theta)$ in (2.6.12), as follows:

$$A = A_1A_2, \tag{2.6.80}$$

$$\psi = A_1\psi_2 + \psi_1A_2, \tag{2.6.81}$$

$$B = A_1B_2 + B_1A_2 - i\bar{\psi}_1\gamma\,^5\psi_2, \tag{2.6.82}$$

$$G = A_1G_2 + G_1A_2 - \bar{\psi}_1\psi_2, \tag{2.6.83}$$

$$V^\mu = A_1V^\mu{}_2 + V_1^\mu A_2 + \bar{\psi}_1\gamma\,^5\gamma^\mu\psi_2, \tag{2.6.84}$$

$$\chi = -\frac{i}{\sqrt{2}}\left[(\partial_\mu A_1)\gamma^\mu\psi_2 + (\gamma^\mu\psi_1)\partial_\mu A_2\right] + (A_1\chi_2 + \chi_1A_2) - \frac{i}{\sqrt{2}}\left[(\gamma\,^5\psi_1)B_2\right.$$

$$\left. + B_1(\gamma\,^5\psi_2)\right] + \frac{1}{\sqrt{2}}\gamma\,^5\gamma_\mu(\psi_1V^\mu{}_2 + V_1^\mu\psi_2) - \frac{1}{\sqrt{2}}(\psi_1G_2 + G_1\psi_2),$$

$$\tag{2.6.85}$$

$$\mathscr{D} = -\frac{1}{2}\left(\partial_\mu A_1 \partial^\mu A_2 + \partial_\mu A_2 \partial^\mu A_1\right) + \sqrt{2}\left(\overline{\psi}_1 \chi_2 + \overline{\chi}_1 \psi_2\right) + B_1 B_2 + G_1 G_2$$

$$- V_1^\mu V_{2\mu} + (A_1 \mathscr{D}_2 + \mathscr{D}_1 A_2) - \frac{1}{\sqrt{2}}\left[(\frac{\partial_\mu}{i}\overline{\psi}_1)\gamma^\mu \psi_2 - \overline{\psi}_1 \gamma^\mu \frac{\partial_\mu}{i}\psi_2\right],$$

$$(2.6.86)$$

in general, up to total derivatives.

2.6.2 Chiral Superfields

In reference to (2.6.5)–(2.6.11), if we set $\chi = 0$, $\mathscr{D} = 0$, and V_μ of the form $V_\mu = \partial_\mu \xi$, for some ξ, then we have the following closed system of transformations of fields among themselves:

$$\delta A = -i\overline{\epsilon}\gamma^5 \psi, \tag{2.6.87}$$

$$\delta \psi = -\frac{1}{2}(B + i\gamma^5 G)\epsilon - \frac{1}{2}\gamma^\mu (\gamma^5 \partial_\mu A + iV_\mu)\epsilon, \tag{2.6.88}$$

$$\delta B = i\partial_\mu \overline{\epsilon}\gamma^\mu \psi, \tag{2.6.89}$$

$$\delta G = -\partial_\mu \overline{\epsilon}\gamma^5 \gamma^\mu \psi, \tag{2.6.90}$$

$$\delta V_\mu = \partial_\mu \overline{\epsilon}\psi, \tag{2.6.91}$$

where note that for $V_\mu = \partial_\mu \xi$, $[\gamma^\sigma, \gamma^\mu]\partial_\sigma \partial_\mu \xi = 0$ in (2.6.10).

Upon multiplying (2.6.88) by $(1 \pm \gamma^5)/2$ and using

$$\gamma^5(1 \pm \gamma^5) = \pm(1 \pm \gamma^5),$$

$\{\gamma^5, \gamma^\mu\} = 0$, the system of (2.6.87)–(2.6.91) may be conveniently rewritten as

$$\delta \psi_{R/L} = -\left(\frac{B \pm iG}{2}\right)\epsilon_{R/L} - \gamma^\mu \left(\frac{\mp \partial_\mu A + iV_\mu}{2}\right)\epsilon_{L/R}, \tag{2.6.92}$$

$$\delta\left(\frac{B \pm iG}{2}\right) = i\overline{\epsilon}\gamma\partial\psi_{R/L}, \tag{2.6.93}$$

$$\delta\left(\frac{\mp \partial_\mu A + iV_\mu}{2}\right) = i\overline{\epsilon}\partial_\mu \psi_{R/L}. \tag{2.6.94}$$

These introduce two systems of equations. One by choosing $\psi_R = 0$, $B + iG = 0$, $-\partial_\mu A + iV_\mu = 0$. Thus upon setting $-iG = B \equiv i\mathscr{F}$, $iV_\mu = \partial_\mu A$, we obtain the closed system of transformations in (2.6.20)–(2.6.22). Hence using the fact that in (2.6.2) we now have

$$\frac{1}{4}(i\,\overline{\theta}\gamma^5\theta B - \overline{\theta}\theta G) = -\frac{1}{4}(\overline{\theta}\gamma^5\theta - \overline{\theta}\theta)\mathscr{F} = \frac{1}{2}\overline{\theta}\theta_L\,\mathscr{F},$$

we generate a so-called left-chiral superfield as given in (2.6.23).

The second system is introduced by choosing

$$\psi_L = 0, \ B - iG = 0, \ \partial_\mu A + iV_\mu = 0.$$

Hence by setting, $i\,G = B \equiv -i\,\mathscr{F}$, $i\,V_\mu = -\partial_\mu A$, where, needless to say, $\mathscr{F}(x)$, $A(x)$ may be chosen as a new set of functions. We may then generate a right-chiral superfield as given in (2.6.33), with the transformation rules in (2.6.30)–(2.6.32).

Consider the operation defined below

$$\exp\left[\frac{i}{4}\,\overline{\theta}\gamma^5\gamma^\mu\theta\,\partial_\mu\right]\left(A(x) + i\,\overline{\theta}\psi_L(x) + \frac{1}{2}\overline{\theta}\theta_L\,\mathscr{F}(x)\right). \tag{2.6.95}$$

There are two ways of evaluating this expression. One way is to expand the exponential using, in the process, the first identity in (2.2.24) thus obtaining

$$\exp\left[\frac{i}{4}\,\overline{\theta}\gamma^5\gamma^\mu\theta\,\partial_\mu\right] = 1 + \frac{i}{4}\,\overline{\theta}\gamma^5\gamma^\mu\theta\,\partial_\mu - \frac{(\overline{\theta}\gamma^5\theta)^2}{32}\,\Box, \tag{2.6.96}$$

showing that the expression in (2.6.95) is nothing but $\Phi_L(x, \theta)$ in (2.6.23), since, in particular,

$$\overline{\theta}\gamma^5\gamma^\mu\theta\,\overline{\theta}\theta_L = 0, \ \overline{\theta}\gamma^5\gamma^\mu\theta\,\overline{\theta}_a = \overline{\theta}\gamma^5\theta\,(\overline{\theta}\gamma^\mu)_a,$$

as given in (2.2.22), (2.2.23). On the other hand the exponential term is nothing but a translation operator of x^μ by $i\,\overline{\theta}\gamma^5\gamma^\mu\theta/4$. That is, we have obtained the simple expression for $\Phi_L(x, \theta)$ in (2.6.27), and similarly for $\Phi_R(x, \theta)$ in (2.6.33)/(2.6.34).

By applying the right-hand superderivative D^R, defined in (2.6.28), to $\Phi_L(x, \theta)$ in (2.6.27), and using, in the process, the chain rule,

$$\frac{\partial}{\partial\theta_a}f(\hat{x}) = \frac{\partial f(\hat{x})}{\partial\hat{x}^\mu}\frac{\partial}{\partial\theta_a}\hat{x}^\mu,$$

and similarly with respect to x^μ, for a function f of $\hat{x}$, a left-chiral superfield is easily seen to be annihilated by the right-hand superderivative. Similarly, a right-hand chiral field $\Phi_R(x, \theta)$ in (2.6.34), is annihilated by the left-hand superderivative D^L, which is also defined in (2.6.28). These are, respectively, given in (2.6.28), (2.6.35).

2.6.3   (Scalar-) Vector Superfields

The general structure of a (scalar-) vector superfield was given in (2.6.37).

(Scalar-) Vector Superfields: Abelian Case

Under the transformation in (2.6.38), $\mathscr{V}$ transforms as in (2.6.40). The gauge-parameter superfield has the structure

$$\Lambda(x,\theta) = a(x) - \frac{i}{\sqrt{2}}\,\overline{\theta}\xi_L(x) - \frac{i}{2}\,\overline{\theta}\theta_L\,b(x) + \frac{i}{4}\,\overline{\theta}\gamma^5\gamma^\mu\theta\,\partial_\mu a(x)$$

$$+\frac{1}{4\sqrt{2}}\,\overline{\theta}\gamma^5\theta\,\overline{\theta}\gamma\partial\xi_L(x) - \frac{1}{32}\,(\overline{\theta}\gamma^5\theta)^2\,\square\,a(x), \tag{2.6.97}$$

From (2.6.38)–(2.6.40), and Problem 2.16, we obtain the supergauge transformations of the field components of $\mathscr{V}$ in (2.6.37)

$$S' = S - \mathrm{Im}\,a, \tag{2.6.98}$$

$$\tilde{\psi}' = \tilde{\psi} + \frac{i}{2\sqrt{2}}\,\gamma^5\xi, \tag{2.6.99}$$

$$B' = B - \mathrm{Im}\,b, \tag{2.6.100}$$

$$G' = G - \mathrm{Re}\,b, \tag{2.6.101}$$

$$V_\mu' = V_\mu + \partial_\mu\,\mathrm{Re}\,a, \tag{2.6.102}$$

$$\chi' = \chi, \tag{2.6.103}$$

$$\mathscr{D}' = \mathscr{D}. \tag{2.6.104}$$

Hence the field components $(S, \tilde{\psi}, B, G)$ of $\mathscr{V}$ may be *gauged* away, by choosing $\mathrm{Im}\,a, \xi, b$, appropriately, to cancel the just mentioned field components. This supergauge is referred to as the Wess-Zumino gauge. In this supergauge, the vector superfield $\mathscr{V}$ then reduces to the expression given in (2.6.41) with the vector field components V_μ having the standard gauge transformation of abelian gauge theory in (2.6.102).

As mentioned in the text, the advantage in working in this gauge is that since now $\mathscr{V}$ has no θ-independent part, $\exp(-2e\mathscr{V})$ in (2.6.39) reduces from an infinite series to a *polynomial*. Also the parameter gauge function now takes the simple form given in (2.6.43). Finally we note that χ and $\mathscr{D}$ are invariant.

(Scalar-) Vector Superfields: Non-Abelian Case

To investigate the nature of the right-hand side of (2.6.45), we may invoke the Baker-Campbell-Hausdorff formula which for three matrices A, B and C, with the latter two involving infinitesimal matrix elements, reads

$$e^C e^A e^B = \exp\left((A + (B + C) + \frac{1}{2}[A, B - C] + \frac{1}{12}[A, [A, B + C]]\right.$$
$$\left. + \mathscr{O}\big([A, [A, [A, B]]], [A, [A, [A, C]]]\big)\right),$$
(2.6.105)

where $\mathscr{O}\big([A, [A, [A, B]]], [A, [A, [A, C]]]\big)$ denotes commutators involving *more* than two factors of the matrix A, which will be taken into account below.[35] To this end, we set

$$A = -2g\mathscr{V}, \qquad B = -ig\Lambda, \qquad C = ig\Lambda^\dagger.$$
(2.6.106)

We note, in particular that $\left(A + (B + C)\right)$ may be written as

$$\left(A + (B + C)\right) = -2gt_\alpha\left(\mathscr{V}^\alpha - \frac{i}{2}(\Lambda^{\alpha\dagger} - \Lambda^\alpha)\right),$$
(2.6.107)

(compare with (2.6.40)), and the Wess-Zumino supergauge now consists, from (2.6.40), (2.6.105), (2.6.107), and finally, (2.6.97)–(2.6.101) (see also Problem 2.16), to *choose*

$$\text{Im}\, a_E \quad \text{to cancel} \quad S_E, \tag{2.6.108}$$

$$-\frac{i}{2\sqrt{2}}\gamma^5\xi_E \quad \text{to cancel} \quad \dot{\tilde{\psi}}_E, \tag{2.6.109}$$

$$\text{Im}\, b_E \quad \text{to cancel} \quad B_E, \tag{2.6.110}$$

$$\text{Re}\, b_E \quad \text{to cancel} \quad G_E. \tag{2.6.111}$$

The (scalar-) vector superfield $\mathscr{V}$, then *reduces* explicitly to the structure given in (2.6.41) with $\mathscr{V} = t_E \mathscr{V}_E$ involving at least *two* powers of θ's. That is, all commutators involving more than two factors of $\mathscr{V}$ in the exponential on the right-hand side of (2.6.105) (i.e., in $\mathscr{O}$) vanish. Also from (2.6.106),

$$B + C = ig(\Lambda^\dagger - \Lambda),$$

---

[35]For a proof of the Baker-Campbell-Hausdorff formula and further generalizations, see, e.g., [24, p. 975].

now contains at least two powers of θ's which makes the term $[A, [A, B + C]]$ in (2.6.105) also vanish.

Therefore from (2.6.45), (2.6.105), (2.6.108)–(2.6.111), we have

$$\mathscr{V}' = \left(\mathscr{V} - \frac{i}{2}(\Lambda^{\dagger} - \Lambda)\right) - ig\left[\mathscr{V}, \frac{1}{2}(\Lambda^{\dagger} + \Lambda)\right], \tag{2.6.112}$$

where the second term corresponds to $(1/2)[A, B - C]$ in the exponential on the right-hand side of (2.6.105). In the Wess-Zumino gauge, $\mathscr{V}$ is given in (2.6.41), and the gauge parameter Λ now to be used in this supergauge is given in (2.6.43) with a real. That is, in particular, with

$$a = a_E\, t_E, \qquad [t_D, t_E] = i f_{DEC}\, t_C,$$

we have

$$\frac{1}{2}(\Lambda^{\dagger} + \Lambda) = a - \frac{1}{32}\,(\bar{\theta}\gamma^{\,5}\theta)^2\,\square\, a, \tag{2.6.113}$$

$$\frac{i}{2}(\Lambda^{\dagger} - \Lambda) = \frac{1}{4}\,\bar{\theta}\gamma^{\,5}\gamma^{\mu}\theta\partial_{\mu}a, \tag{2.6.114}$$

and the second term in (2.6.113), does not contribute to the commutator in (2.6.112) since $\mathscr{V}$, in (2.6.41), starts with second order in θ. Thus we obtain the expression in (2.6.46) with infinitesimal $a(x)$. This leads, in the super Wess-Zumino supergauge, the explicit rules spelled out in (2.6.47)–(2.6.49), for infinitesimal $a(x)$.

2.6.4   Pure Vector Superfields

In view of deriving the explicit expression of the pure vector superfield $\mathscr{V}_{\mu}$, in (2.6.57), we first apply the superderivative D_b to $\exp[-2g\mathscr{V}]$ with the latter given in (2.6.58) in the Wess-Zumino supergauge. Using, in the process, the expansion of the product $\theta_a\bar{\theta}_b$ in the Box 2.1, given in the beginning of the section, together with the orthogonality relations between the product of any of two of $\bar{\theta}\gamma^{\,5}\theta, \bar{\theta}\theta, \bar{\theta}\gamma^{\,5}\gamma^{\mu}\theta$, we obtain

$$\frac{1}{g}D_b\, e^{-2g\mathscr{V}} = (\gamma^{\,5}\gamma^{\mu}\theta)_b V_{\mu} - \frac{i}{4}\,\bar{\theta}\gamma^{\,5}\gamma^{\mu}\theta\,(\gamma^{\sigma}\theta)_b\partial_{\sigma}V_{\mu}$$

$$+ \frac{i}{2\sqrt{2}}\{\bar{\theta}\theta\,(\gamma^{\,5}\chi)_b - \bar{\theta}\gamma^{\,5}\theta\,\chi_b + \bar{\theta}\gamma^{\,5}\gamma_{\lambda}\theta\,(\gamma^{\,\lambda}\chi)_b\}$$

$$+ \frac{1}{8\sqrt{2}}\,(\bar{\theta}\gamma^{\,5}\theta)^2\,(\gamma^{\sigma}\gamma^{\,5}\partial_{\sigma}\chi)_b + \frac{1}{2}\,\bar{\theta}\gamma^{\,5}\theta\,(\gamma^{\,5}\theta)_b[\mathscr{D} + gV^{\nu}V_{\nu}]. \tag{2.6.115}$$

We provide enough details of the underlying derivation. Multiplying the latter equation by

$$\left[\mathscr{C} \gamma^{\rho} \frac{(1 - \gamma^5)}{2} \right]_{ab} \exp(2g\,\mathscr{V}),$$

from the left, leads to

$$\frac{1}{g} \left(\mathscr{C} \gamma^{\rho} \frac{1 - \gamma^5}{2} \right)_{ab} e^{2g\mathscr{V}} D_b\, e^{-2g\mathscr{V}} =$$

$$-(\mathscr{C} \gamma^{\rho} \frac{1 - \gamma^5}{2} \gamma^{\mu} \theta)_a V_{\mu} + \frac{i}{4} \bar{\theta} \gamma^5 \theta\, (\mathscr{C} \gamma^{\rho} \frac{1 - \gamma^5}{2} \gamma^{\sigma} \gamma^{\mu} \theta)_a \partial_{\sigma} V_{\mu}$$

$$- \frac{i}{2\sqrt{2}} \left\{ (\bar{\theta}\theta - \bar{\theta}\gamma^5\theta)(\mathscr{C}\gamma^{\rho}\frac{1 - \gamma^5}{2}\chi)_a - \bar{\theta}\gamma^5\gamma_{\lambda}\theta\,(\mathscr{C}\gamma^{\rho}\frac{1 - \gamma^5}{2}\gamma^{\lambda}\chi)_a \right\}$$

$$- \frac{1}{8\sqrt{2}} (\bar{\theta}\gamma^5\theta)^2 (\mathscr{C}\gamma^{\rho}\frac{1 - \gamma^5}{2}\gamma^{\sigma}\gamma^5\partial_{\sigma}\chi)_a - \frac{1}{2}\bar{\theta}\gamma^5\theta\,(\mathscr{C}\gamma^{\rho}\frac{1 - \gamma^5}{2}\theta)_a[\mathscr{D} + gV^{\nu}V_{\nu}]$$

$$- \frac{g}{2}\bar{\theta}\gamma^5\theta\,(\mathscr{C}\gamma^{\rho}\frac{1 - \gamma^5}{2}\gamma^{\sigma}\gamma^{\mu}\theta)_a V_{\mu}V_{\sigma} - \frac{ig}{4\sqrt{2}} (\bar{\theta}\gamma^5\theta)^2 (\mathscr{C}\gamma^{\rho}\frac{1 - \gamma^5}{2}\gamma^{\sigma}[V_{\sigma}\chi - \chi V_{\sigma}])_a.$$

$$(2.6.116)$$

Now we apply $-D_a/2$ to the above equation, and use, in the process, the following properties,

$$\theta_a \mathscr{C}_{ab} = \bar{\theta}_b, \qquad\qquad (\gamma^5\theta)_a \mathscr{C}_{ab} = (\bar{\theta}\gamma^5)_b, \qquad\qquad (2.6.117)$$

$$(\gamma^{\lambda}\theta)_a \mathscr{C}_{ab} = -(\bar{\theta}\gamma^{\lambda})_b, \quad (\gamma^5\gamma^{\lambda}\theta)_a \mathscr{C}_{ab} = -(\bar{\theta}\gamma^{\lambda}\gamma^5)_b, \qquad (2.6.118)$$

to obtain

$$\mathscr{V}^{\rho}(x,\theta) = V^{\rho}(x) + \frac{i}{\sqrt{2}}\bar{\theta}\gamma^{\rho}\chi(x) + \bar{\theta}\gamma^5\gamma_{\lambda}\theta\,A^{\lambda\rho}(x) + \bar{\theta}\gamma^5\theta\,\bar{\theta}\,B^{\rho}(x) - (\bar{\theta}\gamma^5\theta)^2 C^{\rho}(x),$$

$$(2.6.119)$$

where

$$A^{\lambda\rho} = \frac{i}{16} \mathrm{Tr}\left[\left(\gamma^{\sigma}\gamma^{\rho}\gamma^{\mu}\gamma^{\lambda}\frac{1 - \gamma^5}{2} \right) + \left(\gamma^{\rho}\gamma^{\sigma}\gamma^{\mu}\gamma^{\lambda}\frac{1 + \gamma^5}{2} \right) \right]\partial_{\sigma}V_{\mu}$$

$$- \frac{g}{8} \mathrm{Tr}\left[\gamma^{\rho}\gamma^{\sigma}\gamma^{\mu}\gamma^{\lambda}\frac{1 + \gamma^5}{2} \right] V_{\mu}V_{\sigma} + \frac{1}{4}\eta^{\rho\lambda}[\mathscr{D} + gV^{\nu}V_{\nu}], \qquad (2.6.120)$$

$$B^{\rho} = \frac{1}{2\sqrt{2}} (\eta^{\rho\sigma} - \frac{1}{2}\gamma^5\gamma^{\rho}\gamma^{\sigma})\partial_{\sigma}\chi + \frac{ig}{2\sqrt{2}}\gamma^{\rho}\gamma^{\sigma}\frac{1 + \gamma^5}{2}(V_{\sigma}\chi - \chi V_{\sigma}),$$

$$(2.6.121)$$

$$C^\rho = -\frac{i}{16}\,\partial^\rho[\mathscr{D} + gV^\nu V_\nu] + \frac{ig}{32}\,\mathrm{Tr}\left[\gamma^\lambda\gamma^\rho\gamma^\sigma\gamma^\mu\frac{1-\gamma^5}{2}\right]\partial_\lambda(V_\mu V_\sigma)$$

$$+\frac{1}{64}\,\mathrm{Tr}\left[\gamma^\lambda\gamma^\rho\gamma^\sigma\gamma^\mu\frac{1-\gamma^5}{2}\right]\partial_\lambda\partial_\sigma V_\mu. \tag{2.6.122}$$

The identities

$$\mathrm{Tr}[\gamma^\sigma\gamma^\rho\gamma^\mu\gamma^\lambda] = 4(\eta^{\sigma\rho}\eta^{\mu\lambda} - \eta^{\sigma\mu}\eta^{\rho\lambda} + \eta^{\sigma\lambda}\eta^{\rho\mu}),$$

$$\mathrm{Tr}[\gamma^\sigma\gamma^\rho\gamma^\mu\gamma^\lambda\gamma^5] = -i4\,\varepsilon^{\sigma\rho\mu\lambda}, \tag{2.6.123}$$

and $\varepsilon^{\lambda\rho\sigma\mu}\partial_\lambda\partial_\sigma V_\mu = 0$, lead to the following expressions for $A^{\lambda\rho}$, and C^ρ,

$$A^{\lambda\rho}(x) = \frac{i}{4}\,\partial^\lambda V^\rho(x) - \frac{i}{4}\,G^{\lambda\rho}(x) - \frac{1}{8}\,\varepsilon^{\rho\sigma\mu\lambda}\,G_{\sigma\mu}(x) + \frac{1}{4}\,\eta^{\lambda\rho}\mathscr{D}(x), \tag{2.6.124}$$

$$G_{\sigma\mu}(x) = \partial_\sigma V_\mu(x) - \partial_\mu V_\sigma(x) - i\,g\,[V_\sigma(x), V_\mu(x)], \tag{2.6.125}$$

$$C^\rho(x) = -\frac{i}{4}\,\partial_\lambda A^{\lambda\rho}(x) - \frac{1}{32}\,\Box\, V^\rho(x). \tag{2.6.126}$$

These equations then immediately lead to the expression for $\mathscr{V}^\rho$ in (2.6.61) for the non-abelian case, expressed completely in terms of $A^{\lambda\rho}$, defined in (2.6.124), and B^ρ, defined in (2.6.121), and in (2.6.59), (2.6.60) for the abelian case.

In (2.6.66), $\mathscr{V}_\mu$ is also expressed as a function of $\hat{x}^\mu$, with the latter defined in (2.6.26).

2.6.5 Spinor Superfields

Under the supersymmetric gauge transformation (2.6.3), the superfield $\mathscr{W}_a$ in (2.6.67) transforms as [see also (2.6.45)]

$$\mathscr{W}_a \to \frac{1}{2\sqrt{2}\,ig}\,\mathscr{C}_{bc}D_b^R D_c^R\,e^{ig\Lambda}\,e^{2g\mathscr{V}}e^{-ig\Lambda^\dagger}D_a^L\,e^{ig\Lambda^\dagger}e^{-2g\mathscr{V}}\,e^{-ig\Lambda}, \tag{2.6.127}$$

with Λ, defined in (2.6.97), being a left-chiral superfield means that $D^R e^{ig\Lambda} = e^{ig\Lambda}D^R$, and $D^L e^{ig\Lambda^\dagger} = e^{ig\Lambda^\dagger}D^L$. Hence we may rewrite the above as

$$\mathscr{W}_a \to \frac{1}{2\sqrt{2}\,ig}\,e^{ig\Lambda}\left(\mathscr{C}_{bc}\,D_b^R D_c^R\,e^{2g\mathscr{V}}\,D_a^L e^{-2g\mathscr{V}}\right)e^{-ig\Lambda}$$

$$+\frac{1}{2\sqrt{2}\,ig}\,e^{ig\Lambda}\left(\mathscr{C}_{bc}\,D_b^R D_c^R\,D_a^L\right)e^{-ig\Lambda}. \tag{2.6.128}$$

The expression within the second set of round brackets $\mathscr{C}_{bc} D_b^R D_c^R D_a^L$ may be rewritten as $-\overline{D}D^R D_a^L$. According to Problem 2.14,

$$\overline{D}D^R D_a^L = -D_a^L \overline{D}D^R + 2\mathrm{i}\,(\gamma^\mu D_a^R)\partial_\mu,$$

which is expressed in terms of the right-hand superderivative D^R on its extreme right-hand side. Since nothing stands to the right of $\mathrm{e}^{-\mathrm{i}g\Lambda}$ in the above equation, the second term is zero in (2.6.128). This establishes the transformation rule of $\mathscr{W}_a$ in (2.6.68).

The multiplicative operator factor to $\mathrm{e}^{2g\mathscr{V}} D_a^L \mathrm{e}^{-2g\mathscr{V}}$ in (2.6.67), may be rewritten simply as

$$\frac{1}{2\sqrt{2}\,\mathrm{i}\,g}\,\mathscr{C}_{cd}D_c\left(\frac{1+\gamma^5}{2}D\right)_d. \tag{2.6.129}$$

A direct application of this operator to the explicit expression already obtained in (2.6.116), not including the overall factor $\mathscr{C}\gamma^\rho(1-\gamma^5)/2g$, now yields the expression for $\mathscr{W}_{Aa}$ in (2.6.69), where the index A, as a capital letter, corresponds to the group generators t_A labels, a (standard) notation used before in Sect. 2.6.3, while the lower case index a denotes a spinor index. The abelian counterpart is a special case of the structure given above.

2.7 Supersymmetric Maxwell-Field Theory

In reference to the spinor superfield $\mathscr{W}_a$, in the Wess-Zumino supergauge, associated with an abelian gauge field, we may write from (2.6.69), (2.6.70),

$$\mathscr{W}_a = \exp\left[\frac{\mathrm{i}}{4}\overline{\theta}\gamma^5\gamma^\mu\theta\partial_\mu\right]\left(\chi_a + \frac{1}{\sqrt{2}}(\gamma^\mu\gamma^\nu\theta_L)_a F_{\mu\nu} - \mathrm{i}(\gamma\partial\chi)_a\overline{\theta}\theta_L - \mathrm{i}\sqrt{2}\,\mathscr{D}(\theta_L)_a\right), \tag{2.7.1}$$

$$F_{\mu\nu} = \partial_\mu V_\nu - \partial_\nu V_\mu. \tag{2.7.2}$$

We consider the $\mathscr{F}$ term of the expression $(1/8)(\mathscr{C}\mathscr{W})_a\mathscr{W}_a$, and, in turn, carry out the integration

$$(1/8)\int(\mathrm{d}\theta)\,\delta^{(2)}(\theta_R)(\mathscr{C}\mathscr{W})_a\mathscr{W}_a.$$

To this end, note that since $\overline{\theta}\gamma^\mu\theta = 0$, we may write

$$\overline{\theta}\gamma^5\gamma^\mu\theta = -\overline{\theta}\gamma^\mu\gamma^5\theta - \overline{\theta}\gamma^\mu\theta = -2\overline{\theta}\gamma^\mu\theta_R, \tag{2.7.3}$$

and the contribution of the translational operator in (2.7.1) may be neglected in finally carrying out the above mentioned integral.

We note, in particular, that

$$\mathscr{C}_{ab}\left[\chi_b + \frac{1}{\sqrt{2}}(\gamma^\mu\gamma^\nu\theta_L)_b\, F_{\mu\nu} - i(\gamma\partial\chi)_b\overline{\theta}\theta_L - i(\theta_L)_b\,\mathscr{D}\right]$$

$$= -\overline{\chi}_a - \frac{1}{\sqrt{2}}\left(\overline{\theta}\gamma^\nu\gamma^\mu\frac{1-\gamma^5}{2}\right)_a F_{\mu\nu} - i\partial_\mu(\overline{\chi}\,\gamma^\mu)_a\overline{\theta}\theta_L + i\sqrt{2}\,\mathscr{D}\left(\overline{\theta}\frac{1-\gamma^5}{2}\right)_a.$$

$$(2.7.4)$$

From this equation and (2.7.1), we obtain

$$\frac{1}{8}(\mathscr{C}\mathscr{W})_a\mathscr{W}_a\bigg|_{\mathscr{F}}(x,\theta) = \frac{i}{8}\left(\overline{\chi}\gamma^\mu\partial_\mu\chi - (\partial_\mu\overline{\chi}\gamma^\mu)\chi\right)\overline{\theta}\theta_L + \frac{1}{4}\mathscr{D}^2\overline{\theta}\theta_L$$

$$- \frac{1}{16}\left(\overline{\theta}\gamma^\nu\gamma^\mu\gamma^\alpha\gamma^\beta\frac{1-\gamma^5}{2}\theta\right)F_{\mu\nu}F_{\alpha\beta}.$$

$$(2.7.5)$$

$$\left(\overline{\theta}\gamma^\nu\gamma^\mu\gamma^\alpha\gamma^\beta\frac{1-\gamma^5}{2}\theta\right) = \left(\gamma^\nu\gamma^\mu\gamma^\alpha\gamma^\beta\frac{1-\gamma^5}{2}\right)_{ba}\overline{\theta}_b\theta_a$$

$$= \frac{1}{4}\left(\gamma^\nu\gamma^\mu\gamma^\alpha\gamma^\beta\frac{1-\gamma^5}{2}\right)_{ba}\left[\delta_{ab}\overline{\theta}\theta + \gamma^5_{ab}\overline{\theta}\gamma^5\theta + (\gamma^5\gamma_\mu)_{ab}\overline{\theta}\gamma^5\gamma^\mu\theta\right]$$

$$= \frac{1}{4}\text{Tr}\left[\gamma^\nu\gamma^\mu\gamma^\alpha\gamma^\beta(1-\gamma^5)\right]\frac{1}{2}[\overline{\theta}\theta - \overline{\theta}\gamma^5\theta]$$

$$= \left(\eta^\mu\eta^\beta - \eta^\alpha\eta^\mu\beta + \eta^\beta\eta^\mu\alpha + i\varepsilon^{\nu\mu\alpha\beta}\right)\overline{\theta}\theta_L,$$

$$(2.7.6)$$

where we have used the symmetry property of γ^5, $(\gamma^5)^2 = 1$, the trace of an odd number of gamma matrices is zero, and the first entry in the table in the beginning of the last section to carry out the indicated expansion of $\overline{\theta}_b\theta_a$.

In Problem 2.18, it is shown that $\varepsilon^{\nu\mu\alpha\beta}F_{\mu\nu}F_{\alpha\beta}$ is a total derivative. Accordingly using the anti-symmetry property of $F_{\mu\nu}$, the fact that $(\partial_\mu\overline{\chi})\gamma^\mu\chi = -\overline{\chi}\gamma^\mu\partial_\mu\chi$, the Lagrangian density of the supersymmetric Maxwell field in the Wess-Zumino supergauge emerges to have the structure

$$\mathscr{L}(x) = -\frac{1}{4}F^{\mu\nu}(x)\,F_{\mu\nu}(x) - \overline{\chi}(x)\frac{\gamma\partial}{2i}\,\chi(x) + \frac{1}{2}\mathscr{D}^2(x),$$

$$(2.7.7)$$

to be integrated over Minkowski spacetime, where the particle associated with the field χ is referred to as the photino. The presence of the $1/2$ factor in the second term in the Lagrangian density should not be surprising, as for a Majorana spinor, this term may be rewritten as $[-(1/2i)\mathscr{C}^{-1}_{ba}\chi_a(\gamma\partial)_{bc}\chi_c]$, where we have used the property $\mathscr{C}^{-1}_{ab} = -\mathscr{C}^{-1}_{ba}$, and the spinor field χ appears twice in the same way as in the kinetic term of the Lagrangian density of a real scalar field is $[-(1/2)\partial_\mu\phi\partial^\mu\phi]$ involving a factor of $1/2$. See also Problem 2.19.

2.8 Supersymmetric Yang and Mills-Field Theory

In reference to the spinor superfield $\mathscr{W}_a$, in the Wess-Zumino supergauge, associated with a non-abelian gauge field, we have from (2.6.69), (2.6.70),

$$\mathscr{W}_{Aa} = \exp\left[\frac{i}{4}\overline{\theta}\gamma^5\gamma^\mu\theta\partial_\mu\right]\left(\chi_{Aa}+\frac{1}{\sqrt{2}}(\gamma^\mu\gamma^\nu\theta_L)_a\,G_{A\mu\nu}-i(\gamma\nabla\chi)_{Aa}\overline{\theta}\theta_L-i\sqrt{2}\,\mathscr{D}_A(\theta_L)_a\right), \tag{2.8.1}$$

$$G_{A\mu\nu} = \partial_\mu V_{A\nu}-\partial_\nu V_{A\mu}+gf_{ABC}\,V_{B\mu}V_{C\nu}, \quad (\gamma\nabla\chi)_{Aa}=\left(\gamma^\mu(\delta_{AC}\partial_\mu+gf_{ABC}V_{B\mu})\chi_C\right)_a. \tag{2.8.2}$$

The analysis carried out for the supersymmetric version of Maxwell theory may be now taken over with almost no labor with $F^{\mu\nu}$, in particular, replaced by $G_A^{\mu\nu}$, and $(\gamma\partial\chi)_a$ by $(\gamma\Delta\chi)_{Aa}$. The Langrangian density of the supersymmetric Yang-Mills field, in the Wess-Zumino supergauge, follows to be

$$\mathscr{L} = -\frac{1}{4}\,G_E^{\mu\nu}G_{E\mu\nu} - \frac{1}{2i}\,\overline{\chi}_E\gamma^\nu[\delta_{EC}\partial_\nu + gf_{EAC}V_{A\nu}]\chi_C + \frac{1}{2}\mathscr{D}_E\mathscr{D}_E, \tag{2.8.3}$$

to be integrated over Minkowski spacetime, with sums over repeated indices understood. Note also from Problem 2.18, that $\varepsilon^{\nu\mu\alpha\beta}G_{\mu\nu}G_{\alpha\beta}$ is a total derivative.[36]

2.9 Spin 0: Spin 1/2 Supersymmetric Interacting Theories

Consider the following left-chiral superfield $\Phi(x,\theta)$, as given in (2.6.23), and its adjoint:

$$\Phi_L(x,\theta) = A(x) + i\overline{\theta}\psi_L(x) + \frac{1}{2}\overline{\theta}\theta_L\,\mathscr{F}(x) + \frac{i}{4}\overline{\theta}\gamma^5\gamma^\mu\theta\,\partial_\mu A(x)$$

$$- \frac{1}{4}\overline{\theta}\gamma^5\theta\,\overline{\theta}\gamma\partial\psi_L(x) - \frac{1}{32}(\overline{\theta}\gamma^5\theta)^2\square A(x), \tag{2.9.1}$$

---

[36] Although such a term is a total divergence and does not contribute to equations of motion, it is not completely void of applications. It is referred to as the θ term in the literature. We will not, however, go into this here.

$$\Phi_L^\dagger(x,\theta) = A^\dagger(x) - i\,\overline{\theta}\,\psi_R(x) + \frac{1}{2}\overline{\theta}\theta_R\,\mathscr{F}^\dagger(x) - \frac{i}{4}\overline{\theta}\gamma\,^5\gamma^\mu\theta\,\partial_\mu A^\dagger(x)$$

$$-\frac{1}{4}\overline{\theta}\gamma\,^5\theta\,\overline{\theta}\gamma\partial\psi_R(x) - \frac{1}{32}(\overline{\theta}\gamma\,^5\theta)^2\Box A^\dagger(x). \tag{2.9.2}$$

It is easily verified that

$$\Phi_L^\dagger(x,\theta)\Phi_L(x,\theta)\Big|_{\mathscr{D}} = -\frac{1}{8}(\overline{\theta}\gamma\,^5\theta)^2\Big[-\partial_\mu A^\dagger\partial^\mu A + \mathscr{F}^\dagger\mathscr{F} - \overline{\psi}\frac{\gamma^\mu\partial_\mu\psi_L}{2i} + \frac{\partial_\mu\overline{\psi}\gamma^\mu}{2i}\psi_R\Big], \tag{2.9.3}$$

where in view of defining an action integral, we have effectively replaced an expression like $A^\dagger\Box A$ by $-\partial_\mu A^\dagger\partial^\mu A$. The last term in (2.9.3) may be rewritten as $-\overline{\psi}\,\gamma\partial\psi_R/2i$ which may be combined[37] with the third term, thus introducing the simple free Lagrangian density

$$\mathscr{L}_0 = -\partial_\mu A^\dagger\partial^\mu A + \mathscr{F}^\dagger\mathscr{F} - \overline{\psi}\frac{\gamma\partial}{2i}\psi. \tag{2.9.4}$$

Interaction may be introduced in a simple way. To this end, consider the $\mathscr{F}$ terms of the following readily derived products

$$\Phi_L(x,\theta)\Phi_L(x,\theta)\Big|_{\mathscr{F}} = \Big(A\mathscr{F} + \frac{1}{2}\overline{\psi}\psi_L\Big)\overline{\theta}\theta_L, \tag{2.9.5}$$

$$\Phi_L(x,\theta)\Phi_L(x,\theta)\Phi_L(x,\theta)\Big|_{\mathscr{F}} = \frac{3}{2}\Big(AA\mathscr{F} + A\overline{\psi}\psi_L\Big)\overline{\theta}\theta_L. \tag{2.9.6}$$

We consider the combination

$$W[\Phi] = \frac{m}{2}\,\Phi_L(x,\theta)\Phi_L(x,\theta) + \frac{2}{3}\lambda\,\Phi_L(x,\theta)\Phi_L(x,\theta)\Phi_L(x,\theta), \tag{2.9.7}$$

where m and λ are real, $m > 0$, referred to as a superpotential, and consider the following contribution to the action integral

$$\int(dx)(d\theta)\big[-\delta^{(2)}(\theta_R)W[\Phi] + h.c\big], \tag{2.9.8}$$

thus generating the Lagrangian density

$$\mathscr{L} = \mathscr{L}_0 - \Big(m\,(A\mathscr{F} + \frac{1}{2}\overline{\psi}\psi_L) + 2\lambda\,(AA\mathscr{F} + A\overline{\psi}\psi_L) + h.c\Big). \tag{2.9.9}$$

---

[37]Recall that $\psi_R + \psi_L = \psi$.

The field equations of $\mathscr{F}$ and $\mathscr{F}^\dagger$ are readily obtained. They simply lead to the constraints:

$$\mathscr{F} = mA^\dagger + 2\lambda A^\dagger A^\dagger, \quad \mathscr{F}^\dagger = mA + 2\lambda AA. \tag{2.9.10}$$

That is, they do not propagate. They are not, however, void of physical relevance as we will see in the next section. By eliminating these auxiliary fields, as they are called, the Lagrangian density in (2.9.9) takes the form

$$\mathscr{L} = -\partial_\mu A^\dagger \partial^\mu A - \overline{\psi}\frac{\gamma\partial}{2i}\psi - \frac{m}{2}\overline{\psi}\psi - |mA + 2\lambda AA|^2 - 2\lambda A\overline{\psi}\psi_L - 2\lambda A^\dagger\overline{\psi}\psi_R, \tag{2.9.11}$$

which may also be rewritten in the more convenient form

$$\mathscr{L} = -\partial_\mu A^\dagger \partial^\mu A - \overline{\psi}\frac{\gamma\partial}{2i}\psi - |mA + 2\lambda AA|^2 - \frac{(m + 4\lambda A)}{2}\overline{\psi}\psi_L - \frac{(m + 4\lambda A^\dagger)}{2}\overline{\psi}\psi_R. \tag{2.9.12}$$

Finally, we may also express the field A in terms of a real and imaginary part

$$A = \frac{1}{\sqrt{2}}(\varphi_1 + i\varphi_2), \tag{2.9.13}$$

from which the following expression for the Lagrangian density emerges

$$\mathscr{L} = -\frac{1}{2}\partial_\mu\varphi_1\partial^\mu\varphi_1 - \frac{1}{2}\partial_\mu\varphi_2\partial^\mu\varphi_2 - \overline{\psi}\frac{\gamma\partial}{2i}\psi - \frac{m}{2}\overline{\psi}\psi - \frac{m^2}{2}(\varphi_1^2 + \varphi_2^2)$$
$$- \lambda^2(\varphi_1^2 + \varphi_2^2)^2 - \sqrt{2}\,\lambda m\varphi_1(\varphi_1^2 + \varphi_2^2) - \sqrt{2}\,\lambda\,\varphi_1\overline{\psi}\psi - i\sqrt{2}\,\lambda\,\varphi_2\overline{\psi}\gamma^5\psi. \tag{2.9.14}$$

It is referred to as a Wess and Zumino [42] Lagrangian density involving an interaction.

2.10   Supersymmetry and Improvement of the Divergence Problem

Supersymmetry improves significantly the divergence problem in quantum field theory. We illustrate this by computing the one point function $\langle 0|\varphi_1(x)|0\rangle$ and the two-point function $\langle 0|\big(\varphi_1(x')\varphi_1(x)\big)_+|0\rangle$ of the scalar field $\varphi_1(x)$ for the Wess-

Zumino Lagrangian density with interaction given in $(2.9.14)$[38] to second order in λ. The Lagrangian density in question, with the fields coupled to external sources K_1, K_2, may be written in detail as

$$\mathcal{L} = -\frac{1}{2}\partial_\mu \varphi_1 \partial^\mu \varphi_1 - \frac{1}{2}\partial_\mu \varphi_2 \partial^\mu \varphi_2 - \overline{\psi}\frac{\gamma \partial}{2i}\psi - \frac{m}{2}\overline{\psi}\psi - \frac{m^2}{2}(\varphi_1^2 + \varphi_2^2)$$
$$- \lambda^2 (\varphi_1^2 + \varphi_2^2)^2 - \sqrt{2}\,\lambda m \varphi_1 (\varphi_1^2 + \varphi_2^2) - \sqrt{2}\,\lambda\,\varphi_1\overline{\psi}\psi$$
$$- i\sqrt{2}\,\lambda\,\varphi_2\overline{\psi}\gamma^5\psi + K_1\varphi_1 + K_2\varphi_2 + \overline{\eta}\,\psi. \qquad (2.10.1)$$

In particular, the following vacuum expectation values ($\langle 0_+| . |0_-\rangle \equiv \langle\,.\,\rangle$), in the presence of external sources, are readily obtained:

$$(-\Box + m^2)\langle \varphi_1(x)\rangle = -4\lambda^2\langle \varphi_1^3(x)\rangle - 4\lambda^2\langle \varphi_1(x)\varphi_2(x)^2\rangle - 3\sqrt{2}\,\lambda m \langle \varphi_1^2(x)\rangle$$
$$- \sqrt{2}\,\lambda m \langle \varphi_2^2(x)\rangle - \sqrt{2}\,\lambda\langle \overline{\psi}(x)\psi(x)\rangle + K_1(x)\langle 0_+|0_-\rangle, \qquad (2.10.2)$$

$$(-\Box + m^2)\langle \big(\varphi_1(x')\varphi_1(x)\big)_+\rangle = -4\lambda^2\langle \big(\varphi_1(x')\varphi_1^3(x)\big)_+\rangle$$
$$- 4\lambda^2\langle \big(\varphi_1(x')\varphi_1(x)\varphi_2(x)^2\big)_+\rangle - 3\sqrt{2}\,\lambda m\langle \big(\varphi_1(x')\varphi_1^2(x)\big)_+\rangle$$
$$- \sqrt{2}\,\lambda m\langle \big(\varphi_1(x')\varphi_2^2(x)\big)_+\rangle - \sqrt{2}\,\lambda\langle \big(\varphi_1(x')\overline{\psi}(x)\psi(x)\big)+\rangle$$
$$- i\,\delta^{(4)}(x - x')\langle 0_+|0_-\rangle + K_1(x)\langle 0_+|\varphi_1(x')0_-\rangle. \qquad (2.10.3)$$

$$\langle \psi(x)\rangle = \Big(\frac{i\gamma\partial + m}{-\Box + m^2}\Big)\Big\{-2\sqrt{2}\lambda\langle \varphi_1(x)\psi(x)\rangle$$
$$-2\sqrt{2}\lambda\,i\langle \varphi_2(x)(\gamma^5\psi(x))\rangle + (\mathscr{C}\overline{\eta}^{\mathsf{T}})\langle 0_+|0_-\rangle\Big\}. \qquad (2.10.4)$$

Equation $(2.10.3)$ is obtained by functional differentiation $(-i)\delta/\delta K_1(x')$ of $(2.10.2)$. For a derivation of $(2.10.4)$ see Problem 2.19. In the absence of interaction, we recognize $(2.10.4)$ as the equation of the vacuum expectation value of a spinor field in the external source $\eta = \mathscr{C}\overline{\eta}^{\mathsf{T}}$.

Using the notation $\big|_j$ to represent jth order in λ, in the absence of sources, and

$$\Delta_+(x) = \int \frac{(dk)}{(2\pi)^4}\frac{e^{ik\cdot x}}{k^2 + m^2}, \qquad (2.10.5)$$

---

[38]$(.)_+$ denotes the time ordered product: $\big(\varphi_1(x')\varphi_1(x)\big)_+ = \varphi_1(x')\varphi_1(x)\,\Theta(x'^0 - x^0) + \varphi_1(x)\varphi_1(x')\,\Theta(x^0 - x'^0)$.

where $\Delta_+(0)$ is quadratically divergent, and[39]

$$\langle 0|(\phi_1^2(x))_+|0\rangle\big|_0 = \langle 0|(\phi_2^2(x))_+|0\rangle\big|_0 = -i\,\Delta_+(0), \tag{2.10.6}$$

$$\langle 0|(\overline{\psi}_a(x)\psi_a(x))_+|0\rangle\big|_0 = 4m\,i\,\Delta_+(0). \tag{2.10.7}$$

From (2.10.2), the above equations show that the quadratic divergence cancels out in the one-point function to first order:

$$\langle 0|(\phi_1(x))|0\rangle\big|_1 = \left(3\,i\sqrt{2}\,\lambda + i\sqrt{2}\,\lambda - 4i\sqrt{2}\,\lambda\right)\frac{\Delta_+(0)}{m^2} = 0. \tag{2.10.8}$$

From (2.10.3), and similarly for φ_2,

$$(-\Box + m^2)\langle\,(\varphi_j(x')\varphi_j(x))_+\rangle\big|_1 = 0, \qquad j = 1, 2. \tag{2.10.9}$$

Also $\langle 0|\overline{\psi}(x)\psi(x)|0\rangle\big|_1 = 0$, since the latter is linear in the scalar fields. Finally, we obviously have

$$\langle 0|(\phi_1^3(x))_+|0\rangle\big|_0 = 0, \qquad \langle 0|(\phi_1(x)\phi_2^2(x))_+|0\rangle\big|_0 = 0,$$

and we see from (2.10.2) that all the divergences cancel out in the one-point function to second order as well.

To investigate the nature of divergence of the two-point function, we proceed as follows. Upon taking the functional derivative $(-i)\delta/\delta K_1(x'')$ of (2.10.3), we have

$$(-\Box + m^2)\langle\,(\varphi_1(x'')\varphi_1(x')\varphi_1(x))_+\rangle\big|_1 = -(3\sqrt{2}\,\lambda m)\langle\,(\varphi_1(x'')\varphi_1(x')\varphi_1^2(x))_+\rangle\big|_0$$

$$-(\sqrt{2}\,\lambda m)\langle\,(\varphi_1(x'')\varphi_1(x')\varphi_2^2(x))_+\rangle\big|_0 - (\sqrt{2}\,\lambda)\langle\,(\varphi_1(x'')\varphi_1(x')\overline{\psi}(x)\psi(x))_+\rangle\big|_0. \tag{2.10.10}$$

This gives

$$\langle\,(\varphi_1(x'')\varphi_1(x')\varphi_1(x))_+\rangle\big|_1 = +(3\sqrt{2}\lambda m)\left\{\frac{\Delta(x''-x')\,\Delta_+(0)}{m^2} + 2\frac{1}{-\Box + m^2}\right.$$

$$\times\Delta(x-x'')\Delta(x-x')\bigg\} + (\sqrt{2}\lambda m)\frac{\Delta(x''-x')\Delta_+(0)}{m^2} - (4\sqrt{2}\lambda m)\frac{\Delta(x''-x')\Delta_+(0)}{m^2}. \tag{2.10.11}$$

---

[39]In the absence of external sources we use the notation $\langle 0|\,.\,|0\rangle$ for $\langle 0_+|\,.\,|0_-\rangle$.

Hence upon using the useful equations

$$\frac{1}{-\Box+m^2}\,\delta^{(4)}(z-x) = \Delta_+(z-x), \tag{2.10.12}$$

$$\Delta_+(x-x'')\,\Delta_+(x-x') = \int(dz)\,\delta^{(4)}(z-x)\,\Delta_+(z-x'')\,\Delta_+(z-x'), \tag{2.10.13}$$

we obtain

$$\langle\,(\varphi_1(x')\varphi_1^2(x))_+\rangle\big|_1 = 0 + 6\sqrt{2}\,\lambda m\int(dz)\,\Delta_+(z-x)\,\Delta_+(z-x)\,\Delta_+(z-x'). \tag{2.10.14}$$

This will be used for the third term on the right-hand of (2.10.3) of the two-point function equation and is to be multiplied by $-3\sqrt{2}\,\lambda m$. Similarly taking the functional derivative, this time of (2.10.2) with respect to K_2 twice, we readily obtain

$$\langle\,(\varphi_1(x')\varphi_2^2(x))_+\rangle = 2\sqrt{2}\,\lambda m\int(dz)\,\Delta_+(z-x)\,\Delta_+(z-x)\,\Delta_+(z-x'), \tag{2.10.15}$$

corresponding to the fourth term on the right-hand side of (2.10.3) and is to be multiplied by $(-\sqrt{2}\,\lambda m)$. Finally in Problem 2.20 it is shown that

$$\langle\,(\varphi_1(x')\overline{\psi}(x)\psi(x))_+\rangle = 8\sqrt{2}\,\lambda\,\Delta_+(x-x')\,\Delta_+(0) - 4\sqrt{2}\,\lambda\Delta(x-x')\,\Delta_+(x-x')$$

$$- 12\sqrt{2}\,m^2\lambda\int(dz)\,\Delta_+(z-x)\,\Delta_+(z-x)\,\Delta_+(z-x'), \tag{2.10.16}$$

corresponding to the fifth term on the right-hand side of (2.10.3) and is to be multiplied by $(-\sqrt{2}\,\lambda)$. Finally the first two terms on the right-hand side of (2.10.3), to second order, are directly obtained to be given, respectively, by

$$12\lambda^2\Delta_+(x-x')\,\Delta_+(0), \qquad 4\lambda^2\Delta_+(x-x')\,\Delta_+(0). \tag{2.10.17}$$

All told, (2.10.3) leads from (2.10.14)–(2.10.17) to

$$\langle\,(\varphi_1(x')\varphi_1(x))_+\rangle\big|_2 = [\,12+4-16\,]\,\lambda^2\Delta_+(0)\int(dz)\,\Delta_+(z-x)\,\Delta_+(z-x')$$

$$+ [\,-36-4+24\,]\,\lambda^2 m^2\int(dz_1)(dz_2)\,\Delta_+(z_2-x)\,\Delta_+(z_1-z_2)\,\Delta_+(z_1-z_2)\,\Delta_+(z_1-x')$$

$$+ 8\lambda^2 m^2\int(dz)\,\Delta_+(z-x)\,\Delta_+(z-x')\,\Delta_+(z-x'). \tag{2.10.18}$$

The quadratic divergence, involving the factor $\Delta_+(0)$, cancels out exactly. On the other hand, the remaining integrals are at most logarithmically divergent.

2.11 Spontaneous Symmetry Breaking

To investigate the concept of spontaneous supersymmetry breaking,[40] it is easiest to consider the Wess-Zumino Lagrangian density obtained in Sect. 2.9. To this end, define the superpotential ($m > 0$)

$$W[A] = \frac{m}{2}A^2 + \frac{2}{3}\lambda A^3,$$ (2.11.1)

which leads to the expressions

$$W'[A] \equiv \frac{\partial W[A]}{\partial A} = (mA + 2\lambda A^2), \qquad W''[A] = (m + 4\lambda A),$$ (2.11.2)

and set

$$\mathscr{F} = mA^\dagger + 2\lambda A^\dagger A^\dagger, \quad \mathscr{F}^\dagger = mA + 2\lambda AA.$$ (2.11.3)

Upon comparison with (2.9.12), we may then write

$$\mathscr{L} = -\partial_\mu A^\dagger \partial^\mu A - \overline{\psi}\frac{\gamma\partial}{2i}\psi - \frac{1}{2}W''[A]\overline{\psi}\psi_L - \frac{1}{2}W''^\dagger[A]\overline{\psi}\psi_R - V[A],$$ (2.11.4)

$$V[A] \equiv \left|W'[A]\right|^2 = \mathscr{F}^\dagger\mathscr{F} = |mA + 2\lambda A^2|^2 \geq 0.$$ (2.11.5)

For $A = 0$, we have $V = 0$ which provides a minimum, since $W''[0] = m > 0$, which is a condition that supersymmetry is not spontaneously broken, at least in the tree approximation, since V is non-negative. This minimum corresponds to $\mathscr{F} = 0$.

Given left-chiral superfields $\Phi_i(x, \theta)$, $i = 1, \ldots, n$ [see (2.6.23)], with theta independent parts $A_i(x)$:

$$\Phi_i(x, \theta) = A_i(x) + i\overline{\theta}\psi_{iL}(x) + \frac{1}{2}\overline{\theta}\theta_L \mathscr{F}_i(x) + \frac{i}{4}\overline{\theta}\gamma^5\gamma^\mu\theta\,\partial_\mu A_i(x)$$

$$- \frac{1}{4}\overline{\theta}\gamma^5\theta\,\overline{\theta}\gamma\partial\psi_{iL}(x) - \frac{1}{32}(\overline{\theta}\gamma^5\theta)^2\square A_i(x).$$ (2.11.6)

---

[40]Here the reader is advised to review the content of the introduction to this chapter.

suppressing the label L for left-handed, we note that $(\Phi_i - A_i)(\Phi_j - A_j)(\Phi_k - A_k)$, and higher order products, involve no $\mathscr{F}$ terms. On the other-hand

$$(\Phi_i - A_i)\Big|_{\mathscr{F}} = \frac{1}{2}\overline{\theta}\theta_{\mathrm{L}}\,\mathscr{F}_i, \qquad (2.11.7)$$

$$(\Phi_i - A_i)(\Phi_j - A_j)\Big|_{\mathscr{F}} = -\overline{\theta}\psi_{i\mathrm{L}}\,\overline{\theta}\psi_{j\mathrm{L}} = \frac{1}{2}\overline{\theta}\theta_{\mathrm{L}}\,\overline{\psi}_i\psi_{j\mathrm{L}}. \qquad (2.11.8)$$

Accordingly one may expand a given function $W[\Phi_1,\dots,\Phi_n] \equiv W[\phi]$ of the left-chiral superfields about $(A_1,\dots,A_n)$, and its $\mathscr{F}$ term terminates after two differentiations, giving

$$W[\Phi]\Big|_{\mathscr{F}} = \left(\sum_i W_i[A]\,\mathscr{F}_i + \frac{1}{2}\sum_{ij} W_{ij}[A]\,\overline{\psi}_i\psi_{j\mathrm{L}}\right)\frac{1}{2}\overline{\theta}\theta_{\mathrm{L}}, \qquad (2.11.9)$$

$$W_i \equiv \frac{\partial W[A]}{\partial A_i}, \qquad W_{ij} \equiv \frac{\partial^2 W[A]}{\partial A_i \partial A_j}. \qquad (2.11.10)$$

We may then introduce a Lagrangian density involving the n superfields given by [see (2.11.4)]

$$\mathscr{L} = -\partial_\mu A_i^\dagger \partial^\mu A_i + \mathscr{F}_i^\dagger \mathscr{F}_i - \overline{\psi}_i \frac{\gamma\partial}{2\mathrm{i}}\psi_i$$

$$- \left(W_i[A]\,\mathscr{F}_i + \frac{1}{2}W_{ij}[A]\,\overline{\psi}_i\psi_{j\mathrm{L}} + h.c.\right). \qquad (2.11.11)$$

with a summation over repeated indices understood. The auxiliary fields then satisfy the equations

$$\mathscr{F}_i = W_i[A]^\dagger, \quad \mathscr{F}_i^\dagger = W_i[A]. \qquad (2.11.12)$$

From (2.6.20)–(2.6.22), only the $\delta\psi_{i\mathrm{L}}$ in a chiral superfield in (2.11.6), for a given i, may (or may not) develop a non-vanishing vacuum expectation values:[41]

$$\langle\mathrm{vac}|\delta\psi_{i\mathrm{L}}|\mathrm{vac}\rangle = (-\mathrm{i})\,\langle\mathrm{vac}|\mathscr{F}_i|\mathrm{vac}\rangle\epsilon_{\mathrm{L}}, \qquad (2.11.13)$$

where $\mathscr{F}_i$ is a scalar field. Since $\delta\psi_{i\mathrm{L}} = [\psi_{i\mathrm{L}}, \delta\overline{\epsilon}Q]/\mathrm{i}$, we have

$$\mathrm{i}\,\langle\mathrm{vac}|\delta\psi_{i\mathrm{L}}|\mathrm{vac}\rangle = \langle\mathrm{vac}|\psi_{i\mathrm{L}}\,\delta\overline{\epsilon}Q|\mathrm{vac}\rangle - \langle\mathrm{vac}|(Q^\dagger\gamma^0)\delta\epsilon\,\psi_{i\mathrm{L}}|\mathrm{vac}\rangle. \qquad (2.11.14)$$

---

[41]Note that $\partial_\mu\langle\mathrm{vac}|A_i(x)|\mathrm{vac}\rangle = \partial_\mu\langle\mathrm{vac}|A_i(0)|\mathrm{vac}\rangle = 0$, invoking translational invariance of the vacuum state.

If the vacuum expectation value $\langle \text{vac}|\mathcal{F}_i|\text{vac}\rangle$ in (2.11.13) does not vanish for the given i in question, the above equation implies that $Q_a|\text{vac}\rangle \neq 0$, at least for an a, since otherwise the vacuum expectation of the expression in (2.11.14) will vanish. We also obtain from (2.3), in the introduction to this chapter, $\langle \text{vac}|H|\text{vac}\rangle > 0$, implying spontaneous symmetry breaking.[42] In the tree approximation, i.e., by neglecting radiative corrections, the vacuum expectation value of $\mathcal{F}_i$ becomes simply $\mathcal{F}_i$ itself.

Guided by the criterion of renormalizability, thus avoiding having interaction terms with coupling constants with negative dimensionalities, we may introduce the general superpotential

$$W[A] = \lambda_i A_i + \alpha_{ij} A_i A_j + \lambda_{ijk} A_i A_j A_k, \tag{2.11.15}$$

with the Lagrangian density given by

$$\mathcal{L} = -\partial_\mu A_i^\dagger \partial^\mu A_i - \overline{\psi}_i \frac{\gamma \partial}{2} \psi_i - \frac{1}{2} W_{ij} \overline{\psi}_i \psi_{jL} - \frac{1}{2} W_{ij}^\dagger \overline{\psi}_i \psi_{jR} - W_i^\dagger W_i, \quad \mathcal{F}_i = W_i^\dagger. \tag{2.11.16}$$

The non-vanishing of W_i, i.e., $\mathcal{F}_i \neq 0$ for some i, at the absolute minimum, signals the breaking of supersymmetry, making the minimum of the scalar potential $V[A] = W_i^\dagger W_i$ positive definite.

An example of spontaneous symmetry breaking, with three chiral superfields is given with superpotential defined by[43]

$$W[A] = \alpha A_1 A_3 + \beta A_2 (A_3^2 - M^2). \tag{2.11.17}$$

leading to a scalar potential $V[A] = W_i^\dagger[A] W_i[A]$,

$$V[A] = \alpha^2 |A_3|^2 + \beta^2 |A_3^2 - M^2|^2 + |\alpha A_1 + 2\beta A_2 A_3|^2. \tag{2.11.18}$$

To find the minimum of this potential, we may set $A_1 = -(2\beta/\alpha)A_2 A_3$, since the last term above is non-negative. For definiteness suppose that $\alpha^2 \geq 2\beta^2 M^2$. The case with a reversed inequality may be similarly handled. Note that upon writing $(A_3 = (a_3 + i b_3)/\sqrt{2})$, the remaining first two terms in V may be rewritten as

$$\frac{1}{4}\beta^2 (a_3^2 + b_3^2)^2 + \frac{1}{2}(\alpha^2 - 2\beta^2 M^2)a_3^2 + \frac{1}{2}(\alpha^2 + 2\beta^2 M^2)b_3^2 + \beta^2 M^4. \tag{2.11.19}$$

Clearly, the minimum for this will occur for $A_3 = 0$, thus giving $V_{\min} = \beta^2 M^2$. But the value of A_2 in (2.11.18) is undetermined. At the minimum of the potential, the

---

[42]See also (2.4) and below it in the introduction to this chapter.
[43]This model is referred to as the O'Raifeartaigh Model [28].

auxiliary fields $\mathscr{F}_i$ satisfy

$$\mathscr{F}_1 = W_1^{\dagger} = 0, \quad \mathscr{F}_2 = W_2^{\dagger} = -\beta M^2, \quad \mathscr{F}_3 = W_3^{\dagger} = 0. \tag{2.11.20}$$

Because at least one of the auxiliary fields is non-zero, the spontaneous symmetry breaking encountered below is referred to as $\mathscr{F}$-type supersymmetry breaking.

By introducing the vacuum expectation values of the fields, we may, in particular, make the substitution $A_2 \rightarrow A_2 + \langle A_2 \rangle$, working at the tree level, where now the vacuum expectation value of the field A_2 is zero, and we have denoted $\langle 0|A_2|0 \rangle$ simply by $\langle A_2 \rangle$, and will be taken to be real. Thus the scalar potential in (2.11.18) takes the form

$$V = \alpha^2 |A_3|^2 + \beta^2 |A_3^2 - M^2|^2 + |\alpha A_1 + 2\beta A_2 A_3 + 2\beta \langle A_2 \rangle A_3|^2, \tag{2.11.21}$$

where, at the tree level, the vacuum expectation values of all the fields are zero. The quadratic part of V in the fields is given by $(A_j = (a_j + i\,b_j)/\sqrt{2})$

$$V\big|_{\text{quad}} = \frac{1}{2}\Big(\alpha^2(a_1^2 + b_1^2) + a_3^2\big[\alpha^2 - 2\beta^2 M^2 + 4\beta^2 (\langle A_2 \rangle)^2\big]$$

$$+ b_3^2\big[\alpha^2 + 2\beta^2 M^2 + 4\beta^2(\langle A_2 \rangle)^2\big] + 4\beta\alpha\,(a_1 a_3 + b_1 b_3)\langle A_2 \rangle\Big). \tag{2.11.22}$$

Since the latter does not involve the field A_2, the masses of the associated real fields $a_2\, b_2$ are zero. For $\langle A_2 \rangle \neq 0$, one may diagonalize the mass matrix extracted from (2.11.22) in the basis (a_1, b_1, a_3, b_3) to find the mass spectrum of the scalar fields. For $\langle A_2 \rangle = 0$, in particular, we have also two scalar fields with masses α and two additional scalar fields with masses, respectively, $\sqrt{\alpha^2 - 2\beta^2 M^2}$, and $\sqrt{\alpha^2 + 2\beta^2 M^2}$. All in all we have six real scalar fields.

To find the masses of the fermion fields, we express the superpotential in terms of the shifted fields with zero vacuum expectation values

$$W[A] = \alpha A_1 A_3 + \beta \langle A_2 \rangle (A_3^2 - M^2) + \beta A_2 (A_3^2 - M^2). \tag{2.11.23}$$

The quadratic part of the Lagrangian density (2.11.16), for the example in question, in the fermion fields is then given by

$$\mathscr{L}\big|_{\text{quad}} = -\sum_{j=1}^{3} \overline{\psi}_j \frac{\gamma\partial}{2i}\psi_j - \frac{1}{2}\,(2\beta\langle A_2 \rangle)\overline{\psi}_3\psi_3 - \frac{1}{2}\alpha\,\overline{\psi}_1\psi_3 - \frac{1}{2}\alpha\,\overline{\psi}_3\psi_1,$$

$$\tag{2.11.24}$$

with the fermion field ψ_2 obviously being massless. The above expression generates the fermion mass matrix involving the other fermion fields given by

$$\mathbf{M} = \begin{pmatrix} 0 & \alpha \\ \alpha & 2\beta\langle A_2\rangle \end{pmatrix}, \quad \mathbf{M}^\dagger\mathbf{M} = \begin{pmatrix} \alpha^2 & 2\beta\langle A_2\rangle \\ 2\beta\langle A_2\rangle & \alpha^2 + 4\beta^2(\langle A_2\rangle)^2 \end{pmatrix}. \tag{2.11.25}$$

Upon diagonalizing the latter matrix, we obtain, in addition to the massless fermion field, two fermions with masses

$$\sqrt{\beta^2(\langle A_2\rangle)^2 + \alpha^2} \pm \beta\langle A_2\rangle, \tag{2.11.26}$$

respectively.

When a symmetry is broken spontaneously, massless particles (Goldstones) appear in the theory. In supersymmetry theory, the generator of supersymmetry transformations is fermionic, and the Goldstone now is a fermion, referred to as a Goldstino. In the present model, it is associated with the massless fermion field we observed above.

2.12 Supersymmetric Gauge Theories

Now we have all the ingredients to define the Lagrangian density of gauge theories. To this end, with a left-chiral field as defined in (2.6.23), (2.9.1), and its adjoint, as given in (2.9.2), we have seen in (2.9.3), (2.9.5), (2.9.6), (2.7.5), respectively, that

$$\Phi_L^\dagger(x,\theta)\Phi_L(x,\theta)\Big|_{\mathscr{D}} = -\frac{1}{8}(\overline{\theta}\gamma^5\theta)^2\Big[-\partial_\mu A^\dagger\partial^\mu A + \mathscr{F}^\dagger\mathscr{F} - \overline{\psi}\frac{\gamma\partial}{2i}\psi\Big], \tag{2.12.1}$$

$$\Phi_L(x,\theta)\Phi_L(x,\theta)\Big|_{\mathscr{F}} = [A\mathscr{F} + \frac{1}{2}\overline{\psi}\psi_L]\overline{\theta}\theta_L, \tag{2.12.2}$$

$$\Phi_L(x,\theta)\Phi_L(x,\theta)\Phi_L(x,\theta)\Big|_{\mathscr{F}} = \frac{3}{2}[AA\mathscr{F} + A\overline{\psi}\psi_L]\overline{\theta}\theta_L. \tag{2.12.3}$$

$$\frac{1}{8}(\mathscr{C}\mathscr{W})_a\mathscr{W}_a\Big|_{\mathscr{F}}(x,\theta)$$

$$= \Big[-\frac{1}{4}G_E^{\mu\nu}G_{E\mu\nu} - \frac{1}{2i}\overline{\chi}_E\gamma^\nu[\delta_{EC}\partial_\nu + gf_{EAC}V_{A\nu}]\chi_C + \frac{1}{2}\mathscr{D}_E\mathscr{D}_E\Big]\overline{\theta}\theta_L. \tag{2.12.4}$$

On the other hand, according to (2.6.58) (in the Wess-Zumino gauge), we also have

$$e^{-2g\mathscr{V}} = 1 + \frac{g}{2}\overline{\theta}\gamma^5\gamma^\mu\theta\,V_\mu - \frac{ig}{\sqrt{2}}\overline{\theta}\gamma^5\theta\,\overline{\theta}\chi + \frac{g}{8}(\overline{\theta}\gamma^5\theta)^2[\mathscr{D} + gV^\nu V_\nu]. \tag{2.12.5}$$

The contributions to the following expression

$$\Phi_{L}^{\dagger}(x,\theta)\, e^{-2g\,\mathscr{V}}\Phi_{L}(x,\theta)\Big|_{\mathscr{D}}, \tag{2.12.6}$$

involving, the $\mathscr{D}$ terms are easily worked out and are spelled out as follows:

$$\Phi_{L}^{\dagger}(x,\theta)\,\frac{(\overline{\theta}\gamma^{5}\theta)^{2}}{8}[\mathscr{D}+g\,V^{\nu}V_{\nu}]\,\Phi_{L}(x,\theta)\Big|_{\mathscr{D}}$$
$$= \frac{(\overline{\theta}\gamma^{5}\theta)^{2}}{8}\,g\,A^{\dagger}[\mathscr{D}+g\,V^{\nu}V_{\nu}]A, \tag{2.12.7}$$

$$\Phi_{L}^{\dagger}(x,\theta)\left[-\frac{i\,g}{\sqrt{2}}\,\overline{\theta}\gamma^{5}\theta\,\overline{\theta}\chi\right]\Phi_{L}(x,\theta)\Big|_{\mathscr{D}}$$
$$= \frac{(\overline{\theta}\gamma^{5}\theta)^{2}}{8}\,g\,\sqrt{2}\left[\overline{\chi}A^{\dagger}\psi_{L}+\overline{\chi}A\psi_{R}\right], \tag{2.12.8}$$

$$\Phi_{L}^{\dagger}(x,\theta)\left[\frac{g}{2}\,\overline{\theta}\gamma^{5}\gamma^{\mu}\theta\,V_{\mu}\right]\Phi_{L}(x,\theta)\Big|_{\mathscr{D}}$$
$$= -\frac{(\overline{\theta}\gamma^{5}\theta)^{2}}{8}\,g\left[\overline{\psi}\,V_{\mu}\gamma^{\mu}\psi_{L}+i(\partial_{\mu}A^{\dagger})V^{\mu}A-iA^{\dagger}V^{\mu}(\partial_{\mu}A)\right], \tag{2.12.9}$$

where we may add to $\overline{\psi}\,V_{\mu}\gamma^{\mu}\psi_{L}$ its hermitian conjugate with the L $\rightarrow$ R label exchanged, in the last equation, and divide by 2 as follows

$$\frac{1}{2}\left[\overline{\psi}\,V_{\mu}\gamma^{\mu}\psi_{L}+(\overline{\psi}\,V_{\mu}\gamma^{\mu}\psi_{L})^{\dagger}\right]$$
$$= \frac{1}{2}\left[\overline{\psi}\,V_{C\mu}t_{C}\,\gamma^{\mu}\psi_{L}-\overline{\psi}\,V_{C\mu}[t_{C}^{\dagger}]^{\mathsf{T}}\,\gamma^{\mu}\psi_{R}\right], \tag{2.12.10}$$

Finally the term $\Phi_{L}^{\dagger}\Phi_{L}\big|_{\mathscr{D}}$ contributing to the expression in (2.12.6) is given in (2.12.1).

The Lagrangian density, up to the contribution of a superpotential, may be then taken as

$$\mathscr{L} = \left[-\partial_{\mu}A^{\dagger}\partial^{\mu}A+\mathscr{F}^{\dagger}\mathscr{F}-\overline{\psi}\frac{\gamma\partial}{2i}\psi\right]-\sqrt{2}\,g\left[\overline{\chi}A^{\dagger}\psi_{L}+\overline{\chi}A\psi_{R}\right]$$
$$+\frac{1}{2}\,g\left[\overline{\psi}\,V_{C\mu}t_{C}\,\gamma^{\mu}\psi_{L}-\overline{\psi}\,V_{C\mu}[t_{C}^{\dagger}]^{\mathsf{T}}\,\gamma^{\mu}\psi_{R}\right]$$
$$-g\,A^{\dagger}[\mathscr{D}+g\,V^{\nu}V_{\nu}]A+i\,g\,(\partial_{\mu}A^{\dagger})V^{\mu}A-i\,g\,A^{\dagger}V^{\mu}(\partial_{\mu}A)$$
$$+\left[-\frac{1}{4}\,G_{E}^{\mu\nu}\,G_{E\mu\nu}-\frac{1}{2i}\,\overline{\chi}_{E}\,\gamma^{\nu}\big[\delta_{EC}\partial_{\nu}+gf_{EAC}\,V_{A\nu}\big]\chi_{C}+\frac{1}{2}\mathscr{D}_{E}\mathscr{D}_{E}\right]. \tag{2.12.11}$$

Upon setting

$$\nabla_\mu \psi = \left(\partial_\mu \psi - i g\left[(V_{C\mu}tc)\psi_L - (V_{C\mu}[t_C^\dagger]^T)\psi_R\right]\right), \qquad [t_C^\dagger]^T = t_C^*,$$

$$(2.12.12)$$

$$\nabla_\mu A = \partial_\mu A - i g \, V_{C\mu}tc \, A, \qquad (\nabla_\mu A)^\dagger = \partial_\mu A^\dagger + i g A^\dagger V_{C\mu}tc, \qquad (2.12.13)$$

$$\nabla_{\mu EC}\chi_C = \left(\delta_{EC}\partial_\mu + g f_{EDC} V_{D\mu}\right)\chi_C, \qquad (2.12.14)$$

and taking into account the contribution of a superpotential, as given in (2.11.9), (2.11.10), the following expression emerges for the Lagrangian density of supersymmetric gauge theories in the Wess-Zumino gauge:

$$\mathscr{L}_{\text{Tot}} = -\frac{1}{4}G_E^{\mu\nu}G_{E\mu\nu} - \frac{\overline{\chi}_E\gamma^\mu(\nabla_\mu\chi)_E}{2i} + \frac{1}{2}\mathscr{D}_E\mathscr{D}_E - g A_i^\dagger(t_E)_{ij}\mathscr{D}_E A_j$$

$$- \frac{\overline{\psi}_i\gamma^\mu(\nabla_\mu\psi)_i}{2i} - (\nabla_\mu A)_i^\dagger(\nabla^\mu A)_i + \mathscr{F}_i^\dagger\mathscr{F}_i$$

$$- \sqrt{2}\,g\left(\overline{\chi}_E A_i^\dagger(t_E)_{ij}\psi_{jL} + h.c.\right) - \left(\mathscr{W}_i'[A]\mathscr{F}_i + h.c.\right) - \frac{1}{2}\left(\overline{\psi}_i\mathscr{W}_{ij}''[A]\psi_{jL} + h.c.\right),$$

$$(2.12.15)$$

involving several fields, with a summation over repeated indices understood.

2.13 Incorporating Supersymmetry in the Standard Model and Couplings Unification

The generation of the superpartners of the particles of the standard model is implicitly taken care of by promoting the field of the latter particles to superfields. For example, we may define the left-chiral superfield associated with the electron, as in (2.6.23), by

$$E(x,\theta) = A(x) + i\overline{\theta}\psi_L^{(e)}(x) + \frac{1}{2}\overline{\theta}\theta_L \mathscr{F}(x) + \frac{i}{4}\overline{\theta}\gamma^5\gamma^\mu\theta\,\partial_\mu A(x)$$

$$- \frac{1}{4}\overline{\theta}\gamma^5\theta\,\overline{\theta}\gamma\partial\psi_L^{(e)}(x) - \frac{1}{32}(\overline{\theta}\gamma^5\theta)^2\Box A(x), \qquad (2.13.1)$$

where $\psi_L^{(e)}$ is the left-hand part of the electron field, and A, here, is the field associated with its superpartner of spin 0, referred to as the selectron, and $\mathscr{F}$ is an auxiliary field. Similarly, we may introduce the superfield of a gauge boson, say, of a W vector boson, in the Wess-Zumino gauge, as in (2.6.41), by

$$\mathscr{V}^{(W)}(x,\theta) = -\frac{1}{4}\overline{\theta}\gamma^5\gamma^\mu\theta\, W_\mu(x) + \frac{i}{2\sqrt{2}}\overline{\theta}\gamma^5\theta\,\overline{\theta}\chi(x) - \frac{1}{16}(\overline{\theta}\gamma^5\theta)^2\mathscr{D}(x),$$

(2.13.2)

where W_μ represent the W vector boson field, in question, suppressing other indices, and χ, here, denotes its superpartner, a Majorana spinor, referred to as the wino, and $\mathscr{D}$ denotes an auxiliary field. In general, the superpartner of a gauge field is called a gaugino.

By continuing, the above process, we may generate the superfields associated, in particular, with the quarks and leptons of the standard model, conveniently, denoted in the following manner:

$$\begin{pmatrix} N_i \\ E_i \end{pmatrix} \equiv L_i, \quad (E)_i^c; \quad \begin{pmatrix} U_i \\ D_i \end{pmatrix} \equiv Q_i, \quad (U)_i^c, \quad (D)_i^c,$$

(2.13.3)

where the subscript i specifies the various generations, with the N_i denoting the left-chiral superfields associated with the left handed neutrino fields, the U_i denoting the left-chiral superfields associated with the left-handed u, c, t quarks, and the D_i denoting the left-chiral superfields associated with the left-handed d, s, b quarks, respectively, for $i = 1, 2, 3$. $(E)_i^c$; $(U)_i^c$, $(D)_i^c$ are associated with the left-handed positron, and the left-handed anti-quarks, respectively. The superpartner of a lepton is referred to as a slepton, and of the quark as a squark. The superparticles of the Gluons, eight of them, are referred as gluinos.

To consider the interaction of the above fermion fields, with the Higgs bosons, recall how the $\mathscr{F}$ term of a superpotential $W\big|_{\mathscr{F}}$, in Sects. 2.11, 2.12, was defined. To this end the following elementary properties, involving left-chiral superfields Φ_L and their adjoints should be noted:

$$\delta^{(2)}(\theta_R)\,(\Phi_{jL} - A_j)^\dagger = 0,$$

(2.13.4)

$$\delta^{(2)}(\theta_R)\,(\Phi_{iL} - A_i)(\Phi_{jL} - A_j)^\dagger = 0,$$

(2.13.5)

$$\delta^{(2)}(\theta_R)\,(\Phi_{i_1L} - A_{i_1})(\Phi_{i_2L} - A_{i_2})(\Phi_{jL} - A_j)^\dagger = 0,$$

(2.13.6)

from which we may infer that for describing Yukawa interactions of the quark and leptons, in a supersymmetric setting, the superpotential must be a functions of the (left-) chiral fields but not of their adjoints. By attempting to define their interactions with only one Higgs left-chiral superfield doublet, say with weak hypercharge $Y = +1$, we will encounter a difficulty as will be evident below. To this end, one defines two Higgs doublets of left-chiral superfields

$$H_1 = \begin{pmatrix} H_1^0 \\ H_1^- \end{pmatrix}, \qquad H_2 = \begin{pmatrix} H_2^+ \\ H_2^0 \end{pmatrix}, \tag{2.13.7}$$

respectively, of weak hypercharges $Y = \mp 1$. We note that using the identity $Y = 2(Q - T_3)$, the weak hypercharges associated with the superfields in (2.13.3) are respectively, $Y = -1, 2; 1/3, -4/3, 2/3$. Accordingly, SU(3) × SU(2) × U(1) singlets may be introduced to define the Lagrangian density of the Yukawa interactions of the fermions with the Higgs bosons and the superpotential will involve linear combinations of $\mathscr{F}$ parts of the following anti-symmetric terms:

$$\frac{1}{\sqrt{2}}(E_i H_1^0 - N_i H_1^-)(E_j)^c, \quad \frac{1}{\sqrt{2}}(D_i H_1^0 - U_i H_1^-)(D_j)^c, \quad \frac{1}{\sqrt{2}}(D_i H_2^+ - U_i H_2^0)(U_j)^c, \tag{2.13.8}$$

with coefficients depending on i and j, where we note that the anti-symmetric combinations of the two doublets, in each term, give rise to SU(2) singlets. Clearly, if we have only one Higgs doublet, say H_1, then the last term will involve the adjoints of the superfield components of H_1. Similarly, if we have the Higgs doublet H_2 the first two terms will involve the adjoints of the superfield components of H_2. In particular, the two Higgs fields now give masses to u-type and d-type quarks consistent with supersymmetry. The presence of two Higgs doublets turns out to be also important in the elimination of higgsino related anomalies. In addition to the Yukawa couplings in (2.13.8), we may also consider a mass-like term involving the two Higgs doublets in the form $\propto (H_2^0 H_1^0 - H_2^+ H_1^-)\big|_{\mathscr{F}}$.[44] Interactions with gauge fields occur through the kinetic energy terms.

The theory under consideration, with just two Higgs doublets, with minimal number of couplings and minimal number of fields is referred to as the Minimal Supersymmetric Model (MSSM).

An important contribution of this model is to the problem of the unification of gauge couplings at high energies, that is, beyond the MSSM model. This is considered next.

The beta functions of the MSSM are given in Table 2.1 together with the corresponding ones for the SM for comparison,[45] where

$$\mu^2 \frac{\mathrm{d}}{\mathrm{d}\mu^2} \frac{1}{\alpha_\#(\mu^2)} = \beta_\#, \qquad \alpha_\# = \frac{g_\#^2}{4\pi}, \tag{2.13.9}$$

---

[44]Here it is worth noting that selecting renormalizable supersymmetric interactions terms which conserve baryon and lepton numbers, may be equivalently achieved by imposing a symmetry known as the conservation of R-parity, but we will not go into it here. See, e.g., [11, 14].

[45]See Appendix B of this chapter for a discussion of the SM in this context.

Table 2.1 Table depicting the expressions for $12 \times \pi \times$ beta functions, with the beta functions, as introduced in (2.13.1), for the SM, and MSSM models as functions of the number of generations n_g, and the number of complex Higgs doublets n_H

$12\pi\beta_\#$	SM	MSSM
$12\pi\beta_s$	$(33 - 4n_g)$	$(27 - 6n_g)$
$12\pi\beta$	$(22 - 4n_g - \frac{1}{2}n_H)$	$(18 - 6n_g - \frac{3}{2}n_H)$
$12\pi\beta'$	$-(\frac{20}{3}n_g + \frac{1}{2}n_H)$	$-(10n_g + \frac{3}{2}n_H)$

with B.C.:

$$\alpha_s(\mu^2)\Big|_{\mu^2=M^2} = \alpha(\mu^2)\Big|_{\mu^2=M^2} = \frac{5}{3}\alpha'(\mu^2)\Big|_{\mu^2=M^2}, \qquad (2.13.10)$$

at a unifying energy scale to be determined below. The number of complex Higgs bosons doublets are taken to be $n_H = 1$ for the SM, and $n_H = 2$ for the MSSM models.[46]

The following two equations immediately follow:

$$\mu^2 \frac{d}{d\mu^2}\left(\frac{1}{\alpha_s(\mu^2)} - \frac{1}{\alpha(\mu^2)}\right) = \frac{1}{\pi}, \qquad (2.13.11)$$

$$\mu^2 \frac{d}{d\mu^2}\left(\frac{1}{\alpha_s(\mu^2)} - \frac{3}{5}\frac{1}{\alpha'(\mu^2)}\right) = \frac{12}{5\pi}. \qquad (2.13.12)$$

Upon subtracting (2.13.12) from 12/5 times (2.13.11), and using the unifying boundary conditions (2.13.10), at a unifying energy specified by a mass parameter M, as well as the defining equations[47]

$$\frac{1}{\alpha'} = \frac{\cos^2\theta_W}{\alpha_{em}}, \quad \frac{1}{\alpha} = \frac{\sin^2\theta_W}{\alpha_{em}}, \qquad (2.13.13)$$

we obtain at an energy scale specified by the mass M_Z of the Z vector boson,

$$\sin^2\theta_W\Big|_{M_Z^2} = \frac{1}{5} + \frac{7}{15}\frac{\alpha_{em}(M_Z^2)}{\alpha_s(M_Z^2)}. \qquad (2.13.14)$$

---

[46]For details concerning these beta functions see [13, 22, 23]. Their computations parallel very closely to those computed in Sect. 6.6, and already used in Sect. 6.18 in Vol. I.

[47]See Appendix B of this chapter for a review and for the relevant analysis in the standard model.

On the other hand, the second defining equation in (2.13.11), (2.13.13) and the equality in (2.13.14) lead upon integration to

$$\ln\left(\frac{M^2}{M_Z^2}\right) = \frac{\pi}{5\alpha_{em}(M_Z^2)}\left[1 - \frac{8}{3}\frac{\alpha_{em}(M_Z^2)}{\alpha_s(M_Z^2)}\right]. \tag{2.13.15}$$

Using the experimentally input,[48] $\alpha_s(M_Z^2) = 0.1184 \pm 0.0007$, $1/\alpha_{em}(M_Z^2) = 127.916 \pm 0.015$, give, $\sin^2\theta_W|_{M_Z^2} \simeq 0.231$, which compares remarkably well with the experimental result.[49] (2.13.15) gives rise to a scale $M \simeq 2.2 \times 10^{16}$ GeV. Moreover,

$$\mu^2 \frac{d}{d\mu^2}\frac{1}{\alpha_s(\mu^2)} = \frac{(27 - 6n_g)}{12\pi}, \tag{2.13.16}$$

gives, upon integration for $n_g = 3$, the expression

$$\frac{1}{\alpha_s(M^2)} = \frac{1}{\alpha_s(M_Z^2)} + \frac{3}{4\pi}\ln\left(\frac{M^2}{M_Z^2}\right)$$

$$= \frac{3}{5\alpha_s(M_Z^2)} + \frac{3}{20\alpha_{em}(M_Z^2)}. \tag{2.13.17}$$

The latter gives $1/\alpha_s(M^2) \approx 24.3$, $1/\alpha_{em}(M^2) \approx 65$.

As a measure of the accuracy of the approach of the coupling parameters to eventual unification in (2.13.10), one may introduce the following critical parameter[50]

$$\Delta = \frac{\alpha^{-1}(M_Z^2) - \alpha_s^{-1}(M_Z^2)}{(3/5)\alpha'^{-1}(M_Z^2) - \alpha^{-1}(M_Z^2)}, \tag{2.13.18}$$

which experimentally takes the value $\simeq 0.74$. On the other hand, by integrating the renormalization group equations (2.13.9)/Table 2.1 from $\mu^2 = M_Z^2$ to $\mu^2 = M^2$ and using the boundary conditions in (2.13.10), we obtain for the theoretical expression for (2.13.18)

$$\Delta_{theor} = \frac{\beta - \beta_s}{(3/5)\beta' - \beta} = \frac{-3 - 9}{-(3/5) \times 33 + 3} = 0.714, \tag{2.13.19}$$

which compares well with the experimental value. This is unlike the value of 0.5 obtained for the non-supersymmetric version as discussed in Appendix B of the

---

[48]Beringer [5].
[49]See, e.g., [5].
[50]Peskin [29].

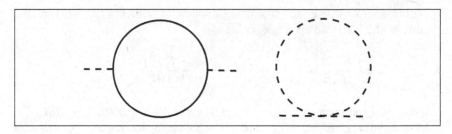

Fig. 2.2 Quadratic divergent contributions to a scalar particle self-mass squared from the two graphs, with the *solid* and the *dashed lines* denoting a spin 1/2 and a scalar particle, respectively

chapter. Also with the lifetime of the proton $\propto M^4$, one obtains the very welcome additional power of $\sim 10^4$ for the lifetime to that of the non-supersymmetric theory. The recent experimental large lower bound[51] for the lifetime of the proton justifies, however, the need of further investigations of unifications schemes.

In Sect. 2.10, we have seen, by a particular example, how supersymmetry eliminates quadratic divergences in the radiative corrections to the mass squared, through the study of the propagator, of a scalar particle. Due to opposite signs of the statistics of the fermion and the scalar particle, and with the particular relationship existing between the respective dimensionless couplings, as imposed by supersymmetry, such quadratic divergences cancel out. As a matter of fact, in a supersymmetric theory, some radiative corrections may not only be finite but may be completely absent in perturbation theory.[52]

A fundamental energy scale arises in the SM from the vacuum expectation value of the Higgs boson field $\simeq 246\,\mathrm{GeV}$,[53] which sets up the scale for the masses of the particles in the theory, including its own mass. This energy scale is much smaller in comparison to the Planck energy scale $\sim 10^{19}\,\mathrm{GeV}$ or lower, at which gravitation is expected to be significant. It is even much smaller than a grand unified energy scale, say $\sim 10^{16}\,\mathrm{GeV}$. The question then arises as to what amounts for the enormous energy scale difference between a grand unified energy scale and the energy scale characteristic of the SM? This is known as the hierarchy problem. What kind of new physics arises in this huge range of energy? As a *scalar* particle, the self-mass squared δM_H^2 of the Higgs boson, in the non-supersymmetric SM model, is quadratically divergent,[54] as inferred, by simple power counting, from such diagrams as shown in Fig. 2.2.

[51]See [27].

[52]See [20, 34]. The underlying theorems are referred to as non-renormalization theorems.

[53]See, e.g., (6.14.34) in Chap. 6 of Vol. I [26].

[54]See also [39].

With an ultraviolet cut-off taken of the order, say, $M \sim 10^{16} - 10^{19}$ GeV, an unnatural cancelation,[55] referred to as fine-tuning, has to occur between the bare mass squared of the Higgs boson and radiative corrections of the order M^2 in order to give a net finite mass for the Higgs boson not much different from the energy scale characterizing the SM.[56] This unnatural cancelation of enormously large numbers has been termed a facet of the hierarchy problem. We have seen in Sect. 6.14 of Vol. I, that at the tree level approximation, the electroweak theory provides very good agreement with experiments for the masses of the gauge bosons. A supersymmetric removal of an unwanted quadratic divergence is a positive contribution to the hierarchy problem. Supersymmetry is of significance in dealing with the hierarchy problem, as in supersymmetric field theories cancelations of such large quadratic corrections, a priori, generally, occur between loops involving particles and loops involving their supersymmetric counterparts in a supersymmetric version of a non-supersymmetric field theory, similar to the two loops shown in Fig. 2.2. Moreover, this cancelation is, possibly, up to divergences of logarithmic type which are tolerable, thus protecting a scalar particle from acquiring a large bare mass.

Since no superparticles are expected to have the same masses as their particle counterparts, otherwise some of them would have been discovered so far, supersymmetry is to be broken. For supersymmetry to provide a solution to the hierarchy problem, the relationship between the dimensionless coupling constants, in the unbroken supersymmetric theory, must be maintained without spoiling renormalizability, and without re-introducing quadratic divergences that supersymmetry was here to eliminate. This may be done by introducing supersymmetry breaking terms in the Lagrangian density with mass terms and coupling terms with positive mass dimensionalities, referred to as "soft" supersymmetry breaking terms.[57] This, in turn, introduces new physics beyond the SM, specified by an energy scale which may be denoted by M_s. With the quadratic divergence now removed, the self mass squared $\delta M_H{}^2$, will, by dimensional reasoning, be proportional to $M_s{}^2$, up to the well known logarithmic corrective factors of perturbation theory. Clearly, such a mass scale M_s cannot be too large, say of the order of a TeV or so, otherwise the hierarchy problem would re-emerge all over, or perhaps give rise to a little hierarchy problem.

---

[55]In QED the self-mass of the electron is only logarithmically divergent and the shift between the bare and the physical mass is not huge in comparison with a cut-off mass scale, say, of the order of the Planck energy.

[56]Aad, et al. [1], Chatrchyan, et al. [6].

[57]"Soft" supersymmetry breaking terms have been classified systematically by Girardello and Grisaru [18]. Such terms are spelled out in detail in [38, p. 68].

2.14   Spinors in Curved Spacetime: Geometrical Intricacies

With the detailed geometrical aspects developed of general relativity in Sect. 1.1, we now consider further generalizations of geometrical nature to describe supergravity. To this end, in the last three sections of this chapter, we consider a local version of supersymmetry, where the parameter $\epsilon(x)$, involved in supersymmetric transformations, depends on the particular spacetime point into consideration, and hence takes into account the underlying local structure of spacetime. Not surprisingly, this local version of supersymmetry is called supergravity. In the present section we develop the action of the gravitational field, associated with the spin 2 particle massless particle- the so-called graviton, and consider the role of a spinor in such a formalism. In the next section, the action of the Rarita-Schwinger field of spin 3/2 is developed, as the superpartner of the graviton. In the final section, the full action of pure supergravity is derived.[58] Here we follow a rather standard notation and, in order not to confuse Lorentz indices with general curved spacetime ones, we use Latin indices for the former and Greek indices for the latter. Also since spinor components will, in general, be suppressed throughout this section and the following two sections, this turns out to be quite convenient.

Let us recall once more the definition of the parallel transfer of the *components* $V^{\mu}(x)$ of a vector field from a point x to a point infinitesimally close point $x + dx$ in reference to some given curve as given in (1.1.28)

$$V^{\mu}(x) \;\rightarrow\; V^{\mu}(x) - \Gamma_{\nu\rho}{}^{\mu}(x)\, V^{\rho}(x)\, dx^{\nu} \;\equiv\; V_{\parallel}^{\mu}(x + dx), \tag{2.14.1}$$

where we will see that in our study of supergravity, we have to consider a connection $\Gamma_{\nu\rho}{}^{\mu}$ which is non-symmetric in $(\nu\rho)$. We may, similarly, define the parallel transfer of a spinor field by

$$\psi(x) \;\rightarrow\; \psi_{\parallel}(x + dx) = \psi(x) + \Omega_{\mu}(x)\, \psi(x)\, dx^{\mu}, \tag{2.14.2}$$

where $\Omega_{\nu}(x)$ is a matrix the expression of which is self consistently derived. Then $\overline{\psi}(x)\psi(x)$ would remain invariant under such an infinitesimal parallel transform if

$$\psi^{\dagger}(x)\gamma^{0}\psi(x) = \psi_{\parallel}^{\dagger}(x + dx)\, \gamma^{0}\, \psi_{\parallel}(x + dx)$$

$$= \left(\psi^{\dagger}(x) + \psi^{\dagger}(x)\, \Omega_{\mu}^{\dagger}(x)\, dx^{\mu}\right) \gamma^{0}\left(\psi(x) + \Omega_{\nu}(x)\psi(x)\, dx^{\nu}\right), \tag{2.14.3}$$

---

[58]The reader is advised to review the material in the introductory section Sect. 1.1 of Chap. 1 and, in particular, in the tetrad field treatment of the underlying geometry of general relativity.

to first order in dx^μ, from which we obtain

$$\gamma^0 \, \Omega_\mu^\dagger(x) \, \gamma^0 = - \Omega_\mu(x). \tag{2.14.4}$$

On the other hand, starting from the special relativistic expression of the components $\overline{\psi}(x)\gamma^a\psi(x)$ of the spinor field, locally defined, in a local Lorentz coordinate system at the point x, we may introduce the corresponding expression in the curvilinear coordinate spacetime by $\overline{\psi}(x)\gamma^\mu(x)\psi(x)$, where $\gamma^\mu(x) = e_a^\mu(x)\gamma^a$, with $e_a^\mu(x)$ denoting tetrad (vierbein) fields , to obtain from (2.14.1) for its parallel transfer

$$\overline{\psi}(x)\gamma^\mu(x)\psi(x) \; \rightarrow \; \overline{\psi}(x)\gamma^\mu(x)\psi(x) - \Gamma_{\nu\rho}{}^\mu(x) \, \overline{\psi}(x)\gamma^\rho(x)\psi(x) \, dx^\nu$$

$$\equiv \; \left(e_a{}^\mu(x+dx) \, \overline{\psi}_{\|}(x+dx) \, \gamma^a \, \psi_{\|}(x+dx) \right). \tag{2.14.5}$$

We may compare the two expressions on the right-hand sides of (2.14.5) and use, in the process, the expression for $\psi_{\|}(x+dx)$ on the right-hand of (2.14.2) to obtain as a matrix equation

$$[\Omega_\nu, \gamma_b] = e_{b\,\mu}\left(\partial_\nu e_a^\mu + \Gamma_{\nu\rho}{}^\mu \, e_a{}^\rho \right) \gamma^a, \tag{2.14.6}$$

where we have also used (2.14.4), and the fact that $e^b{}_\mu \, e_a{}^\mu = \delta^b{}_a$, upon multiplying first the resulting equation by $e^b{}_\mu$. The solution for $\Omega_\nu(x)$ is obtained from the double commutator,

$$[\gamma^a, [\gamma^b, \gamma^c]] = 4\left(\gamma^b\eta^{ac} - \gamma^c\eta^{ab}\right),$$

and is given by

$$\Omega_\nu(x) = (\omega_\nu)^{ab}(x) \, \frac{1}{8}\,[\gamma_a, \gamma_b], \qquad (\omega_\nu)^{ab} = -\,(\omega_\nu)^{ba}, \tag{2.14.7}$$

$$(\omega_\nu)_{ab}(x) + e_{b\mu}(x)\left(\partial_\nu e_a{}^\mu(x) + \Gamma_{\nu\rho}{}^\mu(x) \, e_a{}^\rho(x)\right) = 0, \tag{2.14.8}$$

where $(\omega_\nu)_{ab}$ is referred to as the spin connection.

From (2.14.2), (2.14.7), the covariant derivative of the spinor field $\mathscr{D}_\nu\psi$, should satisfy

$$\mathscr{D}_\nu\psi(x) \, dx^\nu = \left(\psi(x+dx) - \psi_{\|}(x+dx)\right), \tag{2.14.9}$$

$$\mathscr{D}_\nu\psi(x) = \partial_\nu\psi(x) - \Omega_\nu(x)\psi(x). \tag{2.14.10}$$

Multiplying (2.14.8) by $e^b{}_\sigma$, we may write the generalized connection as

$$\Gamma_{\nu\sigma}{}^\rho = e^{a\rho}\left(\partial_\nu e_{a\sigma} + (\omega_\nu)_{ab}\, e^b{}_\sigma\right). \tag{2.14.11}$$

The Riemann curvature tensor may be expressed in terms of the generalized connection and its derivative. On the other hand, the spin connection may be expressed from (2.14.8) as

$$(\omega_v)^{ab} = -e^b{}_\mu \left(\partial_v e^{a\mu} + \Gamma_{v\rho}{}^\mu(x) e^{a\rho} \right), \tag{2.14.12}$$

and the curvature $R_{\mu v}{}^{ab}$, with mixed indices, may be expressed in terms of the spin connection and its derivative as follows

$$R_{\mu v}{}^{ab}(\omega) = \partial_v(\omega_\mu)^{ab} + (\omega_v)^{ac}(\omega_\mu)_c{}^b - \partial_\mu(\omega_v)^{ab} - (\omega_\mu)^{ac}(\omega_v)_c{}^b, \tag{2.14.13}$$

where $R_{\mu v}{}^{ab}$ is anti-symmetric in (μ, v) and, independently, anti-symmetric in (a, b). The gravitational action, associated with the spin 2 particle, may be then spelled out to be

$$\int (\mathrm{d}x)\mathscr{L}_{(2)} = -\frac{1}{2\kappa^2} \int (\mathrm{d}x)\, e\, e_a{}^\mu e_b{}^v R_{\mu v}{}^{ab}(\omega), \qquad e = \det[e^a{}_\mu], \tag{2.14.14}$$

with coupling parameter $2\kappa^2$, and recall that the metric may be expressed as $g_{\mu v} = e^a{}_\mu e_{av}$.

At this stage we may define the torsion as the anti-symmetric part of the connection $\Gamma_{\mu v}{}^\rho$ in its indices (μ, v):

$$T_{\mu v}{}^\rho = \frac{1}{2}\left(\Gamma_{\mu v}{}^\rho - \Gamma_{v\mu}{}^\rho \right), \tag{2.14.15}$$

not to be confused with an energy-momentum tensor, and in the next section we will derive the general expression for the torsion $T_{ab}{}^\mu$ by considering variations of the above action together with the action associated with the spin 3/2 field, *with respect* to $(\omega_\mu)^{ab}$. Accordingly, we consider the corresponding variation of the action in (2.14.14) first. To this end,

$$\delta R_{\mu v}{}^{ab}(\omega) = \partial_v(\delta\omega_\mu)^{ab} + (\delta\omega_v)^{ac}(\omega_\mu)_c{}^b + (\omega_v)^{ac}(\delta\omega_\mu)_c{}^b$$
$$- \partial_\mu(\delta\omega_v)^{ab} - (\delta\omega_\mu)^{ac}(\omega_v)_c{}^b - (\omega_\mu)^{ac}(\delta\omega_v)_c{}^b. \tag{2.14.16}$$

This, in particular, requires to find the derivative $\partial_v(e\, e_a^\mu\, e_b^v)$, in carrying out a partial integration of the action in (2.14.14). We note that (2.14.8) leads to

$$\partial_\mu e_{av} - \Gamma_{\mu v}{}^\rho e_{a\rho} + (\omega_\mu)_{ab}\, e^b{}_v = 0, \quad \partial_v e_a{}^\mu + \Gamma_{v\sigma}{}^\mu e_a{}^\sigma + (\omega_v)_a{}^c e_c{}^\mu = 0. \tag{2.14.17}$$

Also using the variation of a determinant A derived in Appendix A of Chap. 1

$$\delta \det[A] = \det[A]\, \mathrm{Tr}[A^{-1}\delta A],$$

and with $e = \det [e^a{}_\mu]$, we have

$$e^\mu_a e^a{}_\nu = \delta^\mu{}_\nu, \qquad \delta e = e\, e_a{}^\sigma\, \delta e^a{}_\sigma, \qquad \partial_\nu e = e\, e^{a\sigma} \partial_\nu e_{a\sigma} = e\, \Gamma_{\nu\sigma}{}^\sigma. \qquad (2.14.18)$$

where we have used (2.14.11), to write $\Gamma_{\nu\sigma}{}^\sigma = e^{a\sigma} \partial_\nu e_{a\sigma}$.

From (2.14.17), (2.14.18), the following identity then readily follows

$$\partial_\nu \left(e\, e_a{}^\mu\, e_b{}^\nu\right) + e\, (\omega_\nu)_a{}^c\, e_c{}^\mu\, e_b{}^\nu + e\, (\omega_\nu)_b{}^c\, e_c{}^\nu\, e_a{}^\mu$$

$$= e\left\{\left(\Gamma_{\nu\lambda}{}^\lambda - \Gamma_{\lambda\nu}{}^\lambda\right) e_a{}^\mu\, e_b{}^\nu - \Gamma^\mu_{\nu\lambda} e_a{}^\lambda\, e_b{}^\nu\right\}. \qquad (2.14.19)$$

Finally upon comparison of the above equation with the following one obtained from (2.14.16)

$$e\, e_a{}^\mu\, e_b{}^\nu\, \delta R_{\mu\nu}{}^{ab}(\omega) = 2e\, e_a{}^\mu\, e_b{}^\nu\left\{\partial_\nu(\delta\omega_\mu)^{ab} + (\omega_\nu)^{ac}(\delta\omega_\mu)_c{}^b - (\delta\omega_\mu)^{ac}(\omega_\nu)_c{}^b\right\}$$

$$= 2e\left\{e_a{}^\mu\, e_b{}^\nu\, \partial_\nu - e_c{}^\mu\, e_b{}^\nu(\omega_\nu)_a{}^c - e_a{}^\mu\, e_c{}^\nu(\omega_\nu)_b{}^c\right\}\delta(\omega_\mu)^{ab}, \qquad (2.14.20)$$

where we have, in the process, took the advantage of the anti-symmetry property of the spin connection $(\omega_\mu)^{ab} = -(\omega_\mu)^{ba}$, and by using the definition of the torsion in (2.14.15), the following key equation emerges

$$\delta_\omega \int (dx)\mathcal{L}_{(2)} = \frac{1}{\kappa^2} \int (dx)\, e\left(e_b{}^\mu T_{\sigma a}{}^\sigma - e_a{}^\mu T_{\sigma b}{}^\sigma + T_{ab}{}^\mu\right)\delta(\omega_\mu)^{ab}. \qquad (2.14.21)$$

This equation will be used in conjunction with the corresponding one for the spin 3/2 field to determine the torsion.

Before closing this section, we also derive an expression relating the curvature and the torsion. To this end, from (2.14.11), we may infer that

$$\Gamma_{\beta\sigma}{}_a = \left(\partial_\beta e_{a\sigma} + (\omega_\beta)_{ac}\, e^c{}_\sigma\right), \qquad (2.14.22)$$

and

$$\partial_\alpha \Gamma_{\beta\sigma}{}_a = \partial_\alpha \partial_\beta e_{a\sigma} + (\partial_\alpha(\omega_\beta)_{ac})\, e^c{}_\sigma + (\omega_\beta)_{ac}\partial_\alpha e^c{}_\sigma$$

$$= \partial_\alpha \partial_\beta e_{a\sigma} + (\partial_\alpha(\omega_\beta)_{ac})\, e^c{}_\sigma + (\omega_\beta)_{ac}\left(\Gamma_{\alpha\sigma}{}^\lambda e^c{}_\lambda - (\omega_\alpha)^c{}_d e^d{}_\sigma\right), \qquad (2.14.23)$$

where in writing the second expression we have used the first equation in (2.14.17) with the index a there raised, or

$$\left(\partial_\alpha \Gamma_{\beta\sigma\ a} - (\omega_\beta)_{ac}\, e^c{}_\lambda \Gamma_{\alpha\sigma}{}^\lambda\right) = \partial_\alpha \partial_\beta e_{a\sigma} - (\omega_\beta)_{ac}(\omega_\alpha)^c{}_d e^d{}_\sigma + (\partial_\alpha(\omega_\beta)_{ac})\, e^c{}_\sigma.$$

(2.14.24)

Upon multiplying the latter equation by $\varepsilon^{\rho\alpha\beta\sigma}$, upon the exchange of some of the indices, and using the anti-symmetry relation of the spin connection, we obtain

$$\varepsilon^{\rho\alpha\beta\sigma}\left(\partial_\alpha \Gamma_{\beta\sigma\ a} + (\omega_\alpha)_{ac}\, e^c{}_\lambda \Gamma_{\beta\sigma}{}^\lambda\right) = \varepsilon^{\rho\alpha\beta\sigma}\left(\partial_\beta(\omega_\alpha)_{ca} + (\omega_\beta)_c{}^b(\omega_\alpha)_{ba}\right)e^c{}_\sigma.$$

(2.14.25)

From the expression of $R_{\mu\nu}{}^{ab}(\omega)$ in (2.14.13), we infer that the right-hand of (2.14.25) may be written as $\left(\varepsilon^{\rho\alpha\beta\sigma}R_{\alpha\beta\sigma\ a}\right)/2$. Hence using the definition of the torsion in (2.14.15), we obtain the following identity

$$\frac{1}{2}\varepsilon^{\rho\alpha\beta\sigma}R_{\alpha\beta\sigma\ a} = \varepsilon^{\rho\alpha\beta\sigma}\left(\partial_\alpha T_{\beta\sigma\ a} + (\omega_\alpha)_a{}^b\, T_{\beta\sigma\ b}\right),$$

(2.14.26)

and note that the $1/2$ factor on the left-hand side of the above equation does not cancel out because the definition of torsion in (2.14.15) already includes a $1/2$ factor. This identity will turn up to be useful later on.

2.15 Rarita-Schwinger Field and Induced Torsion: More Geometry

The Rarita-Schwinger Lagrangian density of a Majorana massless spin $3/2$ field in Minkowski spacetime may be written as[59]

$$\mathcal{L} = -\frac{1}{2i}\,\overline{\psi}_a[\eta^{ab}\gamma^c - (\eta^{ac}\gamma^b + \eta^{bc}\gamma^a) - \gamma^a\gamma^c\gamma^b]\partial_c\psi_b,$$

(2.15.1)

involving Lorentz indices a, b, c, and where a multiplicative factor of $1/2$ is included due to the Majorana character of the field, with a derivative acting to the right. Problem 2.21 allows us to rewrite the above Lagrangian density as[60]

$$\mathcal{L} = \frac{1}{2}\,\varepsilon^{abcd}\,\overline{\psi}_d\gamma_a\gamma^5\partial_b\psi_c,$$

(2.15.2)

---

[59]See Appendix II for the Rarita-Schwinger Lagrangian density at the end of this volume.
[60]Here we are using the Latin alphabet for Lorentz indices for convenience.

where, as mentioned above, the additional $1/2$ factor is due to the Majorana character of the field and avoids double counting. In curved spacetime, the Lagrangian density is taken as

$$\mathcal{L}_{3/2} = \frac{1}{2}\left(\varepsilon^{\alpha\beta\sigma\rho}\right)\overline{\psi}_\rho\gamma_\alpha\gamma^5\mathcal{D}_\beta\psi_\sigma, \qquad \mathcal{D}_\beta = \partial_\beta - \frac{1}{8}(\omega_\beta)^{ab}[\gamma_a, \gamma_b], \qquad (2.15.3)$$

where due to the nature of $\varepsilon^{\alpha\beta\sigma\rho}$ as a density, the above Lagrangian density does not involve an additional multiplicative $e = \det[e^a{}_\mu]$ factor. To consider its variation with respect to the spin connection, we use the following properties involving gamma matrices ($\varepsilon_{0123} = -1$, $\eta_{00} = -1$)

$$\gamma_a[\gamma_b, \gamma_c] = 2\left(i\varepsilon_{abcd}\gamma^5\gamma^d + \eta_{ab}\gamma_c - \eta_{ac}\gamma_b\right), \qquad \gamma_a\gamma^5 = -\gamma^5\gamma_a, \qquad (\gamma^5)^2 = I, \qquad (2.15.4)$$

to write

$$\delta_\omega\mathcal{L}_{3/2} = -\frac{i}{8}\varepsilon^{\alpha\beta\sigma\rho}\varepsilon_{cabd}\,e^c{}_\alpha\,\overline{\psi}_\sigma\gamma^d\psi_\rho\,\delta(\omega_\beta)^{ab}, \qquad (2.15.5)$$

where note that $\overline{\psi}_\sigma\gamma^d\psi_\rho = -\overline{\psi}_\rho\gamma^d\psi_\sigma$, while $\overline{\psi}_\sigma\gamma^5\gamma^d\psi_\rho = \overline{\psi}_\rho\gamma^5\gamma^d\psi_\sigma$, that is why the second and the third terms within the brackets in (2.15.4) do not contribute in (2.15.5). To simplify the above expression, we note, in passing, that

$$\varepsilon^{\alpha\beta\sigma\rho}e^a{}_\alpha\,e^b{}_\beta\,e^c{}_\sigma\,e^d{}_\rho = \varepsilon^{abcd}\,e, \qquad (2.15.6)$$

and hence by multiplying the latter by $e_b{}^{\beta'}e_c{}^{\mu'}e_d{}^{\nu'}$ and eventually make the replacement, $(\beta', \mu', \nu') \to (\beta, \mu, \nu)$, and multiplying the final expression, which is now independent of the indices (b, c, d), as they were summed over, by ε_{abcd} give, upon relabeling the indices, the identity

$$e^c{}_\alpha\varepsilon^{\alpha\beta\sigma\rho}\varepsilon_{cabd} = -\left(\delta_a^\beta\delta_b^\sigma\delta_d^\rho + \delta_a^\sigma\delta_b^\rho\delta_d^\beta + \delta_a^\rho\delta_b^\beta\delta_d^\sigma\right.$$
$$\left. - \delta_a^\beta\delta_d^\sigma\delta_b^\rho - \delta_a^\sigma\delta_d^\rho\delta_b^\beta - \delta_a^\rho\delta_d^\beta\delta_b^\sigma\right)e. \qquad (2.15.7)$$

We may then re-express (2.15.5), by conveniently relabeling the indices, as

$$\delta_\omega\mathcal{L}_{3/2} = -\frac{i}{4}e\left(e_b{}^\mu\overline{\psi}_\sigma\gamma^\sigma\psi_a - e_a{}^\mu\overline{\psi}_\sigma\gamma^\sigma\psi_b + \overline{\psi}_a\gamma^\mu\psi_b\right)\delta(\omega_\mu)^{ab}. \qquad (2.15.8)$$

Hence from this equation and the one associated with the graviton in (2.14.21), and from the variation,

$$\delta_\omega\int (dx)\left(\mathcal{L}_{(2)} + \mathcal{L}_{3/2}\right) = 0, \qquad (2.15.9)$$

we obtain the following expression for the torsion in (2.14.15)

$$T_{ab}{}^{\mu} = \mathrm{i}\,\frac{\kappa^2}{4}\,\overline{\psi}_a \gamma^{\mu}\psi_b,\tag{2.15.10}$$

where the i factor ensures the reality of the torsion.

2.16   From Geometry to Supergravity: The Full Theory

The action of the full theory considered together in Sects. 2.14 and 2.15, is given from (2.14.14) and (2.15.3) to be

$$\mathscr{A} = \int (\mathrm{d}x)\left[-\frac{1}{2\kappa^2}\,e\,e_a{}^{\mu}e_b{}^{\nu}R_{\mu\nu}{}^{ab}(\omega) + \frac{1}{2}\,\varepsilon^{\alpha\beta\sigma\rho}\,\overline{\psi}_{\rho}\gamma_{\alpha}\gamma^5 \mathscr{D}_{\beta}\psi_{\sigma}\right].\tag{2.16.1}$$

We have a Lagrangian density depending on the tetrad fields $e^a{}_{\mu}$, the fields $\overline{\psi}_{\mu}$, ψ_{μ}, and the spin connection $(\omega_{\mu})^{ab}$. We have already considered the variation of this action with respect to the spin connection in (2.15.9) and determined the torsion in (2.15.10). Since the spin connection will depend on the tetrad and the Rarita-Schwinger fields, the variations of the actions with respect to the latter may be *restricted to variations of the explicit dependence of the Lagrangian density* only to these latter variables. This is because a variation of the action $\mathscr{A}$ with respect to, say, a tetrad $e^a{}_{\mu}$, may, by using the chain rule, be symbolically written as

$$\delta_e \mathscr{A} = \frac{\delta\omega}{\delta e}\,\delta_{\omega}\mathscr{A} + \delta_e \mathscr{A}\Big|_{\text{explicit}},\tag{2.16.2}$$

and the first expression on the right-hand side of this equation already vanishes. A similar remark applies to the operations δ_{ψ} and $\delta_{\overline{\psi}}$.

Now we develop the algebra involving supersymmetric transformations. To this end, we define the following transformations:

$$\delta e^a{}_{\mu} = -\frac{\mathrm{i}\kappa}{2}\,\overline{\epsilon}\,\gamma^a\psi_{\mu},\tag{2.16.3}$$

$$\delta\psi_{\mu} = \frac{1}{\kappa}\,\mathscr{D}_{\mu}\epsilon = \frac{1}{\kappa}\left(\partial_{\mu} - \frac{1}{8}\,(\omega_{\mu})_{ab}[\gamma^a,\gamma^b]\right)\epsilon,\tag{2.16.4}$$

and self consistently establish the supersymmetry of the action $\mathscr{A}$ in (2.16.1), involving the Lagrangian densities associated with the graviton and the gravitino, under the *transformation rules* in (2.16.3), (2.16.4).

To the above end, from (2.14.18), we know that $\delta e = e\,e_a{}^{\sigma}\,\delta e^a{}_{\sigma}$, also

$$e^b{}_{\nu}\,\delta e_b{}^{\mu} = -e_b{}^{\mu}\,\delta e^b{}_{\nu},\quad e_a{}^{\nu}e^b{}_{\nu}\,\delta e_b{}^{\mu} = \delta e_a{}^{\mu} = -e_a{}^{\nu}e_b{}^{\mu}\,\delta e^b{}_{\nu}.\tag{2.16.5}$$

Hence (2.16.3) leads to

$$\delta\left(e\, e_a{}^\mu\, e_b{}^\nu\right) = -\frac{i\kappa e}{2}\left(e_a{}^\mu e_b{}^\nu\,\bar{\epsilon}\,\gamma^\sigma\psi_\sigma - e_b{}^\nu\,\bar{\epsilon}\,\gamma^\mu\psi_a - e_a{}^\mu\,\bar{\epsilon}\,\gamma^\nu\psi_b\right). \qquad (2.16.6)$$

Using the facts that

$$R_{\mu\nu}{}^{ab} = R_{\nu\mu}{}^{ba}, \qquad \text{with} \qquad R = e_a{}^\mu e_b{}^\nu R_{\mu\nu}{}^{ab}, \qquad R_\mu{}^a = e_b{}^\nu R_{\mu\nu}{}^{ab},$$

we obtain

$$\delta_e \mathcal{L}_{(2)} = -i\,\frac{e}{2\kappa}\left(R_{\mu a} - \frac{1}{2}e_{a\mu}R\right)\bar{\epsilon}\,\gamma^\mu\psi^a. \qquad (2.16.7)$$

The corresponding expression, associated with the spin $3/2$ part is much more involved. In particular [61]

$$\delta_e \mathcal{L}_{(3/2)} = \frac{1}{2}\,\varepsilon^{\alpha\beta\sigma\rho}\,\overline{\psi}_\rho\gamma_a\gamma^5\mathcal{D}_\beta\psi_\sigma\,\delta e^a{}_\alpha$$

$$= -\frac{i\kappa}{4}\,\varepsilon^{\alpha\beta\sigma\rho}\left(\bar{\epsilon}\,\gamma^a\psi_\alpha\right)\overline{\psi}_\rho\gamma_a\gamma^5\mathcal{D}_\beta\psi_\sigma. \qquad (2.16.8)$$

Now we invoke the Fierz identity in (A-2.3)

$$(\gamma^a)_{AB}(\gamma_a)_{CD} = -\delta_{AD}\,\delta_{CB} - \frac{1}{2}\,(\gamma^a)_{AD}\,(\gamma_a)_{CB}$$

$$- \frac{1}{2}\,(\gamma^5\gamma^a)_{AD}\,(\gamma^5\gamma_a)_{CB} + (\gamma^5)_{AD}\,(\gamma^5)_{CB}, \qquad (2.16.9)$$

where here a denotes a Lorentz index, as before, and in order not to confuse spinor indices with the other indices we have used capital letters for them. The Fierz identity above allows us to rewrite (2.16.8) as

$$\delta_e \mathcal{L}_{(3/2)} = \frac{i\kappa}{8}\,\varepsilon^{\alpha\beta\sigma\rho}\left(\bar{\epsilon}\,\gamma^a\gamma^5\mathcal{D}_\beta\psi_\sigma\right)\overline{\psi}_\rho\gamma_a\psi_\alpha$$

$$= \frac{1}{2\kappa}\,\varepsilon^{\alpha\beta\sigma\rho}\left(\bar{\epsilon}\,\gamma^a\gamma^5\mathcal{D}_\beta\psi_\sigma\right)T_{\rho\alpha a} = \frac{1}{2\kappa}\,\varepsilon^{\alpha\beta\sigma\rho}\left(\bar{\epsilon}\,\gamma^a\gamma^5\mathcal{D}_\beta\psi_\alpha\right)T_{\sigma\rho a}, \qquad (2.16.10)$$

where only the second term on the right-hand side of (2.16.9) contributes to the latter. In detail, the above equation may be rewritten from (2.15.3) as

$$\delta_e \mathcal{L}_{(3/2)} = -\frac{1}{2\kappa}\,\varepsilon^{\alpha\beta\sigma\rho}T_{\sigma\rho a}\left[(\partial_\beta\overline{\psi}_\alpha)\gamma^5\gamma^a\epsilon + \frac{1}{8}\overline{\psi}_\alpha\gamma^5[\gamma_c,\gamma_d]\gamma^a\epsilon\,(\omega_\beta)^{cd}\right], \qquad (2.16.11)$$

---

[61] We consider the variation of the fields $\overline{\psi}_\mu$, ψ_ν separately.

where we have used the Majorana properties

$$\bar{\epsilon}\gamma_a\gamma^5\psi_\mu = -\bar{\psi}_\mu\gamma^5\gamma_a\epsilon, \qquad \bar{\epsilon}\gamma^5\gamma_a[\gamma_b,\gamma_c]\psi_\mu = -\bar{\psi}_\mu\gamma^5[\gamma_b,\gamma_c]\gamma_a\epsilon. \tag{2.16.12}$$

On the other hand, using the transformation rule in (2.16.4), we have ($\{\gamma^5, \gamma_\alpha\} = 0$)

$$\delta_\psi \mathcal{L}_{(3/2)} = \frac{1}{2\kappa}\varepsilon^{\alpha\beta\sigma\rho}\dot{\overline{\psi}}_\rho\gamma_\alpha\gamma^5\mathcal{D}_\beta\mathcal{D}_\sigma\,\epsilon = \frac{1}{2\kappa}\varepsilon^{\alpha\beta\sigma\rho}\overline{\psi}_\sigma\gamma^5\gamma_\alpha\,\mathcal{D}_\beta\mathcal{D}_\rho\,\epsilon. \tag{2.16.13}$$

Similarly, the transformation rule (2.16.4) leads to

$$\delta_{\overline{\psi}}\mathcal{L}_{(3/2)} = -\frac{1}{2\kappa}\varepsilon^{\alpha\beta\sigma\rho}\left(\partial_\beta\overline{\psi}_\sigma\gamma^5 + \frac{1}{8}\overline{\psi}_\sigma\gamma^5(\omega_\beta)^{ab}[\gamma_a,\gamma_b]\right)\gamma_\alpha\,\mathcal{D}_\rho\epsilon. \tag{2.16.14}$$

Integrating the latter by parts reduces it to

$$\delta_{\overline{\psi}}\mathcal{L}_{(3/2)} = \frac{1}{2\kappa}\varepsilon^{\alpha\beta\sigma\rho}\overline{\psi}_\sigma\gamma^5\left(\mathcal{D}_\beta\,\gamma_\alpha\,\mathcal{D}_\rho\,\epsilon\right). \tag{2.16.15}$$

Using the first equation in (2.14.17) as well as the definition of $\mathcal{D}_\beta$, the above equation may be rewritten as ($\gamma_\alpha = e^c{}_\alpha\gamma_c$)

$$\delta_{\overline{\psi}}\mathcal{L}_{(3/2)} = \frac{1}{2\kappa}\varepsilon^{\alpha\beta\sigma\rho}\overline{\psi}_\sigma\gamma^5\left(\Gamma_{\beta\alpha}{}^\lambda\,\gamma_\lambda + \gamma_\alpha\mathcal{D}_\beta\right)\mathcal{D}_\rho\epsilon. \tag{2.16.16}$$

Therefore this equation and (2.16.13) give

$$(\delta_\psi + \delta_{\overline{\psi}})\mathcal{L}_{(3/2)} = \frac{1}{2\kappa}\varepsilon^{\alpha\beta\sigma\rho}\overline{\psi}_\sigma\gamma^5\left(\Gamma_{\beta\alpha}{}^\lambda\,\gamma_\lambda\,\mathcal{D}_\rho\epsilon + 2\,\gamma_\alpha\mathcal{D}_\beta\mathcal{D}_\rho\epsilon\right)$$

$$= \frac{1}{2\kappa}\varepsilon^{\alpha\beta\sigma\rho}\overline{\psi}_\sigma\gamma^5\left(T_{\beta\alpha}{}^\lambda\gamma_\lambda\,\mathcal{D}_\rho\epsilon + \gamma_\alpha[\mathcal{D}_\beta,\mathcal{D}_\rho]\epsilon\right), \tag{2.16.17}$$

where we have used anti-symmetry property in the indices (β, σ), to replace $2\mathcal{D}_\beta\mathcal{D}_\rho$ by the commutator $[\mathcal{D}_\beta, \mathcal{D}_\rho]$, as well as the definition of the torsion in (2.14.15).

In Problem 2.22, it is shown that

$$[\mathcal{D}_\mu,\mathcal{D}_\nu]\epsilon = \frac{1}{8}R_{\mu\nu}{}^{ab}[\gamma_a,\gamma_b]\epsilon. \tag{2.16.18}$$

This equation together with the ones in (2.15.4) and (2.15.7), allow one to rewrite (2.16.17) as

$$
(\delta_\psi + \delta_{\overline{\psi}}) \, \mathcal{L}_{(3/2)} = \frac{1}{2\kappa} \, \varepsilon^{\alpha\beta\sigma\rho} \, \overline{\psi}_\sigma \gamma^5 \left(T_{\beta\alpha}{}^\lambda \gamma_\lambda \, \mathcal{D}_\rho \epsilon + \frac{1}{2} \gamma_a R_{\rho\beta\alpha}{}^a \epsilon \right)
$$

$$
- \, i \frac{e}{4\kappa} \left(\overline{\epsilon} \, \gamma^c \psi_c \, R_{ab}{}^{ab} - 2 \overline{\epsilon} \, \gamma^\mu \psi_b \, R_{\mu a}{}^{ba} \right), \qquad (2.16.19)
$$

where we have used the Majorana relation $\overline{\psi}^a \gamma^\mu \epsilon = -\overline{\epsilon} \, \gamma^\mu \psi^a$. The second term above, on the right-hand of the equation, is simply

$$
\frac{i\,e}{2\kappa} \left(R_{\mu a} - \frac{1}{2} e_{a\mu} R \right) \overline{\epsilon} \, \gamma^\mu \psi^a,
$$

which cancels $\delta \mathcal{L}_{(2)}$ in (2.16.7). Hence from (2.16.11), (2.16.19), (2.14.26) and the definition of $\mathcal{D}_\beta \epsilon$ in (2.15.3), the following expression emerges for the total supersymmetric variation

$$
\delta_{\text{SUSY}} \left(\mathcal{L}_{(2)} + \mathcal{L}_{(3/2)} \right) = -\frac{1}{2\kappa} \, \varepsilon^{\alpha\beta\sigma\rho} \Big[\overline{\psi}_\alpha \gamma^5 \gamma^a T_{\sigma\rho a} \, \partial_\beta \epsilon
$$

$$
+ \overline{\psi}_\alpha \gamma^5 \gamma_a \epsilon \left(\partial_\beta T_{\sigma\rho}{}^a + (\omega_\beta)^a{}_c T_{\sigma\rho}{}^c \right) + (\partial_\beta \overline{\psi}_\alpha) \gamma^5 \gamma^a \epsilon \, T_{\sigma\rho a}
$$

$$
- \overline{\psi}_\alpha \gamma^5 \frac{1}{8} \left[\gamma_a, [\gamma_c, \gamma_d] \right] \epsilon \, (\omega_\beta)^{cd} T_{\sigma\rho}{}^a \Big], \qquad (2.16.20)
$$

upon the exchange of some of the indices. The terms explicitly dependent on ω_β, within the square brackets in the above equation, are given in the expression

$$
\overline{\psi}_\alpha \gamma^5 \left(-\frac{1}{8} \left[\gamma_a, [\gamma_c, \gamma_d] \right] (\omega_\beta)^{cd} + \gamma_c (\omega_\beta)^c{}_a \right) T_{\sigma\rho}{}^a \epsilon, \qquad (2.16.21)
$$

and on account of the identity

$$
\frac{1}{4} \left[\gamma_a, [\gamma_c, \gamma_d] \right] = \left(\eta_{ad} \gamma_c - \eta_{ac} \gamma_d \right), \qquad (2.16.22)
$$

the expression in (2.16.21) vanishes identically. On the other hand, the remaining terms in (2.16.20) are precisely given by

$$
\delta_{\text{SUSY}} \left(\mathcal{L}_{(2)} + \mathcal{L}_{(3/2)} \right) = -\frac{1}{2\kappa} \, \varepsilon^{\alpha\beta\sigma\rho} \, \partial_\beta \left(\overline{\psi}_\alpha \gamma^5 \gamma^a \epsilon \, T_{\sigma\rho a} \right), \qquad (2.16.23)
$$

as a total partial derivative, establishing the supersymmetry of the action $\mathcal{A}$ in (2.16.1).

Before closing the section, we note that by considering the combination

$$e_a{}^\rho e_b{}^\sigma (T_{\rho\sigma\nu} + T_{\nu\sigma\rho} - T_{\nu\rho\sigma}),$$

we obtain from (2.14.8)/(2.14.11) and the definition of the torsion in (2.14.15), the explicit expression for the spin connection

$$(\omega_\nu)_{ab} = e_a{}^\rho e_b{}^\sigma \left(T_{\rho\sigma\nu} + T_{\nu\sigma\rho} - T_{\nu\rho\sigma} \right)$$

$$+ \frac{1}{2} \left[e_a{}^\rho e_b{}^\sigma e^c{}_\nu \left(\partial_\sigma e_{c\rho} - \partial_\rho e_{c\sigma} \right) + e_a{}^\sigma \left(\partial_\nu e_{b\sigma} - \partial_\sigma e_{b\nu} \right) - e_b{}^\sigma \left(\partial_\nu e_{a\sigma} - \partial_\sigma e_{a\nu} \right) \right].$$

$$(2.16.24)$$

Appendix A: Fierz Identities Involving the Charge Conjugation Matrix

The following Fierz identities involving the charge conjugation matrix are useful:

$$(\gamma^5 \gamma^\mu \mathscr{C})_{ba} (\gamma_\mu)_{ck} - (\gamma^5 \gamma^\mu \mathscr{C})_{ca} (\gamma_\mu)_{bk} + (\gamma^5 \gamma^\mu \mathscr{C})_{cb} (\gamma_\mu)_{ak}$$

$$= 2 \left[\mathscr{C}_{ba} \gamma^5_{ck} - \mathscr{C}_{ca} \gamma^5_{bk} + \mathscr{C}_{cb} \gamma^5_{ak} \right.$$

$$\left. - (\gamma^5 \mathscr{C})_{ba} \delta_{ck} + (\gamma^5 \mathscr{C})_{ca} \delta_{bk} - (\gamma^5 \mathscr{C})_{cb} \delta_{ak} \right]. \qquad (A\text{-}2.1)$$

$$(\gamma^5 \gamma^\mu \mathscr{C})_{ba} (\gamma^5 \gamma_\mu \mathscr{C})_{dc} - (\gamma^5 \gamma^\mu \mathscr{C})_{ca} (\gamma^5 \gamma_\mu \mathscr{C})_{db} + (\gamma^5 \gamma^\mu \mathscr{C})_{da} (\gamma^5 \gamma_\mu \mathscr{C})_{cb}$$

$$= 2 \left[- \mathscr{C}_{ba} \mathscr{C}_{dc} + \mathscr{C}_{ca} \mathscr{C}_{db} - \mathscr{C}_{da} \mathscr{C}_{cb} \right.$$

$$\left. + (\gamma^5 \mathscr{C})_{ba} (\gamma^5 \mathscr{C})_{dc} - (\gamma^5 \mathscr{C})_{ca} (\gamma^5 \mathscr{C})_{db} + (\gamma^5 \mathscr{C})_{da} (\gamma^5 \mathscr{C})_{cb} \right].$$

$$(A\text{-}2.2)$$

Due to the obvious anti-symmetry in the indices a, b, c, d in the second Fierz identity, there is an overall constant factor relating its right-hand with its left-hand side which is easily worked out to be 2. To obtain the first identity simply multiply the second by $-(\gamma^5 \mathscr{C})_{dk}$, and recall the anti-symmetry of the matrix $\gamma^5 \gamma^\mu \mathscr{C}$, $\{\gamma^5, \gamma^\mu\} = 0$.

Another interesting derivation of the above identities follows by using the classic Fierz identity given below and deriving, in the process, as well the identity following it.

The classic Fierz identity is given by

$$(\gamma^\mu)_{ab}(\gamma_\mu)_{cd} = -\delta_{ad}\delta_{cb} - \frac{1}{2}(\gamma^\mu)_{ad}(\gamma_\mu)_{cb} - \frac{1}{2}(\gamma^5\gamma^\mu)_{ad}(\gamma^5\gamma_\mu)_{cb} + (\gamma^5)_{ad}(\gamma^5)_{cb}.$$

$$(A\text{-}2.3)$$

In terms of the charge conjugation matrix, the following identity is useful (see Problem 2.5)

$$(\gamma^5\gamma^\mu)_{ad}(\gamma_\mu\mathscr{C})_{ck} + (\gamma^5\gamma^\mu)_{cd}(\gamma_\mu\mathscr{C})_{ka} + (\gamma^5\gamma^\mu)_{kd}(\gamma_\mu\mathscr{C})_{ca} = 0. \qquad (A\text{-}2.4)$$

Appendix B: Couplings Unification in the Non-supersymmetric Standard Model

The standard model is based on the symmetry of the product groups SU(3)×SU(2)× U(1). Here we provide only a summary with key points of couplings unification. The purpose of this appendix is to recall the unification of couplings in the non-supersymmetric SM at high energies investigated in Vol. I for comparison with the supersymmetric version given in Sect. 2.13.[62] The effective couplings of the theory, at an energy scale μ, satisfy the following renormalization group equations to the leading orders:

$$\mu^2\frac{d}{d\mu^2}\frac{1}{\alpha_s(\mu^2)} = \beta_s, \qquad \left(\alpha_s = \frac{g_s^2}{4\pi} \text{ for SU(3)}\right), \qquad (B\text{-}2.1)$$

$$\mu^2\frac{d}{d\mu^2}\frac{1}{\alpha(\mu^2)} = \beta, \qquad \left(\alpha = \frac{g^2}{4\pi} \text{ for SU(2)}\right), \qquad (B\text{-}2.2)$$

$$\mu^2\frac{d}{d\mu^2}\frac{1}{\alpha'(\mu^2)} = \beta', \qquad \left(\alpha' = \frac{g'^2}{4\pi} \text{ for U(1)}\right). \qquad (B\text{-}2.3)$$

and if one neglects the small contribution of the Higgs boson,

$$\beta_s = +\frac{1}{12\pi}(33 - 4n_g), \qquad (B\text{-}2.4)$$

$$\beta = +\frac{1}{12\pi}(22 - 4n_g), \qquad (B\text{-}2.5)$$

$$\beta' = -\frac{1}{12\pi}\frac{(20n_g)}{3}. \qquad (B\text{-}2.6)$$

---

[62]For details of the renormalization group analysis of the standard model discussed here see: Chap. 6 in Vol I [26].

In particular, the fine-structure coupling α_e is given in terms of the $SU(2)$ coupling and the Weinberg angle θ_W,

$$\alpha_{em} = \alpha \sin^2 \theta_W, \tag{B-2.7}$$

Moreover, for the coupling α' we have

$$\frac{1}{\alpha'} = \frac{\cos^2 \theta_W}{\alpha_{em}}, \tag{B-2.8}$$

The unification energy scale M is defined as the energy at which the following effective couplings become equal

$$\alpha_s(M^2) = \frac{5}{3}\alpha'(M^2) = \alpha(M^2), \tag{B-2.9}$$

The solutions of the renormalization groups equations give rise to an energy scale $M \simeq 1.1 \times 10^{15}$ GeV. To assess the approach of the eventual equalities of the couplings in (B-2.9), one may define the critical parameter[63]

$$\Delta = \frac{\alpha^{-1}(M_Z{}^2) - \alpha_s^{-1}(M_Z{}^2)}{(3/5)\alpha'^{-1}(M_Z{}^2) - \alpha^{-1}(M_Z{}^2)}, \tag{B-2.10}$$

where, by convention, M_Z is taken to be the mass of the neutral Z vector boson. Experimentally,[64] $\alpha_s(M_Z^2) = 0.1184 \pm 0.0007$, $1/\alpha_{em}(M_Z^2) = 127.916 \pm 0.015$, $\sin^2 \theta_W|_{M_Z^2} \simeq 0.23$, which give $\Delta_{exp} \simeq 0.74$. On the other hand, by integrating the renormalization groups equations (B-2.1)–(B-2.3) from $\mu^2 = M_Z^2$ to $\mu^2 = M^2$ and using the boundary conditions in (B-2.9), we obtain for the theoretical expression for (B-2.10)

$$\Delta_{theor} = \frac{\beta - \beta_s}{(3/5)\beta' - \beta} = \frac{-11/12\pi}{-22/12\pi} = 0.5, \tag{B-2.11}$$

and the departure is significant. If we include the small contribution of the Higgs boson given in Table 2.1 in Sect. 2.13, we obtain $\Delta_{theor} \simeq 0.53$. As shown in Sect. 2.13, the excellent agreement between the theoretical and experimental values of Δ in a supersymmetric version of the standard model is quite impressive.

---

[63]Peskin [29].

[64]See, e.g., [5] .

Problems

2.1 Verify that $\bar{\epsilon} K \gamma^\mu K^{-1} \epsilon = \overline{(K^{-1}\epsilon)} \gamma^\mu (K^{-1}\epsilon)$.

2.2 Verify the Super-Poincaré Transformations in (2.1.14), (2.1.15).

2.3   (i) Derive the identities in (2.2.22).
 (ii) Derive the identities in (2.2.23), (2.2.24).
(iii) Derive the identities in (2.2.25), (2.2.26).

2.4 Derive the identities in (2.2.27), and in (2.2.28).

2.5 Derive the key identity (A-2.4) involving the charge conjugation matrix, using the classic Fierz identity in (A-2.3).

2.6 Upon writing $\psi = \psi_+ + i\psi_-$, where $\psi_- = -i(\psi - \mathscr{C}\bar{\psi}^\mathsf{T})/2$, $\psi_+ = (\psi + \mathscr{C}\bar{\psi}^\mathsf{T})/2$, show that $\psi_\pm$ are Majorana spinors. That is, an arbitrary spinor may be written as a simple linear combination of two Majorana spinors.

2.7 Derive the anti-commutation rules of the superderivatives in (2.3.10), (2.3.11).

2.8 Derive the identities in (2.3.12).

2.9 Derive the identity in (2.3.13).

2.10 Verify the relations (2.3.18)–(2.3.23).

2.11 Derive the expression of the matrix in (2.4.17).

2.12 Derive the expression for the matrix M in (2.4.18)–(2.4.20).

2.13 Show that the superdeterminant of the matrix Y defined in (2.4.31) is as given in that equation.

2.14 Prove the basic identity involving superderivatives: $\bar{D} D^\mathrm{R} D_a^\mathrm{L} = D_a^\mathrm{L} \bar{D} D^\mathrm{R} - 2i(\gamma^\mu D_a^\mathrm{R})\partial_\mu$.

2.15 For two Majorana spinors θ, ξ, with all the components anti-commuting, derive the following useful identities in the chiral representation: [Below $\xi_1, \xi_2 / \xi_3, \xi_4$ are the respective components of $\xi_\mathrm{R}/\xi_\mathrm{L}$.]

$$\text{(i)} \quad \bar{\theta}\xi_\mathrm{L} = \theta_3\xi_4 - \theta_4\xi_3, \qquad \bar{\theta}\xi_\mathrm{R} = \theta_2\xi_1 - \theta_1\xi_2,$$

$$\bar{\theta}\gamma^\mu\gamma^\nu\theta_\mathrm{L} = -\bar{\theta}\theta_\mathrm{L}\eta^{\mu\nu}, \qquad \bar{\theta}\gamma^5\gamma^\mu\theta = -2\theta^\mathsf{T}(\mathscr{C}^{-1}\gamma^\mu)\theta_\mathrm{L}.$$

$$\text{(ii)} \ (\bar{\theta}\xi_\mathrm{L})^\dagger = \bar{\theta}\xi_\mathrm{R}, \qquad (\bar{\theta}\gamma^\mu\xi_\mathrm{L})^\dagger = -\bar{\theta}\gamma^\mu\xi_\mathrm{R}.$$

2.16 Show that the explicit structures of $(i/2)(\Lambda^\dagger - \Lambda)$, where Λ is the gauge parameter (left-chiral) superfield in (2.6.97), is given by

$$\frac{i}{2}(\Lambda^\dagger - \Lambda) = \text{Im}(a) - \frac{1}{2\sqrt{2}}\overline{\theta}\xi + \frac{i}{8\sqrt{2}}\overline{\theta}\gamma^5\theta\,\overline{\theta}\gamma\partial\gamma^5\xi + \frac{i}{4}\overline{\theta}\gamma^5\theta\,\text{Im}(b)$$

$$-\frac{1}{4}\overline{\theta}\theta\,\text{Re}(b) + \frac{1}{4}\overline{\theta}\gamma^5\gamma^\mu\theta\,\partial_\mu\text{Re}(a) - \frac{1}{32}(\overline{\theta}\gamma^5\theta)^2\,\Box\,\text{Im}(a).$$

2.17 Show that the pure vector superfield $\mathscr{V}_\mu(x)$, given in (2.6.61), may be re-expressed as a function of $\hat{x}^\mu$, with the latter defined in (2.6.26), and is given by (2.6.66).

2.18 Show that $\varepsilon^{\mu\nu\alpha\beta}G_{\mu\nu}G_{\alpha\beta}$ is a total differential.

2.19 Derive the field equation of the Majorana spinor in (2.10.4).

2.20 Derive (2.10.16) involving the spinor fields.

2.21 Show that the Lagrangian density of a massless spin $3/2$ may be rewritten as $\mathscr{L} = \frac{1}{2}\varepsilon^{abcd}\overline{\psi}_d\gamma_a\gamma^5\partial_b\psi_c$.

2.22 Show that $[\mathscr{D}_\mu, \mathscr{D}_\nu]\epsilon = (1/8)R_{\mu\nu}{}^{ab}[\gamma_a, \gamma_b]\epsilon$, where

$$\mathscr{D}_\mu = \partial_\mu - \frac{1}{8}(\omega_\mu)^{ab}[\gamma_a, \gamma_b],$$

$R_{\mu\nu}{}^{ab}$ is given in (2.14.13), and ϵ is a spinor.

Recommended Reading

Baer, H., & Tata, X. (2006). *Weak scale supersymmetry: From superfields to scattering events.* Cambridge: Cambridge University Press.

Binetruy, P. (2006). *Supersymmetry, experiment, and cosmology.* Oxford: Oxford University Press.

Dine, M. (2007). *Supersymmetry and string theory: Beyond the standard model.* Cambridge: Cambridge University Press.

Manoukian, E. B. (2012). The explicit pure vector superfield in gauge theories. *Journal of Modern Physics, 3,* 682–685.

Manoukian, E. B. (2016). *Quantum field theory I: Foundations and abelian and non-abelian gauge theories.* Dordrecht: Springer.

Weinberg, S. (2000). *The quantum theory of fields. III: Supersymmetry.* Cambridge: Cambridge University Press.

References

1. Aad, G., et al. (2012). Observation of a new particle in the search for the standard model Higgs Boson with the ATLAS detector at the LHC. *Physics Letters B, 716*, 1–29.
2. Bailin, D., & Love, A. (1994). *Supersymmetric gauge field theory and string theory*. Bristol: Institute of Physics Publishing.
3. Bare, H., & Tata, X. (2006). *Weak scale supersymmetry: From superfields to scattering events*. Cambridge: Cambridge University Press.
4. Berezin, F. A. (1987). *Introduction to superanalysis*. Dordrecht: Reidel.
5. Beringer, J., et al. (2012). Particle data group. *Physical Review D, 86*, 010001.
6. Chatrchyan, S., et al. (2012). Observation of a new boson at mass 125 GeV with the CMS experiment at LHC. *Physics Letters B, 716*, 30–61.
7. Coleman, S., & Mandula, J. (1967). All possible symmetries of the S matrix. *Physical Review, 150*, 1251–1256.
8. Deser, S. (2000). Infinities in quantum gravities. *Annalen der Physik, 9*, 299–306.
9. Deser, S., Kay, J. H., & Stelle, K. S. (1977). Renormalizability properties of supergravity. *Physical Reviews Letters, 38*, 527–530.
10. Deser, S., & Zumino, B. (1976). Consistent supergravity. *Physics Letters B, 62*, 335–337.
11. Dimopoulos, S., & Georgi, H. (1981). Softly broken supersymmetry and SU(5). *Nuclear Physics B, 193*, 150–162.
12. Dirac, P. A. M. (1970). Can equations of motion be used in high-energy physics? *Physics Today, 23*(4), 29.
13. Einhorn, M. B., & Jones, D. R. T. (1982). The weak mixing angle and unification mass in supersymmetric SU(5). *Nuclear Physics B, 196*, 475–488.
14. Farrar, G. R., & Fayet, P. (1978). Phenomenology of the production, decay, and detection of new hadronic states associated with supersymmetry. *Physics Letters B, 76*, 575–579
15. Fayet, P. (1977). Spontaneously broken supersymmetric theories of weak, electromagnetic, and strong interactions. *Physics Letters B, 69*, 489–494.
16. Ferrara, S. (Ed.). (1987). *Supersymmetry* (Vols. 1 & 2). New York: Elsevier Publishers, B. V.
17. Freedman, D. Z., van Nieuwenhuizen, P., & Ferrara, S. (1976). Progress toward a theory of supergravity. *Physical Review B, 13*, 3214–3218.
18. Girardello, L., & Grisaru, M. T. (1982). Soft breaking of supersymmetry. *Nuclear Physics B, 194*, 65–76.
19. Gol'fand, A., & Likhtman, E. P. (1971). Extension of the Poincaré group generators and violation of P invariance. *JETP Letters, 13*, 323–326. (Reprinted In Ferrara, S. (Ed.), *Supersymmetry* (Vols. 1 & 2). New York: Elsevier Publishers, B. V, 1987).
20. Grisaru, M. T., Siegel, N., & Roček, M. (1979). Improved methods for supergraphs. *Nuclear Physics B, 159*, 429–450.
21. Howe, P. S., & Stelle, K. S. (2003). Supersymmetry counterterms revisited. *Physics Letters B, 554*, 190–196. hep–th/0211279v1.
22. Jones, D. R. T. (1974). Two-loop diagrams in Yang-Mills theory. *Nuclear Physics B, 75*, 531–538.
23. Jones, D. R. T. (1975). Asymptotic behaviour of supersymmetric Yang-Mills theories in the two-loop approximation. *Nuclear Physics B, 87*, 127–132.
24. Manoukian, E. B. (2006). *Quantum theory: A wide spectrum*. AA Dordrecht: Springer.
25. Manoukian, E. B. (2012). The explicit pure vector superfield in gauge theories. *Journal of Modern Physics, 3*, 682–685.
26. Manoukian, E. B. (2016). *Quantum field theory I: Foundations and abelian and non-abelian gauge theories*. Dordrecht: Springer.
27. Miura, M. (2010). Search for nucleon decays in super-Kamiokande. ICHEP, Paris, Session, 10.
28. O'Raifeartaigh, L. (1975). Spontaneous symmetry breaking for chiral scalar superfields. *Nuclear Physics B, 96*, 331–352.

29. Peskin, M. (1997).   Beyond the standard model.   In N. Ellis & M. Neubert (Eds.), *1996 European school of high-energy physics*. Genève: CERN-97-03.
30. Salam, A., & Strathdee, J. (1974). Supersymmetry and non-abelian gauges. *Physics Letters B, 51*, 353–355. (Reprinted In Ferrara, S. (Ed.), *Supersymmetry* (Vols.1 & 2). New York: Elsevier Publishers, B. V, 1987)
31. Salam, A., & Strathdee, J. (1974). Supergauge transformations. *Nuclear Physics B, 76*, 477–482.
32. Salam, A., & Strathdee, S. (1975). Feynman rules for superfields. *Nuclear Physics B, 86*, 42–152.
33. Salam, A., & Strathdee, J. (1975). Superfields and Fermi-Bose symmetry. *Physical Review D, 8*(11), 1521–1535.
34. Seiberg, N. (1993). Naturalness versus supersymmetric non-renormalization theorems. *Physics Letters B*, 318, 469–475.
35. Seiberg, N., & Witten, E. (1994).   Electric-magnetic duality, monopole condensation, and confinement in N=2 supersymmetric Yang-Mills theory.   *Nuclear Physics B, 426*, 19–52. [Erratum: *Nuclear Physics B, 430*, pp. 485–486, 1994]; hep-th/9407087.
36. Stelle, K. S. (2001).   Revisiting supergravity and super Yang-Mills renormalization.   In J. Lukierski & J. Rembielinski (Eds.), *Proceedings of 37th Karpacz Winter School of Theoretical Physics*, Feb 2001. hep-th/0203015v1.
37. Stelle, K. S. (2012).   String theory, unification and quantum gravity.   In 6th Aegean Summer School, "Quantum Gravity and Quantum Cosmology", 12–17 Sept, 2011, Chora, Naxos Island. hep-th/1203.4689v1.
38. Terning, J. (2006). *Modern supersymmetry: Dynamics and duality*. Oxford: Clarendon Press.
39. Veltman, M. J. G. (1981).   The infrared-ultraviolet connection. *Acta Physica Polonica B, 12*, 437.
40. Volkov, D. V., & Akulov, V. P. (1973). Is the neutrino a goldstone particle. *Physics Letters B, 46*, 109–110. Reprinted In Ferrara, S. (Ed.), *Supersymmetry* (Vols. 1 & 2). New York: Elsevier Publishers, B. V, 1987).
41. Weinberg, S. (2000). *The quantum theory of fields, III: Supersymmetry*. Cambridge: Cambridge University Press.
42. Wess, J., & Zumino, B. (1974).   Supergauge transformations in four dimensions.   *Nuclear Physics, B*70, 39–50. Reprinted In Ferrara, S. (Ed.), *Supersymmetry* (Vols. 1 & 2). New York: Elsevier Publishers, B. V, 1987).

Chapter 3
Introduction to String Theory

A string, whether open or closed, as it moves in spacetime, it sweeps out a two-dimensional surface called a worldsheet. String Theory is a *quantum field theory*[1] *which operates on this two-dimensional worldsheet* with remarkable consequences in spacetime itself, albeit in higher dimensions. The strings are supposedly very small in extension and may "appear" almost point-like if they are indeed very small, say, of the order of Planck length. Since no experiments can probe distances of the order of the Planck length, such a string in present day experiments is considered to be point-like. What is remarkable about string theory is that particles that are needed to describe the dynamics of elementary particles arise naturally in the mass spectra of oscillating strings, and are not, a priori, assumed to exist or put in by hand in the underlying theories. Particles are identified as vibrational modes of strings, and a single vibrating string may describe several particles depending on its vibrational modes. Strings describing bosonic particles are referred to as bosonic strings, while those involving fermionic ones as well are referred to as superstrings. Accordingly, superstrings are expected to be more relevant to the real world than bosonic strings. We will witness the emergence of the graviton, in addition to a whole spectrum of other excitation modes from the theory and thus, hopefully, string theory may provide a framework for the unification of general relativity and quantum theory. The ultimate goal of String Theory is actually to provide a unified description of all the fundamental interactions in Nature and, in particular, give rise to a consistent theory of quantum gravity. If string theory has to do with quantum gravity, it must involve the three fundamental constants: Newton's gravitational constant G_N, the quantum unit of action $\hbar$, and the speed of light c. The unit of length emerging from these fundamental constants is the Planck length $\ell_P = \sqrt{G_N \hbar / c^3} \sim 10^{-33}$ cm mentioned earlier.

---

[1]For an overall view of quantum field theory since its birth in 1926 see Chap. 1 of Vol. I [29]. This introduction is partly based on the latter.

© Springer International Publishing Switzerland 2016

E.B. Manoukian, *Quantum Field Theory II*, Graduate Texts in Physics,
DOI 10.1007/978-3-319-33852-1_3

The *dimensionality* of the spacetime in which the strings live are *predicted* by the underlying theory as well and are necessarily of higher dimensions than four for consistency with Lorentz invariance of spacetime at the quantum level, consisting of a dimensionality of 26 for the bosonic strings and a spacetime dimensionality of 10 for the superstrings. We will see, in particular, how the knowledge of the number of degrees of freedom associated with massless fields lead self consistently to the determination of the dimensionality of the underlying spacetime of a string theory. The extra dimensions are expected to curl up into a space that is too small to be detectable with present available energies. For example the surface of a hollow extended cylinder with circular base is two dimensional, with one dimension along the cylinder, and another one encountered as one moves on its circumference. If the radius of the base of the cylinder is relatively small, the cylinder will appear as one dimensional when viewed from a large distance (low energies). Accordingly, the extra dimensions in string theory are expected to be small and methods, referred to as compactifications,[2] have been developed to deal with them thus ensuring that the "observable" dimensionality of spacetime is four.

Unlike loop quantum gravity, which provides a background independent formulation of spacetime with the latter emerging from the theory itself, as discussed earlier, the strings in string theories are assumed to move in a pre-determined spacetime, and thus spacetime plays a passive role in them.[3] There are also several superstring theories, and a theory, referred to as M-Theory,[4] based on nonperturbative methods, is envisaged to unify the existing superstrings theories into one single theory, instead of several ones, and be related to them by various limiting and/or transformation rules, referred to as dualities,[5] and is of 11 dimensional spacetime. M-Theory is believed to be approximated by 11 dimensional supergravity,[6] and the spacetime structure may emerge from the theory as well. Bosonic strings involve tachyonic states. This is unlike the situation in superstring theories in which supersymmetry plays a key role in their definitions, and a process referred to as a GSO projection method, ensuring the equality of the degrees of freedom of bosonic and fermionic states, as required by supersymmetry, in turn implies that no tachyonic states appear in the theory.[7]

String theory was accidentally discovered through work carried out by Veneziano in 1968 when he attempted to write down consistent explicit expressions of meson-meson scattering amplitudes in strong interactions physics.[8] This was, of course before the discovery of QCD. With the many excited states of mesons and baryons

---

[2] Such an idea was used by Kaluza and Klein in their attempt to unify gravity and electromagnetism in a 5 dimensional generalization of general relativity.

[3] See also [23].

[4] Townsend [44], Witten [46], Duff [10].

[5] Duff [10], Schwarz [37].

[6] Cremmer et al. [8].

[7] The GSO method of projection was proposed in [13, 14].

[8] Veneziano [45], see also [9, 24].

(resonances), it was observed experimentally that there exists a linear relationship between spin J and the mass M squared of a resonance given by

$$J = \alpha(M^2), \quad \text{with a universal slope :} \quad \frac{dJ}{dM^2} = \alpha', \quad \alpha' \cong 1 \, \text{GeV}^{-2}, \quad (3.1)$$

defining so-called Regge trajectories. Veneziano postulated and wrote down a scattering amplitude of meson-meson scattering: $p_1(m_1) + p_2(m_2) \rightarrow p_3(m_3) + p_4(m_4)$, involving the beta function given by

$$A(s,t) = \frac{\Gamma(-\alpha(s))\Gamma(-\alpha(t))}{\Gamma(-\alpha(s) - \alpha(t))}, \quad (3.2)$$

where s, t are two of so-called Mandelstam variables defined by

$$s = -(p_1 + p_2)^2, \quad t = -(p_1 - p_4)^2, \quad u = -(p_1 - p_3)^2, \quad s + t + u = \sum_{i=1}^{4} m_i^2, \quad (3.3)$$

where $p_1 + p_2 = p_3 + p_4$. The amplitude has some desirable properties, it satisfies the crossing relation $s \leftrightarrow t$, and because of the singular behavior of the *Gamma* function

$$\Gamma(z) \simeq \frac{(-1)^n}{n!} \frac{1}{(z+n)}, \quad \text{for} \quad z \simeq -n, \quad \text{with} \quad n \text{ non-negative integers}, \quad (3.4)$$

we note from (3.2), that for $\alpha(s) \simeq n \geq 0$, that as a function of s

$$A(s,t) \simeq \frac{(-1)^n}{n!} \frac{\Gamma(-\alpha(t))}{\Gamma(-\alpha(s) - n)} \frac{1}{n - \alpha(s)}. \quad (3.5)$$

The above amplitude involves the exchange of an infinite number of particles for the various values of the non-negative integer n corresponding to arbitrary (integer) spins. This is unlike the situation in conventional field theory as QED or the standard model, where a tree approximation involves the exchange of a finite number of particles. String theory shares this property of the Veneziano amplitude. As a matter of fact we will see in Sect. 3.5.1 how the Veneziano amplitude may be derived from string theory. In string theory, two strings with given vibrational modes, identifying two given particles, may combine forming one string with an arbitrary number of different vibrational modes associated with a myriad number of particles, defining generalized 3-vertices (see Sect. 3.5). The combined string may again split into two strings with associated vibrational modes, identified appropriately with two more particles, describing a scattering process of 2 particles → 2 particles. Thus interactions involve string worldsheets of various topologies. Due to the assumed non-zero extensions of strings, it is hoped that they provide, naturally, an ultraviolet cut-off $\Lambda \sim (\ell_P)^{-1}$ and render all processes involving strings ultraviolet finite. This is unlike conventional quantum field theory interactions where all the quantum fields are multiplied locally at the same spacetime points, like multiplying distributions at the same point, and are, in this sense, quite troublesome.

Nambu [30], Nielsen [33] and Susskind [39] have shown that the famous expression of the amplitude postulated by Veneziano may be interpreted as a quantum theory of scattering of relativistic strings. Although, a priori, this was assumed to describe a strong interaction process, Yoneya [48], and Scherk and Schwarz [36] made use of the fact that string theory (involving closed strings) contains a spin 2 massless state, which was identified with the elusive graviton, in addition to a whole spectrum of other excitation modes, to propose that string theory provides a framework for the unification of general relativity and quantum mechanics.

As early as 1971, Neveu and Schwarz [32], and Ramond [35] included fermions in their analyses, which eventually led to the notion of superstrings, and during a short period of time, several types[9] of superstrings were introduced in the literature.

Other extended objects are also encountered in string theory called branes which, in general, are of higher spatial dimensions than one, with the string defined as a one dimensional brane. For example, an open string, satisfying a particular boundary condition, referred to as a Dirichlet boundary condition, specifies a hypersurface, referred to as a D brane, on which the end points of the open string reside. On the other hand, the graviton corresponds to a vibrational mode of closed strings, and since the latter, having no ends, may not be restricted to a brane and moves away from it. This might explain the weakness of the gravitational field, if our universe is a 3 dimensional brane embedded in a higher dimensional spacetime. Massless particles encountered in string theory are really the physically relevant ones because of the large unit of mass $(\ell_P)^{-1} \sim 10^{19}$ GeV in attributing masses to the spectrum of massive particles. A systematic analysis of all the massless field excitations encountered in both bosonic and superstrings are investigated in Manoukian [26–28], in their respective higher dimensional spacetimes, and include the determinations of the degrees of freedom associated with them.[10] As we will see later a massless particle may acquire mass if, for example, the end points of the open string are attached to two different branes, instead of a single brane.

Later we will learn the remarkable fact that Einstein's general relativity (Sect. 3.6), as well as Yang-Mill field theory (Sect. 3.7) may be obtained from string theory.

---

[9]Green and Schwarz [16, 17], Gross et al. [19, 20].

[10]Note that in four dimensional spacetime the number of degrees of freedom (spin states) of non-scalar fields is always two. This is not true in higher dimensional spacetime. For example, the degrees of freedom associated with a massless vector particle is 8 in 10 dimensions, while for the graviton is 35, as shown later in the text. In 4 dimensions, their degrees of freedom are, of course, two.

Interesting high energy scattering amplitudes have been computed in string theory over the years,[11] which provide a hint that space may not be probed beyond the Planck length—a result shared with "loop quantum gravity". It is also worth mentioning that the Bekenstein-Hawking Entropy relation has been also derived in string theory.[12]

In recent years much work has been done indicating that general relationships may exist between field theories and string theories, and consequently considerable attention was given trying to make such a statement more and more precise, with the ultimate hope of providing, in turn, a consistent and acceptable quantum theory of gravitation relevant to our world, but much work still remains to be done. In particular, much study has been made to study the equivalence between certain four dimensional gauge theories and superstring theories, referred to as the AdS/CFT correspondence which we now briefly discuss. Such a correspondence has been also referred to as a Gauge/Gravity duality, as well as Maldacena Duality, a duality which was first proposed by Maldacena.[13] Here AdS stands for Anti-de-Sitter, and CFT stands for conformal field theory which we define below. Although we will not go into theoretical developments of this correspondence principle in the text, it is worth noting and, in turn, briefly discuss that the aim of such a work is to show, for example, the existence of an equivalence relation between a certain supersymmetric SU(N) Yang-Mills field theory in 4 dimensional Minnkowski spacetime, and a superstring theory in 5 dimensional AdS space, having one additional dimension to the Minkowskian one, and with the 5 dimensions of the AdS space supplemented by the 5 dimensions of a 5-sphere, making up the 10 dimensions of a superstring theory. The interest in here is that it deals with a connection between string theory (involving gravity) and a supersymmetric gauge theory.

To the above end, we first define the conformal group as applied in 4 dimensional Minkowski spacetime. This involves a scale transformation $x^{\mu} \to \lambda x^{\mu}$, a so-called special conformal transformation

$$\frac{x'^{\mu}}{x'^{2}} = \frac{x^{\mu}}{x^{2}} + a^{\mu}, \tag{3.6}$$

for a given four vector a^{μ}, in addition to the Poincaré ones, generating all in all *15 generators* which act on the fields as a set of *field transformations*: 4 for the special transformation, one for the scaling, 6 for Lorentz and 4 translations. The field theory (FT) in question is a supersymmetric SU(N) Yang-Mills field theory with an $\mathcal{N} = 4$ supersymmetry.

---

[11]See, e.g., [3, 4, 41].

[12]See, e.g., [22, 38].

[13]Maldacena [25]. See also [2, 21, 47].

D dimensional AdS space, may be defined in terms of the set of all coordinate variables $(z^0, z^1, \ldots, z^{D-1}, z^D) = z$ satisfying a quadratic equation

$$z^2 = \sum_{k=1}^{D-1} (z^k)^2 - (z^0)^2 - (z^D)^2 = -R^2, \qquad (3.7)$$

for a given constant R^2, embedded in a $D + 1$ dimensional space with interval squared

$$ds^2 = \sum_{j=1}^{D-1} dz^{j\,2} - dz^{0\,2} - dz^{D\,2}. \qquad (3.8)$$

The invariance group for theories defined on AdS$_D$ which preserve distances squared $(z_1 - z_2)^2$ via linear transformations: $(z_1 - z_2) \rightarrow (z_1' - z_2') = L(z_1 - z_2)$, involves operators as that of a D-dimensional flat space. That is, it involves $D(D-1)/2$ generators which correspond to Lorentz transformations, and D generators which correspond to translations, making a total of $D(D+1)/2$ generators. This number will coincide with the number of 15 generators of conformal transformations discussed above only if $D = 5$. The latter implies that the space AdS has to be of 5 dimensions, and consequently it may be denoted by AdS$_5$. That is, due to the additional number of generators in the conformal group, the AdS space is of one dimensional higher than that of the 4 dimensional Minkowski space. In order to have a superstring theory in 10 dimensional spacetime, we need extra 5 dimensions to that of AdS$_5$ just described. This is achieved by the 5-sphere S$_5$ which matches the $\mathcal{N} = 4$ transformations of the supersymmetric field theory. We thus generate the ten dimensional spacetime AdS$_5 \times$ S$_5$ for the superstring theory.[14] The latter space carries as isometries, i.e., *distance preserving maps*, the symmetries of the Yang-Mills field theory which reside in 4 dimensional Minkowski spacetime. Thus one expects that the conformal supersymmetric Yang-Mills field theory (with $\mathcal{N} = 4$ spinor supercharges) is the same as a 10 dimensional superstring theory (which includes gravity) on AdS$_5 \times$ S^5. This was conjectured in a remarkable classic paper by Maldacana,[15] and needless to say, there has been much interest in it as it deals with a connection that may exist between string theory (involving gravity) and supersymmetric gauge theories as mentioned earlier.

The above brings us into contact with the holographic principle, in analogy to holography in capturing 3 dimensional images of objects on a two dimensional plate.[16] The conformal quantum field theory is like a hologram capturing infor-

---

[14]The underlying superstring theory is of type II. Such superstrings as well as other ones, of course, will be studied in this chapter.

[15]*loc. cit.*

[16]Recall that the *two* dimensional holographic plate which registers the interference of reflected light off an object and an unperturbed Laser beam stores information of the shape of the *three*

mation about the higher dimensional quantum gravity theory, i.e., the boundary of AdS space may be regarded as the spacetime for a conformal field theory. In this case the SU(N) theory provides a holographic description of gravitational field. This is in analogy to black hole entropy, with information encoded in it, being proportional to the area rather than to the volume of the region enclosed by the horizon. Perhaps holography is a basic property of string theory and one expects that much has to be done before developing a realistic quantum gravity, and in turn provide a background independent formulation for string theory. The holographic principle was first proposed by 't Hooft.[17] We will not, however, elaborate further on AdS/CFT correspondence and the holographic principle in the text, and the interested reader may refer to the references just provided.

This chapter is involved with a systematic presentation of the fundamentals of string theory. The first section deals with the development of action integrals for relativistic particles and superparticles and of their interpretations as a preparation to study strings. Bosonic strings are treated in Sect. 3.2, which includes the definition of the action integral, underlying boundary conditions, details of the light-cone gauge, open and closed strings, oriented and unoriented strings. Quantum aspects of bosonic strings are developed and the mass spectra of the string vibrational modes are analyzed with emphasis put on massless particles. The central problem of their compactification and the associated concept of a duality, referred to as T duality, are finally analyzed. A detailed study is carried out of all the massless fields that emerge from the theory of bosonic strings and their properties, such as their associated degrees of freedom, are spelled out. This is followed by a study of superstrings. Here Dirac fields are included in the action integral, and the Dirac equation in two dimensions plays a key role. Boundary conditions are examined in detail and the concept of the so-called GSO projection method, to ensure supersymmetry, is invoked. The fundamental types of superstrings are analyzed, and the concept of compactification and the various duality relationships that are envisaged to exist between these types and that of M-theory are briefly discussed, giving the hope of having one unified theory instead of several superstring theories. Detailed studies are also carried out of all the massless bosonic and fermionic fields that emerge from the theory of superstrings and their properties are spelled out. D branes, as extended objects, specified by Dirichlet boundary conditions, and to which the ends of open strings are restricted to move is the subject matter of Sect. 3.4. Here, in particular, we show how a massless vector field may acquire mass, when the ends of the open string, in question, are attached to two different D branes mentioned earlier. Anti-symmetric tensor fields, which are obtained in the mass spectra of vibrating strings, due to their very nature, interact, in turn, with such extended objects providing a justification of the existence of branes. Section 3.5 deals with the introduction of

---

dimensional object. As one shines a Laser beam on it an image of the three dimensional object emerges.

[17]'t Hooft [42], see also especially [43], as well as the analysis, with further interpretations, by Susskind [40]. See also [6].

vertices in string theory and in developing expressions for scattering amplitudes. In particular, the classic expression of the Veneziano amplitude, mentioned above, is derived from string theory.

Simple demonstrations are given in Sects. 3.6 and 3.7, showing how Einstein's theory of general relativity and Yang-Mills field theory may emerge from string theory, respectively. No attempt will be made in this introductory presentation, however, to consider supergravity in the light of superstring theory.

3.1   The Relativistic Particle and the Relativistic Superparticle

In this section we set up the action integrals for a relativistic particle as well as of a relativistic superparticle.

3.1.1   Action of the Relativistic Particle

A relativistic particle, as it moves in spacetime, traces a curve referred to as its worldline (Fig. 3.1). In Minkowski spacetime, generalized, for convenience for later studies, to D dimensions, with one time variable x^0, and $D - 1$ space variables $x^1, \ldots, x^{D-1}$, as set up in a given coordinate system, the interval ds may be expressed as

$$\mathrm{d}s = \mathrm{c}\,\mathrm{d}t\,\sqrt{1 - \frac{1}{\mathrm{c}^2}\dot{\mathbf{X}}^2},\qquad(3.1.1)$$

where the X^μ, in capital letters, with $\mu = 0, 1, \ldots, D - 1$, $\mathbf{X} = (X^1, \ldots, X^{D-1})$, are the coordinate labels of points on the string in the "laboratory" system one has set up, and we have inserted the speed of light constant c for further analysis. Here $\dot{\mathbf{X}} = \mathrm{d}\mathbf{X}/\mathrm{d}t$, and as usual $X^0 = \mathrm{c}\,t$. The metric is defined by $\eta_{\mu\nu} = \mathrm{diag}[-1, 1, \ldots, 1]$.

For low speeds, we may expand the expression in (3.1.1) as follows,

$$\mathrm{d}s \simeq -\frac{1}{m\mathrm{c}}\left[\frac{m\dot{\mathbf{X}}^2}{2} - m\,\mathrm{c}^2\right]\mathrm{d}t.\qquad(3.1.2)$$

We recognize the terms within the square brackets as the non-relativistic kinetic energy minus a constant potential energy. Accordingly, we may define the action

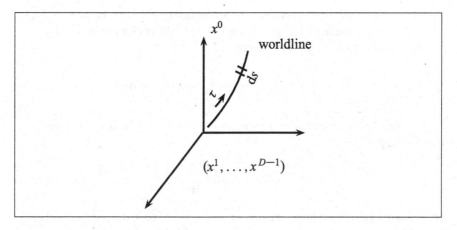

Fig. 3.1 Worldline of a relativistic particle

for the relativistic particle by

$$W = -mc \int ds = -mc \int \sqrt{-\eta_{\mu\nu} dX^\mu dX^\nu} = -mc \int d\tau \sqrt{-\eta_{\mu\nu} \frac{dX^\mu}{d\tau} \frac{dX^\nu}{d\tau}},$$
(3.1.3)

where in the last integral we have parameterized the integrand by the proper time τ.

By varying the action with respect to X^μ, we obtain

$$\frac{d}{d\tau} \left[\left(-\frac{dX^\nu}{d\tau} \frac{dX_\nu}{d\tau} \right)^{-1/2} \frac{d}{d\tau} X^\mu \right] = 0.$$
(3.1.4)

The above equation may be more conveniently rewritten as a simultaneous set of two equations by introducing, in the process, an additional variable, as a function of τ, as follows. Set

$$mc\, a(\tau) = \left(-\frac{dX^\nu}{d\tau} \frac{dX_\nu}{d\tau} \right)^{1/2},$$
(3.1.5)

to obtain

$$\frac{d}{d\tau} \left[\frac{1}{a(\tau)} \frac{d}{d\tau} X^\mu(\tau) \right] = 0.$$
(3.1.6)

Therefore, instead of working with the action in (3.1.3), involving the square root expression, one may consider, anew, the following action integral for the relativistic particle

$$W = \frac{1}{2} \int d\tau \left[\frac{1}{a(\tau)} \frac{dX^\mu}{d\tau} \frac{dX_\mu}{d\tau} - m^2 c^2 a(\tau) \right],$$
(3.1.7)

By varying the action with respect to X^μ we obtain (3.1.6). On the other hand, by varying $a(\tau)$, we obtain (3.1.5) which when substituted in (3.1.6) gives (3.1.4).

The action in (3.1.7) may be re-expressed as

$$W = \frac{1}{2} \int \left[\frac{1}{a(\tau)\,d\tau} \, dX^\mu dX_\mu - m^2 c^2 a(\tau)\,d\tau \right]. \tag{3.1.8}$$

The action is invariant under re-parametrizations $\tau \to \tau'$, with the function $a(\tau)$ transforming as

$$a'(\tau') = a(\tau)\frac{d\tau}{d\tau'}. \tag{3.1.9}$$

Indeed under an infinitesimal variation $\delta\tau = \tau - \tau' = \lambda(\tau)$, $d\tau'/d\tau = 1 - d\lambda/d\tau$, the following variations emerge

$$\delta X^\mu(\tau) = X^\mu(\tau) - X'^\mu(\tau) = -\dot{X}^\mu(\tau)\,\lambda(\tau), \quad \dot{X}^\mu = \frac{dX^\mu}{d\tau}, \tag{3.1.10}$$

$$\delta a(\tau) = a(\tau) - a'(\tau) = -\frac{d}{d\tau}\left(a(\tau)\lambda(\tau)\right), \tag{3.1.11}$$

and $\delta W = 0$, up to a total derivative. Accordingly, one may choose a gauge such that $a(\tau)$ is some arbitrary constant, say, κ, having the dimension of $(\text{mass})^{-1}$, to rewrite (3.1.7) in this gauge as

$$W = \frac{1}{2\kappa} \int d\tau \left[\dot{X}^\mu \dot{X}_\mu - m^2 c^2 \kappa^2 \right], \quad \dot{X}^\mu = \frac{dX^\mu}{d\tau}. \tag{3.1.12}$$

For a massless particle $m = 0$, $1/\kappa$ may be chosen as any convenient mass scale parameter. On the other hand, for $m \neq 0$, set $\kappa = 1/m$.

3.1.2 The Relativistic Superparticle

We note that the integral (3.1.12) is a one dimensional integral. Consider the latter for $m = 0$. To develop an action with worldline supersymmetry to describe a superparticle, we note that in one dimension, one may introduce one theta. Accordingly, superfields may be defined involving a single power of theta as follows

$$\Phi^\mu(\tau,\theta) = X^\mu(\tau) + \frac{i}{\sqrt{2}} \, \theta \, \psi^\mu(\tau), \qquad \theta^2 = 0, \tag{3.1.13}$$

where $\psi^0(\tau), \ldots, \psi^{D-1}(\tau)$ denote D fermion fields. The i factor in the equation is to ensure the reality of Φ^μ since complex conjugation reverses the order of anti-commuting factors in a product and $\{\psi^\mu, \theta\} = 0$.

We define one dimensional supersymmetry transformations

$$\tau' = \tau + \frac{i}{2}\epsilon\theta, \quad \theta' = \theta + \epsilon. \tag{3.1.14}$$

Accordingly, from (3.1.14), the infinitesimal variation of the superfield is given by

$$\delta\Phi^\mu = \epsilon\left(\frac{\partial}{\partial\theta} + \frac{i}{2}\theta\frac{\partial}{\partial\tau}\right)\Phi^\mu, \tag{3.1.15}$$

leading from (3.1.13) to

$$\delta X^\mu = \frac{i}{\sqrt{2}}\epsilon\psi^\mu, \quad \delta\psi^\mu = \frac{1}{\sqrt{2}}\epsilon\dot{X}^\mu, \quad \left(\dot{X}^\mu = \frac{dX^\mu}{d\tau}\right). \tag{3.1.16}$$

We recall the definition of the supercovariant derivative [see (2.3.9)], which now takes the simple form

$$D = \left(\frac{\partial}{\partial\theta} - i\frac{\theta}{2}\frac{\partial}{\partial\tau}\right),$$

and leads to

$$D\Phi^\mu = i\left(\frac{1}{\sqrt{2}}\psi^\mu - \frac{1}{2}\theta\dot{X}^\mu\right), \quad D^2\Phi^\mu = \frac{1}{2}\left(\frac{1}{\sqrt{2}}\theta\dot{\psi}^\mu - i\dot{X}^\mu\right), \tag{3.1.17}$$

$$(D\Phi^\mu)D(D\Phi_\mu) = -\frac{1}{4}\theta\left(\dot{X}^\mu\dot{X}_\mu + i\psi^\mu\dot{\psi}_\mu\right) + \frac{1}{2\sqrt{2}}\psi^\mu\dot{X}_\mu, \tag{3.1.18}$$

where we have used the anti-commutativity of ψ^μ and θ to write the last equation. This equation allows us to define a Lagrangian $\propto 2(D\Phi^\mu)D(D\Phi^\mu)$, giving an action

$$W = \frac{1}{2\kappa}\int d\tau\left(\dot{X}^\mu\dot{X}_\mu + i\psi^\mu\dot{\psi}_\mu\right), \tag{3.1.19}$$

where we have used, by now, the elementary integrals

$$\int d\theta = 0, \quad \int d\theta\,\theta = 1.$$

Again the i factor, within the integrand in (3.1.19), ensures the reality of the action. This action should be compared with the one in (3.1.12) for $m = 0$.

Next we consider strings.

3.2   Bosonic Strings

This section deals with the intricacies of bosonic strings.

3.2.1   The Bosonic String Action

An open or closed string as it moves in spacetime, it sweeps out a two-dimensional surface referred to as its worldsheet . Some worldsheets generated by an open string and a closed one are shown in Fig. 3.2:

The points on the worldsheet of a string are parametrized by coordinates $\xi^0 = \tau$, $\xi^1 = \sigma$, which correspond, respectively, to timelike and spacelike directions. That is, $\xi^1 = \sigma$ is the spacial coordinate *along* the string, i.e., it provides a partition of the string for somebody "sitting" on it, while $\xi^0 = \tau$, describes its propagation in time. In a coordinate system set up ("laboratory system"), a point on the string is labeled by $X^\mu(\tau, \sigma), \mu = 0, 1, \ldots, D-1$.

We recall that the action for a relativistic particle was given by $-mc \int ds$, corresponding to the curve traced by the particle, where ds is a measure of "length" in spacetime. The proportionality constant $-mc$, was fixed by considering the non-relativistic limit of the action, and, of course, by checking the correct dimensionality of the action. To obtain the expression for the action of a string, we do the same thing

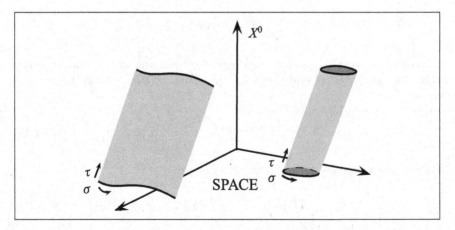

Fig. 3.2 Examples of worldsheets of an open string and a closed one. No specific dynamics of the strings are considered in these diagrams

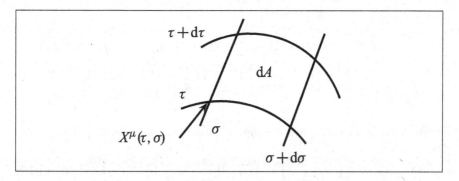

Fig. 3.3 $\int dA$ corresponds to the area swept by the string in spacetime

and write the action as Const. $\int dA$, corresponding to the area swept by the string in spacetime, where dA is such an infinitesimal area, and investigate the nature of the proportionality constant (Fig. 3.3).

To see how a two-dimensional surface may be generated in Minkowski spacetime, consider some event with coordinate label $X^\mu(\lambda^1, \lambda^2)$, in some coordinate system, parametrized by a pair of variables λ^1, λ^2. Suppose λ^2 is kept fixed and λ^1 is varied. $X^\mu(\lambda^1, \lambda^2)$ will then trace a curve, and a tangent vector to it may be then defined:

$$X^\mu(\lambda^1, \lambda^2)\Big|_{\lambda^2 \text{ fixed}}, \quad \text{tangent vector}: \quad d_1 X^\mu = \frac{\partial X^\mu}{\partial \lambda^1} d\lambda^1. \tag{3.2.1}$$

Similarly, for λ^1 kept fixed and λ^2 is varied, $X^\mu(\lambda^1, \lambda^2)$ will then trace a curve, and a tangent vector to it may be then defined:

$$X^\mu(\lambda^1, \lambda^2)\Big|_{\lambda^1 \text{ fixed}}, \quad \text{tangent vector}: \quad d_2 X^\mu = \frac{\partial X^\mu}{\partial \lambda^2} d\lambda^2. \tag{3.2.2}$$

In analogy to an oriented surface in 3 dimensional Euclidean space

$$(d_1 \mathbf{X} \times d_2 \mathbf{X})^i = \varepsilon^{ijk} d_1 X^j d_2 X^k, \tag{3.2.3}$$

an oriented two-dimensional surface element may be defined in a D dimensional Minkowski space by

$$d\sigma^{\mu_1 \cdots \mu_{D-2}} = \epsilon^{\mu_1 \cdots \mu_{D-2} \mu \nu} d_1 X_\mu d_2 X_\nu, \tag{3.2.4}$$

where $\epsilon^{\mu_1 \cdots \mu_D}$ is totally anti-symmetric with $\epsilon^{01 \cdots (D-1)} = 1$.

Due to the antisymmetry of $\epsilon^{\mu_1 \cdots \mu_D}$, one and only one of its indices must be zero, i.e., $\epsilon_{\mu_1 \cdots \mu_D} = -\epsilon^{\mu_1 \cdots \mu_D}$, where we recall that we are working in Minkowski spacetime. Using the identity

$$\epsilon_{\mu_1 \cdots \mu_{D-2} \mu \nu} \, \epsilon^{\mu_1 \cdots \mu_{D-2} \sigma \lambda} = (D-2)! \, (\delta_\mu{}^\lambda \delta_\nu{}^\sigma - \delta_\mu{}^\sigma \delta_\nu{}^\lambda), \tag{3.2.5}$$

the following equality then emerges

$$d\sigma_{\mu_1\dots\mu_{D-2}}\, d\sigma^{\mu_1\dots\mu_{D-2}} = (D-2)!(d_1X^\mu d_2X_\mu\, d_1X^\nu d_2X_\nu - d_1X^\mu d_1X_\mu\, d_2X^\nu d_2X_\nu)$$

$$= (D-2)!\left[\left(\frac{\partial X^\mu}{\partial\lambda^1}\frac{\partial X_\mu}{\partial\lambda^2}\right)^2 - \left(\frac{\partial X^\mu}{\partial\lambda^1}\right)\left(\frac{\partial X_\mu}{\partial\lambda^1}\right)\left(\frac{\partial X^\nu}{\partial\lambda^2}\right)\left(\frac{\partial X_\nu}{\partial\lambda^2}\right)\right](d\lambda^1)^2\,(d\lambda^2)^2.$$

$$(3.2.6)$$

Therefore as a measure of a two-dimensional surface area, we may define

$$dA = d\lambda^1\, d\lambda^2 \left[\left(\frac{\partial X^\mu}{\partial\lambda^1}\frac{\partial X_\mu}{\partial\lambda^2}\right)^2 - \left(\frac{\partial X^\mu}{\partial\lambda^1}\right)\left(\frac{\partial X_\mu}{\partial\lambda^1}\right)\left(\frac{\partial X^\nu}{\partial\lambda^2}\right)\left(\frac{\partial X_\nu}{\partial\lambda^2}\right)\right]^{1/2}, \qquad (3.2.7)$$

provided the expression under the square-root is non-negative.

With the worldsheet parametrized by (τ,σ), we may use (3.2.7), to write the action for a string as

$$W = -\frac{T}{c}\int d\tau\, d\sigma\, \sqrt{(\dot{X}^\mu X'_\mu)^2 - (\dot{X}^\mu \dot{X}_\mu)(X'^\nu X'_\nu)}, \qquad (3.2.8)$$

where T/c is a proportionality constant whose nature will be investigated below, and

$$\dot{X}^\mu(\tau,\sigma) = \frac{\partial X^\mu(\tau,\sigma)}{\partial\tau}, \qquad X'^\mu(\tau,\sigma) = \frac{\partial X^\mu(\tau,\sigma)}{\partial\sigma}. \qquad (3.2.9)$$

Since the scalar product of a timelike vector with itself has the opposite sign of the scalar product of a spacelike vector with itself, the expression under the square root sign in the integrand in (3.2.8) is non-negative. The presence of the minus sign multiplying the integral in (3.2.8) will become clear below.

Clearly, the dimension of T in (3.2.8) is energy per unit length. As for the situation with a particle, let us examine a non-relativistic limit of this action. To this end, consider a string for which $\partial X^\mu/\partial\sigma = (0, \partial\mathbf{X}/\partial\sigma) \equiv (0, \mathbf{X}')$, i.e., $\partial t/\partial\sigma = 0$, meaning that the string moves in such a way that all the clocks set at each point on it are always synchronized. This leads with, $\mathbf{v} = \dot{\mathbf{X}}$, to

$$\dot{X}^\mu X'_\mu = \left(\frac{dt}{d\tau}\right)\mathbf{v}\cdot\mathbf{X}', \quad X'^\mu X'_\mu = \mathbf{X}'^2, \quad \dot{X}^\mu \dot{X}_\mu = \left(\frac{dt}{d\tau}\right)^2(\mathbf{v}^2 - c^2), \quad (3.2.10)$$

where t, is the coordinate time, i.e., $t = X^0/c$ as determined in the "laboratory" system. Hence for $\mathbf{v}^2 \ll c^2$, the action becomes

$$W \simeq -T\int dt\, d\sigma \left(1 - \frac{1}{2c^2}[\mathbf{v}^2 - (\mathbf{v}\cdot\mathbf{N})^2]\right)|\mathbf{X}'|, \qquad (3.2.11)$$

where $\mathbf{N} = \mathbf{X}'/|\mathbf{X}'|$, specifies a unit vector along the vector $\mathbf{X}'$ at a point σ on the string. Then we note that the transverse velocity $\mathbf{v}_{\text{trans}} = [\mathbf{v} - \mathbf{N}(\mathbf{v} \cdot \mathbf{N})]$, satisfies

$$(\mathbf{v}_{\text{trans}})^2 = [\mathbf{v} - \mathbf{N}(\mathbf{v} \cdot \mathbf{N})]^2 = [\mathbf{v}^2 - (\mathbf{v} \cdot \mathbf{N})^2]. \tag{3.2.12}$$

Thus the action takes the limiting form

$$W \simeq \int \mathrm{d}t \left[\int \mathrm{d}\sigma \left(\frac{T|\mathbf{X}'|}{c^2} \right) \frac{\mathbf{v}_{\text{trans}}^2}{2} - \int \mathrm{d}\sigma \, T|\mathbf{X}'| \right]. \tag{3.2.13}$$

Note the minus sign within the square brackets above. We see that the string moves with a transverse velocity with an associated mass per unit length m_σ, and a potential energy, given, respectively, by

$$m_\sigma = \frac{T|\mathbf{X}'|}{c^2}, \quad \text{Pot. En.} = \int \mathrm{d}\sigma \, T|\mathbf{X}'| = mc^2, \quad \int \mathrm{d}\sigma \, m_\sigma = m. \tag{3.2.14}$$

Since $|\mathbf{X}'|\mathrm{d}\sigma = |\partial \mathbf{X}/\partial \sigma|\mathrm{d}\sigma$ is an infinitesimal change of length of the string, the factor T is interpreted as the tension in the string, as the energy responsible of stretching the string per unit length. In turn, we see that the origin of the mass of the string is due to its inherit tension. The above equation also justifies the introduction of the minus sign multiplying the integral in (3.2.8).

The action in (3.2.8) is called the Nambu-Goto action [15, 31]. Further analysis of this action is given next.

3.2.2 The String Sigma Model Action: Curved Nature of the Worldsheet

We will re-express the Nambu-Goto action in (3.2.8) in a way that it does not involve the square root sign in the integrand by following a procedure similar to the one carried out for a point particle in Sect. 3.1.1. From now on we work with natural units, with, in particular, $c = 1$.

The interval (squared) connecting two points $X^\mu(\tau, \sigma)$ and $X^\mu(\tau + \mathrm{d}\tau, \sigma + \mathrm{d}\sigma)$ on a string worldsheet, may be written as (in our earlier notation $\mathrm{d}s = \sqrt{\mathrm{d}S^2}$)

$$\mathrm{d}S^2 = -\eta_{\mu\nu} \, \mathrm{d}X^\mu \, \mathrm{d}X^\nu = -\eta_{\mu\nu} \frac{\partial X^\mu}{\partial \xi^\alpha} \frac{\partial X^\nu}{\partial \xi^\beta} \, \mathrm{d}\xi^\alpha \, \mathrm{d}\xi^\beta = -h_{\alpha\beta} \, \mathrm{d}\xi^\alpha \, \mathrm{d}\xi^\beta, \tag{3.2.15}$$

thus introducing the induced metric on the surface of a worldsheet

$$h_{\alpha\beta} = \eta_{\mu\nu} \frac{\partial X^\mu}{\partial \xi^\alpha} \frac{\partial X^\nu}{\partial \xi^\beta} \equiv \partial_\alpha X^\mu \, \partial_\beta X_\mu, \quad \text{with} \ \ \xi^0 = \tau, \ \ \xi^1 = \sigma. \tag{3.2.16}$$

Note that μ, ν denote spacetime indices, taking values $0, 1, \ldots, (D-1)$, while ξ^0, ξ^1 are defined above.

The inverse $h^{\alpha\beta}$ of $h_{\alpha\beta}$ is defined by $h^{\alpha\gamma} h_{\gamma\beta} = \delta^\alpha{}_\beta$, giving the following connecting formulae between the components

$$h_{10} = h_{01} = -h\,h^{01}, \quad h_{11} = h\,h^{00}, \quad h_{00} = h\,h^{11}, \quad h \equiv \det[h_{\alpha\beta}].$$
$$(3.2.17)$$

The equation of motion obtained by varying X^μ in the Nambu-Goto action in (3.2.8) easily follows by noting, in the process, that for a function $f(X)$

$$\delta\sqrt{f(X)} = \frac{1}{2\sqrt{f(X)}}\,\delta(\sqrt{f(X)})^2,$$
$$(3.2.18)$$

and remembering the definitions of the partial derivatives in (3.2.9). The field equations resulting from varying the Nambu-Goto action in (3.2.8), with respect to X^μ, and using the notations $\partial_0 = \partial/\partial\tau = \partial/\partial\xi^0$, $\partial_1 = \partial/\partial\sigma = \partial/\partial\xi^1$, are given by

$$\partial_0 \frac{1}{\sqrt{-h}}\Big([\partial_0 X \cdot \partial_1 X]\,\partial_1 X^\mu - [\partial_1 X \cdot \partial_1 X]\,\partial_0 X^\mu\Big)$$
$$+ \partial_1 \frac{1}{\sqrt{-h}}\Big([\partial_0 X \cdot \partial_1 X]\,\partial_0 X^\mu - [\partial_0 X \cdot \partial_0 X]\,\partial_1 X^\mu\Big) = 0.$$
$$(3.2.19)$$

Here we have used a standard notation $\partial_\alpha X^\mu\,\partial_\beta X_\mu = \partial_\alpha X \cdot \partial_\beta X$, and the fact that from (3.2.16), the expression under the square root in (3.2.8) is equal to $-h$. Using the expression of the connecting formulae of induced metric $h_{\alpha\beta}$ with its inverse in (3.2.17), we may rewrite (3.2.19) in the compact form[18]

$$\partial_\alpha(\sqrt{-h}\,h^{\alpha\beta}\,\partial_\beta X^\mu) = 0.$$
$$(3.2.20)$$

By multiplying (3.2.16) by $h^{\alpha\beta}$, the resulting equality $2 = h^{\alpha\beta}\,\partial_\alpha X \cdot \partial_\beta X$, allows one to rewrite (3.2.16) as

$$\partial_\alpha X \cdot \partial_\beta X = \frac{1}{2} h_{\alpha\beta}(h^{\gamma\delta}\partial_\gamma X \cdot \partial_\delta X).$$
$$(3.2.21)$$

Instead of working with the action in (3.2.8), involving the square root expression, one may consider, *anew*, an equivalent action integral depending in addition to X^μ, the $h_{\alpha\beta}$, and vary these variables independently. This is easy to do provided one uses the following formula for the variation of a determinant $h = \det h_{\mu\nu}$, given and

---

[18]See Problem 3.2.

proved in Appendix A, (A-1.9) of Chap. 1 which leads to $\left(h^{\mu\nu}\,\delta h_{\mu\nu} = -h_{\mu\nu}\,\delta h^{\mu\nu}\right)$

$$\delta\sqrt{-h} = -\frac{1}{2}\sqrt{-h}\,h_{\alpha\beta}\,\delta h^{\alpha\beta}. \tag{3.2.22}$$

The action of the bosonic string may be then rewritten as

$$W = -\frac{T}{2}\int (d\xi)\sqrt{-h}\,h^{\alpha\beta}\,\partial_\alpha X\cdot\partial_\beta X, \qquad (d\xi) = d\xi^0\,d\xi^1 = d\tau\,d\sigma. \tag{3.2.23}$$

The overall $1/2$ factor above is because of the reality of the components X^μ and hence due to the symmetry of the product $h^{\alpha\beta}\,\partial_\alpha X\cdot\partial_\beta X$.

By considering the variation of the action (3.2.23) with respect to the X^μ leads to (3.2.20). On the other hand, to carry out the variation with respect to $h^{\alpha\beta}$ we may use the identity

$$\delta(\sqrt{-h}\,h^{\alpha\beta}) = \sqrt{-h}\left(\delta^\alpha{}_\gamma\,\delta^\beta{}_\delta - \frac{1}{2}h_{\gamma\delta}\,h^{\alpha\beta}\right)\delta h^{\gamma\delta}, \tag{3.2.24}$$

which follows from (3.2.22), and immediately gives (3.2.21).

The energy-momentum tensor $T_{\alpha\beta}$ is defined by[19]

$$T_{\alpha\beta} = -\frac{2}{T}\frac{1}{\sqrt{-h}}\frac{\delta W}{\delta h^{\alpha\beta}}, \tag{3.2.25}$$

which from (3.2.23), (3.2.21) gives

$$T_{\alpha\beta} = \partial_\alpha X\cdot\partial_\beta X - \frac{1}{2}h_{\alpha\beta}\,h^{\gamma\delta}\partial_\gamma X\cdot\partial_\delta X = 0, \tag{3.2.26}$$

by a direct application of (3.2.24). The equality to zero on the extreme right-hand side of (3.2.26) is a consequence of (3.2.21) which now follows from the action in (3.2.23). That the energy-momentum tensor is zero is of no surprise, since the action in (3.2.23) does not include an additional term corresponding to a dynamical equation (no kinetic energy expression) for $h^{\alpha\beta}$.

The action W in (3.2.23) is referred to as the "String Sigma Model Action" and has the names of many authors associated with it, notably Polyakov, Deser, Zumino, Brink, Di Vecchia and many others.

In the next section we will see that a particular gauge may be found to simplify the expression of the action considerably.

---

[19]See also (1.2.22).

3.2.3 Parametrization Independence and the Light-Cone Gauge

The string sigma model action in (3.2.23) may be simplified quite a bit. We will see below and in Problem 3.3 that due to its parametrization independence, via coordinate transformations, followed by a Weyl scale transformation: $h_{\alpha\beta} \rightarrow h_{\alpha\beta} \exp[\phi]$, under which the following combination remains invariant

$$\sqrt{-h}\, h^{\alpha\beta} \rightarrow \sqrt{-h}\, h^{\alpha\beta}, \qquad (3.2.27)$$

the induced metric $[h_{\alpha\beta}]$, on the worldsheet of a string, may be reduced to a flat space metric, i.e., to the Minkowski one in two dimensions: $[\eta_{\alpha\beta}] = \text{diag}[-1, 1]$.

For a Minkowski metric in 2D, the field equations in (3.2.20) read

$$\Box X^\mu = 0, \quad \Box \equiv \eta^{\alpha\beta} \partial_\alpha\, \partial_\beta, \quad [\eta^{\alpha\beta}] = \text{diag}[-1, 1], \qquad (3.2.28)$$

and the energy-momentum tensor in (3.2.26) takes the form

$$T_{\alpha\beta} = \partial_\alpha X \cdot \partial_\beta X - \frac{1}{2}\, \eta_{\alpha\beta} \partial^\gamma X \cdot \partial_\gamma X = 0. \qquad (3.2.29)$$

providing a constraint in the theory. We note that the trace $T^\alpha{}_\alpha$ is identically equal to zero.

The re-parametrization, via coordinate transformations, and Weyl scaling to bring the worldsheet induced metric $h_{\alpha\beta}$ to the Minkowski form, as a gauge choice, is referred to as the conformal gauge.

It is convenient to re-express the field equations in terms of light-cone variables:

$$X^+ = \frac{X^0 + X^{D-1}}{\sqrt{2}}, \quad X^- = \frac{X^0 - X^{D-1}}{\sqrt{2}}, \quad X^i, \ i = 1, \ldots, D-2, \qquad (3.2.30)$$

to read

$$\Box X^+ = 0, \quad \Box X^- = 0, \quad \Box X^i = 0, \ i = 1, \ldots, D-2. \qquad (3.2.31)$$

By a re-parametrization via a coordinate transformation $\xi \rightarrow \rho$, leading to $h_{\alpha\beta} \rightarrow \hat{h}_{\alpha\beta}$, the latter two metrics are related by

$$\hat{h}_{\alpha\beta}(\rho) = h_{\alpha'\beta'}(\xi(\rho)) \frac{\partial \xi^{\alpha'}}{\partial \rho^\alpha} \frac{\partial \xi^{\beta'}}{\partial \rho^\beta}. \qquad (3.2.32)$$

In matrix form, this equation takes the form ($\hat{h}_{10} = \hat{h}_{01}$, $h_{10} = h_{01}$)

$$
\begin{pmatrix} \hat{h}_{00} & \hat{h}_{01} \\ \hat{h}_{10} & \hat{h}_{11} \end{pmatrix} = \begin{pmatrix} \frac{\partial \xi^0}{\partial \rho^0} & \frac{\partial \xi^1}{\partial \rho^0} \\ \frac{\partial \xi^0}{\partial \rho^1} & \frac{\partial \xi^1}{\partial \rho^1} \end{pmatrix} \begin{pmatrix} h_{00} & h_{01} \\ h_{10} & h_{11} \end{pmatrix} \begin{pmatrix} \frac{\partial \xi^0}{\partial \rho^0} & \frac{\partial \xi^0}{\partial \rho^1} \\ \frac{\partial \xi^1}{\partial \rho^0} & \frac{\partial \xi^1}{\partial \rho^1} \end{pmatrix},
\tag{3.2.33}
$$

where the first matrix on the right-hand side is the transpose of the third matrix. The matrices $\hat{h}_{\beta\beta}$, $h_{\alpha\beta}$ have the same signatures as the Minkowski metric with one of their eigenvalues being negative and the other positive. In Problem 3.3, it is shown that $h_{\alpha\beta}$ may be brought to a diagonal form

$$
\hat{h}_{\alpha\beta} = \begin{pmatrix} -\lambda & 0 \\ 0 & \lambda \end{pmatrix}, \quad \text{where} \quad \lambda \equiv \exp[\phi] > 0.
$$

We can relate the metric $[\hat{h}_{\alpha\beta}] = \text{diag}[-e^\phi, e^\phi]$ to the Minkowski one, via a transformation rule as follows:

$$
e^\phi \, \eta_{\alpha\beta} = \eta_{\alpha'\beta'} \frac{\partial \zeta^{\alpha'}}{\partial \rho^\alpha} \frac{\partial \zeta^{\beta'}}{\partial \rho^\beta}.
\tag{3.2.34}
$$

Upon multiplying the latter equation by minus one, we obtain the following set of equations

$$
e^\phi = \frac{\partial \zeta^0}{\partial \rho^0} \frac{\partial \zeta^0}{\partial \rho^0} - \frac{\partial \zeta^1}{\partial \rho^0} \frac{\partial \zeta^1}{\partial \rho^0},
\tag{3.2.35}
$$

$$
0 = \frac{\partial \zeta^0}{\partial \rho^0} \frac{\partial \zeta^0}{\partial \rho^1} - \frac{\partial \zeta^1}{\partial \rho^0} \frac{\partial \zeta^1}{\partial \rho^1},
\tag{3.2.36}
$$

$$
-e^\phi = \frac{\partial \zeta^0}{\partial \rho^1} \frac{\partial \zeta^0}{\partial \rho^1} - \frac{\partial \zeta^1}{\partial \rho^1} \frac{\partial \zeta^1}{\partial \rho^1}.
\tag{3.2.37}
$$

Clearly,

$$
\frac{\partial \zeta^0}{\partial \rho^0} = \frac{\partial \zeta^1}{\partial \rho^1},
\tag{3.2.38}
$$

$$
\frac{\partial \zeta^1}{\partial \rho^0} = \frac{\partial \zeta^0}{\partial \rho^1},
\tag{3.2.39}
$$

satisfy these equations exactly. Linear combinations of the above two equations lead to

$$\left(\frac{\partial}{\partial\rho^0} - \frac{\partial}{\partial\rho^1}\right)\zeta^+ = 0, \qquad \zeta^+ = \frac{\zeta^0 + \zeta^1}{\sqrt{2}}, \qquad (3.2.40)$$

$$\left(\frac{\partial}{\partial\rho^0} + \frac{\partial}{\partial\rho^1}\right)\zeta^- = 0, \qquad \zeta^- = \frac{\zeta^0 - \zeta^1}{\sqrt{2}}. \qquad (3.2.41)$$

Using the facts that $[(\partial/\partial\rho^0) \pm (\partial/\partial\rho^1)]\rho^\mp = 0$, we may infer that $\zeta^\pm$ is each a linear combination of functions of $\rho^\pm$. Since in turn $\zeta^0 = (\zeta^+ + \zeta^-)/\sqrt{2}$, we note that ζ^0 is a linear combination of functions of ρ^+ and ρ^-. That is, ζ^0 satisfies the free field equation

$$\frac{\partial}{\partial\rho^+}\frac{\partial}{\partial\rho^-}\zeta^0 = 0 \quad \text{or} \quad \left[\left(\frac{\partial}{\partial\rho^1}\right)^2 - \left(\frac{\partial}{\partial\rho^0}\right)^2\right]\zeta^0 = 0. \qquad (3.2.42)$$

This allows one to choose ζ^0 as one of the fields in (3.2.31), corresponding to a flat metric. The light-cone gauge consists in choosing ζ^0 to coincide with X^+, or more precisely as a linear function of the latter: $\zeta^0 = c_1 + c_2 X^+$ where c_1, c_2, are constants, and express X^+ as

$$X^+(\zeta^0, \zeta^1) = a + b\zeta^0. \qquad (3.2.43)$$

Here a, b are constants. From now on and for convenience we will use the notation $\tau \equiv \xi^0$ for ζ^0, $\sigma \equiv \xi^1$ for ζ^1, and use the notation $X^-(\xi)$, $X^i(\xi)$ for $i = 1, \ldots, D-2$, with the understanding that all the fields X^+, X^-, X^i now satisfy the field equations (3.2.31) with respect to the newly defined coordinate variables ξ^0, ξ^1, in a two dimensional flat, i.e., in Minkowski, space.

In the conformal gauge , the bosonic action in (3.2.23) takes the form

$$W = -\frac{T}{2}\int(\text{d}\xi)\,\eta^{\alpha\beta}(\partial_\alpha X)\cdot(\partial_\beta X), \quad (\text{d}\xi) = \text{d}\xi^0\,\text{d}\xi^1 \equiv \text{d}\tau\,\text{d}\sigma. \qquad (3.2.44)$$

Equation (3.2.43) may be used to express X^+ as

$$X^+ = x^+ + \ell^2 p^+ \tau, \qquad (3.2.45)$$

defining the so-called light-cone gauge. Here x^+, ℓ, p^+ are independent of σ and τ. In a quantum setting, as mentioned earlier in this subsection, we choose units such that the unit of action $\hbar = 1$. This gives, the dimension of tension T to be [length]$^{-2}$. We set

$$\ell^2 = 1/(\pi T) = 2\alpha', \qquad (3.2.46)$$

where the first equality will be justified in (3.2.60) in the next section, and the significance of the parameter α' will also emerge later, with $1/\sqrt{\alpha'}$ defining a mass scale.[20] Clearly, the coordinates $(\xi^0, \xi^1) \equiv (\tau, \sigma)$ in the action are to be taken to be dimensionless.

For future reference, we note that the constraint $T_{00} = 0$, with $T_{\alpha\beta}$ given in (3.2.29), leads to the equation

$$\partial_\sigma X^+ \cdot \partial_\sigma X^- + \partial_\tau X^+ \cdot \partial_\tau X^- = \frac{1}{2}\left[(\partial_\sigma X^i)^2 + (\partial_\tau X^i)^2\right], \qquad (3.2.47)$$

with a sum over $i = 1, \ldots, D-2$, understood. By using the expression for X^+ in (3.2.45), the above equation simplifies to

$$\ell^2 p^+ \partial_\tau X^- = \frac{1}{2}\left[(\partial_\sigma X^i)^2 + (\partial_\tau X^i)^2\right], \qquad (3.2.48)$$

which allows us to solve for X^- in terms of the so-called transverse components X^i.

$$* \; ** $$

A simple model may be provided to obtain further insight on the significance of the parameter α' in (3.2.46). Consider a straight string of length L rotating about its midpoint with its end points moving at the speed of light. At a point r from its center, the speed, as a function of r, is given by $v(r) = r/(L/2)$. Now for a point particle of mass m moving with speed v, its energy is given by $p^0 = m/(1-v^2)^{1/2}$. Accordingly, the energy of the string E and its angular momentum J are given, respectively, by

$$E = 2\int_0^{L/2} T dr/\sqrt{1-v^2(r)} = TL\pi/2,$$

$$J = 2\int_0^{L/2} Tr\, v(r)\, dr/\sqrt{1-v^2(r)} = TL^2\pi/8,$$

from which $J = \alpha' E^2$, where, as a proportionality factor, $\alpha' = 1/(2\pi T)$ as given in (3.2.46) expressed in terms of the string tension T.

$$* \; ** $$

---

[20] For an almost "point-like" string, an intrinsic length ℓ of the order of the Planck length ($\sim 10^{-33}$ cm) sets up particularly a very large value for the mass scale $1/\sqrt{\alpha'}$.

3.2.4 Boundary Conditions and Solutions of Field Equations

In reference to the bosonic action in (3.2.44), in the conformal gauge, we choose dimensionless coordinates such that $\tau_i \leq \tau \leq \tau_f$, and $0 \leq \sigma \leq \pi$, for both open and closed strings with an obvious periodicity condition for closed strings over the coordinate σ. The response of this action W for the bosonic string to the variation of the fields δX^μ is given by

$$\delta W = -T \int (\mathrm{d}\xi)\, \eta^{\alpha\beta} (\partial_\alpha \delta X) \cdot \partial_\beta X$$

$$= T\left[\int (\mathrm{d}\xi)(\delta X \cdot \Box X) - \int (\mathrm{d}\xi)\, \partial^\alpha (\delta X \cdot \partial_\alpha X)\right]. \tag{3.2.49}$$

The first integral gives the field equations (3.2.28). With $\delta X(\tau_i, \sigma) = 0$ and $\delta X(\tau_f, \sigma) = 0$, the second integral becomes

$$-\int \mathrm{d}\tau \left[\delta X \cdot \partial_\sigma X\big|_{\sigma=\pi} - \delta X \cdot \partial_\sigma X\big|_{\sigma=0} \right], \tag{3.2.50}$$

the vanishing of which leads us to consider the following boundary conditions:
    For *closed* strings , we impose the periodicity condition

$$X^\mu(\tau, \sigma) = X^\mu(\tau, \sigma + \pi). \tag{3.2.51}$$

This implies the equality of the two "surface terms" in (3.2.50).
    For *open* strings we may choose

$$\partial_\sigma X^\mu(\tau, \sigma)\big|_{\sigma=0} = 0, \quad \partial_\sigma X^\mu(\tau, \sigma)\big|_{\sigma=\pi} = 0, \qquad \text{Neumann B.C.}, \tag{3.2.52}$$

referred to as Neumann boundary conditions (B.C.) as stated. In this case the end points of the string are not fixed.
    In the other extreme case, we may choose the following boundary conditions for an open string

$$\delta X^\mu(\tau, \sigma)\big|_{\sigma=0,\pi} = 0, \quad X^\mu\big|_{\sigma=0} = \underline{X}^\mu, \quad X^\mu\big|_{\sigma=\pi} = \overline{X}^\mu, \quad \text{Dirichlet B.C.}, \tag{3.2.53}$$

where $\underline{X}^\mu, \overline{X}^\mu$ are constants and the position of the end points do not change with τ. That is, the end points of the string are fixed. Such underlying boundary conditions are referred to as Dirichlet boundary conditions as stated.

One may also consider mixed boundary conditions such as

$$\delta X^\mu(\tau,\sigma)\big|_{\sigma=0,\pi} = 0, \quad \text{for} \quad \mu = p+1,\ldots,D-1, \quad \text{i.e.,}$$

$$X^{p+1}\big|_{\sigma=0} = \underline{X}^{p+1},\ldots,X^{D-1}\big|_{\sigma=0} = \underline{X}^{D-1},$$

$$X^{p+1}\big|_{\sigma=\pi} = \overline{X}^{p+1},\ldots,X^{D-1}\big|_{\sigma=\pi} = \overline{X}^{D-1}$$

are fixed;

$$\text{and} \quad \partial_\sigma X^\mu(\tau,\sigma)\big|_{\sigma=0,\pi} = 0, \quad \text{for} \quad \mu = 0,1,\ldots,p. \tag{3.2.54}$$

Here we note that the condition $\delta X^\mu(\tau,\sigma)\big|_{\sigma=0,\pi} = 0$, means to hold for all τ. That is, $\partial_\tau X^\mu(\tau,\sigma)\big|_{\sigma=0,\pi} = 0$, hold true for $\mu = p+1,\ldots,D-1$. This implies that

$$\underline{X}^{p+1},\ldots,\underline{X}^{D-1},\overline{X}^{p+1},\ldots,\overline{X}^{D-1},$$

are constants, as stated, and the end points of a string are restricted to move only in the other directions as they have no velocity components in the former directions. This spacetime "background", where the end points of the string are restricted, defines an extended object referred to as a D brane of spatial dimensionality p and is denoted by D p-brane.

According to this last boundary condition, the end of the strings are restricted to the brane and free to move and "slide" only on this object. For $p = D - 1$, the D brane is referred to as spacetime filling . The letter D stands for Dirichlet and the nomenclature "brane" is derived from the word membrane. The subject of D branes will be taken up in Sect. 3.4.

In the present section, we develop the solutions of closed strings, and for open strings satisfying Neumann boundary conditions.

The canonical momentum conjugate to X^μ is given by

$$P_\mu = \frac{\partial \mathscr{L}}{\partial \dot{X}^\mu} = T\dot{X}_\mu, \quad \mathscr{L} = -\frac{T}{2}\eta^{\alpha\beta}(\partial_\alpha X)\cdot(\partial_\beta X), \quad \dot{X}^\mu = \frac{\partial X^\mu}{\partial \tau}, \tag{3.2.55}$$

where $\mathscr{L}$ is the Lagrangian density corresponding to the bosonic action (3.2.44).

Open Strings

We work in the light-cone gauge, with Neumann boundary conditions. That is, we consider the following system which we spell out as

$$X^+ = x^+ + \ell^2 p^+ \tau, \quad \Box X^i = 0, \quad \partial_\sigma X^i\big|_{\sigma=0,\pi} = 0, \quad i = 1,\ldots,D-2,$$

$$\tag{3.2.56}$$

$$\ell^2 p^+ \, \partial_\tau X^- \;=\; \frac{1}{2} \left[(\partial_\sigma X^i)^2 + (\partial_\tau X^i)^2 \right], \tag{3.2.57}$$

where we have used (3.2.45), and (3.2.48). [In passing, we note that $\partial_\sigma X^+ = 0$, and we will see later, that $\partial_\sigma X^-(\tau,\sigma)\big|_{\sigma=0,\pi} = 0$ (see below (3.2.68)).]

We may carry out a Fourier transform as a mode expansion for the transverse fields satisfying the above conditions to obtain

$$X^i(\tau,\sigma) \;=\; x^i + \ell^2 p^i \tau + i\ell \sum_{n\neq 0} \frac{\alpha^i(n)}{n} \, e^{-in\tau} \cos n\sigma, \tag{3.2.58}$$

where recall that τ is dimensionless. We note from (3.2.55) and the above equation that

$$P^i(\tau,\sigma) \;=\; T\ell^2 p^i + T\ell \sum_{n\neq 0} \alpha^i(n)\, e^{-in\tau} \cos n\sigma, \tag{3.2.59}$$

$$\int_0^\pi d\sigma \, P^i \;=\; \pi\, T\, \ell^2 p^i = p^i, \tag{3.2.60}$$

That is, p^i denotes the ith component of the momentum of the string in the specified directions, and $x^i + \ell^2 p^i \tau$ as the ith component of its center of mass position, where we have used the definition in (3.2.46).

We note that (3.2.59) allows us to introduce a constant term $\alpha^i(0)$ and re-express $P^i(\tau,\sigma)$ as follows:

$$\alpha^i(0) \;=\; \ell p^i, \qquad P^i(\tau,\sigma) \;=\; T\ell \sum_{n=-\infty}^{\infty} \alpha^i(n)\, e^{-in\tau} \cos n\sigma. \tag{3.2.61}$$

Now we solve for X^- from (3.2.57) which is expressed in terms of derivatives of the fields X^i. To this end,

$$(X'^i)^2 + (\dot X^i)^2 \;=\; \ell^2 \sum_{n,m} e^{-i(n+m)\tau} \, \alpha^i(n)\, \alpha^i(m) \cos(n+m)\sigma, \tag{3.2.62}$$

which upon setting $(n+m) = N$, this gives from (3.2.57)

$$\ell^2 p^+ \, \partial_\tau X^- \;=\; \frac{\ell^2}{2} \sum_N e^{-iN\tau} \left(\sum_m \alpha^i(N-m)\, \alpha^i(m) \right) \cos N\sigma$$

$$=\frac{\ell^2}{2} \sum_m \alpha^i(-m)\, \alpha^i(m) + \frac{\ell^2}{2} \sum_{N\neq 0} e^{-iN\tau} \left(\sum_m \alpha^i(N-m)\, \alpha^i(m) \right) \cos N\sigma. \tag{3.2.63}$$

We may now readily integrate this equation for X^- to obtain

$$X^-(\tau,\sigma) = x^- + \ell^2 p^- \tau + \frac{i}{2p^+} \sum_{N \neq 0} \frac{e^{-iN\tau}}{N} \left(\sum_m \alpha^i(N-m)\,\alpha^i(m) \right) \cos N\sigma,$$

(3.2.64)

with

$$2\ell^2 p^+ p^- = \sum_{m=-\infty}^{\infty} \alpha^i(-m)\,\alpha^i(m).$$

(3.2.65)

Or using the first equation in (3.2.61), this leads to the important equation

$$-p^2 = -\eta_{\mu\nu} p^\mu p^\nu = 2p^+ p^- - p^i p^i = \frac{1}{\ell^2} \sum_{m \neq 0} \alpha^i(-m)\,\alpha^i(m).$$

(3.2.66)

We may also rewrite X^- in the form of X^i in (3.2.58) as

$$X^-(\tau,\sigma) = x^- + \ell^2 p^- \tau + i\ell \sum_{N \neq 0} \frac{\hat{\alpha}(N)}{N} e^{-iN\tau} \cos N\sigma,$$

(3.2.67)

$$\hat{\alpha}(N) = \frac{1}{2\ell p^+} \sum_{m=-\infty}^{\infty} \alpha^i(N-m)\,\alpha^i(m).$$

(3.2.68)

Clearly, $\Box X^- = 0$, $\partial_\sigma X^-|_{\sigma=0,\pi} = 0$.

Before discussing the physical significance of the solutions of the field equations just derived for open strings in the next section, we turn to the corresponding analysis for closed strings.

Closed Strings

The field equations $\Box X^i = 0$, with $\Box = \partial_\sigma^2 - \partial_\tau^2$, imply by the chain rules

$$\partial_\sigma = -\partial/\partial_{(\tau-\sigma)} + \partial/\partial_{(\tau+\sigma)}, \quad \partial_\tau = \partial/\partial_{(\tau-\sigma)} + \partial/\partial_{(\tau+\sigma)},$$

that $X^i(\tau,\sigma)$ may be written as a sum of a right-mover and a left-mover (see Problem 3.4)

$$X^i(\tau,\sigma) = X^i_R(\tau-\sigma) + X^i_L(\tau+\sigma), \qquad i = 1,\dots,D-2.$$

(3.2.69)

For a closed string, the latter two movers may be written quite generally in Fourier mode expansions

$$X_R^i(\tau - \sigma) = x_R^i + \ell\,\alpha^i(0)(\tau - \sigma) + \frac{i\ell}{2}\sum_{n\neq 0}\frac{\alpha^i(n)}{n}\,e^{-2in(\tau-\sigma)}, \tag{3.2.70}$$

$$X_L^i(\tau + \sigma) = x_L^i + \ell\,\bar{\alpha}^i(0)(\tau + \sigma) + \frac{i\ell}{2}\sum_{n\neq 0}\frac{\bar{\alpha}^i(n)}{n}\,e^{-2in(\tau+\sigma)}, \tag{3.2.71}$$

where recall that τ and σ are dimensionless. The above two equations give

$$X^i(\tau,\sigma) = x^i + \ell\left(\alpha^i(0) + \bar{\alpha}^i(0)\right)\tau + \ell\left(\bar{\alpha}^i(0) - \alpha^i(0)\right)\sigma$$
$$+ \frac{i\ell}{2}\sum_{n\neq 0}\left(\frac{\alpha^i(n)}{n}\,e^{-2in(\tau-\sigma)} + \frac{\bar{\alpha}^i(n)}{n}\,e^{-2in(\tau+\sigma)}\right), \tag{3.2.72}$$

where $x^i = x_R^i + x_L^i$. The periodicity condition in (3.2.51) is satisfied only if

$$\alpha^i(0) = \bar{\alpha}^i(0), \tag{3.2.73}$$

in order to make the third term in (3.2.72) vanish.

The canonical momentum conjugate to X^i in (3.2.55) then takes the form

$$P^i(\tau,\sigma) = T\ell\sum_{n=-\infty}^{\infty}\left[\alpha^i(n)\,e^{-2in(\tau-\sigma)} + \bar{\alpha}^i(n)\,e^{-2in(\tau+\sigma)}\right]. \tag{3.2.74}$$

The definition of the momentum of string $p^i = \int_0^\pi d\sigma\, P^i$, along the specified directions leads, upon using the definition (3.2.46), to the identification

$$\ell p^i = \alpha^i(0) + \bar{\alpha}^i(0), \qquad \frac{\ell}{2}p^i = \alpha^i(0) = \bar{\alpha}^i(0). \tag{3.2.75}$$

where in writing the second equation we have used (3.2.73).

The solutions of the field equations

$$\Box X^i = 0, \qquad i = 1,\ldots,D-2,$$

for a closed string satisfying the periodicity condition in (3.2.51), now takes the form

$$X^i(\tau,\sigma) = x^i + \ell^2 p^i\tau + \frac{i\ell}{2}\sum_{n\neq 0}\left[\frac{\alpha^i(n)}{n}\,e^{-2in(\tau-\sigma)} + \frac{\bar{\alpha}^i(n)}{n}\,e^{-2in(\tau+\sigma)}\right]. \tag{3.2.76}$$

The right-mover and a left-mover may be spelled out as

$$X_R^i(\tau - \sigma) = x_R^i + \frac{1}{2}\ell^2 p^i(\tau - \sigma) + \frac{i\ell}{2} \sum_{n \neq 0} \frac{\alpha^i(n)}{n} e^{-2in(\tau-\sigma)}, \qquad (3.2.77)$$

$$X_L^i(\tau + \sigma) = x_L^i + \frac{1}{2}\ell^2 p^i(\tau + \sigma) + \frac{i\ell}{2} \sum_{n \neq 0} \frac{\bar{\alpha}^i(n)}{n} e^{-2in(\tau+\sigma)}. \qquad (3.2.78)$$

From (3.2.75), we know, in particular, that $\alpha^i(0) = \bar{\alpha}^i(0)$ giving the expressions

$$\partial_\tau X^i = +\ell \sum_n [\alpha^i(n)\, e^{-2in(\tau-\sigma)} + \bar{\alpha}^i(n)\, e^{-2in(\tau+\sigma)}], \qquad (3.2.79)$$

$$\partial_\sigma X^i = -\ell \sum_n [\alpha^i(n)\, e^{-2in(\tau-\sigma)} - \bar{\alpha}^i(n)\, e^{-2in(\tau+\sigma)}], \qquad (3.2.80)$$

for all n. Upon substituting the latter equations in (3.2.48), the following expression for X^- emerges

$$X^-(\tau,\sigma) = x^- + \ell^2 p^- \tau + \frac{i\ell}{2} \sum_{N \neq 0} \left[\frac{\beta(N)}{N} e^{-2iN(\tau-\sigma)} + \frac{\bar{\beta}(N)}{N} e^{-2iN(\tau+\sigma)} \right], \qquad (3.2.81)$$

where the explicit expressions for the coefficients $\beta^i(N)$, $\bar{\beta}^i(N)$ may be obtained as in (3.2.67), for the open string, but will not be needed here.

The equation analogous to the open string case in (3.2.65) easily follows from (3.2.48), (3.2.79), (3.2.80) to be (see Problem 3.5)

$$2p^+ p^- = \frac{(\alpha^i(0) + \bar{\alpha}^i(0))^2}{\ell^2} + \frac{2}{\ell^2} \sum_{n \neq 0} [\alpha^i(-n)\alpha^i(n) + \bar{\alpha}^i(-n)\bar{\alpha}^i(n)]. \qquad (3.2.82)$$

Upon using (3.2.75), this leads to the important equation

$$-p^2 = \eta_{\mu\nu} p^\mu p^\nu = 2p^+ p^- - p^i p^i = \frac{2}{\ell^2} \sum_{n \neq 0} [\alpha^i(-n)\alpha^i(n) + \bar{\alpha}^i(-n)\bar{\alpha}^i(n)], \qquad (3.2.83)$$

which is the analogous equation obtained in (3.2.66) for the open strings.

The physical interpretation of our findings for open and closed strings is given next.

3.2.5 Quantum Aspect, Critical Spacetime Dimension and the Mass Spectrum

We first consider open strings.

Open Strings

The quantum aspect of string theory emerges upon introducing the commutation relations of the fields $X^i(\tau,\sigma)$ in (3.2.58) and the canonical momenta $P^i(\tau,\sigma)$ in (3.2.61) conjugate to them:

$$[X^i(\tau,\sigma), P^j(\tau,\sigma')] = i\delta^{ij}\delta(\sigma - \sigma'), \qquad (3.2.84)$$

$$[X^i(\tau,\sigma), X^j(\tau,\sigma')] = 0, \qquad (3.2.85)$$

$$[P^i(\tau,\sigma), P^j(\tau,\sigma')] = 0. \qquad (3.2.86)$$

The Hermiticity of the quantum field $X^i(\tau,\sigma)$ requires that

$$\alpha^i(-n) = (\alpha^i(n))^\dagger. \qquad (3.2.87)$$

The corresponding commutation relations of the Fourier coefficients $\alpha^i(n)$ in (3.2.58), are then given by

$$[\alpha^i(n), \alpha^j(m)] = n\,\delta(m, -n)\,\delta^{ij}, \qquad (3.2.88)$$

where $\delta(m, -n)$ is a Kronecker delta.

The mass-shell condition $p^2 = -M^2$ in (3.2.66) may be spelled out as

$$M^2 = \frac{1}{\ell^2}\sum_{m\neq 0}\alpha^i(-m)\,\alpha^i(m) = \frac{1}{\ell^2}\sum_{m=1}^{\infty}\left(\alpha^i(-m)\,\alpha^i(m) + \alpha^i(m)\,\alpha^i(-m)\right),$$
$$(3.2.89)$$

with a sum over $i = 1,\ldots,D-2$ understood. Having been careful in keeping the order of the products $\alpha^i(-m)\alpha^i(m) + \alpha^i(m)\alpha^i(-m)$ intact, we may use (3.2.87), and the commutation relations (3.2.88), to rewrite (3.2.89) as

$$M^2 = \frac{2}{\ell^2}\left(\sum_{m=1}^{\infty}\alpha^i(m)^\dagger\,\alpha^i(m) + \frac{D-2}{2}\sum_{m=1}^{\infty}m\right). \qquad (3.2.90)$$

In making a transition from a classical description to a quantum one, the constant $\sum_{m=1}^{\infty} m$ is interpreted as the unique analytical continuation of the Riemann zeta function[21]

$$\zeta(s) = \sum_{m=1}^{\infty} m^{-s}, \qquad \text{for } \mathrm{Re}\, s > 1, \tag{3.2.91}$$

to $s = -1$ having the value $\zeta(-1) = -1/12$. That this is the correct interpretation also follows from the fact that Lorentz invariance requires this value to be as just given, thus providing another *different* way of determining it. The corresponding constraint imposed by the commutation relations of the Lorentz generators will be spelled out below.

Using the definition of the parameter α' in (3.2.46), we obtain

$$M^2 = \frac{1}{\alpha'} \left(\sum_{m=1}^{\infty} \alpha^i(m)^\dagger \alpha^i(m) - \frac{D-2}{24} \right), \tag{3.2.92}$$

with $(\alpha')^{-1/2}$ providing a fundamental mass scale.

We may define ground-states, annihilated by $\alpha^i(m)$, and labeled by the momenta, say,

$$|0;p'\rangle: \quad p' = (p^+, p^1, \ldots, p^{D-2}),$$
$$\alpha^i(m)\,|0;p'\rangle = 0, \quad i = 1, 2, \ldots, D-2, \quad m = 1, 2, \ldots . \tag{3.2.93}$$

The particle of lowest mass corresponds to

$$M^2\,|0;p'\rangle = -\frac{1}{\alpha'}\left(\frac{D-2}{24}\right), \tag{3.2.94}$$

which for $D > 2$ corresponds to a particle of negative mass squared—a tachyon—not a very welcome particle. The first excited state corresponds to $(i = 1, \ldots, D, -2)$

$$M^2 \alpha^i(1)^\dagger\,|0;p'\rangle = \frac{1}{\alpha'}\left(1 - \frac{D-2}{24}\right)\alpha^i(1)^\dagger\,|0;p'\rangle. \tag{3.2.95}$$

Now a vector particle in 4 dimensions which has only two degrees of freedom is a massless particle. In D dimensions, a vector particle, with $D-2$ degrees of freedom is a massless particle—the *photon*. A detailed analysis of the degrees of freedom of massless fields in higher dimensions are derived in Sect. 3.2.9. Accordingly, the right-hand side of (3.2.95) must be zero and leads to the critical dimension of

---

[21]For details and interesting applications of the Riemann zeta function, see, e.g., [11, 12]. The analytical continuation to $s = -1$ is all what is needed here.

spacetime for bosonic strings to be $D = 26$.[22] That is we learn that *the number of degrees of freedom and the masslessness of the particle* (described by a vector field) *determine the dimensionality of the underlying spacetime of the theory.*

The next states are massive and are obtained by the applications of $\alpha^i(2)^\dagger$ and $\alpha^i(1)^\dagger \alpha^j(1)^\dagger$, respectively, to $| 0; p')$, with the mass given by $1/\sqrt{\alpha'}$, which may be enormous, and the analysis may be carried out further to higher massive states. In such cases, only the massless excitation may be expected to be physically relevant at present available energies acquiring masses by some mechanism. Such a mechanism to generate a massive vector boson will be described later when dealing with branes.

The mass squared operator in (3.2.92) may be now rewritten as

$$
M^2 = \frac{1}{\alpha'} \left(\sum_{m=1}^{\infty} m\, a^i(m)^\dagger a^i(m) - 1 \right), \quad [a^i(m), a^j(n)^\dagger] = \delta^{ij}\delta(m,n).
$$

$$(3.2.96)$$

The number of states $\rho(n)$ for a given mass squared $M^2 = (n-1)/\alpha'$ are given through the expression of the trace of the latter:

$$
\alpha'\, \mathrm{Tr}\,[M^2] = \sum_{n=0}^{\infty} n\,\rho(n).
$$

$$(3.2.97)$$

These may be obtained from the taking derivatives of the following generating function as follows

$$
G[q] = \sum_{n=0}^{\infty} \rho(n)\, q^n, \quad \rho(n) = \frac{1}{n!} \left(\frac{d}{dq} \right)^n G[q]\big|_{q=0}, \; n \geq 1, \; \rho(0) \equiv G[0],
$$

$$(3.2.98)$$

with $\rho(0) = 1$.

Introducing the notation, $N = \sum_{n=1}^{\infty} n\, a^i(n)^\dagger a^i(n)$, and making use of the fact that the commutator

$$
[a^i(n)^\dagger a^i(n), a^j(m)^\dagger a^j(m)] = 0, \quad \text{for} \quad i \neq j,
$$

---

[22]In reference to (3.2.89), if we set

$$
\sum_{m=1}^{\infty} \left(\alpha^i(-m)\alpha^i(m) + \alpha^i(m)\alpha^i(-m) \right) = 2 \left(\sum_{m=1}^{\infty} \alpha^i(m)^\dagger \alpha^i(m) - a \right),
$$

then the Lorentz algebra satisfied by the Lorentz generators $J^{\mu\nu}$, implies that

$$
\frac{m}{2}(26 - D) + \frac{2}{m}\left((1-a) - \frac{(26-D)}{24} \right) = 0,
$$

for all $m \neq 0$, providing the solutions $a = 1, D = 26$, consistent with the corresponding ones obtained through (3.2.90), (3.2.95).

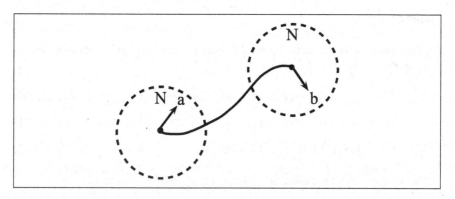

Fig. 3.4 Additional degrees of freedom may be associated with an open string, by labeling each end of the string by N indices, referred to as Chan-Paton degrees of freedom

we may write the generating function as

$$G[q] = \text{Tr}[q^N] = \Pi_{n=1}^{\infty} \Pi_{i=1}^{24} \text{Tr}[q^{n a^i(n)^\dagger a^i(n)}]$$

$$= \Pi_{n=1}^{\infty} \Pi_{i=1}^{24} \left(1 + q^n + q^{2n} + q^{3n} + \cdots\right) = \Pi_{n=1}^{\infty} \left(\frac{1}{1 - q^n}\right)^{24}, \qquad (3.2.99)$$

where we have used the fact that $a^\dagger a (a^\dagger)^m |0\rangle = m(a^\dagger)^m |0\rangle$, for a one dimensional problem, with $|0\rangle$ being the Fock vacuum, in computing the trace for all m. We recall that one has the normalization condition $G[0] \equiv \rho(0) = 1$, and it is easily seen that $\rho(1) = 24$, as expected.

In order to generate several gauge fields, as is necessarily required in particle physics, the end of the open strings may be assumed to carry certain degrees of freedom referred to as Chan-Paton degrees of freedom. That is, the ends of the strings may exist in any of these "charge" states. One may label these states by a pair of indices $a, b = 1, 2, \ldots, N$ corresponding to the two ends of a string: The ends of the strings will have associated with them the matrices of a gauge group (Fig. 3.4).

The ground-state would then carry two indices a, b: $|0; p', ab\rangle$, specifying, as well, the states of the end points of the string. As the indices a, b take each N possible values, this provides a way to generate N^2 gauge fields associated with the states $a^i(1) |0; p', ab\rangle$ corresponding to a U(N) gauge symmetry. Here we recall that the group U(N) consists of all N by N unitary complex matrices and may be generated by N^2 independent matrices.

Closed Strings

The commutation relations of the fields X^i in (3.2.76), and the canonical momenta P^i conjugate to them in (3.2.74) read

$$[X^i(\tau, \sigma), P^j(\tau, \sigma')] = i \delta^{ij} \delta(\sigma - \sigma'), \qquad (3.2.100)$$

$$[X^i(\tau, \sigma), X^j(\tau, \sigma')] = 0, \qquad (3.2.101)$$

$$[P^i(\tau, \sigma), P^j(\tau, \sigma')] = 0. \qquad (3.2.102)$$

The Hermiticity of the quantum fields $X^i(\tau, \sigma)$ also requires that

$$\alpha^i(-n) = (\alpha^i(n))^\dagger, \qquad \bar{\alpha}^i(-n) = (\bar{\alpha}^i(n))^\dagger. \qquad (3.2.103)$$

The ensuing commutation rules of $\alpha^i(-n), \bar{\alpha}^i(n)$ are then given by

$$[\alpha^i(n), \alpha^j(m)] = n \delta(m, -n) \delta^{ij}, \quad [\bar{\alpha}^i(n), \bar{\alpha}^j(m)] = n \delta(m, -n) \delta^{ij}, \qquad (3.2.104)$$

$$[\alpha^i(m), \bar{\alpha}^j(n)] = 0. \qquad (3.2.105)$$

The mass shell condition given through (3.2.83) with $M^2 = -p^2$ contains both sums

$$\sum_{n \neq 0} \alpha^i(-n) \alpha^i(n), \quad \text{and} \quad \sum_{n \neq 0} \bar{\alpha}^i(-n) \bar{\alpha}^i(n).$$

We will see that there exists a constraint on them to be set to be equal. To this end, we note that the constraint $T_{01} = 0$ in (3.2.26), in particular, gives

$$\partial_\tau X \cdot \partial_\sigma X = 0, \qquad (3.2.106)$$

which, from (3.2.30), (3.2.45), means that

$$-\ell^2 p^+ \partial_\sigma X^- + \partial_\tau X^i \partial_\sigma X^i = 0. \qquad (3.2.107)$$

From (3.2.81),

$$\partial_\sigma X^- = -\ell \sum_{N \neq 0} [\beta(N) e^{-2iN(\tau - \sigma)} - \bar{\beta}(N) e^{-2iN(\tau + \sigma)}], \qquad (3.2.108)$$

which when integrated out over σ from 0 to π gives zero. That is, (3.2.107) leads to

$$\frac{1}{\pi} \int_0^\pi d\sigma \, \partial_\tau X^i \partial_\sigma X^i = 0. \qquad (3.2.109)$$

This integral is explicitly worked out from (3.2.79), (3.2.80) leading to

$$\sum_{m\neq0}[\alpha^i(-m)\,\alpha^i(m) - \tilde{\alpha}^i(-m)\,\tilde{\alpha}^i(m)] = \sum_{m\neq0}[\alpha^i(m),\tilde{\alpha}^i(m)]\,e^{-4\mathrm{i}m\tau} = 0,$$

(3.2.110)

where we have finally used (3.2.105), and recall from (3.2.75) that $\alpha^i(0) = \tilde{\alpha}^i(0)$.

The mass shell condition, as obtained from (3.2.83), together with the above constraint then give

$$M^2 = \frac{2}{\ell^2}\sum_{m\neq0}[\alpha^i(-m)\,\alpha^i(m) + \tilde{\alpha}^i(-m)\,\tilde{\alpha}^i(m)],$$

(3.2.111)

$$\sum_{m\neq0}\alpha^i(-m)\,\alpha^i(m) - \sum_{m\neq0}\tilde{\alpha}^i(-m)\,\tilde{\alpha}^i(m) = 0.$$

(3.2.112)

Having been careful in keeping the orders of the products $\alpha^i(-m)\,\alpha^i(m)$, $\tilde{\alpha}^i(-m)\,\tilde{\alpha}^i(m)$, intact, we note by using (3.2.103), (3.2.104) and the unique analytical continuation of the zeta function in (3.2.91) to $s = -1$, giving the value $\zeta(-1) = -1/12$, also dictated independently by Lorentz invariance, the mass shell condition (3.2.111), and the constraint (3.2.112) for closed bosonic strings become

$$M^2 = \frac{2}{\alpha'}\sum_{m=1}^{\infty}\left(\alpha^i(m)^\dagger\,\alpha^i(m) + \tilde{\alpha}^i(m)^\dagger\,\tilde{\alpha}^i(m) - \frac{D-2}{12}\right),$$

(3.2.113)

$$\sum_{m=1}^{\infty}\alpha^i(m)^\dagger\,\alpha^i(m) = \sum_{m=1}^{\infty}\tilde{\alpha}^i(m)^\dagger\,\tilde{\alpha}^i(m),$$

(3.2.114)

where we have finally introduced the parameter α' defined in (3.2.46).

Again for $D > 2$, we have a tachyon state satisfying

$$M^2\,|\,0;p'\rangle = -\frac{1}{\alpha'}\frac{D-2}{6},$$

(3.2.115)

with a negative mass squared.

With the constraint (3.2.114), the first excited states are given by products of the form

$$\alpha^i(1)^\dagger\,\tilde{\alpha}^j(1)^\dagger\,|\,0;p'\rangle,$$

(3.2.116)

corresponding to

$$\sum_{m=1}^{\infty}\alpha^i(m)^\dagger\,\alpha^i(m) = \sum_{m=1}^{\infty}\tilde{\alpha}^i(m)^\dagger\,\tilde{\alpha}^i(m) \to 1.$$

We have at our disposal two (commuting[23]) vectors $\alpha^i(1)^\dagger$ and $\bar\alpha^j(1)^\dagger$. We can enumerate the underlying particles described by states described in (3.2.116), by considering the following linear combinations:

$$\alpha^i(1)^\dagger \bar\alpha^j(1)^\dagger = \sum_{a=1,2,3,k,\ell} C_a{}^{ikj\ell} \alpha^k(1)^\dagger \bar\alpha^\ell(1)^\dagger,$$

where

$$C_1{}^{ikj\ell} = \frac{1}{2}\left[\delta^{ik}\delta^{j\ell} + \delta^{i\ell}\delta^{jk} - \frac{2}{D-2}\delta^{ij}\delta^{k\ell} \right], \qquad (3.2.117)$$

$$C_2{}^{ikj\ell} = \frac{1}{2}\left[\delta^{ik}\delta^{j\ell} - \delta^{i\ell}\delta^{jk} \right], \qquad (3.2.118)$$

$$C_3{}^{ikj\ell} = \frac{1}{D-2}\delta^{ij}\delta^{k\ell}, \qquad (3.2.119)$$

and $i, j, k, \ell = 1, 2, \ldots, (D-2)$. We may thus generated the following three states:

$$\sum_{k,\ell=1}^{D-2} C_a{}^{ikj\ell} \alpha^k(1)^\dagger \bar\alpha^\ell(1)^\dagger \mid 0;p'\rangle, \qquad a = 1, 2, 3, \qquad (3.2.120)$$

satisfying the eigenvalue equations

$$M^2 \sum_{k,\ell=1}^{D-2} C_a{}^{ikj\ell} \alpha^k(1)^\dagger \bar\alpha^\ell(1)^\dagger \mid 0;p'\rangle,$$

$$= \frac{4}{\alpha'}\left(1 - \frac{D-2}{24}\right) \sum_{k,\ell=1}^{D-2} C_a{}^{ikj\ell} \alpha^k(1)^\dagger \bar\alpha^\ell(1)^\dagger \mid 0;p'\rangle. \qquad (3.2.121)$$

For $a = 1$, corresponding to

$$\sum_{k,\ell=1}^{D-2} C_1{}^{ikj\ell} \alpha^k(1)^\dagger \bar\alpha^\ell(1)^\dagger \mid 0;p'\rangle,$$

we are dealing with a symmetric traceless tensor, which has

$$\frac{1}{2}(D-1)(D-2) - 1 = \frac{1}{2}D(D-3), \qquad (3.2.122)$$

---

[23]See (3.2.105).

independent components, where the first term denotes the number of independent components of a symmetric tensor and the minus one is due to the fact that one of the elements on the diagonal may be determined in terms of the sum of the other $(D-3)$ ones on the diagonal. This necessarily implies that the particle associated with it is *massless* and represents the graviton in D-dimensional spacetime as analyzed in detail in Sect. 3.2.9. From (3.2.121) we then conclude that the critical dimension of spacetime, again, is $D = 26$. Thus we learn once more that *the number of degrees of freedom and the masslessness of the particle* (described by a symmetric tensor field) *determine the dimensionality of the underlying spacetime of the theory.*

For $a = 2$, corresponding to

$$\sum_{k,\ell=1}^{D-2} C_2{}^{ikj\ell} \alpha^k(1)^\dagger \bar{\alpha}^\ell(1)^\dagger \mid 0;p'\rangle,$$

we are dealing with an anti-symmetric tensor, which has

$$\frac{1}{2}(D-2)(D-3), \tag{3.2.123}$$

independent components as also analyzed in detail in Sect. 3.2.9. The field associated with it is usually referred to as the Kalb-Ramond field .

Finally for $a = 3$, corresponding to

$$\sum_{k,\ell=1}^{D-2} C_3{}^{ikj\ell} \alpha^k(1)^\dagger \bar{\alpha}^\ell(1)^\dagger \mid 0;p'\rangle,$$

we have a scalar, that is with one degree of freedom, and is referred to as the dilation. Thus the total number of states of the massless cases is

$$\frac{1}{2}D(D-3) + \frac{1}{2}(D-2)(D-3) + 1 = (D-2)^2, \tag{3.2.124}$$

as it should be. Incidentally, the generation of the three types of the (massless) fields out of two vectors of $D-2$ dimensions is written in a group theoretical notation as

$$(D-2) \otimes (D-2) = \left(\frac{1}{2}D(D-3)\right) \oplus \left(\frac{1}{2}(D-2)(D-3)\right) \oplus 1,$$

spelling out explicitly the various degrees of freedom of the (massless) fields discovered above.

Massive particle states may be also considered by satisfying, in the process, the constraint in (3.2.114).

The expression for the mass squared in (3.2.113) together with the constraint (3.2.114) may be rewritten as

$$\frac{\alpha'}{4} M^2 = (N_R - 1) = (N_L - 1), \tag{3.2.125}$$

$$N_R = \sum_{m=1}^{\infty} \alpha^i(m)^\dagger \alpha^i(m), \quad N_L = \sum_{m=1}^{\infty} \bar{\alpha}^i(m)^\dagger \bar{\alpha}^i(m). \tag{3.2.126}$$

3.2.6   Unoriented String Theories and Chan-Paton Degrees of Freedom

Consider the worldsheet parity transformation, defined by

$$\Omega X^i(\tau,\sigma) \Omega^{-1} = X^i(\tau, \pi - \sigma). \tag{3.2.127}$$

After the above transformation is carried out, the direction of increase of σ for the given τ will in general be different.

From the expression of the worldsheet fields $X^i(\tau,\sigma)$ in (3.2.76), for the closed bosonic string, one may infer that their two Fourier coefficients are interchanged under a worldsheet parity transformation

$$\Omega \alpha^i(n) \Omega^{-1} = \bar{\alpha}^i(n), \tag{3.2.128}$$

or equivalently one interchanges its right- and left-movers. The states of the string theories considered earlier, however, are not invariant under the parity operation, assuming the invariance of the ground-state $\Omega| 0; p' \rangle = | 0; p' \rangle$. For example, the anti-symmetric state

$$\sum_{k,\ell=1}^{D-2} C_2^{ikj\ell} \alpha^k(1)^\dagger \bar{\alpha}^\ell(1)^\dagger | 0; p' \rangle,$$

in (3.2.120)/(3.2.118), gives the eigenvalue $\Omega = -1$. By restricting these theories only to Ω-invariant states, one generates theories which are referred to as *unoriented* string theories, while the earlier ones as *oriented* ones. In general, one may define the orientation of a worldsheet as the direction of increasing σ for fixed τ. Referring to the zero mass modes of the closed strings just mentioned, the antisymmetric state, corresponding to the Kalb-Ramond field just given above is then projected out of the spectrum in the process of defining an unoriented bosonic string.

The situation dealing with open unoriented strings is a bit more involved. The transformation in (3.2.127) interchanges the end points of the string. Here we note that the expression of the fields for the open strings in (3.2.58) imply that under worldsheet parity transformation

$$\Omega : \alpha^i(n) \leftrightarrow (-1)^n \alpha^i(n). \tag{3.2.129}$$

As in the oriented open strings, we may consider the end of the open strings to carry Chan-Paton degrees of freedom. That is, the ends of the strings may exist in any of these "charge" states (see Fig. 3.4). One may label these states by a pair of indices $a, b = 1, 2, \ldots, N$ corresponding to the two ends of a string. The strings ends will have associated with them matrices of a gauge group. The ground-state would then carry two indices a, b: $|0; p', ab\rangle$, specifying, as well, the states of the end points of the string.

We consider the action of the worldsheet parity operation on the states which interchanges the Chan-Paton labels via a unitary operator γ

$$\Omega : |0; p', ab\rangle \rightarrow \gamma_{aa'} |0; p', b'a'\rangle \gamma^{-1}_{b'b} = \gamma_{ab'} |0; p', a'b'\rangle \gamma^{-1}_{a'b}. \tag{3.2.130}$$

By setting $\Omega^2 = 1$, we have for a subsequent transformation,

$$\Omega^2 : |0; p', ab\rangle \rightarrow \gamma_{ab'} \gamma_{a'b''} |0; p', a''b''\rangle \gamma^{-1}_{a''b'} \gamma^{-1}_{a'b}$$

$$= \gamma_{ab'} (\gamma^\mathsf{T})^{-1}_{b'a''} |0; p', a''b''\rangle \gamma^\mathsf{T}_{b''a'} \gamma^{-1}_{a'b} = |0; p', ab\rangle, \tag{3.2.131}$$

and we may set,

$$\gamma = \pm \gamma^\mathsf{T}. \tag{3.2.132}$$

To introduce gauge fields, suppose that we have a group with generators denoted by $\Lambda = [\Lambda_{ab}]$. We may define states

$$|0; p', \Lambda\rangle = \sum_{a,b=1}^{N} \Lambda_{ab} |0; p', ab\rangle. \tag{3.2.133}$$

The transformation property of the states $|0; p', ab\rangle$ in (3.2.130) may be transferred to the expansion coefficients Λ_{ab} in (3.2.133), instead of the states $|0; p', ab\rangle$, via the equation

$$\Omega : |0; p', \Lambda\rangle \rightarrow \sum_{a,b,a',b'=1}^{N} \gamma_{a'b} \Lambda_{a'b'} \gamma^{-1}_{ab'} |0; p', ab\rangle, \tag{3.2.134}$$

by appropriate relabelings. This gives the transformation rule

$$\Omega : \Lambda \rightarrow \Lambda' = \gamma^{-1} \Lambda^\mathsf{T} \gamma. \tag{3.2.135}$$

Referring to the massless vector field state in (3.2.95), we note from (3.2.129) that $a^i(1)^\dagger \rightarrow -a^i(1)^\dagger$. Hence in order that the states $a^i(1)^\dagger \mid 0; p', \Lambda\rangle$ be invariant under the worldsheet parity transformation, i.e., $\Omega\, a^i(1)^\dagger \mid 0; p', \Lambda\rangle = a^i(1)^\dagger \mid 0; p', \Lambda\rangle$, it is necessary that $\Lambda' = -\Lambda$. This implies from (3.2.135) that

$$\gamma\, \Lambda + \Lambda^\mathsf{T} \gamma = 0. \tag{3.2.136}$$

An obvious solution, corresponding to $\gamma = \gamma^\mathsf{T}$ in (3.2.132), is $\gamma = I$, which, from (3.2.136), implies that the Hermitian Λ matrices are antisymmetric. That is, the Λ matrices, generate the group SO(N) with group elements $O = e^{-\Lambda}$, with $O\,O^\mathsf{T} = I$ and det $O = e^{-\mathrm{Tr}\Lambda} = 1$. The number of gauge fields generated then is $N(N-1)/2$.

A solution, corresponding to $\gamma = -\gamma^\mathsf{T}$ in (3.2.132), is provided by the group $U\mathrm{Sp}(N)$ (read symplectic group), with N restricted to be even, which we define as involving generators given by $N \times N$ Hermitian matrices Λ satisfying (3.2.136) with the unitary matrix γ given by

$$\gamma = \mathrm{i} \begin{pmatrix} 0 & I_{N/2} \\ -I_{N/2} & 0 \end{pmatrix}. \tag{3.2.137}$$

We may then infer from (3.2.136), (3.2.137) that the matrices Λ are of the form

$$\begin{pmatrix} A & B \\ C & D \end{pmatrix}, \quad B = B^\mathsf{T}, \quad C = C^\mathsf{T}, \quad D = -A^\mathsf{T}.$$

Thus the number of independent elements of A, B, C, in general, are: $N^2/4$, $N(N/2+1)/4$, $N(N/2+1)/4$, respectively. Adding up, this leads to the generation of $N(N+1)/2$ gauge fields.

3.2.7   Compactification and T-Duality: Closed Strings

We consider the compactification of one of the extra spatial component, say, X^{25}, of a bosonic string into a circle of radius R. In this respect, the light-cone variables may be defined by

$$X^\pm = (X^0 \pm X^{24})/\sqrt{2}, \quad X^i, \, i = 1, \ldots, 23,$$

and the compactified dimension, X^{25} will be denoted by X. We will see, that there arises a duality, referred to as T-duality in the analysis, which relates two theories with compactified extra dimension into a circle of radius R and one into a circle of radius α'/R. In particular, the application of T-duality to open strings, as will be seen in the next section, introduces necessarily the concept of D branes which were briefly mentioned in Sect. 3.2.4. We first consider closed strings.

For the closed string, the boundary condition for the extra dimension compactified into a circle of radius R becomes

$$X(\tau, \sigma + \pi) = X(\tau, \sigma) + 2\pi R w, \tag{3.2.138}$$

replacing the boundary condition for $X^{25} \equiv X$ in (3.2.51), where w denotes the winding number, indicating that the string can wind around the compactified dimension any number of times, with w taking positive integer values for winding in a certain direction and negative ones in the opposite one. The value $w = 0$ corresponds to no winding.

For a closed string, we have obtained in (3.2.72), the general expression for X^i, and in particular for $X^{25} \equiv X$ we have

$$X(\tau, \sigma) = x + \ell \left(\alpha(0) + \bar{\alpha}(0) \right) \tau + \ell \left(\bar{\alpha}(0) - \alpha(0) \right) \sigma$$
$$+ \frac{i\ell}{2} \sum_{n \neq 0} \left(\frac{\alpha(n)}{n} e^{-2in(\tau-\sigma)} + \frac{\bar{\alpha}(n)}{n} e^{-2in(\tau+\sigma)} \right). \tag{3.2.139}$$

From the expression of the momentum canonically conjugate to X, as given in (3.2.74) and the first relation in (3.2.75), we may infer that

$$\ell p^{25} = \alpha(0) + \bar{\alpha}(0), \tag{3.2.140}$$

and on the other-hand, we may infer from (3.2.138), (3.2.139) that

$$2Rw = \ell \left(\bar{\alpha}(0) - \alpha(0) \right), \tag{3.2.141}$$

where now, in general, $\bar{\alpha}(0) \neq \alpha(0)$, corresponding to $i = 25$. The Fourier mode expansion of X then reads

$$X(\tau, \sigma) = x + \ell^2 p^{25} \tau + 2Rw\sigma + \frac{i\ell}{2} \sum_{n \neq 0} \left[\frac{\alpha(n)}{n} e^{-2in(\tau-\sigma)} + \frac{\bar{\alpha}(n)}{n} e^{-2in(\tau+\sigma)} \right], \tag{3.2.142}$$

instead of (3.2.76) for this field. From (3.2.139), we may rewrite $X(\tau, \sigma)$ as the sum of its right-mover and left-mover

$$X_R(\tau - \sigma) = x_R + \ell \alpha(0)(\tau - \sigma) + \frac{i\ell}{2} \sum_{n \neq 0} \frac{\alpha(n)}{n} e^{-2in(\tau-\sigma)}, \tag{3.2.143}$$

$$X_L(\tau + \sigma) = x_L + \ell \bar{\alpha}(0)(\tau + \sigma) + \frac{i\ell}{2} \sum_{n \neq 0} \frac{\bar{\alpha}(n)}{n} e^{-2in(\tau+\sigma)}, \tag{3.2.144}$$

$$p_R \equiv \frac{\alpha(0)}{\ell} = \frac{1}{2} \left(p^{25} - \frac{Rw}{\alpha'} \right), \quad p_L \equiv \frac{\bar{\alpha}(0)}{\ell} = \frac{1}{2} \left(p^{25} + \frac{Rw}{\alpha'} \right), \tag{3.2.145}$$

where in writing the two equations in (3.2.145), we have used (3.2.140), (3.2.141), and finally (3.2.46).

Using the facts that

$$\partial_\sigma X = 2Rw - \ell \sum_{n\neq 0} \left[\alpha(n)\,e^{-2in(\tau-\sigma)} - \tilde{\alpha}(n)\,e^{-2in(\tau+\sigma)}\right], \tag{3.2.146}$$

$$\partial_\tau X = \ell^2 p^{25} + \ell \sum_{n\neq 0} \left[\alpha(n)\,e^{-2in(\tau-\sigma)} + \tilde{\alpha}(n)\,e^{-2in(\tau+\sigma)}\right], \tag{3.2.147}$$

the expression in (3.2.82), as obtained from (3.2.48), becomes simply replaced by

$$2p^+p^- = \frac{4}{\ell^4}R^2 w^2 + (p^{25})^2 + \frac{(\alpha^i(0) + \tilde{\alpha}^i(0))^2}{\ell^2}$$

$$+\frac{2}{\ell^2} \sum_{n\neq 0} \left[\alpha^i(-n)\,\alpha^i(n) + \tilde{\alpha}^i(-n)\,\tilde{\alpha}^i(n) + \alpha(-n)\,\alpha(n) + \tilde{\alpha}(-n)\,\tilde{\alpha}(n)\right],$$

$$\tag{3.2.148}$$

where a sum over $i = 1, \ldots, 23$ is understood (see also Problem 3.7). The Mass-squared operator in 25 dimensions, i.e., for $M^2 = 2p^+p^- - \sum_{i=1}^{23} p^i p^i$, becomes

$$M^2 = (p^{25})^2 + \frac{R^2 w^2}{\alpha'^2} + \frac{2}{\alpha'}(N_L + N_R - 2), \tag{3.2.149}$$

$$N_L = \sum_{n=1}^{\infty} \left(\tilde{\alpha}^i(n)^\dagger \tilde{\alpha}^i(n) + \tilde{\alpha}(n)^\dagger \tilde{\alpha}(n)\right), \tag{3.2.150}$$

$$N_R = \sum_{n=1}^{\infty} \left(\alpha^i(n)^\dagger \alpha^i(n) + \alpha(n)^\dagger \alpha(n)\right), \tag{3.2.151}$$

where recall from (3.2.75) that

$$\frac{\alpha^i(0) + \tilde{\alpha}^i(0)}{\ell} = p^i, \quad \text{for } i = 1, 2, \ldots, 23,$$

and the -2 term in $(N_L + N_R - 2)$ in (3.2.149) corresponds to $-(D-2)/12$ with $D = 26$. Also due to the additional factor $2Rw$ in (3.2.146), the constraint in (3.2.114) becomes, by following the procedure in deriving (3.2.110) (see Problem 3.8), replaced by

$$N_R - N_L = Rw\, p^{25}. \tag{3.2.152}$$

The mass-squared operator in (3.2.149), including the constraint in (3.2.152) may be expressed in a convenient way as

$$\frac{\alpha'}{4} M^2 = \alpha' p_{\mathrm{L}}^2 + (N_{\mathrm{L}} - 1) = \alpha' p_{\mathrm{R}}^2 + (N_{\mathrm{R}} - 1), \tag{3.2.153}$$

where, in the process of writing this equation, we have used (3.2.145).

Imposing single valuedness for the translation of a state in the x direction by $2\pi R$, via the operator $\exp[i\,xp]$, leads to the quantization of the momentum eigenvalues to $p^{25} = k/R$ for integers k, called the Kaluza-Klein excitation number. The expression for the mass operator in (3.2.149), may be then spelled out as

$$M^2 = \frac{k^2}{R^2} + \frac{w^2 R^2}{\alpha'^2} + \frac{2}{\alpha'} (N_{\mathrm{L}} + N_{\mathrm{R}} - 2). \tag{3.2.154}$$

The mass-squared operator is invariant under the simultaneous interchange $w \leftrightarrow k$, and $R \leftrightarrow \widetilde{R} = \alpha'/R$ referred to as T-duality. For the critical value $R^* = \sqrt{\alpha'}$, the radius R^* is transformed into itself and is called the self-dual radius. Each radius smaller than R^* is equivalent to some radius larger than R^* for the interchange of the winding number w, and the Kaluza-Klein excitation number k.

Ground states of the compactified closed string may be determined by considering, in the process, states specified by $|\,k,w,p'\,\rangle$, where $p' = (p^+, p^1, \ldots, p^{23})$, subject to the constraint (3.2.152). For $N_{\mathrm{L}} = 0 = N_{\mathrm{R}}$, the constraint is satisfied if $(k = 0$ or $w = 0)$, or $(k = w = 0)$, remembering that $p^{25} = k/R$. In the latter case, $|\,0,0,p'\,\rangle$ corresponds to a tachyonic state with mass squared $-4/\alpha'$. For the former cases, one has the states $|\,0,w,p'\,\rangle$, and $|\,k,0,p'\,\rangle$ corresponding to scalar fields, with masses squared

$$\frac{1}{\alpha'} \left(w^2 \frac{R^2}{\alpha'} - 4 \right), \qquad \frac{1}{\alpha'} \left(k^2 \frac{\alpha'}{R^2} - 4 \right), \tag{3.2.155}$$

respectively, and may or may not be tachyonic depending on the size of the radius R.

For $k = 0 = w$, and $N_{\mathrm{L}} = 1 = N_{\mathrm{R}}$, thus, in particular, satisfying the constraint (3.2.152), we are led to consider the states:

$$a(1)^\dagger \bar{a}(1)^\dagger \,|\,0,0,p'\,\rangle, \tag{3.2.156}$$

$$a^i(1)^\dagger \bar{a}(1)^\dagger \,|\,0,0,p'\,\rangle, \qquad \bar{a}^i(1)^\dagger a(1)^\dagger \,|\,0,0,p'\,\rangle, \tag{3.2.157}$$

$$\bar{a}^i(1)^\dagger a^j(1)^\dagger \,|\,0,0,p'\,\rangle, \tag{3.2.158}$$

with all corresponding to massless fields in 25 dimensions. Here $\alpha(n) = \sqrt{n}\,a(n)$, and similarly defined for the other operators. The first defines a state of a massless scalar field. The two in the second equations, define two vector field states. The last one defines states of a graviton, an anti-symmetric second rank tensor field, and a scalar field—the dilaton. Before we had *no* vector fields for the closed string, now we encounter *two* vector fields at this stage.

For $k = w = \pm 1$, $N_R - N_L = 1$, with the lowest-mass states for $N_R = 1$, $N_L = 0$, we are led to consider the states

$$a(1)^\dagger \,|\pm 1, \pm 1, p'\rangle, \quad a^i(1)^\dagger \,|\pm 1, \pm 1, p'\rangle \tag{3.2.159}$$

Similarly, for $k = -w = \mp 1$, $N_R - N_L = -1$, with the lowest-mass states for $N_R = 0$, $N_L = 1$, we are led to consider the states

$$\bar{a}(1)^\dagger \,|\mp 1, \pm 1, p'\rangle, \quad \bar{a}^i(1)^\dagger \,|\mp 1, \pm 1, p'\rangle |1, p'\rangle, \tag{3.2.160}$$

with *all* these eight states in (3.2.159), (3.2.160) corresponding to masses-squared equal to $(1/R - R/\alpha')^2$. Thus for the self-dual radius $R = R^* = \sqrt{\alpha'}$,

$$a(1)^\dagger \,|\pm 1, \pm 1, p'\rangle, \quad \bar{a}(1)^\dagger \,|\mp 1, \pm 1, p'\rangle,$$

describe four massless scalar fields. On the other hand, the states

$$a^i(1)^\dagger \,|\pm 1, \pm 1, p'\rangle, \quad \bar{a}^i(1)^\dagger \,|\mp 1, \pm 1, p'\rangle,$$

describe four massless vector fields. Hence with the states in (3.2.157), we have 6 massless vector fields as the gauge fields of the symmetry group $SU(2) \times SU(2)$, with $k = w$ for one set, and $k = -w$, for a second. Here we recall that $SU(2)$ denotes the group of 2 by 2 matrices of determinant one, involving three generators. The product $SU(2) \times SU(2)$ means that the two groups act independently, with each involving three generators, they correspond to a total of 6 gauge fields.

3.2.8 Compactification, T-Duality, Open Strings and Emergence of D Branes

From the expression of the fields X^i in (3.2.58), for an open bosonic string, we may write the compactified dimension as $X^{25} \equiv X$

$$X(\tau, \sigma) = x + \ell^2 p^{25} \tau + i\ell \sum_{n \neq 0} \frac{\alpha(n)}{n} e^{-in\tau} \cos n\sigma, \tag{3.2.161}$$

satisfying Neumann boundary condition in (3.2.52)

$$\partial_\sigma X(\tau, \sigma)\big|_{\sigma = 0, \pi} = 0. \tag{3.2.162}$$

We may rewrite $X(\tau, \sigma)$, in terms of a right-mover and a left-mover as $X = X_R + X_L$, with

$$X_R(\tau - \sigma) = x_R + \frac{1}{2}\ell^2 p^{25} (\tau - \sigma) + \frac{i\ell}{2} \sum_{n \neq 0} \frac{\alpha(n)}{n} e^{-in(\tau - \sigma)}, \tag{3.2.163}$$

$$X_L(\tau + \sigma) = x_L + \frac{1}{2}\ell^2 p^{25}(\tau + \sigma) + \frac{i\ell}{2}\sum_{n\neq 0}\frac{\alpha(n)}{n}e^{-in(\tau+\sigma)}, \qquad (3.2.164)$$

where $x = x_R + x_L$.

With compactification into a circle of radius R, one has $p^{25} \equiv p = k/R$, with k, an integer, denoting the Kaluza-Klein excitation number.

The T-dual $\widetilde{X}$ of X, is defined by $\widetilde{X} = X_L - X_R$, giving

$$\widetilde{X}(\tau, \sigma) = \widetilde{x} + \ell^2 p^{25}\sigma + \ell\sum_{n\neq 0}\frac{\alpha(n)}{n}e^{-in\tau}\sin n\sigma, \qquad (3.2.165)$$

satisfying

$$\partial_\tau \widetilde{X}(\tau, \sigma)\big|_{\sigma=0,\pi} = 0, \qquad (3.2.166)$$

giving rise to a Dirichlet boundary condition.[24] That is, the ends of the dual string carry no momentum in the 25th circular direction, and their motion is restricted to the hypersurface $\widetilde{X} = \widetilde{x}$, called a D brane (Fig. 3.5). In this case the spatial

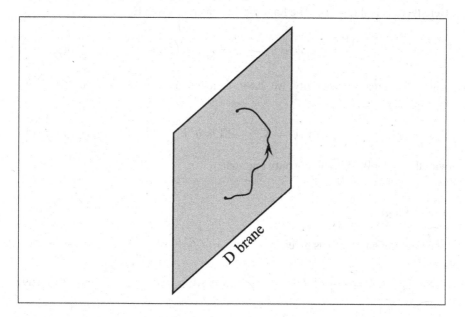

Fig. 3.5 The figure shows end points of an open string being restricted to a structure referred to as a D brane

---

[24]See (3.2.53).

dimensionality of the D brane is 24, and it is referred to as a Dp-brane with $p = 24$. Thus one sees that T-duality necessarily implies the existence of D branes.

T-duality has transformed the string with a Neumann boundary condition into a Dirichlet boundary condition. The subject of D branes will be taken up in Sect. 3.4.

The next section deals with superstring theory. Before moving on to the next section, we may pose and consider in some details all of the massless fields that have emerged from bosonic string theory and study several of their properties including their associated degrees of freedom.

3.2.9　All the Fundamental Massless Fields in Bosonic String Theory

We now consider some technical aspects of all the massless fields encountered in bosonic string theory. We work in arbitrary dimensions D of spacetime.[25]

Scalar Field

Here there is nothing new. The Lagrangian density is given by

$$\mathscr{L} = -\frac{1}{2}\partial^\mu \phi \, \partial_\mu \phi + K\phi, \quad \mu = 0, 1, \ldots, D-1, \tag{3.2.167}$$

where for greater generality we have included an external source. The field equation is

$$-\Box \phi = K, \quad \Box = \partial^i \partial^i - \partial^{02}, \tag{3.2.168}$$

with the scalar field involving only one state.

Vector Field

The Lagrangian density is given by

$$\mathscr{L} = -\frac{1}{4}F^{\mu\nu}F_{\mu\nu}, \quad F^{\mu\nu} = \partial^\mu A^\nu - \partial^\nu A^\mu. \tag{3.2.169}$$

---

[25]This subsection is based on Manoukian [26].

It is invariant under the gauge transformation $A^\mu \to A^\mu + \partial^\mu \Lambda$. We work in the Coulomb gauge

$$\partial^i A^i = 0. \tag{3.2.170}$$

where a sum over $i = 1, 2, \ldots, D-1$ understood. Greek indices $\mu, \nu, \ldots$ go over $0, 1, \ldots, D-1$.

We add an external source contribution to (3.2.169), obtaining

$$\mathscr{L} = -\frac{1}{4} F^{\mu\nu} F_{\mu\nu} + J^\mu A_\mu, . \tag{3.2.171}$$

Due to the constraint in (3.2.170) not all the components of A^μ may be varied independently. We may, however, express A^μ as

$$A^\mu = -A^0 \eta^{\mu 0} + \eta^{\mu i}\left(\delta^{ij} - \frac{\partial^i \partial^j}{\nabla^2}\right)\mathscr{A}^j, \tag{3.2.172}$$

and vary A^0, $\mathscr{A}^j$ to obtain from (3.2.171)

$$-\Box A^i = \left(\delta^{ij} - \frac{\partial^i \partial^j}{\nabla^2}\right) J^j, \tag{3.2.173}$$

$$-\nabla^2 A^0 = J^0. \tag{3.2.174}$$

Clearly only the A^i may propagate. Also $i = 1, \ldots, D-1$, but due to the constraint in (3.2.170), which is also satisfied in (3.2.173), the number of states of the massless vector field is reduced to $D-2$.

Another but equivalent way of counting the number of states of the massless vector field, is to work in the momentum description and use a completeness relation in terms of polarization vectors

$$\delta^{ij} = \frac{k^i}{|\mathbf{k}|} \frac{k^j}{|\mathbf{k}|} + \sum_{\lambda=1}^{D-2} e_\lambda^i e_\lambda^j, \qquad k^i e_\lambda^i = 0, \quad e_\lambda^i e_{\lambda'}^i = \delta_{\lambda\lambda'}, \tag{3.2.175}$$

where

$$\mathbf{e}_1, \ldots, \mathbf{e}_{D-2}, \frac{\mathbf{k}}{|\mathbf{k}|},$$

are $D-1$ mutually pairwise orthonormal vectors spanning a $(D-1)$ Euclidean space, and hence gives rise to the completeness relation in (3.2.175). We may then rewrite (3.2.173) in the momentum description as

$$k^2 A^i(k) = \sum_{\lambda=1}^{D-2} e_\lambda^i (e_\lambda^j J^j(k)), \tag{3.2.176}$$

with the number of states being simply given by $\sum_{\lambda=1}^{D-2} e_\lambda^i e_\lambda^i = \delta^{ii} - k^i k^i / \mathbf{k}^2 = D - 2$, where we have used (3.2.175), and recall that $\delta^{ii} = D - 1$. Additional interesting details are given in Problem 3.9.

Symmetric Traceless Second Rank Tensor Field: The Graviton

The Lagrangian density of a massless symmetric tensor field, which holds in higher dimensions than four as well, in the absence of an external source, may be written as[26]

$$\mathscr{L} = -\frac{1}{2} \partial^\sigma h_{\mu\nu} \partial_\sigma h^{\mu\nu} + \partial_\mu h^{\mu\nu} \partial_\sigma h^\sigma{}_\nu - \partial_\sigma h^{\sigma\mu} \partial_\mu h + \frac{1}{2} \partial^\mu h \partial_\mu h, \qquad (3.2.177)$$

where $h = h^\mu{}_\mu$, $h^{\mu\nu} = h^{\nu\mu}$. The Lagrangian density is invariant under the gauge transformation

$$h^{\mu\nu} \to h^{\mu\nu} + \partial^\mu \Lambda^\nu + \partial^\nu \Lambda^\mu, \qquad (3.2.178)$$

up to a total derivative. We work in the Coulomb-like gauge

$$\partial^i h^{i\nu} = 0, \qquad \nu = 0, 1, \ldots, D-1, \quad i = 1, 2, \ldots, D-1. \qquad (3.2.179)$$

We add a source contribution to (3.2.177), to obtain

$$\mathscr{L} = -\frac{1}{2} \partial^\sigma h_{\mu\nu} \partial_\sigma h^{\mu\nu} + \partial_\mu h^{\mu\nu} \partial_\sigma h^\sigma{}_\nu - \partial_\sigma h^{\sigma\mu} \partial_\mu h + \frac{1}{2} \partial^\mu h \partial_\mu h + T^{\mu\nu} h_{\mu\nu}. \qquad (3.2.180)$$

We may then infer from an earlier analysis (4.7.218)–(4.7.222) in Chap. 4 of Volume I, by working in D dimensional spacetime ($D \neq 2$), that

$$h^{ii} = -\frac{1}{\nabla^2} T^{00}, \qquad h^{00} = -\frac{1}{(D-2)\nabla^2} \Big[(D-3) \frac{\square}{\nabla^2} T^{00} + \pi^{ij} T^{ij} \Big], \qquad (3.2.181)$$

$$h^{0i} = -\frac{1}{\nabla^2} \pi^{ik} T^{0k}, \qquad \pi^{ij} = \Big(\delta^{ij} - \frac{\partial^i \partial^j}{\nabla^2} \Big), \qquad (3.2.182)$$

$$-\square h^{ij} = \frac{1}{D-2} \Big[\frac{D-2}{2} (\pi^{ik} \pi^{jl} + \pi^{il} \pi^{jk}) - \pi^{ij} \pi^{kl} \Big] T^{kl} + \frac{1}{D-2} \pi^{ij} \frac{\square}{\nabla^2} T^{00}. \qquad (3.2.183)$$

---

[26]See Appendix II at the end of this volume.

Clearly, only the h^{ij} may propagate. Now a symmetric matrix h^{ij} has $(D-1)D/2$ independent components. However, for $T^{\mu\nu} = 0$, we have the constraint $h^{ii} = 0$ and the $D-1$ constraints $\partial^i h^{ij} = 0$. Thus the number of states of this massless field is $(D-1)D/2 - 1 - (D-1) = D(D-3)/2$. Note that for $T^{\mu\nu} = 0$, $h^{\mu}{}_{\mu} = 0$.

By using the first equation in (3.2.181), we may rewrite (3.2.183) as

$$-\Box\left(h^{ij} - \frac{\pi^{ij}}{D-2}h^{kk}\right) = \frac{1}{D-2}\left[\frac{D-2}{2}(\pi^{ik}\pi^{jl} + \pi^{il}\pi^{jk}) - \pi^{ij}\pi^{kl}\right]T^{kl}. \tag{3.2.184}$$

We may also use the completeness relation in (3.2.175), the symmetry of T^{kl}, and the elementary identity in the momentum description

$$\sum_{\lambda,\lambda'=1}^{D-2} \delta_{\lambda\lambda'}\left[\frac{(D-2)}{2}(e_\lambda^i e_{\lambda'}^j + e_{\lambda'}^i e_\lambda^j) - \delta_{\lambda\lambda'}\pi^{ij}\right] = 0, \tag{3.2.185}$$

where we note that

$$\sum_{\lambda=1}^{D-2} e_\lambda^i e_\lambda^j = \pi^{ij} = \delta^{ij} - \frac{k^i k^j}{|\mathbf{k}|^2},$$

to rewrite (3.2.184) in the momentum description, as

$$k^2 \bar{h}^{ij}(k) = \sum_{\lambda,\lambda'=1}^{D-2} \epsilon^{ij}(\lambda,\lambda')\, T(\lambda,\lambda';k), \tag{3.2.186}$$

$$\bar{h}^{ij} = h^{ij} - \frac{\pi^{ij}}{D-2}h^{kk}, \qquad T(\lambda,\lambda';k) = \epsilon^{ij}(\lambda,\lambda')\, T^{ij}(k), \tag{3.2.187}$$

$$\epsilon^{ij}(\lambda,\lambda') = \frac{1}{D-2}\left[\frac{(D-2)}{2}(e_\lambda^i e_{\lambda'}^j + e_{\lambda'}^i e_\lambda^j) - \delta_{\lambda\lambda'}\sum_{\kappa=1}^{D-2}e_\kappa^i e_\kappa^j\right]. \tag{3.2.188}$$

It is easily verified that

$$\sum_{\lambda,\lambda'=1}^{D-2} \epsilon^{ij}(\lambda,\lambda')\, \epsilon^{ij}(\lambda,\lambda') = \frac{1}{2}D(D-3), \tag{3.2.189}$$

giving the number of states, as before, as expected. Additional interesting details are given in Problem 3.10.

Anti-Symmetric Second Rank Tensor Field

Consider an anti-symmetric tensor field $A^{\mu\nu} = -A^{\nu\mu}$, and introduce the tensor field

$$F^{\mu\nu\sigma} = \partial^\mu A^{\nu\sigma} + \partial^\nu A^{\sigma\mu} + \partial^\sigma A^{\mu\nu}, \qquad (3.2.190)$$

an anti-symmetric tensor field, defined as a cyclic permutation of the indices. We define the Lagrangian density

$$\mathscr{L} = -\frac{1}{6} F^{\mu\nu\sigma} F_{\mu\nu\sigma}. \qquad (3.2.191)$$

The Lagrangian density is invariant under the gauge transformation

$$A^{\mu\nu} \rightarrow A^{\mu\nu} + \partial^\mu \Lambda^\nu - \partial^\nu \Lambda^\mu. \qquad (3.2.192)$$

We work in the Coulomb-like gauge

$$\partial_i A^{i\nu} = 0, \qquad \nu = 0, 1, \ldots, D-1, \qquad (3.2.193)$$

with a sum over $i = 1, \ldots, D-1$, understood.

We add a source contribution to (3.2.191), to obtain

$$\mathscr{L} = -\frac{1}{6} F^{\mu\nu\sigma} F_{\mu\nu\sigma} + A^{\mu\nu} J_{\mu\nu}. \qquad (3.2.194)$$

With such a constraint, we cannot vary all the components of $A^{\mu\nu}$ independently, we can, however, write

$$A^{\mu\nu} = \eta^{\mu i}\eta^{\nu j}A^{ij} - (\eta^{\mu 0}\eta^{\nu i} - \eta^{\mu i}\eta^{\nu 0})A^{0i}, \qquad (3.2.195)$$

$$A^{ij} = \frac{1}{2}(\pi^{ik}\pi^{jl} - \pi^{il}\pi^{jk})C^{kl}, \quad A^{0i} = \pi^{ij}\phi^j, \quad \pi^{ij} = \delta^{ij} - \frac{\partial^i \partial^j}{\nabla^2}, \qquad (3.2.196)$$

and vary ϕ^j, C^{kl}. These variations give in turn

$$-\nabla^2 A^{0i} = \pi^{ij} J^{0j}, \qquad (3.2.197)$$

$$-\Box A^{ij} = \frac{1}{2}(\pi^{ik}\pi^{jl} - \pi^{il}\pi^{jk})J^{kl}. \qquad (3.2.198)$$

Clearly, only the A^{ij} may propagate. Using the completeness relation in (3.2.175), and the anti-symmetry of J^{ij}, we may rewrite (3.2.198), in the momentum description, as

$$k^2 A^{ij}(k) = \sum_{\lambda, \lambda'=1}^{D-2} \varepsilon^{ij}(\lambda, \lambda') J(\lambda, \lambda'; k), \qquad (3.2.199)$$

$$J(\lambda, \lambda'; k) = \varepsilon^{ij}(\lambda, \lambda') J^{ij}(k), \quad \varepsilon^{ij}(\lambda, \lambda') = \frac{1}{2}(e^i_\lambda e^j_{\lambda'} - e^i_{\lambda'} e^j_\lambda). \tag{3.2.200}$$

The mere fact that $\varepsilon^{ij}(\lambda, \lambda')$ is anti-symmetric in λ, λ', the number of states is given by

$$\sum_{\lambda'=2}^{D-2} \sum_{\lambda=1}^{\lambda'-1} 1 = \frac{1}{2}(D-2)(D-3),$$

as expected. This is also given by

$$\sum_{\lambda'=1}^{D-2} \sum_{\lambda=1}^{D-2} \varepsilon^{ij}(\lambda, \lambda') \varepsilon^{ij}(\lambda, \lambda') = \frac{1}{2}(D-2)(D-3). \tag{3.2.201}$$

Additional interesting details are given in Problem 3.11.

3.3 Superstrings

We extend our earlier analysis of bosonic strings to superstrings. To this end, we must first consider the Dirac equation in 2 dimensional spacetime.

3.3.1 Dirac Equation in Two Dimensions

In two dimensional spacetime, the mass shell condition reads $[p^0 - \sqrt{(p^1)^2 + m^2}]\chi = 0$, which upon multiplying by $p^0 + \sqrt{(p^1)^2 + m^2}$ gives

$$[(p^1)^2 - (p^0 - m)(p^0 + m)]\chi = 0. \tag{3.3.1}$$

Simply set

$$(p^0 - m)\chi = p^1\varphi, \quad \text{hence} \quad (p^0 + m)\varphi = p^1\chi, \tag{3.3.2}$$

thus introducing the object φ. The above two equations may be rewritten as

$$p^1\varphi - (p^0 - m)\chi = 0, \tag{3.3.3}$$

$$-p^1\chi + (p^0 + m)\varphi = 0, \tag{3.3.4}$$

and may be combined in the elegant form

$$\left(\gamma^\alpha \frac{\partial_\alpha}{i} + m\right)\Psi = 0, \qquad \Psi = \begin{pmatrix} \chi \\ \varphi \end{pmatrix}, \tag{3.3.5}$$

$$\gamma^0 = \begin{pmatrix} 1 & 0 \\ 0 & -1 \end{pmatrix}, \quad \gamma^1 = \begin{pmatrix} 0 & 1 \\ -1 & 0 \end{pmatrix}, \quad \gamma^0\gamma^1 = \gamma^5 = \begin{pmatrix} 0 & 1 \\ 1 & 0 \end{pmatrix}, \tag{3.3.6}$$

$$\{\gamma^\alpha, \gamma^\beta\} = -2\,\eta^{\alpha\beta}, \qquad [\eta^{\alpha\beta}] = \mathrm{diag}[-1, 1]. \tag{3.3.7}$$

An equivalent description may be introduced by making a unitary transformation: $G\gamma^\alpha G^\dagger = \rho^\alpha, G\Psi = \psi$ with

$$G = \frac{1}{\sqrt 2}\begin{pmatrix} 1 & 1 \\ i & -i \end{pmatrix}, \qquad G^\dagger = G^{-1} = \frac{1}{\sqrt 2}\begin{pmatrix} 1 & -i \\ 1 & i \end{pmatrix}, \tag{3.3.8}$$

$$\rho^0 = \begin{pmatrix} 0 & -i \\ i & 0 \end{pmatrix}, \quad \rho^1 = \begin{pmatrix} 0 & i \\ i & 0 \end{pmatrix}, \quad \rho^0\rho^1 = \rho^5 = \begin{pmatrix} 1 & 0 \\ 0 & -1 \end{pmatrix}, \tag{3.3.9}$$

$$\left(\rho^\alpha \frac{\partial_\alpha}{i} + m\right)\psi = 0 \qquad \psi = \begin{pmatrix} \psi_1 \\ \psi_2 \end{pmatrix}, \quad \{\rho^\alpha, \rho^\beta\} = -2\,\eta^{\alpha\beta}. \tag{3.3.10}$$

We note that the matrices ρ^0/i, ρ^1/i are real, and we may infer that the Majorana condition of a spinor simply reads

$$\psi_1^* = \psi_1, \qquad \psi_2^* = \psi_2, \tag{3.3.11}$$

signifying the *reality* of the spinor ψ (see Problem 3.13). We note that due to the fact that ρ^5 is diagonal and, as indicated above, the matrices ρ^0/i, ρ^1/i are real, we have a Majorana-chiral, also called Majorana-Weyl, representation.

A complete set of 2 by 2 matrices is

$$I, \rho^0, \rho^1, \rho^5, \tag{3.3.12}$$

where I is the identity matrix. [For example $\rho^\alpha\rho^\beta$, $\alpha, \beta = 0, 1$, may be written as a linear combination of the matrices I and ρ^5.]

The following basic properties should be noted

$$(\rho^0)^\dagger = \rho^0, \quad (\rho^1)^\dagger = -\rho^1, \quad \{\rho^\alpha, \rho^5\} = 0, \quad \rho^\alpha\rho_\alpha = -2, \tag{3.3.13}$$

$$(\rho^\alpha)^\dagger\rho^0 = \rho^0\rho^\alpha, \qquad \rho^\alpha\rho^\beta\rho_\alpha = 0, \qquad [\rho_\alpha, [\rho^\beta, \rho^\gamma]] = 4\,(\rho^\beta\delta_\alpha{}^\gamma - \rho^\gamma\delta_\alpha{}^\beta), \tag{3.3.14}$$

and for two anti-commuting Majorana spinors ψ, ω $(\overline{\psi} = \psi^\dagger \rho^0)$,

$$\overline{\psi}\,\omega = \overline{\omega}\psi, \qquad \overline{\psi}\,\rho^\alpha \omega = -\overline{\omega}\,\rho^\alpha \psi, \quad (\overline{\psi}\,\rho^\alpha \omega)^\dagger = (\overline{\omega}\,\rho^\alpha \psi), \tag{3.3.15}$$

$$\overline{\psi}\,\rho^\alpha \rho^\beta \omega = \overline{\omega}\,\rho^\beta \rho^\alpha \psi, \quad \overline{\psi}\,\rho^\alpha \rho^5 \omega = -\overline{\omega}\,\rho^\alpha \rho^5 \psi, \quad \overline{\psi}\,\rho^5 \omega = -\overline{\omega}\,\rho^5 \psi. \tag{3.3.16}$$

For a Majorana spinor θ with Grassmann, i.e, anti-commuting, components, we have

$$\theta = \begin{pmatrix} \theta_1 \\ \theta_2 \end{pmatrix}, \quad \{\theta_a, \theta_b\} = 0, \quad \overline{\theta}\rho^\alpha \theta = 0, \quad \overline{\theta}\rho^5 \theta = 0, \quad \overline{\theta}\rho^\alpha \rho^5 \theta = 0, \tag{3.3.17}$$

$$\overline{\theta}\theta = 2i\theta_2\theta_1, \qquad \theta_a \overline{\theta}_b = -\frac{1}{2}\overline{\theta}\theta\,\delta_{ab}, \qquad \frac{\partial}{\partial \overline{\theta}_c}\overline{\theta}\theta = 2\theta_c, \tag{3.3.18}$$

$$\overline{\theta}_1 = i\theta_2, \qquad \overline{\theta}_2 = -i\theta_1. \tag{3.3.19}$$

3.3.2 Worldsheet Supersymmetry and the String Action

The basic properties of the theta spinors in (3.3.17), and the completeness of the 2×2 matrices in (3.3.12) allow us to define the general D superfields

$$\Phi^\mu(\xi, \theta) = X^\mu(\xi) + \frac{1}{\sqrt{2}}\overline{\theta}\,\psi^\mu(\xi) + \frac{1}{4}\overline{\theta}\theta\,B^\mu(\xi), \quad \mu = 0, 1, \ldots, D-1, \tag{3.3.20}$$

$\xi = (\tau, \sigma)$, in view of developing a supersymmetric action for a string.

We may infer from (2.6.4), that the variation of the superfields are given by

$$\delta\,\Phi^\mu(\xi, \theta) = \overline{\epsilon}\left[\frac{\partial}{\partial \overline{\theta}} + \frac{i}{2}(\rho^\alpha \theta)\,\partial_\alpha\right]\Phi^\mu(\xi, \sigma), \tag{3.3.21}$$

for an infinitesimal Grassmann parameter $\overline{\epsilon}$. By the application of the last equality in (3.3.18), we obtain, in the process, for the right-hand side of (3.3.21)

$$\frac{1}{\sqrt{2}}\overline{\epsilon}\,\psi^\mu + \frac{i}{2}\overline{\epsilon}\rho^\alpha \theta\,\partial_\alpha X^\mu + \frac{i}{2\sqrt{2}}(\overline{\epsilon}\rho^\alpha \theta)\overline{\theta}\,\partial_\alpha \psi^\mu + \frac{1}{2}\overline{\epsilon}\theta\,B^\mu$$

$$= \frac{1}{\sqrt{2}}\overline{\epsilon}\,\psi^\mu - \frac{i}{2}\overline{\theta}\,\rho^\alpha \epsilon\,\partial_\alpha X^\mu - \frac{i}{4\sqrt{2}}\overline{\theta}\theta\,\overline{\epsilon}\rho^\alpha \partial_\alpha \psi^\mu + \frac{1}{2}\overline{\theta}\epsilon\,B^\mu, \tag{3.3.22}$$

which follows from basic identities in (3.3.17), (3.3.18), and the first identity in (3.3.15).

On the other hand, from (3.3.20), we have for the left-hand side of (3.3.21)

$$\delta\Phi^\mu = \delta X^\mu(\xi) + \frac{1}{\sqrt{2}}\overline{\theta}\,\delta\psi^\mu(x) + \frac{1}{4}\overline{\theta}\theta\,\delta B^\mu(x), \tag{3.3.23}$$

which upon comparison with (3.3.22) gives the supersymmetry transformation rules

$$\delta X^\mu = +\frac{1}{\sqrt{2}}\overline{\epsilon}\,\psi^\mu, \tag{3.3.24}$$

$$\delta\psi^\mu = +\frac{1}{\sqrt{2}}(B^\mu - i\,\rho^\alpha\,\partial_\alpha X^\mu)\epsilon, \tag{3.3.25}$$

$$\delta B^\mu = -\frac{i}{\sqrt{2}}\,\partial_\alpha(\overline{\epsilon}\rho^\alpha\psi^\mu). \tag{3.3.26}$$

From (2.3.9), we may infer that the superderivative now is given by

$$D = \frac{\partial}{\partial\overline{\theta}} - \frac{i}{2}(\rho^\alpha\theta)\partial_\alpha. \tag{3.3.27}$$

The following operation then easily follows

$$D\Phi^\mu = \frac{1}{\sqrt{2}}\psi^\mu - \frac{i}{2}\rho^\alpha\theta\,\partial_\alpha X^\mu + \frac{i}{4\sqrt{2}}\overline{\theta}\theta\,\rho^\alpha\partial_\alpha\psi^\mu + \frac{1}{2}\theta\,B^\mu. \tag{3.3.28}$$

Also

$$(D\Phi^\mu)^\dagger\rho^0 = \frac{1}{\sqrt{2}}\overline{\psi}^\mu + \frac{i}{2}\overline{\theta}\rho^\alpha\partial_\alpha X^\mu - \frac{i}{4\sqrt{2}}\overline{\theta}\theta\,\partial_\alpha\overline{\psi}^\mu\rho^\alpha + \frac{1}{2}\overline{\theta}\,B^\mu. \tag{3.3.29}$$

The term involving $\overline{\theta}\theta$ in the following product is given by

$$(D\Phi^\mu)^\dagger\rho^0 D\Phi_\mu \rightarrow \overline{\theta}\theta\left[\frac{i}{4}\overline{\psi}^\mu\rho^\alpha\partial_\alpha\psi_\mu - \frac{1}{4}\partial^\alpha X^\mu\partial_\alpha X_\mu + \frac{1}{4}B^\mu B_\mu\right], \tag{3.3.30}$$

where we have used the identities [see, in particular, (3.3.17), (3.3.18)]

$$(\partial_\alpha\overline{\psi}^\mu)\rho^\alpha\psi_\mu = -\overline{\psi}^\mu\rho^\alpha\partial_\alpha\psi_\mu, \qquad \overline{\theta}\rho^\alpha\rho^\beta\theta = -\eta^{\alpha\beta}\overline{\theta}\theta, \tag{3.3.31}$$

Also recall from (3.3.14) that $(\rho^\alpha)^\dagger\rho^0 = \rho^0\rho^\alpha$.

Using the facts that

$$\overline{\theta}\theta = 2i\theta_2\theta_1, \qquad \int d\theta_1\,d\theta_2\,(\theta_2\,\theta_1) = 1,$$

we obtain the supersymmetric generalization of the bosonic string in (3.2.44) for the action integral to be

$$W = -\frac{T}{2}\int (d\xi)\left(\partial^\alpha X^\mu\,\partial_\alpha X_\mu + \overline{\psi}^\mu\rho^\alpha\,\frac{\partial_\alpha}{i}\,\psi_\mu - B^\mu B_\mu\right). \qquad (3.3.32)$$

Note that on account that

$$\left(\overline{\psi}^\mu\rho^\alpha\,\frac{\partial_\alpha}{i}\,\psi_\mu\right)^\dagger = -\left(\frac{\partial_\alpha}{i}\overline{\psi}^\mu\right)\rho^\alpha\,\psi_\mu = \overline{\psi}^\mu\rho^\alpha\,\frac{\partial_\alpha}{i}\,\psi_\mu, \qquad (3.3.33)$$

where the last equality follows from the Majorana character of the spinors, the expression in (3.3.32) is Hermitian.

The field equations following from the action (3.3.32) are

$$\Box X^\mu = 0, \qquad \rho^\alpha\,\partial_\alpha\psi^\mu = 0, \qquad B^\mu = 0, \qquad (3.3.34)$$

and the energy-momentum tensor, as a constraint, is given by

$$T_{\alpha\beta} = \partial_\alpha X^\mu\,\partial_\beta X_\mu + \frac{1}{4i}\overline{\psi}^\mu(\rho_\alpha\,\partial_\beta + \rho_\beta\,\partial_\alpha)\psi_\mu$$

$$- \eta_{\alpha\beta}\frac{1}{2}\left(\partial^\gamma X^\mu\,\partial_\gamma X_\mu + \frac{1}{2i}\overline{\psi}^\mu\,\rho^\gamma\partial_\gamma\psi_\mu\right) = 0, \qquad (3.3.35)$$

where we have written the latter in a symmetrized form, and we have set B^μ equal to zero.

The reason why we may set B^μ equal to zero, and will be set equal to zero in the sequel in the action in (3.3.32) as well, is that not only there is no interaction between the fields but the field equation for B, implies that the latter is trivially zero involving no oscillatory modes as the *other* fields. The action in (3.3.32), then reads

$$W = -\frac{T}{2}\int (d\xi)\left(\partial^\alpha X^\mu\,\partial_\alpha X_\mu + \overline{\psi}^\mu\rho^\alpha\,\frac{\partial_\alpha}{i}\,\psi_\mu\right). \qquad (3.3.36)$$

The supersymmetry transformation rules in (3.3.24), (3.3.25) then become

$$\delta X^\mu = +\frac{1}{\sqrt{2}}\,\overline{\epsilon}\psi^\mu, \qquad \delta\psi^\mu = -\frac{i}{\sqrt{2}}\,\rho^\beta\epsilon\,\partial_\beta X^\mu, \qquad \delta\overline{\psi}^\mu = \frac{i}{\sqrt{2}}\,\overline{\epsilon}\,\rho^\beta\,\partial_\beta X^\mu.$$

$$(3.3.37)$$

As a generalization, we consider, for a moment, these transformations with ϵ to be ξ-dependent. The response of the action (3.3.36) to these transformations is given by

$$\delta W = \frac{T}{\sqrt{2}} \int (d\xi)\,(\partial_\alpha \bar{\epsilon})(\rho^\beta \rho^\alpha \psi^\mu\, \partial_\beta X_\mu),\qquad (3.3.38)$$

up to a total derivative, which introduces the supercurrent

$$J^\alpha = \frac{1}{2}\rho^\beta \rho^\alpha \psi^\mu\, \partial_\beta X_\mu = 0,\qquad (3.3.39)$$

as a constraint. The $1/2$ factor is introduced for convenience. It is easily checked that, as consequence of the field equations, the supercurrent is conserved: $\partial_\alpha J^\alpha = 0$. Also from the second identity in (3.3.14) $\rho_\alpha \rho^\beta \rho^\alpha = 0$, the supercurrent satisfies the relation $\rho_\alpha J^\alpha = 0$. In the sequel, we consider, as before, the parameter ϵ to be independent of ξ.

In light-cone coordinates: $X^+, X^-, X^i, \psi^+, \psi^-, \psi^i, i = 1,\dots,D-2$, the integrand in (3.3.36) reads[27]

$$I(\xi) = -2\,\partial^\alpha X^+ \partial_\alpha X^- - \overline{\psi}^+ \frac{\rho^\alpha \partial_\alpha}{i}\,\psi^- - \overline{\psi}^- \frac{\rho^\alpha \partial_\alpha}{i}\,\psi^+$$

$$+\ \partial^\alpha X^i \partial_\alpha X^i + \overline{\psi}^i \frac{\rho^\alpha \partial_\alpha}{i}\,\psi^i.\qquad (3.3.40)$$

The response of this integrand to the variations of $X^+, \psi^+, \overline{\psi}^+$, is explicitly given by

$$\delta_+ I = -2\,(\partial^\alpha \delta X^+)\,\partial_\alpha X^- - (\delta\overline{\psi}^+)\frac{\rho^\alpha \partial_\alpha}{i}\,\psi^- - \overline{\psi}^- \frac{\rho^\alpha \partial_\alpha}{i}\,(\delta\psi^+).\qquad (3.3.41)$$

The supersymmetry transformations in (3.3.37) connecting these fields

$$\delta X^+ = +\frac{1}{\sqrt{2}}\bar{\epsilon}\,\psi^+,\qquad \delta\psi^+ = -\frac{i}{\sqrt{2}}\rho^\beta \epsilon\,\partial_\beta X^+,\qquad \delta\overline{\psi}^+ = \frac{i}{\sqrt{2}}\bar{\epsilon}\rho^\beta\,\partial_\beta X^+,\qquad (3.3.42)$$

then give

$$\delta_+ I = -\sqrt{2}\,\bar{\epsilon}(\partial^\alpha \psi^+)\,\partial_\alpha X^- - \frac{1}{\sqrt{2}}\bar{\epsilon}\,(\rho^\beta\,\partial_\beta X^+)\rho^\alpha \partial_\alpha \psi^-$$

$$+\ \frac{1}{\sqrt{2}}\,\overline{\psi}^- \rho^\alpha \partial_\alpha (\rho^\beta\,\partial_\beta X^+)\,\epsilon.\qquad (3.3.43)$$

---

[27] See (3.2.30) for the definition of light-cone variables.

The interest in this equation is that in the so-called light cone-gauge (3.2.45), for which we take $X^+ = x^+ + \ell^2 p^+ \tau$, $\&_+ I$ is reduced to a total derivative by simultaneously setting $\partial^\alpha \psi^+ = 0$, having the explicit expression

$$\&_+ I\Big|_{\text{light-cone gauge}} = -\frac{1}{\sqrt{2}} \ell^2 p^+ \bar{\epsilon} \partial_\alpha (\rho^0 \rho^\alpha \psi^-), \qquad (3.3.44)$$

where, in the process, we have used the fact that $\partial_\beta X^+ = \delta^0{}_\beta \ell^2 p^+$, and finally that $\partial_\alpha \epsilon = 0$.

Therefore we define the light-cone gauge for which

$$X^+ = x^+ + \ell^2 p^+ \tau, \qquad \partial^\alpha \psi^+ = 0. \qquad (3.3.45)$$

The supercurrent in (3.3.39) written in the form

$$J^\alpha = \frac{1}{2} \rho^\beta \rho^\alpha [-\psi^+ \partial_\beta X^- - \psi^- \partial_\beta X^+ + \psi^i \partial_\beta X^i] = 0, \qquad (3.3.46)$$

may be used for $J^0 = 0$, to solve for ψ^- via the equations

$$\frac{1}{2} \ell^2 p^+ \psi^-_R = -\psi^+_R \partial_- X^-_R + \psi^i_R \partial_- X^i_R,$$

$$\frac{1}{2} \ell^2 p^+ \psi^-_L = -\psi^+_L \partial_+ X^-_L + \psi^i_L \partial_+ X^i_L, \qquad (3.3.47)$$

once the ξ-independent term ψ^+ has been specified and the remaining terms on the right-hand sides of these equations are explicitly evaluated. Here we have introduced the notations and the expressions

$$\partial_\pm = \frac{1}{2} (\partial_\tau \pm \partial_\sigma), \qquad \partial_\pm (\tau \pm \sigma) = 1, \qquad \partial_\pm (\tau \mp \sigma) = 0, \qquad (3.3.48)$$

$$\psi^\mu = \begin{pmatrix} \psi^\mu_R \\ \psi^\mu_L \end{pmatrix}, \qquad \bar{\psi} \cdot \frac{\rho^\alpha \partial_\alpha}{i} \psi = \frac{2}{i} (\psi_L \cdot \partial_- \psi_L + \psi_R \cdot \partial_+ \psi_R), \qquad (3.3.49)$$

and used (3.3.19) as a consequence of the Majorana character of the spinor field. Clearly, the field equations resulting from the action (3.3.36) and (3.3.49), for ψ_R, ψ_L are then

$$\partial_+ \psi^\mu_R = 0, \qquad \partial_- \psi^\mu_L = 0. \qquad (3.3.50)$$

That is, from the field equations we note that ψ^μ_R is a function of $\tau - \sigma$, and is referred to as right moving, and ψ^μ_L is a function of $\tau + \sigma$, and is referred to as left moving.

3.3.3 Boundary Conditions and Solutions of Field Equations

We display some of the equations relevant to superstrings, in the light-cone gauge, which were given in the previous section and include some additional ones discussed below:

$$\Box X^i = 0, \quad X^+ = x^+ + \ell^2 p^+ \tau, \tag{3.3.51}$$

$$\partial_+ \psi_R^\mu = 0, \quad \partial_- \psi_L^\mu = 0, \quad \partial_\alpha \psi^+ = 0, \tag{3.3.52}$$

$$\frac{1}{2}\ell^2 p^+ \psi_R^- = -\psi_R^{+}\partial_- X_R^- + \psi_R^i \partial_- X_R^i, \quad \frac{1}{2}\ell^2 p^+ \psi_L^- = -\psi_L^{+}\partial_+ X_L^- + \psi_L^i \partial_+ X_L^i, \tag{3.3.53}$$

$$\ell^2 p^+ \, \partial_+ X_L^- = \partial_+ X_L^i \, \partial_+ X_L^i + \frac{i}{2} \psi_L^i \partial_+ \psi_L^i - \frac{i}{2} \psi_L^{+}\partial_+ \psi_L^-, \tag{3.3.54}$$

$$\ell^2 p^+ \, \partial_- X_R^- = \partial_- X_R^i \, \partial_- X_R^i + \frac{i}{2} \psi_R^i \partial_- \psi_R^i - \frac{i}{2} \psi_R^{+}\partial_- \psi_R^-, \tag{3.3.55}$$

$$-\frac{i}{2}(\psi_L^{+}\partial_+ \psi_L^- - \psi_R^{+}\partial_- \psi_R^-) + \frac{i}{2}(\psi_L^i \partial_+ \psi_L^i - \psi_R^i \partial_- \psi_R^i)$$

$$-\ell^2 p^+ \partial_1 X^- + \partial_0 X^i \partial_1 X^i = 0. \tag{3.3.56}$$

We recall that (3.3.53) follows from the condition $J^0 = 0$ for the supercurrent in (3.3.46) (see Problem 3.16). On the other hand, (3.3.54)–(3.3.56) follow from the combinations, $T_{00} + T_{01} = 0, T_{00} - T_{01} = 0$, and $T_{01} = 0$, respectively as obtained from the expression of the energy-momentum tensor in (3.3.35) (see Problems 3.17).

From the second expression in (3.3.49), the fermionic part of the action in (3.3.36), is given by

$$W_F = iT \int (d\xi)(\psi_L \cdot \partial_- \psi_L + \psi_R \cdot \partial_+ \psi_R), \quad (d\xi) = d\xi^0 d\xi^1 = d\tau \, d\sigma. \tag{3.3.57}$$

The boundary condition that arises by varying the fields ψ_R, ψ_L is

$$(\psi_L \cdot \delta\psi_L - \psi_R \cdot \delta\psi_R)\big|_{\sigma=\pi} - (\psi_L \cdot \delta\psi_L - \psi_R \cdot \delta\psi_R)\big|_{\sigma=0} = 0. \tag{3.3.58}$$

We consider this boundary condition first for open strings.

Open Strings

For the bosonic part of a string, we choose the Neumann boundary condition

$$\partial_\sigma X^\mu(\tau,\sigma)|_{\sigma=0,\pi} = 0, \tag{3.3.59}$$

in (3.2.52).

For the fermionic boundary condition (3.3.58), we choose

$$\psi^\mu_L(\tau,0) = \psi^\mu_R(\tau,0), \quad \delta\psi^\mu_L(\tau,0) = \delta\psi^\mu_R(\tau,0) \tag{3.3.60}$$

and at the other end of the string, we have the two admissible boundary conditions with no change or change of the relative signs of the ends of the strings:

$$\psi^\mu_L(\tau,\pi) = +\psi^\mu_R(\tau,\pi), \quad \delta\psi^\mu_L(\tau,\pi) = +\delta\psi^\mu_R(\tau,\pi) \quad (R), \tag{3.3.61}$$

$$\psi^\mu_L(\tau,\pi) = -\psi^\mu_R(\tau,\pi), \quad \delta\psi^\mu_L(\tau,\pi) = -\delta\psi^\mu_R(\tau,\pi) \quad (NS), \tag{3.3.62}$$

where (R), (NS) refer, respectively, to Raymond and Neveu-Schwarz boundary conditions .

In view of these boundary conditions, ψ^+ takes a very special role. Since it is independent of (τ,σ), and due to the change in sign of the fields at $\sigma = \pi$, for the (NS) boundary condition, clearly $\psi^+ = 0$, for such a boundary condition. On the other hand, for the (R) boundary condition, we have to consider, in general, $\psi^+ \neq 0$.

We may carry out a Fourier transform as a mode expansion for the transverse fields satisfying the above conditions to obtain[28]

$$\psi^i_{L/R}(\tau,\sigma) = \frac{\ell}{\sqrt{2}} \sum_r d^i(r) e^{-ir(\tau\pm\sigma)}, \tag{3.3.63}$$

$$r = n = 0, \pm 1, \pm 2, \ldots \quad (R), \qquad r = \pm\frac{1}{2}, \pm\frac{3}{2}, \ldots \quad (NS). \tag{3.3.64}$$

From (3.2.58), we also have

$$X^i_{L/R} = \frac{1}{2}x^i + \frac{1}{2}\ell^2 p^i(\tau\pm\sigma) + \frac{i\ell}{2}\sum_{n\neq 0}\frac{\alpha^i(n)}{n}e^{-in(\tau\pm\sigma)}, \tag{3.3.65}$$

---

[28]The coefficients $d^i(R)$ for integers and half-odd integers should not be confused.

and the $\exp[-in(\tau + \sigma)]$ - independent part, of (3.3.54) gives, for an R boundary condition with $\psi^+ \neq 0$, the equation (see Problem 3.18)

$$\frac{\ell^4}{2} p^+ p^- = \frac{\ell^2}{4} \sum_{m=-\infty}^{\infty} \left(\alpha^i(-m)\,\alpha^i(m) + d^i(-m)\,m\,d^i(m) \right) \qquad \text{(R)},$$

(3.3.66)

which may be rewritten in a more useful form as

$$2 p^+ p^- - p^i p^i = \frac{1}{\ell^2} \sum_{m\neq 0} \left(\alpha^i(-m)\,\alpha^i(m) + d^i(-m)\,m\,d^i(m) \right) \quad \text{(R)},$$

(3.3.67)

where we have used the identity $\ell p^i = \alpha^i(0)$.

For a NS boundary condition, the same reasoning, with $\psi^+ = 0$, gives (see Problem 3.18)

$$2 p^+ p^- - p^i p^i = \frac{1}{\ell^2} \left(\sum_{m\neq 0} \alpha^i(-m)\,\alpha^i(m) + \sum_r d^i(-r)\,r\,d^i(r) \right),$$

$$r = \pm\frac{1}{2}, \pm\frac{3}{2}, \dots \quad \text{(NS)}. \qquad (3.3.68)$$

Closed Strings

For the bosonic part of a string, we have the periodic boundary condition $X^\mu(\tau, \sigma) = X^\mu(\tau, \sigma + \pi)$ in (3.2.51). For the fermionic part, one has to specify the movers $\psi^\mu_{R/L}$, relative to an original point, after translating σ by a period π. We enumerate the possible boundary conditions:

$$\psi^\mu_R(\tau, \sigma) = +\psi^\mu_R(\tau, \sigma + \pi), \quad \psi^\mu_L(\tau, \sigma) = +\psi^\mu_L(\tau, \sigma + \pi) \quad \text{(R, R)},$$

(3.3.69)

$$\psi^\mu_R(\tau, \sigma) = +\psi^\mu_R(\tau, \sigma + \pi), \quad \psi^\mu_L(\tau, \sigma) = -\psi^\mu_L(\tau, \sigma + \pi) \quad \text{(R, NS)},$$

(3.3.70)

$$\psi^\mu_R(\tau, \sigma) = -\psi^\mu_R(\tau, \sigma + \pi), \quad \psi^\mu_L(\tau, \sigma) = +\psi^\mu_L(\tau, \sigma + \pi) \quad \text{(NS, R)},$$

(3.3.71)

$$\psi^\mu_R(\tau, \sigma) = -\psi^\mu_R(\tau, \sigma + \pi), \quad \psi^\mu_L(\tau, \sigma) = -\psi^\mu_L(\tau, \sigma + \pi) \quad \text{(NS, NS)},$$

(3.3.72)

with the boundary conditions referred to as indicated on the right-hand sides of these equations. Note that $\delta\psi^\mu_{R/L}$ automatically satisfy the same boundary conditions as $\psi^\mu_{R/L}$.

The corresponding solutions for the transverse fields may be then spelled out as

$$\psi^i_R = \ell \sum_r d^i(r) e^{-2ir(\tau-\sigma)} \text{ (R)}, \qquad \psi^i_L = \ell \sum_r \bar{d}^i(r) e^{-2ir(\tau+\sigma)} \text{ (R)},$$

$$(3.3.73)$$

$$\psi^i_R = \ell \sum_r b^i(r) e^{-2ir(\tau-\sigma)} \text{ (NS)}, \qquad \psi^i_L = \ell \sum_r \bar{b}^i(r) e^{-2ir(\tau+\sigma)} \text{ (NS)},$$

$$(3.3.74)$$

with the corresponding r as given in (3.3.64).

Also we recall from (3.2.70), (3.2.71)

$$X^i_R = \frac{1}{2}x^i + \frac{1}{2}\ell^2 p^i(\tau-\sigma) + \frac{i\ell}{2} \sum_{n\neq 0} \frac{\alpha^i(n)}{n} e^{-2in(\tau-\sigma)}, \qquad (3.3.75)$$

$$X^i_L = \frac{1}{2}x^i + \frac{1}{2}\ell^2 p^i(\tau+\sigma) + \frac{i\ell}{2} \sum_{n\neq 0} \frac{\bar{\alpha}^i(n)}{n} e^{-2in(\tau+\sigma)}, \qquad (3.3.76)$$

$\ell p^i/2 = \alpha^i(0) = \bar{\alpha}^i(0)$, with $x^i_L = x^i_R$. The condition $T_{01} = 0$ in (3.3.56), with $T_{\alpha\beta}$ defined in (3.3.35), when integrated over σ from 0 to π leads to (see Problem 3.19)

$$\sum_n \left(\alpha^i(-n)\alpha^i(n) + d^i(-n) n d^i(n) \right) = \sum_n \left(\bar{\alpha}^i(-n)\bar{\alpha}^i(n) + \bar{d}^i(-n) n \bar{d}^i(n) \right),$$

$$(3.3.77)$$

$$\sum_n \alpha^i(-n)\alpha^i(n) + \sum_r b^i(-r) r b^i(r) = \sum_n \bar{\alpha}^i(-n)\bar{\alpha}^i(n) + \sum_r \bar{b}^i(-r) r \bar{b}^i(r),$$

$$(3.3.78)$$

for (R) and (NS) boundary conditions, respectively, with $r = \pm 1/2, \pm 3/2, \ldots$, in (3.3.78). The equations for the various sectors in (3.3.69)–(3.3.72), similar to the open string equations (3.3.67), (3.3.68), will be given in Sect. 3.3.5.

3.3.4 Quantum Aspect, Critical Spacetime Dimension and the Mass Spectrum: Open Strings

The canonical conjugate momenta to ψ^i_L, ψ^i_R, are from (3.3.57), respectively, and the definition $\partial_\pm = (\partial_\tau \pm \partial_\sigma)/2$ in (3.3.48), given by $iT\psi^i_L/2$, $iT\psi^i_R/2$. The quantum aspect of the theory then emerges upon introducing the anti-commutation relations

$$\{\psi^i_L(\tau,\sigma), \psi^j_L(\tau,\sigma')\} = \frac{2}{T}\delta^{ij}\delta(\sigma-\sigma'), \qquad \qquad (3.3.79)$$

$$\{\psi^i_R(\tau,\sigma), \psi^j_R(\tau,\sigma')\} = \frac{2}{T}\delta^{ij}\delta(\sigma-\sigma'), \qquad \qquad (3.3.80)$$

$$\{\psi^i_L(\tau,\sigma), \psi^j_R(\tau,\sigma')\} = 0. \qquad \qquad (3.3.81)$$

The corresponding anti-commutation relations for the Fourier coefficients in (3.3.63) for the open string, then read

$$\{d^i(r), d^j(r')^\dagger\} = \delta^{ij}\delta(r,r'), \quad \{d^i(r), d^j(r')\} = 0, \quad \{d^i(r)^\dagger, d^j(r')^\dagger\} = 0, \qquad (3.3.82)$$

where we have used the reality condition of the Majorana spinors to set $d^i(-r) = d^i(r)^\dagger$. These operators commute with the Fourier coefficients $\alpha^i(m)$, $\alpha^i(m)^\dagger$.

We investigate, in turn, the nature of NS and R boundary conditions.

NS Boundary Condition

For a NS boundary conditions, $\psi^+ = 0$, and the mass spectrum is directly obtained from (3.3.68) to be : [$\ell^2 = 2\alpha'$ as defined in (3.2.46)]

$$M^2 = \frac{1}{\alpha'}\left(\sum_{m=1}^{\infty} \alpha^i(m)^\dagger\alpha^i(m) + \frac{D-2}{2}\sum_{m=1}^{\infty} m \right.$$
$$\left. + \sum_{r=\frac{1}{2}}^{\infty} d^i(r)^\dagger r\, d^i(r) - \frac{D-2}{2}\sum_{r=\frac{1}{2}}^{\infty} r \right) \quad \text{(NS)}, \qquad (3.3.83)$$

here $r = 1/2, 3/2, \ldots$, where we have used the commutation and anti-commutation rules, respectively, in (3.2.88), (3.3.82).

As for the bosonic string, in making a transition from a classical description to a quantum one, the sum $\sum_{m=1}^{\infty} m$ is interpreted as the analytic continuation of the Riemann zeta function $\zeta(s)$ in (3.2.91) to $s = -1$ giving the value $\zeta(-1) = -1/12$. On the other hand, the zero point energy

$$\sum_{r=1/2, 3/2, \ldots}^{\infty} r \equiv \sum_{m=0}^{\infty}(m+1/2),$$

is interpreted as the analytic continuation of the *Hurwitz* function[29]

$$\zeta(s,a) = \sum_{m=0}^{\infty} (m+a)^{-s}, \quad \mathrm{Re}\, s > 1, \qquad (3.3.84)$$

to $s = -1$, for $a = 1/2$:

$$\zeta(-1,a)\big|_{a=1/2} = -\frac{1}{2}\left(a^2 - a + \frac{1}{6}\right)\bigg|_{a=1/2} = \frac{1}{24}, \qquad (3.3.85)$$

which is all that is needed about the Hurwitz function here. Accordingly, (3.3.83) becomes

$$M^2 = \frac{1}{\alpha'}\left(\sum_{m=1}^{\infty} \alpha^i(m)^\dagger \alpha^i(m) + \sum_{r=1/2}^{\infty} d^i(r)^\dagger r d^i(r) - \frac{D-2}{16}\right), \quad \text{(NS)}.$$

$$(3.3.86)$$

Let $|NS\rangle|0,p'\rangle$ denote the corresponding ground-state. We first examine the first lowest excited state for $r = 1/2$. The application of the operator M^2 to this state gives

$$M^2 d^i(1/2)^\dagger |NS\rangle|0,p'\rangle = \frac{1}{\alpha'}\left(\frac{1}{2} - \frac{D-2}{16}\right) d^i(1/2)|NS\rangle|0,p'\rangle, \qquad (3.3.87)$$

which corresponds to a vector particle with $D - 2$ degrees of freedom and hence is massless—the photon.[30] This gives the critical dimension of spacetime from $(1/2 - (D-2)/16) = 0$ to be 10, with the number of degrees of freedom of the vector particle being equal to 8. Again we learn that the number of degrees of freedom and the masslessness of the particle (described by a vector field) determine the dimensionality of the underlying spacetime of the theory.

There seems to be also a tachyonic state with negative mass squared $-1/(2\alpha')$ which, however, by invoking the necessary supersymmetry restriction that the number of fermion states is equal to the of bosonic states, this state is automatically eliminated as we will see later.

---

[29]For some useful references on the Hurwitz function see [1, 7, 12]. The analytical continuation to $s = -1$, for $a = 1/2$, is all what is needed here.
[30]See Sect. 3.2.9, (3.2.175), (3.2.176).

R Boundary Condition

For the R boundary condition, the mass shell condition follows from (3.3.67) to be

$$M^2 = \frac{1}{\alpha'} \sum_{m=1}^{\infty} \left(\alpha^i(m)^\dagger \alpha^i(m) + d^i(m)^\dagger m \, d^i(m) \right), \qquad (R), \qquad (3.3.88)$$

where the zero point energies cancel out.

A key point which follows from (3.3.63) and distinguishes an R boundary condition from a NS one, is that the former involves the index r taking the value zero in the Fourier mode expansions of the spinor fields [see (3.3.64)] generating the Hermitian coefficients $d^i(0)$ of zero mode. We will see how these coefficients play a major role in defining the ground-states for an R boundary condition.

To investigate the nature of the ground-state of superstrings associated with the R boundary condition, we consider the following well paused problem.

To the above end, given the following anti-commutation relations (a Clifford algebra) satisfied by the set of the Hermitian operators $d^i(0) \equiv d^i, i = 1, \ldots, 2n; n = 4$:

$$\{d^i, d^j\} = \delta^{ij}, \quad i, j = 1, 2 \ldots, 2n, \qquad (d^i)^\dagger = d^i, \qquad (3.3.89)$$

one may define the operators:

$$a_1 = \frac{1}{\sqrt{2}}(d^2 + i d^1), \; a_2 = \frac{1}{\sqrt{2}}(d^4 + i d^3), \; \ldots, \; a_n = \frac{1}{\sqrt{2}}(d^{2n} + i d^{2n-1}),$$
$$(3.3.90)$$

which, as is easily verified, satisfy the anti-commutation relations

$$\{a_I, a_J^\dagger\} = \delta_{IJ}, \quad \{a_I, a_J\} = 0, \quad \{a_I^\dagger, a_J^\dagger\} = 0, \quad I, J = 1, 2, \ldots, n. \qquad (3.3.91)$$

Let $| \, 0 \rangle$ be a state (Fock vacuum) which is annihilated by the operators a_I: $a_I | \, 0 \rangle = 0, I = 1, 2, \ldots, n$. We introduce the following 2^n states:

$$|\lambda_1 \lambda_2, \ldots, \lambda_n \rangle = (a_1^\dagger)^{\lambda_1 + 1/2} (a_2^\dagger)^{\lambda_2 + 1/2} \ldots (a_n^\dagger)^{\lambda_n + 1/2} | \, 0 \rangle, \qquad (3.3.92)$$

where the λ_I's take on the values $\pm 1/2$. The d^i operators may be solved in terms of the a_I operators and their adjoints as follows:

$$d^1 = \frac{1}{i\sqrt{2}}(a_1 - a_1^\dagger), \; d^2 = \frac{1}{\sqrt{2}}(a_1 + a_1^\dagger), \; \ldots$$

$$\ldots, \; d^{2n-1} = \frac{1}{i\sqrt{2}}(a_n - a_n^\dagger), \; d^{2n} = \frac{1}{\sqrt{2}}(a_n + a_n^\dagger). \qquad (3.3.93)$$

With operators Λ^I, as defined below, we have the following eigenvalue equations:

$$\Lambda^I \equiv \frac{d^{2I-1} d^{2I}}{i} = a_I^\dagger a_I - \frac{1}{2}, \qquad \Lambda^I |\lambda_1 \lambda_2, \ldots, \lambda_n\rangle = \lambda_I |\lambda_1 \lambda_2, \ldots, \lambda_n\rangle, \tag{3.3.94}$$

$I = 1, 2, \ldots, n$, where we have used (3.3.91), the following product involving the $d^I, I = 1, 2, \ldots, n$, operators

$$d \equiv (2\,i)^n d^1 \ldots d^{2n} = (-2)^n \Lambda^1 \ldots \Lambda^n, \tag{3.3.95}$$

satisfies the eigenvalue equation

$$d |\lambda_1 \lambda_2, \ldots, \lambda_n\rangle = (-2)^n \lambda_1 \lambda_2 \ldots \lambda_n |\lambda_1 \lambda_2, \ldots, \lambda_n\rangle. \tag{3.3.96}$$

In particular, let k denote the number of λ_I's having the value $+1/2$, then

$$d |\lambda_1 \lambda_2, \ldots, \lambda_n\rangle = (-1)^k |\lambda_1 \lambda_2, \ldots, \lambda_n\rangle, \tag{3.3.97}$$

where we have used the fact that $(-2)^n (1/2)^k (-1/2)^{n-k} = (-1)^k$. That is, thanks to the d operator, the 2^n states can be split into two sets, each with an equal number of 2^{n-1} states $|\psi_1^\kappa\rangle$ and $|\psi_2^\kappa\rangle$, with so-called opposite chirality:

$$d |\psi_1^\kappa\rangle = + |\psi_1^\kappa\rangle, \qquad d |\psi_2^\kappa\rangle = - |\psi_2^\kappa\rangle, \qquad \kappa = 1, 2, \ldots, 2^{n-1}. \tag{3.3.98}$$

Some properties of the d operator are

$$\{d, d^I\} = 0, \qquad (d)^2 = 1. \tag{3.3.99}$$

We note that the states with negative/positive chirality correspond to an odd/even number of creation operators applied to the Fock vacuum, since k in (3.3.97) denotes the number of λ_I taking the value $1/2$ thus providing a non-vanishing power of one for a corresponding creation operator.

In detail for $n = 4$,

$$\Lambda^1 = d^1(0) d^2(0)/i, \qquad \Lambda^2 = d^3(0) d^4(0)/i,$$
$$\Lambda^3 = d^5(0) d^6(0)/i, \qquad \Lambda^4 = d^7(0) d^8(0)/i,$$

are Hermitian operators and commuting in pairs. Clearly, M^2 in (3.3.88) involving the operators $\alpha^i(m)$, $d^i(m)$, for $m \neq 0$, also commutes with the operators $\Lambda^1, \ldots, \Lambda^4$. The non-Hermitian operators $a_1, a_2, a_3, a_4, d^i(m)$, for $m = 1, 2, \ldots$, annihilate the states $|\lambda_1 \lambda_2, \lambda_3, \lambda_4\rangle$ in (3.3.94).

Now we have to look for a fermion, for the R boundary condition, with corresponding degrees of freedom to match the vector particle observed above, in the NS boundary condition, in the light of supersymmetry.

The simultaneous eigenstates of M^2 and of the operators $\Lambda^1, \ldots, \Lambda^4$, define degenerate ground-states, which may be labeled by the eigenvalues of the latter operators. Each of these eigenvalues take on two values (see (3.3.94), with $\lambda_I = \pm 1/2$) generating 16 eigenstates which are annihilated by the non-negative operators M^2 in (3.3.88) leading to a 16-fold degenerate ground-state energy. By referring to (3.3.98), we denote the two sets of these ground states with definite chiralities by $\{|\psi_{\bar{1}}^\kappa\rangle |0; p'\rangle\}$, $\{|\psi_{\bar{2}}^\kappa\rangle |0; p'\rangle\}$, with $|0; p'\rangle$ being the bosonic ground-state, and $\kappa = 1, 2, \ldots, 8$.

We consider the $\exp[-im(\tau + \sigma)]$-independent part of the second equation in (3.3.53) indicated by a bar $|$. To this end we note, in particular, that

$$\partial_+ X_L^- \big| = \frac{1}{2} \ell^2 p^-, \qquad \psi_L^- \big| = \frac{\ell}{\sqrt{2}} d^-, \tag{3.3.100}$$

$$\psi_L^i \partial_+ X_L^i \big| = \frac{\ell^3}{2\sqrt{2}} d^i p^i + \frac{\ell^2}{2\sqrt{2}} \sum_{n=1}^\infty [d^i(n)^\dagger \alpha^i(n) + \alpha^i(n)^\dagger d^i(n)], \tag{3.3.101}$$

where in writing the second equation, we have used the expressions for ψ_L^i, X_L^i in (3.3.63), (3.3.65), respectively, used the fact that the zero point energies of the just mentioned equation cancel out, used the identity $\alpha^i(0) = \ell p^i$ in (3.2.61), and finally, simplified the notation by setting $d^i(0) \equiv d^i$, as before.

The above suggests to define the constant ψ_L^+ by

$$\psi_L^+ = \frac{\ell}{\sqrt{2}} d^+. \tag{3.3.102}$$

where we have used, in the process, the last property $\partial_\alpha \psi^+ = 0$ in (3.3.52).

The zero mode of the second equation in (3.3.53) then simply reads

$$d^- p^+ + d^+ p^- - d^i p^i - \frac{1}{\ell} \sum_{n=1}^\infty [d^i(n)^\dagger \alpha^i(n) + \alpha^i(n)^\dagger d^i(n)] = 0. \tag{3.3.103}$$

This expression now applied to any of the states

$$|\psi_{\bar{1}}^\kappa\rangle |0; p'\rangle, \quad |\psi_{\bar{2}}^\kappa\rangle |0; p'\rangle,$$

gives

$$[d^-p^+ + d^+p^- - d^ip^i] \, | \psi^\kappa_j \rangle \, |0;p'\rangle$$

$$= \sqrt{2\alpha'} \sum_{m=1}^{\infty} \left(d^i(m)^\dagger \alpha^i(m) + \alpha^i(m)^\dagger d^i(m) \right) | \psi^\kappa_j \rangle \, |0;p'\rangle = 0. \qquad (3.3.104)$$

Here we note that the operators $d^i(m)$ annihilate these states since they annihilate the Fock vacuum $|0\rangle$, and the operators $\alpha^i(m)$ annihilate $|0;p'\rangle$, all for $m \geq 1$. Accordingly, the expression on the right-hand side of (3.3.104) does not contribute to the operator in question when acting on such states and vanishes as indicated.

For j equal to one of the values 1 or 2, thus restricting to one of the set of ground states

$$\{| \psi^\kappa_1 \rangle \, |0;p'\rangle \}, \quad \text{or} \quad \{| \psi^\kappa_2 \rangle \, |0;p'\rangle \},$$

i.e., with one of the chiralities, as defined in (3.3.98) for $n = 4$, the above is a Dirac equation

$$[d^-p^+ + d^+p^- - d^ip^i] \, | \psi^\kappa_j \rangle |0;p'\rangle = 0,$$

of a massless chiral field with 8 degrees of freedom. These states are precisely the ones annihilated by the (non-negative) mass operator in (3.3.88). These states are then the lowest lying states with the R boundary condition.

Needless to say, one may multilply (3.3.104) by $i\sqrt{2}$ to have our standard form of gamma matrices.[31] To this end, we have a representation of the gamma matrices with, in particular, $i\sqrt{2}\,d^+ \equiv (\gamma^0 + \gamma^9)/\sqrt{2}$ with $(\gamma^0)^2 = 1, (\gamma^9)^2 = -1, \{\gamma^0, \gamma^9\} = 0$, which make ψ^+_L a Grassmann type, whose square is zero.

To summarise, the massless excitation of an open superstring with an R boundary condition, and a given chirality, is a chiral Fermion with 8 degrees of freedom and matches the number of degrees freedom of the massless vector particle corresponding to a NS boundary condition. What about the tachyon state observed and discussed below (3.3.87) corresponding to the NS boundary condition?

By invoking the necessary condition for the equality of the fermionic and bosonic degrees of freedom as imposed by supersymmetry, we will see that this tachyonic state may be eliminated. For one thing, one may restrict to ground-states with a given chirality for the R boundary condition and thus having 8 degrees of freedom for the massless (chiral) fermion states. We have seen above in (3.3.98), that ground-states, with negative chirality correspond to the application of an odd number of fermion creation operator. If one simultaneously restricts the application of an odd number of fermion operators for the NS boundary condition, not only the condition of supersymmetry is achieved, with the equality of the number of fermionic and

---

[31] Here note that, referring to (3.3.89), $\{i\sqrt{2}\,d^i, i\sqrt{2}d^j\} = -2\delta^{ij}$.

bosonic degrees of freedom, but the tachyonic state is also automatically eliminated. That is supersymmetry eliminates the tachyon. This ingenious method is referred to as the GSO projection[32] method. This is applied next.

Invoking Supersymmetry and the GSO Projection Method

From (3.3.86), with $D = 10$, the mass squared operator for the NS boundary condition may be rewritten as $(\alpha^i(m) = \sqrt{m}\, a^i(m))$

$$M^2 = \frac{1}{\alpha'}\left(\sum_{m=1}^{\infty} m\, a^i(m)^\dagger\, a^i(m) + \sum_{r=1/2}^{\infty} d^i(r)^\dagger\, r\, d^i(r) - \frac{1}{2}\right), \quad \text{(NS)}, \qquad (3.3.105)$$

$$[a^i(m), a^j(n)^\dagger] = \delta^{ij}\,\delta(m,n), \qquad \{d^i(r), d^j(r')^\dagger\} = \delta^{ij}\delta(r,r') \qquad (3.3.106)$$

Following the procedure in deriving (3.2.99), the generating function for determining the number of states, with $M^2 = (N-1/2)/\alpha'$,

$$N = \sum_{m=1}^{\infty} a^i(m)^\dagger\, m\, a^i(m) + \sum_{r=1/2}^{\infty} d^i(r)^\dagger\, r\, d^i(r),$$

removing the contribution of even number of fermions, is given by

$$G_{\text{NS}}[q] = \text{Tr}\left[\frac{1}{2}\left(1 - (-1)^F\right) q^{N-1/2}\right] = \frac{1}{2\sqrt{q}}\,\text{Tr}\left[q^N - (-1)^F q^N\right],$$

$$(3.3.107)$$

where

$$F = \sum_{r=1/2}^{\infty} d^i(r)^\dagger\, d^i(r), \qquad (3.3.108)$$

We note that

$$\text{Tr}[q^N] = \left(\prod_{m=1}^{\infty} \prod_{i=1}^{8} \text{Tr}[q^{m\,a^i(m)^\dagger a^i(m)}]\right)$$

$$\times \left(\prod_{n=1}^{\infty} \prod_{i=1}^{8} \text{Tr}[q^{n'\,d^i(n')^\dagger d^i(n')}]\right), \qquad (3.3.109)$$

---

[32]The GSO method of projection is proposed in [13, 14].

where $n' \equiv (2n - 1)/2$ in the last product. Clearly,

$$\text{Tr}\,[q^{m a^\dagger a}] = 1 + q^m + q^{2m} + \cdots = 1/(1 - q^m),$$

as already encountered in (3.2.99),

$$\text{Tr}\,[q^{n' d^\dagger d}] = 1 + q^{n'}, \quad \text{Tr}\,[(-1)^{d^\dagger d} q^{n' d^\dagger d}] = (-1)^0 1 + (-1)^1 q^{n'} = 1 - q^{n'},$$

due to the Fermi character in the latter two traces, with all carried out in one dimension. This gives

$$G_{\text{NS}}[q] = \frac{1}{2\sqrt{q}} \left[\Pi_{m=1}^{\infty} \left(\frac{1 + q^{m-1/2}}{1 - q^m} \right)^8 - \Pi_{m=1}^{\infty} \left(\frac{1 - q^{m-1/2}}{1 - q^m} \right)^8 \right].$$

(3.3.110)

The ground-states now start with the photon with 8 degrees of freedom, i.e., $G_{\text{NS}}[0] = 8$. The number of states $\rho_{\text{NS}}(n)$ with mass squared equal to n/α', as in (3.2.98), is given by the formula

$$\rho_{\text{NS}}(n) = \frac{1}{n!} \left(\frac{d}{dq} \right)^n G[q]_{\text{NS}} \Big|_{q=0}, \quad n \geq 1, \quad \rho_{\text{NS}}(0) \equiv G_{\text{NS}}[0] = 8.$$

(3.3.111)

For example, it is easily evaluated that $\rho_{\text{NS}}(1) = 128$ and so on for higher states.

The generating function for the number of states $\rho_R(n)$ with mass squared $M^2 = n/\alpha'$, corresponding to an R boundary condition, with expression for M^2 in (3.3.88), with sums over $m \geq 1$, is now even easier to obtain and is given by

$$G_R[q] = 8 \prod_{m=1}^{\infty} \left(\frac{1 + q^m}{1 - q^m} \right)^8,$$

(3.3.112)

with the number 8 multiplying this expression arises because of the multiplicity of 8 of the ground-states $| \psi_{\frac{\kappa}{2}} \rangle \, | \, 0; p' \rangle$ with chosen negative chirality (see (3.3.98)), from which other states are generated by the application of the relevant creation operators.

The reader may amuse himself (herself) to compare the equalities of $\rho_{\text{NS}}(n)$ and $\rho_R(n)$ by explicit differentiations as given in (3.3.111) or equivalently by carrying power expansions of $G_{\text{NS}}[q]$ and $G_R[q]$ in q. [The equalities of these generating functions was proved by C.G.J. Jacobi in 1829, independently, of course, of the problem treated above.] This satisfies the condition for achieving supersymmetry. It turns out to be not only sufficient but also necessary.

3.3.5   Quantum Aspect, Critical Spacetime Dimension and the Mass Spectrum: Closed Strings

For a closed string, the anti-commutation relations for the Fourier coefficients in (3.3.73) and (3.3.74), read

$$\{d^i(r), d^j(r')^\dagger\} = \delta^{ij}\delta(r, r'), \qquad \{\bar{d}^i(r), \bar{d}^j(r')^\dagger\} = \delta^{ij}\delta(r, r'), \qquad (3.3.113)$$

$$\{b^i(r), b^j(r')^\dagger\} = \delta^{ij}\delta(r, r'), \qquad \{\bar{b}^i(r), \bar{b}^j(r')^\dagger\} = \delta^{ij}\delta(r, r'). \qquad (3.3.114)$$

All the other combinations of these coefficients anti-commute in pairs. These fermion operators commute with the Fourier coefficients of the $X^i_{R/L}$ fields in (3.3.75)/(3.3.76).

The key equations for analyzing the mass spectrum of closed strings are the equations for $\partial_+ X^-_L$, $\partial_- X^-_R$, given, respectively, in (3.3.54) and (3.3.55),

$$\ell^2 p^+ \partial_+ X^-_L = \partial_+ X^i_L \partial_+ X^i_L + \frac{i}{2}\psi^i_L \partial_+ \psi^i_L - \frac{i}{2}\psi^+_L \partial_+ \psi^-_L, \qquad (3.3.115)$$

$$\ell^2 p^+ \partial_- X^-_R = \partial_- X^i_R \partial_- X^i_R + \frac{i}{2}\psi^i_R \partial_- \psi^i_R - \frac{i}{2}\psi^+_R \partial_- \psi^-_R, \qquad (3.3.116)$$

together with the defining equations for $X^i_{R/L}$ in (3.3.75), (3.3.76), $\psi^i_{R/L}$ in (3.3.73), (3.3.74),

$$\psi^i_R = \ell \sum_n d^i(n)\, e^{-2in(\tau-\sigma)} \text{ (R)}, \qquad \psi^i_L = \ell \sum_n \bar{d}^i(n)\, e^{-2in(\tau+\sigma)} \text{ (R)}, \qquad (3.3.117)$$

$$\psi^i_R = \ell \sum_r b^i(r)\, e^{-2ir(\tau-\sigma)} \text{ (NS)}, \qquad \psi^i_L = \ell \sum_r \bar{b}^i(r)\, e^{-2ir(\tau+\sigma)} \text{ (NS)}, \qquad (3.3.118)$$

with $n = 0, \pm 1, \pm 2, \ldots; r = \pm 1/2, \pm 3/2, \ldots$.

The various boundary conditions: (R, R), (R, NS), (NS, R), (NS, NS) are spelled out in (3.3.69)–(3.3.72), respectively. These are investigated next. We begin with the last one.

(NS, NS) Boundary Condition

We first consider the left-mover ψ^i_L in (3.3.115), having the expression given in (3.3.118), with a NS boundary condition. For the right-mover ψ^i_R, in (3.3.116), we have two possibilities. We may either choose the one with the R boundary condition in (3.3.117), or the one with a NS boundary condition in (3.3.118). Let us consider the latter condition first corresponding to the case referred to under the above heading. The sum of the just mentioned two equations then readily lead for the zero mode, with

$$\ell^2 p^+ (\partial_- X_R^- + \partial_+ X_L^-) = \ell^2 p^+ \partial_\tau X^-,$$

the equality

$$\ell^4 p^+ p^- = \ell^2 \sum_n \left(\alpha^i(-n)\,\alpha^i(n) + \tilde{\alpha}^i(-n)\,\tilde{\alpha}^i(n) \right)$$

$$+ \ell^2 \sum_r r \left(b^i(-r)\,b^i(r) + \bar{b}^i(-r)\,\bar{b}^i(r) \right), \qquad r = \pm\frac{1}{2}, \pm\frac{3}{2}, \dots .$$

$$(3.3.119)$$

with the constraint (3.3.78)

$$\sum_n \alpha^i(-n)\,\alpha^i(n) + \sum_r b^i(-r)\,r\,b^i(r) = \sum_n \tilde{\alpha}^i(-n)\,\tilde{\alpha}^i(n) + \sum_r \bar{b}^i(-r)\,r\,\bar{b}^i(r).$$

$$(3.3.120)$$

We note that with this boundary condition $\psi^+ = 0$.

Upon using the equality that $\ell p^i/2 = \alpha^i(0) = \tilde{\alpha}^i(0)$, as given in (3.2.75), and that the sums

$$\sum_{m=1}^{\infty} m, \qquad \sum_{r=1/2}^{\infty} r = \sum_{m=0}^{\infty} (m + 1/2),$$

are the analytic continuations of the Riemann zeta function in (3.2.91), and the Hurwitz function in (3.3.84), (3.3.85), as before, (3.3.119) becomes

$$\frac{\alpha'}{2} M^2 = \sum_{n=1}^{\infty} \left(\alpha^i(n)^\dagger \alpha^i(n) + \tilde{\alpha}^i(n)^\dagger \tilde{\alpha}^i(n) \right)$$

$$+ \sum_{r=1/2}^{\infty} r \left(b^i(r)^\dagger b^i(r) + \bar{b}^i(r)^\dagger \bar{b}^i(r) \right) - \frac{D-2}{8}, \qquad (3.3.121)$$

corresponding to the appropriately denoted sector (NS_L, NS_R).

Again a GSO projection eliminating even number of $b^i(r)^\dagger$ operations (and even number of $\bar{b}^i(r)^\dagger$'s) removes the tachyonic state with a mass squared equal to $-(D-2)/(4\alpha')$. The first excited state is then given from the constraint (3.3.120) to be

$$\bar{b}^i(1/2)^\dagger | \, NS\rangle_L \; b^j(1/2)^\dagger | \, NS\rangle_R \, | \, 0; p'\rangle,$$

leading to the application

$$M^2 \bar{b}^i(1/2)^\dagger | \text{NS} \rangle_L \, b^j(1/2)^\dagger | \text{NS} \rangle_R \, | \, 0; p' \rangle$$

$$= \frac{2}{\alpha'} \left(1 - \frac{D-2}{8} \right) \bar{b}^i(1/2)^\dagger | \text{NS} \rangle_L \, b^j(1/2)^\dagger | \text{NS} \rangle_R \, | \, 0; p' \rangle. \qquad (3.3.122)$$

These constitute $(D-2)^2$ states. An identical analysis as in (3.2.116) then tells us that the underlying fields are a scalar field A, the dilaton, with one degree of freedom, a symmetric traceless tensor G^{ij} of $D(D-3)/2$ degrees of freedoms, the graviton, as obtained in (3.2.122), and an antisymmetric field B^{ij} with $(D-2)(D-3)/2$ degrees of freedom, as obtained in (3.2.123),[33] and are all massless fields. That is, from (3.3.122), we may infer that $1 - (D-2)/8 = 0$, *giving* $D = 10$ for the critical spacetime dimension. The degrees of freedom of the fields just mentioned are then, respectively, 1, 35, and 28.

The expression for M^2 for the $(\text{NS}_L, \text{NS}_R)$-sector may be rewritten from (3.3.121) in a convenient form which encompasses the constraint in (3.3.120) as

$$\frac{\alpha'}{4} M^2 = \sum_{n=1}^{\infty} \tilde{\alpha}^i(m)^\dagger \tilde{\alpha}^i(m) + \sum_{r=1/2}^{\infty} r \bar{b}^i(r)^\dagger \, \bar{b}^i(r) - \frac{1}{2}$$

$$= \sum_{n=1}^{\infty} \alpha^i(m)^\dagger \alpha^i(m) + \sum_{r=1/2}^{\infty} r b^i(r)^\dagger b^i(r) - \frac{1}{2} \qquad (3.3.123)$$

(NS, R) Boundary Condition

For the other case mentioned above, we have a left-mover ψ^i_L in (3.3.115), with its expression given in (3.3.118) with an NS boundary condition, together with a right-mover ψ^i_R, in (3.3.116), with a R boundary condition as given in (3.3.117). The mass squared operator corresponding to the (NS_L, R_R), sector is readily worked as in the previous case to be

$$\frac{\alpha'}{4} M^2 = \sum_{n=1}^{\infty} \tilde{\alpha}^i(n)^\dagger \tilde{\alpha}^i(n) + \sum_{r=1/2}^{\infty} r \bar{b}^i(r)^\dagger \, \bar{b}^i(r) - \frac{1}{2}$$

$$= \sum_{n=1}^{\infty} \left(\alpha^i(n)^\dagger \alpha^i(n) + n d^i(n)^\dagger \, d^i(n) \right), \qquad (3.3.124)$$

---

[33] See also (3.2.124).

having used the constraint

$$\sum_{n=1}^{\infty} \bar{\alpha}^i(n)^\dagger \bar{\alpha}^i(n) + \sum_{r=1/2}^{\infty} r \bar{b}^i(r)^\dagger \, \bar{b}^i(r) - \frac{1}{2}.$$

$$= \sum_{n=1}^{\infty} \left(\alpha^i(n)^\dagger \alpha^i(n) + n d^i(n)^\dagger d^i(n) \right). \tag{3.3.125}$$

(R, NS) Boundary Condition

The above case just investigated may be considered simultaneously with the (R_L, NS_R), sector. The corresponding expression for M^2 may be read directly from (3.3.124), by making use of (3.3.117), (3.3.118) for the respective fields $\psi^i_{L/R}$, and is given by

$$\frac{\alpha'}{4} M^2 = \sum_{n=1}^{\infty} \left(\bar{\alpha}^i(n)^\dagger \, \bar{\alpha}^i(n) + n \bar{d}^i(n)^\dagger \, \bar{d}^i(n) \right)$$

$$= \sum_{n=1}^{\infty} \alpha^i(n)^\dagger \alpha^i(n) + \sum_{r=1/2}^{\infty} r b^i(r)^\dagger b^i(r) - \frac{1}{2}, \tag{3.3.126}$$

having used the constraint

$$\sum_{n=1}^{\infty} \left(\bar{\alpha}^i(n)^\dagger \bar{\alpha}^i(n) + n \bar{d}^i(n)^\dagger \bar{d}^i(n) \right)$$

$$= \sum_{n=1}^{\infty} \alpha^i(n)^\dagger \alpha^i(n) + \sum_{r=1/2}^{\infty} r b^i(r)^\dagger b^i(r) - \frac{1}{2}. \tag{3.3.127}$$

Again the GSO projection removes the tachyonic states.

For the (NS_L, R_R), (R_L, NS_R), sectors, their lowest states may be spelled out, respectively, as[34]

$$\bar{b}^i(1/2)^\dagger | \, \psi^\kappa_a \rangle_R \, | \, NS \rangle_L \, | \, 0; p' \rangle, \tag{3.3.128}$$

$$b^i(1/2)^\dagger | \, \psi^\kappa_b \rangle_L \, | \, NS \rangle_R \, | \, 0; p' \rangle. \tag{3.3.129}$$

taken with $a, b = 1, 2; a \neq b$. Here, for example, one is combining a vector (from the index i in b^i) and a spin $1/2$ state $(| \, \psi^\kappa_b \rangle_L)$, to generate a spin $1/2$ (Dirac)

---

[34]Recall that $d^i(0), \bar{d}^i(0) \neq 0$ in (3.3.117).

field of 8 degrees of freedom (the dilatino), and a spin $3/2$ (Rarita-Schwinger) field, with 56 degrees of freedom (the gravitino), with a total of $8 \times 8 = 64$ states, with necessarily zero mass particles. With $a \neq b$, these states are of opposite chiralities in the two sectors. The fundamental massless fermion fields of superstring theory are analyzed in detail in Sect. 3.3.8 *including* the determination of the underlying degrees of freedom.

(R, R) Boundary Condition

To describe the (R, R) sector and investigate the mass spectrum of related super-strings, we consider first the following well paused problem.

The direct product of two matrices $B = [B_{ij}]$ and $A = [A_{ij}]$, $B \times A$, is defined as follows. Write down the matrix A in detail in terms of its matrix elements and multiply each of its elements by the matrix B, i.e.,

$$B \times A = \begin{pmatrix} A_{11}B & A_{12}B & \dots \\ A_{21}B & A_{22}B & \dots \\ \cdot & \cdot & \dots \\ \cdot & \cdot & \dots \\ \cdot & \cdot & \dots \end{pmatrix}. \tag{3.3.130}$$

The latter may, in turn, be rewritten in terms of its resulting matrix elements. For example

$$A_{11}B = \begin{pmatrix} A_{11}B_{11} & A_{11}B_{12} & \dots \\ A_{11}B_{21} & A_{11}B_{22} & \dots \\ \cdot & \cdot & \dots \\ \cdot & \cdot & \dots \\ \cdot & \cdot & \dots \end{pmatrix}, \tag{3.3.131}$$

and so on. An elementary example is given by the direct product of the Pauli matrices

$$\sigma^2 \times \sigma^2 = \begin{pmatrix} 0 & -i\sigma^2 \\ i\sigma^2 & 0 \end{pmatrix} = \begin{pmatrix} 0 & 0 & 0 & -1 \\ 0 & 0 & 1 & 0 \\ 0 & 1 & 0 & 0 \\ -1 & 0 & 0 & 0 \end{pmatrix}. \tag{3.3.132}$$

The direct product of three matrices is defined by

$$C \times B \times A = C \times (B \times A), \tag{3.3.133}$$

and so on for the direct product of more than three matrices. The following properties should be noted:

$$(A \times B \times C)(D \times E \times F) = AD \times BE \times CF, \qquad (3.3.134)$$

$$(A \times B \times C) + (A \times B \times D) = A \times B \times (C + D), \qquad (3.3.135)$$

and if $[A_1, A_2] = 0$, $[B_1, B_2] = 0$, then for the following anti-commutator we have

$$\{A_1 \times B_1 \times C, A_2 \times B_2 \times D\} = A_1 A_2 \times B_1 B_2 \times \{C, D\}. \qquad (3.3.136)$$

Now consider the following 16 by 16 gamma matrices:

$$\begin{aligned}
\Gamma^1 &= i\,\sigma^2 \times \sigma^2 \times \sigma^2 \times \sigma^2 & \Gamma^2 &= i\,I \times \sigma^1 \times \sigma^2 \times \sigma^2 \\
\Gamma^3 &= i\,I \times \sigma^3 \times \sigma^2 \times \sigma^2 & \Gamma^4 &= i\,\sigma^1 \times \sigma^2 \times I \times \sigma^2 \\
\Gamma^5 &= i\,\sigma^3 \times \sigma^2 \times I \times \sigma^2 & \Gamma^6 &= i\,\sigma^2 \times I \times \sigma^1 \times \sigma^2 \\
\Gamma^7 &= i\,\sigma^2 \times I \times \sigma^3 \times \sigma^2 & \Gamma^8 &= i\,I \times I \times I \times \sigma^1
\end{aligned} \qquad (3.3.137)$$

where I is the 2 by 2 unit matrix. From (3.3.134), the square of each of these matrices is equal to minus one. For each pair of matrices Γ^i, Γ^j, $i \neq j$, each matrix at the same respective location in the direct product of the pair commute with the exception of one matrix in the same location of the pair anti-commute. For example, for the pair of matrices Γ^1, Γ^2, only the second matrices from the left, that is σ^2, and σ^1, respectively, anti-commute, and all the other ones at the same respective locations commute, i.e., from (3.3.136), $\{\Gamma^1, \Gamma^2\} = 0$. Hence we have the anti-commutation relations

$$\{\Gamma^i, \Gamma^j\} = -2\,\delta^{ij}, \quad i, j = 1, \ldots, 8. \qquad (3.3.138)$$

From (3.3.134), we have the for the product of the gamma matrices in (3.3.137):

$$\Gamma \equiv \Gamma^1 \Gamma^2 \ldots \Gamma^8 = I \times I \times I \times \sigma^3 = \begin{pmatrix} I_8 & 0 \\ 0 & -I_8 \end{pmatrix}, \qquad (3.3.139)$$

where I_8 denotes the 8 by 8 unit matrix, and I the 2 by 2 one as in (3.3.137). Here we have used, in particular, the identities $\sigma^1 \sigma^2 = i\sigma^3 = -\sigma^2 \sigma^1$. The following basic properties should be noted

$$(\Gamma)^2 = I_{16}, \qquad \{\Gamma^i, \Gamma\} = 0, \qquad \Gamma^1 \ldots \Gamma^k \Gamma = \epsilon_k \Gamma^{k+1} \ldots \Gamma^8, \qquad (3.3.140)$$

where $\epsilon_k = +1$ or -1 depending on k. The main thing about the last expression is that the product of more than 4 gamma matrices with Γ may be expressed as the product of fewer gamma matrices. We will make use of these results to investigate the nature of the (R, R) sector.

The mass operator in this case, as obtained from (3.3.117), is given by

$$\frac{\alpha'}{4} M^2 = \sum_{n=1}^{\infty} \left(\bar{\alpha}^i(n)^\dagger \bar{\alpha}^i(n) + n \bar{d}^i(n)^\dagger \bar{d}^i(n) \right),$$

$$= \sum_{n=1}^{\infty} \left(\alpha^i(n)^\dagger \alpha^i(n) + n d^i(n)^\dagger d^i(n) \right), \qquad (3.3.141)$$

having used the constraint

$$\sum_{n=1}^{\infty} \left(\alpha^i(n)^\dagger \alpha^i(n) + d^i(n)^\dagger n d^i(n) \right) = \sum_{n=1}^{\infty} \left(\bar{\alpha}^i(n)^\dagger \bar{\alpha}^i(n) + \bar{d}^i(n)^\dagger n \bar{d}^i(n) \right), \qquad (3.3.142)$$

where the zero point energies cancel out.

We consider two inequivalent superstrings. One type where the right and left ground-states have opposite chiralities. These are called Type IIA theories, and another type for which they have the same chirality and are referred to as Type IIB theories.

For Type IIA theories we write:

$$[\, |\, \psi_1^\kappa \rangle_L \,] = \left[\, |\, \psi_1^1 \rangle_L, |\, \psi_1^2 \rangle_L, \ldots, |\, \psi_1^8 \rangle_L, 0, \ldots, 0 \right]^T,$$
$$[\, |\, \psi_2^{\kappa'} \rangle_R \,] = \left[0, \ldots, 0, |\, \psi_2^{1'} \rangle_R, |\, \psi_2^{2'} \rangle_R, \ldots, |\, \psi_2^{8'} \rangle_R \right]^T,$$

having each 16 components, by simply extending the values taken by κ, κ'. Their resulting expressions will be denoted respectively by $|\, \psi_1 \rangle_L, |\, \psi_2 \rangle_R$. The matrix Γ in (3.3.139) then gives

$$\Gamma \,|\, \psi_1 \rangle_L = + \,|\, \psi_1 \rangle_L, \qquad \Gamma \,|\, \psi_2 \rangle_R = - \,|\, \psi_1 \rangle_R. \qquad (3.3.143)$$

It doesn't matter if subscript 1 is associated with left-movers, and 2 is associated with right-movers, as long as they have different chiralities.

The lowest state (ground-state) $|\, \psi_1^\kappa \rangle_L \,|\, \psi_2^{\kappa'} \rangle_R$ for the type IIA superstrings just described, corresponding to the (R, R) sector, suppressing the state $|\, 0; p' \rangle$ for simplicity of the notation, may be written as a linear combinations of terms having the structures $(\Gamma^{i_1} \ldots \Gamma^{i_k})^{\kappa \kappa'} \, {}_L\langle \psi_1 \,|\, \Gamma^{i_1} \ldots \Gamma^{i_k} \,|\, \psi_2 \rangle_R$. Using the fact that for $i_1, \ldots i_k$ all unequal

$${}_L\langle \psi_1 \,|\, \Gamma^{i_1} \ldots \Gamma^{i_k} \,|\, \psi_2 \rangle_R = {}_L\langle \psi_1 \,|\, \Gamma^{i_1} \ldots \Gamma^{i_k} (\Gamma)^2 \,|\, \psi_2 \rangle_R$$

$$= (-1)^k \, {}_L\langle \psi_1 \,|\, \Gamma \Gamma^{i_1} \ldots \Gamma^{i_k} \Gamma \,|\, \psi_2 \rangle_R = (-1)^{k+1} \, {}_L\langle \psi_1 \,|\, \Gamma^{i_1} \ldots \Gamma^{i_k} \,|\, \psi_2 \rangle_R, \qquad (3.3.144)$$

where we have used (3.3.143) and the second identity in (3.3.140), we learn, by comparing the first expression with the last, that only an odd number of gamma matrices contribute to the above matrix elements. Finally invoking the last identity in (3.3.140), we may infer that only one or three gamma matrices need to be considered. On the other hand, the contraction of any two indices of the product of three gamma matrices generates a single gamma matrix, which is already accounted for, i.e., we are led to anti-symmetrize the product of the three gamma matrices so that the contraction of any two of its indices is zero.

Hence we conclude that in the (R, R) sector for the Type IIA superstrings, having opposite chiralities of the left and right R ground states, the massless fields are a vector field A^i, corresponding to $_L\langle\psi_1 | \Gamma^i | \psi_2\rangle_R$, with 8 degrees of freedom, and an anti-symmetric field A^{ijk}, corresponding to $_L\langle\psi_1 | \Gamma^{[i}\Gamma^j\Gamma^{k]} | \psi_2\rangle_R$, with 56 degrees of freedom, where $\Gamma^{[i}\Gamma^j\Gamma^{k]}$ means to anti-symmetrize over the i, j, k indices.

For Type IIB superstrings, where the left and right R ground states are taken to have the same chirality, we may write:

$$[| \psi_1\rangle_L] = \left[| \psi_1^1\rangle_L, | \psi_1^1\rangle_L, \ldots, | \psi_1^8\rangle_L, 0 \ldots 0 \right]^T,$$
$$[| \psi_1\rangle_R] = \left[| \psi_1^1\rangle_R, | \psi_1^2\rangle_R, \ldots, | \psi_1^8\rangle_R, 0 \ldots 0 \right]^T,$$

or equivalently as,

$$[| \psi_2\rangle_L] = \left[0, \ldots, 0, | \psi_2^{1'}\rangle_L, | \psi_2^{2'}\rangle_L, \ldots, | \psi_2^{8'}\rangle_L \right]^T,$$
$$[| \psi_2\rangle_R] = \left[0, \ldots, 0, | \psi_2^{1'}\rangle_R, | \psi_2^{2'}\rangle_R, \ldots, | \psi_2^{8'}\rangle_R \right]^T.$$

A similar analysis as in the previous case shows that only an even number of gamma matrices contribute to the matrix elements (no sum over a):

$$_L\langle\psi_a | \Gamma^{i_1} \ldots \Gamma^{i_k} | \psi_a\rangle_R,$$

since the latter equals

$$_L\langle\psi_a | \Gamma^{i_1} \ldots \Gamma^{i_k}\Gamma^2 | \psi_a\rangle_R = (-1)^k {}_L\langle\psi_a | \Gamma^{i_1} \ldots \Gamma^{i_k} | \psi_a\rangle_R,$$

k must be even. This leads to the conclusion that the massless fields are a scalar A, corresponding to $_L\langle\psi_a | \psi_a\rangle_R$, with one degree of freedom, an antisymmetric field A^{ij}, corresponding to $_L\langle\psi_a | \Gamma^{[i}\Gamma^{j]} | \psi_a\rangle_R$, with 28 degrees of freedom, and an anti-symmetric fourth rank (self-dual) tensor field A^{ijkl} corresponding to $_L\langle\psi_a | \Gamma^{[i}\Gamma^j\Gamma^k\Gamma^{l]} | \psi_a\rangle_R$, with 35 degrees of freedom. Here we note that for the fourth rank tensor field A^{ijkl} for all disjoint sets $\{i, j, k, l\}$, $\{i', j', k', l'\}$, with unequal elements, $A^{ijkl} = \pm A^{i'j'k'l'}$ defining a self-duality condition. Additional details on this as well as on all the massless fields in superstring theory will be given in Sects. 3.3.8 and 3.3.9.

The various types of superstrings are considered next.

3.3.6   Types of Superstrings: I, II and Heterotic

In this section we consider the various types of superstrings. Oriented open superstrings were considered in Sect. 3.3.4, and the oriented closed Type IIA, Type IIB superstrins were treated in Sect. 3.3.5. There are also Type I and the so-called heterotic superstrings. The zero-mass modes of all these strings will be summarized as we go along. At low energies, the massless fields are the most significant ones to consider. Details of both massless fermion fields of superstring theory are given in Sect. 3.3.8, including the determination of the associated number of degrees of freedom. Details of all the massless bosonic fields of superstring theory are given in Sect. 3.3.9 (see also Sect. 3.2.9). We urge the reader to consult these subsections. There are no tachyonic states in these strings. We begin with the study of Type I superstrings.

☐ Type I Superstrings

The theory of Type I superstrings is a generalization of the theory of unoriented bosonic strings treated in Sect. 3.2.6 and involves states which are invariant under the worldsheet parity operation Ω. The latter operates on the worldsheet of a string by effectively changing their σ argument to $\pi - \sigma$.

For **closed** strings, one may find the states of the theory directly from those states of Type IIB superstrings, which are Ω - invariant, rather than from the Type IIA superstrings as their R ground-states are different being of opposite chirality. Here we have the transformations:

$$\alpha(n) \leftrightarrow \bar{\alpha}(n), \quad b^i(r) \leftrightarrow \bar{b}^i(r), \quad d^i(r) \leftrightarrow \bar{d}^i(r),$$

$$| \text{NS} \rangle_R \leftrightarrow | \text{NS} \rangle_L, \quad | \psi^\kappa \rangle_R \leftrightarrow | \psi^\kappa \rangle_L.$$

The zero-mass modes are then as follows:

(NS, NS)-sector : a scalar (dilaton) with one state, and a second rank traceless tensor, the graviton, with 35 number of states. The anti-symmetric second rank tensor field being eliminated exactly in the same way as in our study of the unoriented bosonic string given in Sect. 3.2.6.

Under the Ω operation, the sectors $(\text{NS}_L, \text{R}_R)$ and $(\text{R}_L, \text{NS}_R)$ are interchanged, thus one only picks up even combinations of their corresponding states. In particular, we obtain from (3.3.128), (3.3.129), the invariant linear combination for the zero mass mode:

$$\bar{b}^i(1/2)^\dagger | \psi^\kappa_a \rangle_R | \text{NS} \rangle_L | 0; p' \rangle + b^i(1/2)^\dagger | \psi^\kappa_b \rangle_L | \text{NS} \rangle_R | 0; p' \rangle.$$

Hence we may infer that for the zero mass states one has only one chiral spin field (dilatino) with 8 number of states, and only one Rarita-Schwinger field (gravitino) with 56 number of states, instead of having two of each as was the case for Type II.

The number of massless bosonic states obtained so far is $1 + 35 = 36$, while the number of fermionic states obtained was $8 + 56 = 64$. Thus we conclude that only the second anti-symmetric field, with 28 number of states, survives in the (R, R)-sector, with the scalar and fourth rank anti-symmetric tensor fields are eliminated.

For **open** strings, as for the bosonic case in Sect. 3.2.6, we consider that the end points of a string involve Chan-Paton factors. With the underlying gauge group SO(N) with $N = 32 = 2^{D/2}$, $D = 10$, we may infer directly from the open superstrings treated in Sect. 3.3.4 that the zero-mass modes are as follows:

NS-sector: vector particle with $8 \times N(N-1)/2$ states.
R-sector: chiral spinor with $8 \times N(N-1)/2$ states. $\square$

Oriented Open Superstrings
These oriented open superstrings were treated in Sect. 3.3.4 with the following zero-mass modes:

NS-sector: vector particle with eight states.
R-sector: chiral spinor with eight states.

$\square$ Type IIA Superstrings

These oriented closed superstrings were treated in Sect. 3.3.5 with the following zero-mass modes:

(NS, NS)-sector: a scalar (dilaton) with one state, an antisymmetric second rank tensor field, with 28 number of states, and a traceless symmetric second rank tensor (graviton), with 35 number of states.
(NS_L, R_R)-sector: chiral Dirac field (dilatino), with eight number of states, a Rarita-Schwinger field (gravitino), with 56 number of states.
(R_L, NS_R)-sector: chiral Dirac field (dilatino), with 8 number of states, a Rarita-Schwinger field (gravitino), with 56 number of states.
(R, R)-sector: vector field, with 8 number of states, and an anti-symmetric third rank tensor field, with 56 number of states, with the right and left ground-states of opposite chiralities. $\square$

$\square$ Type IIB Superstrings

These oriented closed superstrings were also treated in Sect. 3.3.5 with the following zero-mass modes:

(NS, NS)-sector: a scalar (dilaton) with one state, an antisymmetric second rank tensor field, with 28 number of states, and a traceless symmetric second rank tensor (graviton), with 35 number of states.

$(\mathrm{NS_L}, \mathrm{R_R})$-sector: chiral Dirac field (dilatino), with eight number of states, a Rarita-Schwinger field (gravitino), with 56 number of states.

$(\mathrm{R_L}, \mathrm{NS_R})$-sector: chiral Dirac field (dilatino), with eight number of states, a Rarita-Schwinger field (gravitino), with 56 number of states.

(R, R)-sector: scalar field, with 1 number of states, and an anti-symmetric second rank tensor field, with 28 number of states, and a fourth rank anti-symmetric field, with 35 number of states, with the right and left ground-states of the same chiralities. $\square$

Now we come to a very interesting type of superstrings, referred to as **heterotic** strings. These are closed oriented superstrings for which the right-movers are the ones already considered for the Type II superstrings, in 10 dimensions, while the left-movers are those of the bosonic strings in 26 dimensions, with 16 of the spatial dimensions compactified on a torus, to match the 10 dimensions involving the right-movers. The name heterotic is quite appropriate for these strings as they mix together two theories which are different. The expression for the mass-squared operator is easily obtained from our earlier expressions, as we will see, and is given below.

□ Heterotic Superstrings

Let the index i run from 1 to 8, and the index I run from 10 to 25. The index I specifies the 16 dimensions that are being compactified. The light-cone fields $X^{\pm}$, for example, may be then defined in this case by

$$X^{\pm} = (X^0 \pm X^9)/\sqrt{2}.$$

Clearly, we have two sectors to consider, the $\mathrm{NS_R}$-sector, and the $\mathrm{R_R}$-sector. In reference to the left-movers, as it is obtained from the bosonic string, we may refer to the first equality in (3.2.153)/(A-3.4), for the compactified bosonic part of the string corresponding to the left-movers, together with the expression given in the second equality in (3.3.126) corresponding to the right-movers, to obtain for the

$\mathrm{NS_R}$-sector:

$$\frac{\alpha'}{4} M^2 = \alpha' \mathbf{p}_\mathrm{L}^2 + N_\mathrm{L} - 1 = N_\mathrm{R} - \frac{1}{2} \tag{3.3.145}$$

$$N_\mathrm{L} = \sum_{m=1}^{\infty} \left(\sum_{i=1}^{8} \tilde{\alpha}^i(m)^\dagger \tilde{\alpha}^i(m) + \sum_{I=10}^{25} \tilde{\alpha}^I(m)^\dagger \tilde{\alpha}^I(m) \right), \tag{3.3.146}$$

$$N_\mathrm{R} = \sum_{m=1}^{\infty} \sum_{i=1}^{8} \alpha^i(m)^\dagger \alpha^i(m) + \sum_{r=1/2}^{\infty} \sum_{i=1}^{8} r\, b^i(r)^\dagger b^i(r), \tag{3.3.147}$$

$$\mathbf{p}_\mathrm{L} = (p_\mathrm{L}^{10}, p_\mathrm{L}^{11}, \ldots, p_\mathrm{L}^{25}). \tag{3.3.148}$$

On the other hand, for the R_R-sector, we may refer to the expression in the second equality in (3.3.124) corresponding to the right-mover, to obtain for the

R_R-sector:

$$\frac{\alpha'}{4} M^2 = \alpha' \mathbf{p}_L^2 + N_L - 1 = \widetilde{N}_R, \qquad (3.3.149)$$

$$N_L = \sum_{m=1}^{\infty} \left(\sum_{i=1}^{8} \bar{\alpha}^i(m)^{\dagger} \bar{\alpha}^i(m) + \sum_{I=10}^{25} \bar{\alpha}^I(m)^{\dagger} \bar{\alpha}^I(m) \right), \qquad (3.3.150)$$

$$\widetilde{N}_R = \sum_{m=1}^{\infty} \sum_{i=1}^{8} \left(\alpha^i(m)^{\dagger} \alpha^i(m) + m d^i(m)^{\dagger} d^i(m) \right), \qquad (3.3.151)$$

$$\mathbf{p}_L = (p_L^{10}, p_L^{11}, \ldots, p_L^{25}). \qquad (3.3.152)$$

For the NS $_R$-sector, we have two cases corresponding to $M^2 = 0$:

Case 1: $\alpha' \mathbf{p}_L^2 = 0, N_R = 1/2, N_L = 1$, giving the states

$$b^i(1/2)^{\dagger} \alpha^I(1)^{\dagger} \mid NS \rangle_R \mid 0, \alpha' \mathbf{p}_L^2 = 0; p' \rangle_L,$$

corresponding to a vector, with 8×16 number of states, where we have labeled the bosonic states by $\alpha' \mathbf{p}_L^2$, and $p' = (p^+, p^1, \ldots, p^8)$. We also have the states

$$b^i(1/2)^{\dagger} \alpha^j(1)^{\dagger} \mid NS \rangle_R \mid 0, \alpha' \mathbf{p}_L^2 = 0; p' \rangle_L,$$

corresponding to a scalar with 1 number of states, an anti-symmetric second rank tensor field, with 28 number of states, and a traceless symmetric second rank tensor (graviton), with 35 number of states.

Case 2: $\alpha' \mathbf{p}_L^2 = 1, N_R = 1/2, N_L = 0$, giving the states

$$b^i(1/2)^{\dagger} \mid NS \rangle_R \mid 0, \alpha' \mathbf{p}_L^2 = 1; p' \rangle_L,$$

and with a gauge group SO(32), we will see that the degree of degeneracy $\widehat{d}$ of $\mid 0, \alpha' \mathbf{p}_L^2 = 1; p' \rangle_L$ is 480. That is the number of states of the vector field, just given, is 8×480.

For the R_R-sector, in order that $M^2 = 0$, we have the following cases:

Case 1: $\alpha' \mathbf{p}_L^2 = 0, \widetilde{N}_R = 0, N_L = 1$, giving the states

$$\bar{\alpha}^i(1) \mid \psi^{\kappa} \rangle_R \mid 0, \alpha' \mathbf{p}_L^2 = 0; p' \rangle_L,$$

corresponding to a chiral spinor (dilatino), with 8 number of states, and a Rarita-Schwinger field (gravitino), with 56 number of states, as well as the states

$$\bar{\alpha}^I(1) \mid \psi^{\kappa}\rangle_R \mid 0, \alpha' \mathbf{p}_L^2 = 0; p'\rangle_L,$$

corresponding to a chiral spinor, with 8×16 number of states.
Case 2: $\alpha' \mathbf{p}_L^2 = 1, \widetilde{N}_R = 0, N_L = 0$, giving the state

$$\mid \psi^{\kappa}\rangle_R \mid 0, \alpha' \mathbf{p}_L^2 = 1; p'\rangle_L,$$

with a degree of degeneracy of 480. To obtain this degree of degeneracy, note that the number of vector fields with gauge group SO(32) is $32(32 - 1)/2 = 496$. In Cases 1 and 2 in the NS_R-sector above, we have found $(16 + \hat{d})$ vector fields which must add up to 496 vector fields, from which the degree of degeneracy $\hat{d}$ follows. These 8 components vector fields taken all together are gauge fields of the group SO(32), with the underlying string theory referred to as Heterotic-O, or its T-dual, referred to as Heterotic-E, with a gauge group denoted by $E_8 \times E_8$. The string can wind itself on the torus in nontrivial configurations leading eventually to the total 496 states. The compactification leads to the quantization of $\mathbf{p}_L^2$, [Note that some authors set $\ell^2 = 1$ and hence, from (3.2.46), effectively set $\alpha' = 1/2$.]
□

Type I, including open and closed superstrings, Type IIA, Type IIB together with the oriented open superstrings, and finally the two heterotic (closed) superstrings Heterotic-O, and Heterotic-E, constitute the five basic superstring theories.[35] Relations are expected to exist between them. This is discussed in the next section.

3.3.7 Duality of Superstrings and M-Theory

Type IIA string theory is the T-dual of Type IIB and versa. The reason is that T-duality reverses the relative chiralities of the right- and left-movers. To see this, suppose that X^1 is compactified along a circle of radius R. In reference to (3.3.73), (3.3.75)/(3.3.76), under a T-duality transformation,

$$X_L^1 \rightarrow X_L^1, \quad \psi_L^1 \rightarrow \psi_L^1, \quad X_R^1 \rightarrow -X_R^1, \quad \psi_R^1 \rightarrow -\psi_R^1, \tag{3.3.153}$$

[35]The heterotic string was proposed by Gross [19, 20]. For a lucid presentation of the $E_8 \times E_8$ group see [18], Vol. 1, pp. 344–349, as well as [34], Vol. II, pp. 59–65.

by definition. The last transformation, as applied to the spinor, arises as a consequence of invoking supersymmetry as an immediate extension of the bosonic one. That is, in particular,

$$\bar{d}^{\,1}(0) \;\rightarrow\; \bar{d}^{\,1}(0), \qquad d^{\,1}(0) \;\rightarrow\; -d^{\,1}(0). \tag{3.3.154}$$

Using the notation $d^{\,i}(0) \equiv d^{\,i}$, one may infer from the definition of the chirality operator[36]: $d = (2\,i)^n d^{\,1} d^{\,2} \ldots \, d^{\,2n}$, that under a T-duality transformation, $d \rightarrow -d$, reversing the chirality of right movers. Thus if one starts with Type IIA string theory for which, by definition, the left- and right moving ground-states have opposite chirality, then under a T-duality transformation, the ground-states will have the same chirality thus transforming it to Type IIB and vice versa.

A more complicated argument establishes a similar kind of interchange exists between Heterotic-O and Heterotic-E theories thus they are also related by T-duality. Starting from Type IIB theory, we recall that Type I is obtained by restricting the theory to states which are invariant under the worldsheet parity transformation Ω. Now under a T-duality transformation because of the constraint set, in particular, by the Ω operation, the extra dimension is compactified on an interval instead of a circle. The resulting theory is referred to as Type I$'$ theory. [The latter is sometimes referred to as Type IA and the former as Type IB.]

An interesting theory in development, referred to as M-theory, believed to be approximated by 11 dimensional supergravity in the low energy limit, and is based on a non-perturbative approach which involves an eleventh dimension as well. The reason why an eleventh dimension was not encountered in the earlier sections, is that they were based only on free strings theories as basic ingredients for perturbative analyses. Duality allows this more general theory to be linked to other superstring theories. For example, compactification of M-theory on a circle of small radius leads to Type IIA theory.

A duality, referred to as S-duality, relates *theories with strong coupling to theories with weak coupling*. For example, on may argue that the strong coupling limit of Type IIA theory is described by M-theory, and that the strong coupling limit of Heterotic-O string theory is related to Type I string theory.

Figure 3.6 provides a formal picture that seems to be emerging showing duality transformations between various theories: where T$'$ refers to T-duality with compactification on an interval.

---

[36]See (3.3.95), (3.3.98).

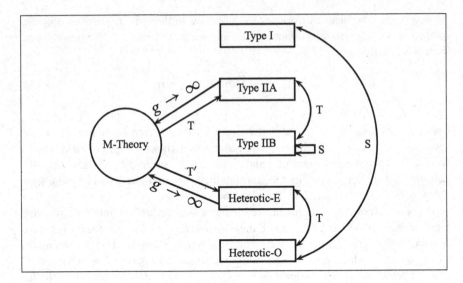

Fig. 3.6 Duality transformations envisaged between various theories

In the compactification process in Heterotic string theory of the bosonic part with 16 extra dimensions, it was implicitly assumed that the compactification was carried out on a torus. The simplicity in toroidal compactification is, in general, the fact that a torus is locally flat, and thus one is involved with the periodic identifications of flat dimensions in the compactification process. A modification of toroidal compactification, is compactification on an orbifold which is a six dimensional space obtained by the identification of points on the toroid which are mapped into one another by some specific discrete symmetries of the torus, defining the point group. This gives the possibly of producing four dimensional theories with only one supersymmetry charge operator instead of several supersymmetry charge operators as encountered in the toroidal one. The orbifold, however, involves singularities at such identified fixed points. A further generalization of this method is by far the more complicated Calabi-Yau compactification which provides a manifold avoiding such singularities. We recall the rather intuitive definition of a manifold as being locally Euclidean and may be covered by patches with mappings which map points in the patches to points in Euclidean space, and if two patches overlap, one has a smooth transformation from one of the coordinates to the other in the corresponding subsets of the Euclidean space. Unfortunately, the Calabi-Yau compactification is not unique.

At this stage we pose and consider all the massless fields that have emerged from superstring theory and study some of their properties such as their associated degrees of freedom.

3.3.8   The Two Fundamental Massless Fermion Fields in Superstring Theory: The Dirac and the Rarita-Schwinger Fields in Ten Dimensions

To derive the field equations of the massless Dirac and the Rarita-Schwinger fields in ten dimensions, we first recall the 8 gamma matrices introduced in (3.3.137)[37]:

$$
\begin{aligned}
\Gamma^1 &= i\,\sigma^2 \times \sigma^2 \times \sigma^2 \times \sigma^2 & \Gamma^2 &= i\,I \times \sigma^1 \times \sigma^2 \times \sigma^2 \\
\Gamma_3 &= i\,I \times \sigma^3 \times \sigma^2 \times \sigma^2 & \Gamma^4 &= i\,\sigma^1 \times \sigma^2 \times I \times \sigma^2 \\
\Gamma_5 &= i\,\sigma^3 \times \sigma^2 \times I \times \sigma^2 & \Gamma^6 &= i\,\sigma^2 \times I \times \sigma^1 \times \sigma^2 \\
\Gamma_7 &= i\,\sigma^2 \times I \times \sigma^3 \times \sigma^2 & \Gamma^8 &= i\,I \times I \times I \times \sigma^1
\end{aligned}
\tag{3.3.155}
$$

$$
\Gamma = \Gamma^1 \Gamma^2 \dots \Gamma^8 = \begin{pmatrix} I_8 & 0 \\ 0 & -I_8 \end{pmatrix}, \quad \{\Gamma, \Gamma^i\} = 0, \;\; i = 1, \dots, 8.
\tag{3.3.156}
$$

Now we add the matrix Γ, multiplied by i, to the set in (3.3.155) which is conveniently denoted by Γ^9, and has the structure

$$
\Gamma^9 = i I \times I \times I \times \sigma^3 \; (\equiv i\Gamma).
\tag{3.3.157}
$$

All of the 16 by 16 matrices $\Gamma^1, \Gamma^2, \dots, \Gamma^9$ are pure imaginary and satisfy the conditions:

$$
\{\Gamma^i, \Gamma^j\} = -2\,\delta^{ij}, \quad i, j = 1, \dots, 9; \qquad \Gamma^1 \Gamma^2 \dots \Gamma^9 = I.
\tag{3.3.158}
$$

The Dirac Equation in Ten Dimensions

A simple way to derive the Dirac equation in ten dimensions, is to follow our procedure in deriving the corresponding cases in two dimensions in Sect. 3.3.1. Upon multiplying the equation: $(p^0 - \sqrt{\mathbf{p}^2 + m^2})\psi_1 = 0$ by $(p^0 + \sqrt{\mathbf{p}^2 + m^2})$ we obtain

$$
\mathbf{p}^2 \psi_1 = (p^{0^2} - m^2)\,\psi_1,
\tag{3.3.159}
$$

where $\mathbf{p}$ is 9 dimensional. From the first relation in (3.3.158), the latter may be rewritten as

$$
-\,\Gamma \cdot \mathbf{p}\; \Gamma \cdot \mathbf{p}\, \psi_1 = (p^{0^2} - m^2)\,\psi_1.
\tag{3.3.160}
$$

---

[37] This subsection is based on [28].

Call the combination $(\boldsymbol{\Gamma} \cdot \mathbf{p} + ip^0)\,\psi_1$ simply $m\psi_2$, i.e., define

$$m\psi_2 = (\boldsymbol{\Gamma} \cdot \mathbf{p} + ip^0)\,\psi_1. \qquad (3.3.161)$$

Upon multiplying the latter by $\boldsymbol{\Gamma} \cdot \mathbf{p}$ and using (3.3.159), (3.3.161), again, give

$$\boldsymbol{\Gamma} \cdot \mathbf{p}\,\psi_2 = ip^0\,\psi_2 + m\psi_1. \qquad (3.3.162)$$

The two equations (3.3.161), (3.3.162), may be now combined into the elegant form

$$\left(\frac{\gamma^\mu \partial_\mu}{i} + m\right)\psi = 0, \quad \mu = 0, 1, \ldots, 9, \qquad (3.3.163)$$

$$\gamma^0 = \begin{pmatrix} 0 & -iI_{16} \\ iI_{16} & 0 \end{pmatrix}, \quad \gamma^i = \begin{pmatrix} 0 & -\Gamma^i \\ -\Gamma^i & 0 \end{pmatrix}, \quad \psi = \begin{pmatrix} \psi_1 \\ \psi_2 \end{pmatrix}. \qquad (3.3.164)$$

as is easily verified. Thus we have an explicit representation of the gamma matrices in 10 dimensions. They are all pure imaginary, and satisfy

$$\{\gamma^\mu, \gamma^\nu\} = -2\,\eta^{\mu\nu}. \qquad (3.3.165)$$

Using, in the process, the second relation in (3.3.158), one may define the chirality matrix by

$$\gamma_C = (-i)\,\gamma^0 \gamma^1 \ldots \gamma^9 = \begin{pmatrix} I_{16} & 0 \\ 0 & -I_{16} \end{pmatrix}, \quad (\gamma_C)^2 = I_{32}, \quad \{\gamma_C, \gamma^\mu\} = 0. \qquad (3.3.166)$$

For $m = 0$ the two equations in (3.3.161), (3.3.162) decouple. From the expression of γ_C in (3.3.166), we may select a spinor with a definite chirality, say one with positive chirality

$$\gamma_C \psi = +\psi, \qquad (3.3.167)$$

referred to as a Weyl spinor of positive chirality. From the structure of ψ in (3.3.164), this amounts in selecting ψ_1 which, from (3.3.161), satisfies the equation ($m = 0$)

$$\boldsymbol{\Gamma} \cdot \mathbf{p}\,\psi_1 = -ip^0\,\psi_1. \qquad (3.3.168)$$

where ψ_1 has 16 components, and this equation projects out further half of the components.

Rarita-Schwinger Equation in Ten Dimensions

The Lagrangian density of the Rarita-Schwinger field for $m = 0$ in 4 dimensions is given in (II.31) in Appendix II at the end of this volume which is conveniently recorded here in 10 dimensions[38]: $(\gamma \partial \equiv \gamma^\mu \partial_\mu, \overleftrightarrow{\partial} \equiv \overrightarrow{\partial} - \overleftarrow{\partial})$

$$\mathscr{L} = -\frac{1}{2i} \overline{\psi}_\mu \left(\gamma \overleftrightarrow{\partial} \eta^{\mu\nu} - (\gamma^\mu \overleftrightarrow{\partial}^\nu + \gamma^\nu \overleftrightarrow{\partial}^\mu) + \gamma^\mu (-\gamma \overleftrightarrow{\partial}) \gamma^\nu \right) \psi_\nu. \qquad (3.3.169)$$

The Lagrangian density is, up to a total derivative, invariant under a gauge transformation $\psi_a^\mu \to \psi_a^\mu + \partial^\mu \Lambda_a$.

We work in a Coulomb-like gauge as before

$$\partial_i \psi_a^i = 0, \qquad (3.3.170)$$

now summing over i from 1 to 9.

We add an external source contribution to the Lagrangian density in (3.3.169) to obtain

$$\mathscr{L} = -\frac{1}{2i} \overline{\psi}_\mu \left(\gamma \overleftrightarrow{\partial} \eta^{\mu\nu} - (\gamma^\mu \overleftrightarrow{\partial}^\nu + \gamma^\nu \overleftrightarrow{\partial}^\mu) + \gamma^\mu (-\gamma \overleftrightarrow{\partial}) \gamma^\nu \right) \psi_\nu$$
$$+ \overline{K}^\mu \psi_\mu + \overline{\psi}^\mu K_\mu. \qquad (3.3.171)$$

Due to the constraint in (3.3.170), the field components may not be varied independently. We may follow our earlier procedure and express the field in terms of fields U_a^i, ρ_a that may be varied independently as follows[39]

$$\psi_a^i = \alpha_{ab}^{ij} U_b^j - \frac{1}{8} \left(\gamma^i - \frac{\gamma^j \partial^j \partial^i}{\nabla^2} \right)_{ab} \rho_b, \qquad (3.3.172)$$

where

$$\alpha^{ij} = \left(\delta^{ij} - \frac{\partial^i \partial^j}{\nabla^2} \right) + \frac{1}{8} \left(\gamma^i - \frac{\gamma^k \partial^k \partial^i}{\nabla^2} \right) \left(\gamma^j - \frac{\gamma^\ell \partial^\ell \partial^j}{\nabla^2} \right), \qquad (3.3.173)$$

satisfying

$$\partial_i \alpha^{ij} = 0, \quad \alpha^{ij} \partial_j = 0, \quad \gamma^i \alpha^{ij} = 0, \quad \alpha^{ij} \gamma^j = 0, \quad \alpha^{ij} \alpha^{jk} = \alpha^{ik}. \qquad (3.3.174)$$

---

[38]For all the details of the four dimensional treatment, see (4.7.175)–(4.7.178) in Chap. 4 of Vol. I [29].

[39]Manoukian [29].

With ψ_a^i given in (3.3.172), we may vary the fields U_a^j, ρ_a, ψ^0 which, as seen below, not only lead to the field equations but to additional (derived) constraints.

To the above end, we note, in particular, that we may write

$$\delta\overline{\psi}_\mu = \delta\overline{\psi}^0 \, \eta_{\mu 0} + \delta\overline{U}^j \, \overleftarrow{\partial}^i \, \eta_{\mu i} - \delta\overline{\rho} \, \frac{1}{8} \left(\gamma^i - \frac{\gamma^j \overleftarrow{\partial}^j \, \overleftarrow{\partial}^i}{\overleftarrow{\nabla}^2} \right) \eta_{\mu i}. \tag{3.3.175}$$

Using the above expression and by varying the Lagrangian density with respect to ψ^0, U_a^j, ρ_a, lead after some labor to the equations ($\nabla^2 \equiv \partial^i \partial^i$)

$$\gamma^j \psi^j = -\frac{i}{\nabla^2} \gamma^j \partial^j \, \gamma^0 \, K_0, \tag{3.3.176}$$

$$\psi^0 = \frac{7}{8} \frac{1}{\nabla^2} \frac{\gamma^\mu \partial_\mu}{i} K^0 - \frac{1}{8 \nabla^2} \left(\frac{\partial^i}{i} + \frac{\gamma^j \partial^j}{i} \gamma^i \right) \gamma^0 K_i, \tag{3.3.177}$$

$$\gamma^\mu \partial_\mu \psi^i = i \alpha^{ij} K_j + \frac{1}{8} \frac{\gamma^\mu \partial_\mu}{\nabla^2} \left(\frac{\partial^i}{i} + \frac{\gamma^j \partial^j}{i} \gamma^i \right) \gamma^0 K_0. \tag{3.3.178}$$

We see from the last equation, that the Coulomb-like condition in (3.3.170) is automatically satisfied. In the absence of the external sources, the above two other equations give also the constraints $\gamma^j \psi^j = 0$, and $\psi^0 = 0$. That is, for a given spinor index, the number of spin states is reduced from a factor of 10 to 7, leading to $8 \times 7 = 56$ degrees of freedom for a spinor of given chirality satisfying the free field equation (see (3.3.168) and (3.3.178), for $K^\mu = 0$, concerning the spinor index). The number of spin states just stated is also rigorously established by performing the contraction of α^{ij} in (3.3.173)

$$\alpha^{ii} = (9-1) + \frac{1}{8} \left(\gamma^i \gamma^i - \frac{1}{\nabla^2} \gamma^j \partial^j \gamma^k \partial^k \right) = 8 + \frac{1}{8}(-9+1) = 7. \tag{3.3.179}$$

Additional interesting details of the Rarita-Schwinger field in 10 dimensions are given in Problem 3.20. We next consider all the massless bosonic fields emerged from superstring theory.

3.3.9   All the Fundamental Massless Bosonic Fields in Superstring Theory

The types of all bosonic massless fields in superstring theory we encountered were: a fourth rank (self-dual) anti-symmetric tensor field, a third rank anti-symmetric tensor field, a second rank anti-symmetric tensor field, a traceless second rank symmetric tensor field, a vector field, and finally a scalar field. In this subsection,[40] we work in 10 dimensions, not only because of the consistency requirement in superstring theory, but also in order to take into account of the self-duality character of the fourth rank anti-symmetric tensor field as spelled out below.

The scalar, the vector, and both second rank tensor fields, mentioned above, were treated in Sect. 3.2.9 in arbitrary dimensions $D > 2$. Accordingly, they will not be considered further in this subsection. But it is interesting to look again in their inherit degrees of freedom. In 4 dimensions, all non-zero spin massless fields have two polarization states (degrees of freedom). In 10 dimensions, the massless vector field has $D - 2 = 8$ degrees of freedom, the massless traceless symmetric second rank tensor field has $D(D - 3)/2 = 35$, while the massless anti-symmetric tensor field has $(D-2)(D-3)/2 = 28$. The scalar field has, of course, 1 degree of freedom in any dimension, while the massless second rank anti-symmetric field has 1 degree of freedom in 4 dimensions. We consider, in turn, the massless third rank anti-symmetric tensor field, followed by the massless (self-dual) anti-symmetric fourth rank tensor field in 10 dimensions.

In this subsection, Latin indices, in the middle alphabet, $i, j, k, \ldots$ take on the values $1, 2, \ldots, 9$, while Latin indices, in the beginning of alphabet, $a, b, c, \ldots$ take on the values $1, 2, \ldots, 8$.

Third Rank Anti-symmetric Tensor Field

The Lagrangian density of a third rank massless anti-symmetric tensor field $A^{\mu\nu\sigma}$, may be defined by

$$\mathscr{L} = -\frac{1}{8} F^{\mu\nu\sigma\lambda} F_{\mu\nu\sigma\lambda}, \tag{3.3.180}$$

where $F^{\mu\nu\sigma\lambda}$ is totally anti-symmetric and is given by

$$F^{\mu\nu\sigma\lambda} = \partial^\mu A^{\nu\sigma\lambda} - \partial^\nu A^{\sigma\lambda\mu} + \partial^\sigma A^{\lambda\mu\nu} - \partial^\lambda A^{\mu\nu\sigma}, \tag{3.3.181}$$

defined as a cyclic permutation. The Lagrangian density is invariant under the gauge transformation

$$A^{\mu\nu\sigma} \rightarrow A^{\mu\nu\sigma} + \partial^\mu \varphi^{\nu\sigma} + \partial^\nu \varphi^{\sigma\mu} + \partial^\sigma \varphi^{\mu\nu}. \tag{3.3.182}$$

We work in a Coulomb-like gauge

$$\partial_i A^{ijk} = 0, \quad \partial_i A^{0ij} = 0, \tag{3.3.183}$$

and similarly defined with respect to the indices j, k.

---

[40]This subsection is based on Manoukian [27].

We add a source contribution to (3.3.180), using the same notation for simplicity, obtaining

$$\mathscr{L} = -\frac{1}{8} F^{\mu\nu\sigma\lambda} F_{\mu\nu\sigma\lambda} + A^{\mu\nu\sigma} J_{\mu\nu\sigma}, \tag{3.3.184}$$

where no constraints are imposed on the external source $J_{\mu\nu\sigma}$ so we may vary its components independently. Due to the constraints in (3.3.183), one cannot vary the components of A^{ijk}, as well as the components of A^{0jk}, independently. We may, however, introduce a field $\mathscr{A}^{ijk}$ that may be varied independently, and set

$$A^{ijk} = \pi^i{}_{[i'} \pi^j{}_{j'} \pi^k{}_{k']} \mathscr{A}^{i'j'k'}, \qquad \pi^{ij} = \delta^{ij} - \frac{\partial^i \partial^j}{\nabla^2}, \tag{3.3.185}$$

and the square brackets in $\pi^i{}_{[i'} \pi^j{}_{j'} \pi^k{}_{k']}$, means an anti-symmetrization over the indices i', j', k'. Similarly, we may introduce a field $\mathscr{A}^{jk}$, and set

$$A^{0jk} = \pi^j{}_{[j'} \pi^k{}_{k']} \mathscr{A}^{j'k'}. \tag{3.3.186}$$

By varying the Lagrangian density in (3.3.184) with respect to $\mathscr{A}^{ijk}$, and $\mathscr{A}^{jk}$, we obtain

$$\Box A^{ijk} = -\pi^i{}_{[i'} \pi^j{}_{j'} \pi^k{}_{k']} J^{i'j'k'}, \tag{3.3.187}$$

$$\nabla^2 A^{0jk} = -\pi^j{}_{[j'} \pi^k{}_{k']} J^{0j'k'}. \tag{3.3.188}$$

Clearly, only A^{ijk} may propagate.

Upon introducing the completeness relation in the momentum description

$$\pi^{ij} = \sum_{a=1}^{8} e_a^i e_a^j, \tag{3.3.189}$$

in terms of polarization vectors e_a^i, where $p^i e_a^i = 0$, $e_{a_1}^i e_{a_2}^i = \delta_{a_1 a_2}$, we may, after some grouping of the polarization vectors, rewrite

$$\pi^i{}_{[i'} \pi^j{}_{j'} \pi^k{}_{k']} = e_{[a_1}^i e_{a_2}^j e_{a_3]}^k e_{[a_1}^{i'} e_{a_2}^{j'} e_{a_3]}^{k'}. \tag{3.3.190}$$

This suggests to introduce the polarizations

$$e_{[a_1}^i e_{a_2}^j e_{a_3]}^k \equiv e_{a_1 a_2 a_3}^{i\,j\,k}. \tag{3.3.191}$$

The number of degrees of freedom is obtained as follows, giving

$$\sum_{a_1,a_2,a_3=1}^{8} e^{i\ j\ k}_{a_1a_2a_3} e^{i\ j\ k}_{a_1a_2a_3} = \frac{1}{3!}\Big(\pi^{i\ i}\pi^{jj}\pi^{kk} - \pi^{ij}\pi^{ij}\pi^{kk} - \pi^{i\ i}\pi^{j\ k}\pi^{kj}$$

$$+ \pi^{ij}\pi^{j\ k}\pi^{ki} - \pi^{ik}\pi^{ij}\pi^{ki} + \pi^{ik}\pi^{j\ i}\pi^{kj}\Big) = \frac{1}{6}(336) = 56 \qquad (3.3.192)$$

degrees of freedom.

Fourth Rank Anti-symmetric Self-Dual Tensor Field

We recall that in this subsection, $i, j, k, \ldots$ go over $1, 2, \ldots, 9$, while $a, b, c, \ldots$ go over $1, 2, \ldots, 8$. A summation over repeated indices is understood throughout.

The Lagrangian density of a fourth rank massless anti-symmetric tensor field $A^{\mu\nu\sigma\rho}$, may be defined by

$$\mathscr{L} = -\frac{1}{10} F^{\mu\nu\sigma\rho}F_{\mu\nu\sigma\rho}, \qquad (3.3.193)$$

where $F^{\mu\nu\sigma\rho}$ is totally anti-symmetric and is given by

$$F^{\mu\nu\sigma\rho} = \partial^\mu A^{\nu\sigma\rho} + \partial^\nu A^{\sigma\rho\mu} + \partial^\sigma A^{\lambda\rho\mu\nu} + \partial^\lambda A^{\rho\mu\nu\sigma} + \partial^\rho A^{\mu\nu\sigma\lambda}, \qquad (3.3.194)$$

defined as a cyclic permutation. The Lagrangian density is invariant under the gauge transformation

$$A^{\mu\nu\sigma\rho} \rightarrow A^{\mu\nu\sigma\rho} + \partial^\mu \varphi^{\nu\sigma\rho} - \partial^\nu \varphi^{\sigma\rho\mu} + \partial^\sigma \varphi^{\rho\mu\nu} - \partial^\rho \varphi^{\mu\nu\sigma}. \qquad (3.3.195)$$

We work in a Coulomb-like gauge

$$\partial_i A^{ijk\ell} = 0, \quad \partial_i A^{0ijk} = 0, \qquad (3.3.196)$$

and similarly defined with respect to the indices j, k, ℓ.

We add a source contribution to (3.3.193), using the same notation for simplicity, obtaining

$$\mathscr{L} = -\frac{1}{10} F^{\mu\nu\sigma\lambda\rho}F_{\mu\nu\sigma\lambda\rho} + A^{\mu\nu\sigma\rho}J_{\mu\nu\sigma\rho}, \qquad (3.3.197)$$

where no constraints are imposed on the external source $J_{\mu\nu\sigma\rho}$, so we may vary its components independently. Due to the constraints in (3.3.196), one cannot vary the components of $A^{ijk\ell}$, as well as the components of A^{0ijk}, independently. We may, however, introduce a field $\mathscr{A}^{ijk\ell}$, that may be varied independently, and set

$$A^{ij\,k\ell} = \Lambda^{i\,j\,k\,\ell}_{i'j'k'\ell'}\,\mathscr{A}^{i'j'k'\ell'},\tag{3.3.198}$$

where $\Lambda^{i\,j\,k\,\ell}_{i'j'k'\ell'}$ is totally anti-symmetric in the indices $\{i,j,k,\ell\}$, as well as in the indices $\{i',j',k',\ell'\}$, satisfying the following properties

$$\Lambda^{i\,j\,k\,\ell}_{i'j'k'\ell'} = \Lambda^{i'j'k'\ell'}_{ij\,k\,\ell},\qquad \Lambda^{i\,j\,k\,\ell}_{i'j'k'\ell'}\,\Lambda^{i'j'k'\ell'}_{i''j''k''\ell''} = \Lambda^{i\,j\,k\,\ell}_{i''j''k''\ell''},\tag{3.3.199}$$

i.e., it is, in particular, a projection operator, and also satisfies orthogonality relations

$$\partial_i\Lambda^{i\,j\,k\,\ell}_{i'j'k'\ell'} = 0 = \Lambda^{i\,j\,k\,\ell}_{i'j'k'\ell'}\,\partial_i,\tag{3.3.200}$$

as well as in all of its indices j,k,ℓ,i',j',k',ℓ'. We will explicitly construct this operator below. We will see how the self-duality condition, spelled out below, may be also defined through this process. Similarly, we may set

$$A^{0ijk} = \Lambda^{i\,j\,k}_{i'j'k'}\,\mathscr{A}^{i'j'k'},\tag{3.3.201}$$

where $\Lambda^{i\,j\,k}_{i'j'k'}$ is totally anti-symmetric in the indices $\{i,j,k\}$, as well as in the indices $\{i',j',k'\}$, and satisfies the properties

$$\Lambda^{i\,j\,k}_{i'j'k'} = \Lambda^{i'j'k'}_{ij\,k},\qquad \Lambda^{i\,j\,k}_{i'j'k'}\,\Lambda^{i'j'k'}_{i''j''k''} = \Lambda^{i\,j\,k}_{i''j''k''},\tag{3.3.202}$$

i.e., it is, in particular, a projection operator, and also satisfies orthogonality relations

$$\partial_i\Lambda^{i\,j\,k}_{i'j'k'} = 0 = \Lambda^{i\,j\,k}_{i'j'k'}\,\partial_i,\tag{3.3.203}$$

as well as in all of its indices j,k,i',j',k'.

By using these properties, and varying the Lagrangian density in (3.3.197) with respect to the fields $\mathscr{A}^{ijk\ell}$, and $\mathscr{A}^{ij\,k}$, we obtain, after some labor, the equations

$$\Box A^{ij\,k\ell} = -\Lambda^{i\,j\,k\,\ell}_{i'j'k'\ell'}\,J^{i'j'k'\ell'},\tag{3.3.204}$$

$$\nabla^2 A^{0j\,k\ell} = -\Lambda^{j\,k\,\ell}_{j'k'\ell'}\,J^{0j'k'\ell'}.\tag{3.3.205}$$

Clearly, only $A^{ij\,k\ell}$ may propagate.

By taking the vacuum expectation value $\langle 0_+|\,.\,|0_-\rangle$ of (3.3.204), and carrying out a Fourier transform, gives

$$\langle\,0_+|\,A^{ij\,k\ell}(x)|0_-\rangle = \int\frac{(\mathrm{d}p)}{(2\pi)^{10}}\frac{1}{p^2 - \mathrm{i}\epsilon}\,\Lambda^{i\,j\,k\,\ell}_{i'j'k'\ell'}\,J^{i'j'k'\ell'}(p)\,\langle\,0_+|0_-\rangle,\tag{3.3.206}$$

and $\Lambda^{i\ j\ k\ \ell}_{i'j'k'\ell'}$, may be explicitly given in terms of mutually orthonormal polarization vectors $e^i_a : p^i e^i_a = 0,\ e^i_a e^i_{a'} = \delta_{aa'},\ a, a' = 1, \ldots, 8$, as follows

$$\Lambda^{i\ j\ k\ \ell}_{i'j'k'\ell'} = e^{\ i}_{[a} e^j_b e^k_c e^\ell_{d]}\ \Gamma^{a\ b\ c\ d}_{a'b'c'd'}\ e^{\ i'}_{[a'} e^{j'}_{b'} e^{k'}_{c'} e^{\ell'}_{d']}, \tag{3.3.207}$$

where $e^{\ i}_{[a} e^j_b e^k_c e^\ell_{d]}$ includes $4! = 24$ terms, with the square brackets in $[a\,b\,c\,d]$ defining an anti-symmetrization over the indices a, b, c, d. $\Gamma^{a\ b\ c\ d}_{a'b'c'd'}$ is explicitly given by

$$\Gamma^{a\ b\ c\ d}_{a'b'c'd'} = \frac{1}{2}\left(\delta^{\ a}_{[a'}\delta^b_{b'}\delta^c_{c'}\delta^d_{d']} + \frac{1}{4!}\,\varepsilon^{abcda'b'c'd'}\right), \tag{3.3.208}$$

where, we recall, $\varepsilon^{abcda'b'c'd'}$ is totally anti-symmetric with $\varepsilon^{12345678} = +1$. The following basic properties should be noted:

$$e^{\ i}_{[a} e^j_b e^k_c e^\ell_{d]}\ e^{\ i}_{[a'} e^j_{b'} e^k_{c'} e^\ell_{d']} = \delta^{\ a}_{[a'}\delta^b_{b'}\delta^c_{c'}\delta^d_{d']}, \tag{3.3.209}$$

$$p^i\,\Lambda^{i\ j\ k\ \ell}_{i'j'k'\ell'} = 0, \tag{3.3.210}$$

and similarly defined with respect to the indices $j, k, \ell, i', j', k', l'$,

$$\Gamma^{a\ b\ c\ d}_{a'b'c'd'}\ \Gamma^{a'\ b'\ c'\ d'}_{a''b''c''d''} = \Gamma^{a\ b\ c\ d}_{a''b''c''d''}, \tag{3.3.211}$$

and most importantly,

$$\frac{1}{4!}\,\varepsilon^{abcda'b'c'd'}\ \Gamma^{a'\ b'\ c'\ d'}_{a''b''c''d''} = \Gamma^{a\ b\ c\ d}_{a''b''c''d''}. \tag{3.3.212}$$

All the properties in (3.3.199), (3.3.200) are now verified.

Now we are ready to discuss the self-duality condition. To this end, the Fourier transform in (3.3.206) reads

$$\langle\, 0_+|A^{ij\,k\ell}(p)|0_-\rangle = \frac{1}{p^2 - i\epsilon}\,\Lambda^{i\ j\ k\ \ell}_{i'j'k'\ell'}\ J^{i'j'k'\ell'}(p)\,\langle\, 0_+|0_-\rangle. \tag{3.3.213}$$

Consider the momentum $\mathbf{p} = (0, 0, \ldots, 0, |\mathbf{p}|)$ with non-zero at the 9th place. The polarization vectors may be written as $e^i_a = \delta^i_a$, and we simply obtain

$$\langle\, 0_+|A^{abcd}(p)\,|0_-\rangle = \frac{1}{p^2 - i\epsilon}\,\Gamma^{a\ b\ c\ d}_{a'b'c'd'}\,J^{a'b'c'd'}(p)\,\langle\, 0_+|0_-\rangle. \tag{3.3.214}$$

The remarkable property in (3.3.212) gives rigorously,

$$\frac{1}{4!}\,\varepsilon^{abcda'b'c'd'}\,\langle\,0_+|A^{a'b'c'd'}(p)|0_-\rangle = \langle\,0_+|A^{abcd}(p)|0_-\rangle, \tag{3.3.215}$$

thus satisfying the self-duality restriction. For example, this gives $A^{1234} = +A^{5678}$, and for all disjoint sets $\{a,b,c,d\}$, $\{a',b',c',d'\}$, with unequal elements, $A^{abcd} = \pm A^{a'b'c'd'}$, where the signs are readily determined.

The property in (3.3.199), together with the one in (3.3.211), suggest to introduce the polarizations

$$e_{[a'}^{i}\,e_{b'}^{j}\,e_{c'}^{k}\,e_{d']}^{\ell}\,\Gamma_{a\,b\,c\,d}^{a'b'c'd'} \equiv e_{abcd}^{ijk\ell}. \tag{3.3.216}$$

The number of independent degrees of freedom may be obtained from the following, giving

$$\sum_{a,b,c,d=1}^{8} e_{abcd}^{ijk\ell}\,e_{abcd}^{ijk\ell} = e_{[a'}^{i}\,e_{b'}^{j}\,e_{c'}^{k}\,e_{d']}^{\ell}\,\Gamma_{a\,b\,c\,d}^{a'b'c'd'}\,e_{[a''}^{i}\,e_{b''}^{j}\,e_{c''}^{k}\,e_{d'']}^{\ell}\,\Gamma_{a\,b\,c\,d}^{a''b''c''d''}$$

$$= \delta_{[a''}^{a'}\,\delta_{b''}^{b'}\,\delta_{c''}^{c'}\,\delta_{d'']}^{d'}\,\Gamma_{a'\,b'\,c'\,d'}^{a''b''c''d''}$$

$$= \frac{1}{2}\sum_{a,b,c,d=1}^{8}\left(\delta_{[a}^{a}\delta_{b}^{b}\delta_{c}^{c}\delta_{d]}^{d} + \frac{1}{4!}\,\delta_{[a}^{a'}\,\delta_{b}^{b'}\,\delta_{c}^{c'}\,\delta_{d]}^{d'}\,\varepsilon^{abcda'b'c'd'}\right)$$

$$= \frac{1}{2}\sum_{a,b,c,d=1}^{8}\left(\delta_{[a}^{a}\delta_{b}^{b}\delta_{c}^{c}\delta_{d]}^{d} + 0\right) = \frac{1}{2}(70) = 35 \tag{3.3.217}$$

degrees of freedom, where we have used, in the process, the expression in (3.3.208), and inserted the summation signs in the last steps to emphasize that we are summing over the repeated indices a, b, c, d. Finally, in the last step, we have used the identity

$$\sum_{a,b,c,d=1}^{8}\delta_{[a}^{a}\delta_{b}^{b}\delta_{c}^{c}\delta_{d]}^{d} = 70. \tag{3.3.218}$$

3.4   D Branes

We have seen (Sect. 3.2.8) that open strings with a compactified dimension with a Neumann boundary conditions, necessarily changes, under T-duality, to a dimension with a Dirichlet boundary condition and implies the existence of extended objects, to which the motion of the end-points are restricted, and are referred to as D branes.

On the other hand, a corresponding Dirichlet boundary condition implies that there must be such objects to which the end points are attached.

The "D" in D branes, stands for Dirichlet, and the word "brane" is derived from the word membrane. If p is the spatial dimension of a D brane, the latter is called a D p-brane. The volume of a brane is termed as its worldvolume. The space outside the brane is called the "bulk". If $p = D - 1$, where D denotes the dimensions of spacetime, then the brane is referred to as space filling. Time flows not only in the brane but in the bulk as well.

Interestingly enough, one may be led to imagine that the universe in which we live is a D3-brane with the bulk being associated with the extra dimensions. Closed strings describing, in particular, gravitons, have no end points and are not restricted to branes, and may move around all over spacetime. One may then infer that gravitons can escape from branes into the bulk providing, perhaps, an explanation of the feebleness of the strength of gravitation in comparison to the other interactions.

3.4.1   Open Strings, D Branes and Massless Particles

Consider a bosonic open string, with field variables $X^\mu : \mu = 0, 1, \ldots, D - 1$, with $X^0, X^1, \ldots, X^p$ satisfying Neumann boundary conditions, i.e.,

$$\left. \frac{\partial X^\mu(\tau, \sigma)}{\partial \sigma} \right|_{\sigma = 0, \pi} = 0, \qquad \mu = 0, 1, \ldots, p, \tag{3.4.1}$$

and $X^{p+1}, \ldots, X^{D-1}$ satisfying Dirichlet conditions

$$X^I(\tau, 0) = X^I(\tau, \pi) = \bar{x}^I, \qquad I = p + 1, \ldots, D - 1. \tag{3.4.2}$$

Here $\bar{x}^I, I = p + 1, \ldots, D - 1$, specify a location of a D p-brane. In particular, we note that $X^{p+1}, \ldots, X^{D-1}$ are coordinates normal to the brane. We set $X^\pm = (X^0 \pm X^1)/\sqrt{2}$.

From (3.2.58),

$$X^i(\tau, \sigma) = x^i + \ell^2 p^i \tau + i\ell \sum_{n \neq 0} \frac{\alpha^i(n)}{n} e^{-in\tau} \cos n\sigma, \quad i = 2, \ldots, p, \tag{3.4.3}$$

and for the field variables satisfying Dirichlet conditions,

$$X^I(\tau, \sigma) = \bar{x}^I + \ell \sum_{n \neq 0} \frac{\alpha^I(n)}{n} e^{-in\tau} \sin n\sigma, \quad I = p + 1, \ldots, D - 1, \tag{3.4.4}$$

with no momentum: $\partial X^I(\tau, 0)/\partial \tau = 0$, $\partial X^I(\tau, \pi)/\partial \tau = 0$ of the end points perpendicular to the brane. From (3.2.96), we may infer that the mass squared pertinent to the brane, $2p^+p^- - p^ip^i$ with a sum over i from 2 to p, is given by

$$\alpha' M^2 = \sum_{n=1}^{\infty} \left(\alpha^i(n)^\dagger \alpha^i(n) + \alpha^I(n)^\dagger \alpha^I(n) \right) - 1, \qquad (3.4.5)$$

with the sum over I from $(p+1), \ldots, (D-1)$.

The ground-state may be labeled as shown in $|0; p^+, p^i\rangle$ with no momentum components perpendicular to the brane. In particular, in terms of a Fourier transform there will be dependence on x^-, x^i. The massless vector particles are given by $\alpha^i(1)^\dagger |0, p^+, p^i\rangle$ with the index i belonging to Lorentz indices pertinent to the brane only, with the number of states being equal to $p - 1 = (p+1) - 2$, i.e., the massless vector particles "live" in the brane. We also have $D - 1 - p$ massless scalar particles, associated with the states $\alpha^I(1)^\dagger |0, p^+, p^i\rangle$, so-called Goldstone bosons, arising from the symmetry breaking of translational invariance by the presence of the brane. Such states correspond to displacements of the brane in the above mentioned normal directions. Clearly, for $p = D - 1$, there are no massless scalar particles and translational invariance holds true in spacetime.

3.4.2 More Than One Brane

We recall (Sect. 3.2.5) that an open (oriented) bosonic string with end points having N possible Chan-Paton states, generates N^2 massless vector particle with an underlying gauge group U(N). For example, for the gauge group U(2), we have four massless vector particles. Now consider two parallel Dp-branes with locations specified by $\bar{x}_1^I, \bar{x}_2^I$, and four possible configurations with ends of string falling on the same brane or one end falling on one brane and the other end on the second one.

These configurations are shown in Fig. 3.7, and one may define four Chan-Paton states $(1, 1), (1, 2), (2, 1), (2, 2)$, e.g., with the first entry in $(1, 2)$ corresponding to the end of the string for $\sigma = 0$, and the second entry for $\sigma = \pi$. For example, for the Chan-Paton state $(1, 2)$ with the bosonic string stretched between the two branes as shown, the field variables, associated with the string, in directions perpendicular to the branes, may be written as

$$X^I(\tau, \sigma) = \bar{x}_1^I + (\bar{x}_2^I - \bar{x}_1^I)\frac{\sigma}{\pi} + \ell \sum_{n \neq 0} \frac{\alpha^I(n)}{n} e^{-in\tau} \sin n\sigma, \qquad (3.4.6)$$

$$\partial_\sigma X^I(\tau, \sigma) = \frac{(\bar{x}_2^I - \bar{x}_1^I)}{\pi} + \ell \sum_{n \neq 0} \alpha^I(n) e^{-in\tau} \cos n\sigma, \quad \alpha^I(0) = \frac{(\bar{x}_2^I - \bar{x}_1^I)}{\ell \pi}, \qquad (3.4.7)$$

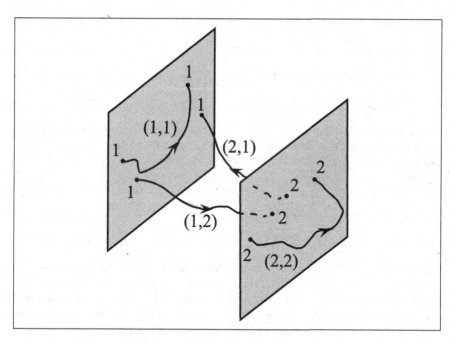

Fig. 3.7 Two parallel Dp-branes. Four possible configurations with ends of string falling on the same brane or one end falling on one brane and the other end on the second one

$$X^I(\tau,\pi) - X^I(\tau,0) = \bar{x}_2^I - \bar{x}_1^I, \tag{3.4.8}$$

$I = p+1,\ldots,D-1$.

The mass-squared $2p^+p^- - p^j p^j$ pertinent to the branes in reference to the Chan-Paton state $(1,2)$, is then given by

$$\alpha' M^2 = \frac{1}{2}\alpha^I(0)\alpha^I(0) + \sum_{n=1}^{\infty}\left(\alpha^j(n)^\dagger\alpha^j(n) + \alpha^I(n)^\dagger\alpha^I(n)\right) - 1, \tag{3.4.9}$$

with sums over $I = p+1,\ldots,D-1$, and $j = 2,\ldots,p$. From the second equality in (3.4.7), we may, in reference to the Chan-Paton states $(1,2)$, $(2,1)$, write[41]

$$\alpha' M^2 = \left[\left(\frac{\bar{x}_2^I - \bar{x}_1^I}{2\pi\sqrt{\alpha'}}\right)^2 - 1\right] + \sum_{n=1}^{\infty}\left(\alpha^j(n)^\dagger\alpha^j(n) + \alpha^I(n)^\dagger\alpha^I(n)\right), \tag{3.4.10}$$

---

[41]Recall that $\ell^2 = 2\alpha'$ [see (3.2.46)].

with the sums over the indices, as spelled out above, understood. In particular, note that $(\bar{x}_2^I - \bar{x}_1^I)^2$ stands for

$$\sum_{I=p+1}^{D-1} (\bar{x}_2^I - \bar{x}_1^I)^2.$$

The ground-states corresponding to the general possible configurations of the end points, may be now labeled as in $|0; p^+, p^i; (a, b)\rangle$ for the Chan-Paton states corresponding to $(a, b) = (1, 1), (1, 2), (2, 1), (2, 2)$.

For

$$\sum_{I=p+1}^{D-1} (\bar{x}_2^I - \bar{x}_1^I)^2 > 4\pi^2\alpha',$$

the state $\alpha^j(1)^\dagger|0; p^+, p^i; (1, 2)\rangle$ corresponds to a massive vector particle but has only $(p - 1)$ of the p polarization states needed to describe a massive particle. The additional degree of freedom needed to define a massive vector particle is obtained from the scalar excitations in the following manner. One may introduce the unit vector with components n^I along $\bar{x}_2^I - \bar{x}_1^I$, and define the state

$$\sum_{I=p+1}^{D-1} n^I \alpha^I(1)^\dagger |0; p^+, p^i; (1, 2)\rangle, \qquad n^I = \frac{\bar{x}_2^I - \bar{x}_1^I}{\sqrt{\sum_{J=p+1}^{D-1} (\bar{x}_2^J - \bar{x}_1^J)^2}}, \qquad (3.4.11)$$

where we emphasize that the index I, as far as the branes are concerned, just labels the scalar particles and does not correspond to a Lorentz index for them. The state $\alpha^j(1)^\dagger|0; p^+, p^i; (1, 2)\rangle$ together the one in (3.4.11) now give the correct number of polarization states of a massive vector particle given by $(p-1)+1 = p$, in reference to the branes. This is interesting. The open string with its both ends on the same brane, as given in part (a) of Fig. 3.8, gives rise to a massless vector particle. On the other hand, in the case, when it is stretched instead between the two branes, as given in part (b) in the figure, gives rise to a massive vector particle. Here what we have is what one may call a "D brane realization" of a Higgs-like mechanism. In this case, two of the vector particles, corresponding to the Chan-Paton states $(1, 1), (2, 2)$, remain massless, while the ones corresponding to $(1, 2), (2, 1)$ are massive.

For

$$\sum_{I=p+1}^{D-1} (\bar{x}_2^I - \bar{x}_1^I)^2 = 4\pi^2\alpha',$$

$|0; p^+, p^i; (1, 2)\rangle$, $|0; p^+, p^i; (2, 1)\rangle$, are massless states, and we also have two massless vector particles, corresponding to the Chan-Paton states $(1, 1), (2, 2)$, with an underlying gauge symmetry of $U(1) \times U(1)$.

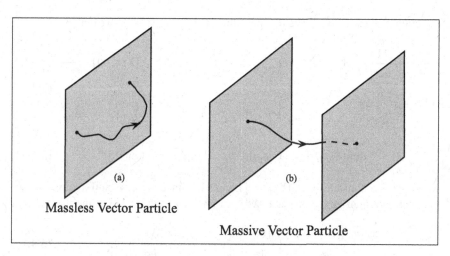

(a)

Massless Vector Particle

(b)

Massive Vector Particle

Fig. 3.8 Consider two parallel Dp-branes with locations specified by $\bar{x}_1^I, \bar{x}_2^I$. For $\sum_{I=p+1}^{D-1} (\bar{x}_2^I - \bar{x}_1^I)^2 > 4\pi^2 \alpha'$, the open string with its both ends on the same brane, as given in part (a) of the figure, gives rise to a massless vector particle. On the other hand, when it is stretched instead between the two branes, as given in part (b), it gives rise to a massive vector particle

Of particular interest is for coincident branes, i.e., for

$$\sum_{I=p+1}^{D-1} (\bar{x}_2^I - \bar{x}_1^I)^2 \to 0,$$

where the gauge symmetry is enhanced from U(1)$\times$U(1) to U(2) with four massless vector particles (including $D - 1 - p$ massless scalar particles).

3.4.3 Anti-symmetric Fields and "Charged" D Branes

The electromagnetic four-vector current associated with a moving charged particle of unit charge is defined by dX^μ/dτ, which is tangent to the worldline traced by the particle in the direction of increasing τ. The interaction Lagrangian in terms of the current and the vector potential A_μ takes the general form $\int d\tau\, A_\mu(X(\tau))\, dX^\mu/d\tau$, where $\dim[A_\mu] = [\text{Length}]^{-1}$. We may, similarly define the interaction of the ant-symmetric fields. In particular, for the second rank anti-symmetric tensor fields $A_{\mu_1 \mu_2}$, we may consider the tangent vector $\partial X^\mu(\tau, \sigma)/\partial \sigma$ along the string, at (τ, σ), in the direction of increasing σ, as well as the tangent vector $\partial X^\mu(\tau, \sigma)/\partial \tau$, in the direction of increasing τ. The action of the field $A_{\mu_1\mu_2}$, may be written as

$$W = -\frac{1}{6\kappa_1^2}\int d^D x \; F^{\mu\mu_1\mu_2}F_{\mu\mu_1\mu_2} + \int d\tau\, d\sigma\; A_{\mu_1\mu_2}(X(\tau,\sigma))\frac{\partial X^{\mu_1}}{\partial \tau}\frac{\partial X^{\mu_2}}{\partial \sigma},$$

$$(3.4.12)$$

$$F_{\mu\mu_1\mu_2} = \partial_\mu A_{\mu_1\mu_2} + \partial_{\mu_1}A_{\mu_2\mu} + \partial_{\mu_2}A_{\mu\mu_1}, \quad A_{\mu_1\mu_2} = -A_{\mu_2\mu_1},$$

$$(3.4.13)$$

$$\dim[A_{\mu_1\mu_2}] = [\text{Length}]^{-2}, \quad \dim[\kappa_1^2] = [\text{Length}]^{D-6}, \quad (3.4.14)$$

where κ_1^2 is a constant, $(\dim[X^\mu] = [\text{Length}])$. Since $A_{\mu_1\mu_2}$ is anti-symmetric, we may anti-symmetrize the product $\partial X^{\mu_1}/\partial\tau\, \partial X^{\mu_2}/\partial\sigma$ in (3.4.12), using the notation

$$\frac{\partial X^{[\mu_1}}{\partial \tau}\frac{\partial X^{\mu_2]}}{\partial \sigma} = \frac{1}{2}\left(\frac{\partial X^{\mu_1}}{\partial \tau}\frac{\partial X^{\mu_2}}{\partial \sigma} - \frac{\partial X^{\mu_2}}{\partial \tau}\frac{\partial X^{\mu_1}}{\partial \sigma}\right). \quad (3.4.15)$$

Upon writing

$$A_{\mu_1\mu_2}(X(\tau,\sigma)) = \int d^D x \; \delta^D(x - X(\tau,\sigma))A_{\mu_1\mu_2}(x),$$

the interaction term in (3.4.12) takes the form

$$\int d^D x\, A_{\mu\nu}(x)J^{\mu\nu}(x), \quad J^{\mu\nu}(x) = \int d\tau\, d\sigma\; \delta^D(x - X(\tau,\sigma))\frac{\partial X^{[\mu}}{\partial \tau}\frac{\partial X^{\nu]}}{\partial \sigma}. \quad (3.4.16)$$

Here we note that the current $J^{\mu\nu}(x)$, and hence also the interaction Lagrangian density, are non-vanishing only for spacetime points x^μ which lie on the two dimensional surface in spacetime described by the totality of the points $X^\mu(\tau,\sigma)$ as τ,σ are varied in their domains of definitions, and traced by a one-dimensional object of a spatial extension.

For a conserved current $\partial_\mu J^{\mu\nu} = 0$, as the source of the $A_{\mu\nu}(x)$ field, one may define a vector charge

$$Q^\nu = \int d^{D-1}\mathbf{x}\; J^{0\nu}(x),$$

indicating the existence of a "charged" one-dimensional object of spatial extension!.

For anti-symmetric $(p + 1)$th-rank $A_{\mu_1\mu_2\ldots\mu_{p+1}}$ tensor fields emerging in string theories, the generalization of the above analysis is immediate.

Equations (3.4.12)–(3.4.14), become replaced by

$$W = -\frac{1}{2(p+2)\,\kappa_p^2} \int d^D x \; F^{\mu\mu_1\cdots\mu_{p+2}} F_{\mu\mu_1\cdots\mu_{p+2}} +$$

$$+ \int d\tau \, d\sigma_1 \ldots d\sigma_p \; A_{\mu_1\cdots\mu_{p+1}}\!\left(X(\tau,\sigma_1,\ldots,\sigma_p)\right) \frac{\partial X^{[\mu_1}}{\partial\tau} \frac{\partial X^{\mu_2}}{\partial\sigma_1} \cdots \frac{\partial X^{\mu_{p+1}]}}{\partial\sigma_p},$$

$$(3.4.17)$$

$$F_{\mu\mu_1\cdots\mu_{(p+2)}} = \partial_\mu A_{\mu_1\cdots\mu_{p+1}} + \text{cycl. perm.}, \quad A_{\mu_1\cdots\mu_{p+1}} = \text{totally anti-symm.},$$

$$(3.4.18)$$

$$\dim[A_{\mu_1\cdots\mu_{p+1}}] = [\text{Length}]^{-(p+1)}, \qquad \dim[\kappa_p^2] = [\text{Length}]^{D-2(p+2)},$$

$$(3.4.19)$$

where κ_p^2 is a constant, and cycl. perm. means following the order of the indices indicated in a c.w. direction around the circle

The interaction part may be rewritten as

$$W_I = \int d^D x \; A_{\mu_1\cdots\mu_{p+1}}(x) \, J^{\mu_1\cdots\mu_{p+1}}(x), \qquad (3.4.20)$$

$$J^{\mu_1\cdots\mu_{p+1}}(x) = \int d\tau \, d\sigma_1 \ldots d\sigma_p \; \delta^D\!\left(x - X(\tau,\sigma_1,\ldots,\sigma_p)\right) \frac{\partial X^{[\mu_1}}{\partial\tau} \frac{\partial X^{\mu_2}}{\partial\sigma_1} \cdots \frac{\partial X^{\mu_{p+1}]}}{\partial\sigma_p}.$$

$$(3.4.21)$$

Again we note that the current $J^{\mu_1\cdots\mu_{p+1}}(x)$, and hence the interaction Lagrangian density, are non-vanishing only for spacetime points x^μ which lie in the $(p+1)$ dimensional region in spacetime described by the totality of the points $X^\mu(\tau,\sigma_1,\ldots,\sigma_p)$ as $(\tau,\sigma_1,\ldots,\sigma_p)$ are varied in their domains of definitions, and generated by a p-dimensional object of spatial extension. For a conserved current $\partial_{\mu_1} J^{\mu_1\cdots\mu_{p+1}}(x) = 0$, as the source of the $A^{\mu_1\cdots\mu_{p+1}}(x)$ field, one may define a tensor charge $Q^{\mu_1\cdots\mu_p} = \int d^{D-1}\mathbf{x}\, J^{0\mu_1\cdots\mu_p}(x)$ with the indication of the existence of a "charged" p-dimensional object of spatial extension—a Dp-brane !

3.5   Interactions, Vertices and Scattering

We consider scattering theory in string theory, where so-called external lines represent single particle excitations of a string, but the internal structure, describing the dynamics, include all the string modes. In a sense, the dynamics involved here is much richer than in conventional field theory where only a finite number of types of particles are exchanged in the scattering process describing the dynamics.

The worldsheets of two open strings may merge together forming a single worldsheet and may then split again as shown in Fig. 3.9.

We have to give some sense of such a process just described. Such diagrams replace familiar Feynman diagrams involving no closed loops and a closed loop, respectively, e.g., as shown in Fig. 3.10. String processes involving "loops" are much harder to compute.

Examples of closed strings corresponding to the above processes are shown in Fig. 3.11.

We present a simple intuitive description of scattering theory with interactions, described by certain vertices. The scattering theory of string theory is not fully developed to the extent of scattering theory in conventional field theory but some interesting results emerge.

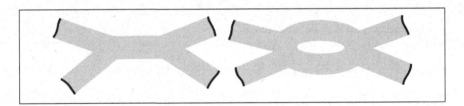

Fig. 3.9 Two of simple examples where the worldsheets of two open strings merge together forming a single worldsheet and then split again

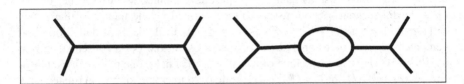

Fig. 3.10 The scattering of the two strings of Fig. 3.9 replace familiar simple Feynman diagrams of the types, e.g., shown in this figure

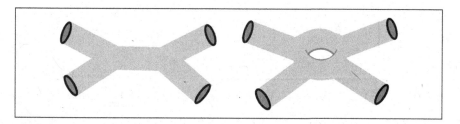

Fig. 3.11 Two of simple examples where the worldsheets of two closed strings merge together forming a single worldsheet and then split again

3.5.1 Open Bosonic Strings

For any given particle state of, say, type Λ, of momentum k^μ, in the open string spectrum, one associates a vertex operator $V_\Lambda(k, \tau)$, using a rather standard notation. This vertex describes the emission of such a particle from the $\sigma = 0$ end of the open string at a proper time τ. The nature of such vertices will be discussed below.

As a very simplistic view of a scattering process, consider a particle of type, say, Λ_1, of momentum k_1, in the spectrum of the string, being attached to the string at the $\sigma = 0$ end of the string in the far past. As time develops, one may, conveniently, introduce a string diagram, not to be confused with its worldsheet, as the semi-infinite rectangular strip, with σ taken along the vertical axis, while τ taken along the horizontal one, with $\tau_1 \leq \tau_2 \leq \cdots \leq \tau_n$, and $\tau_1 \to -\infty$ to $\tau_n \to \infty$. At times $\tau_2, \tau_3, \cdots, \tau_{n-1}$, particles may be emitted (absorbed), at the $\sigma = 0$ boundary of the string, such that with each particle, a vertex operator is associated, as mentioned above. Finally a particle of momentum k_n emerges in the distant future. Such a string diagram may be simply represented as shown in Fig. 3.12.

We may also map the semi-infinite strip of the string diagram into other regions. For example, given $\tau_1 < \tau_2 < \ldots < \tau_n$, we may introduce the variable,

$$\rho = e^{(\tau + i\sigma)}, \quad 0 \leq \sigma \leq \pi, \qquad (3.5.1)$$

which maps the entire semi-infinite strip of the string diagram above to the upper complex ρ-plane, where $z_i = e^{\tau_i}$, as shown in Fig. 3.13, where, in particular, $z_1 \to 0$ for $\tau_1 \to -\infty$, $z_n \to \infty$ for $\tau_n \to \infty$.

One may also map the entire semi-infinite strip of the string diagram above onto a unit disc—a compact space—via the transformation

$$u = \frac{1 + i\rho}{1 - i\rho}, \qquad (3.5.2)$$

leading to Fig. 3.14. Here we note that $u_2, u_3, \ldots, u_{n-1}$ represent the positions of the vertices associated with the external particles. Note that the ordering of the points associated with $z_1, \ldots, z_n$, and $\tau_1, \ldots, \tau_n$, is preserved.

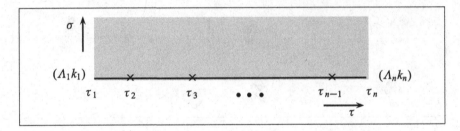

Fig. 3.12 We define a string diagram, not to be confused with the worldsheet of a string, as the semi-infinite rectangular strip, with σ taken along the vertical axis, while τ taken along the horizontal one, with $\tau_1 \le \tau_2 \le \cdots \le \tau_n$, and $\tau_1 \to -\infty$ to $\tau_n \to \infty$

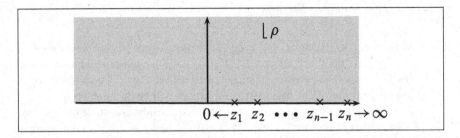

Fig. 3.13 One may map the semi-infinite strip of the string diagram to the upper complex ρ-plane, where $z_i = e^{\tau_i}, \rho = e^{(\tau + i\sigma)}, 0 \le \sigma \le \pi$

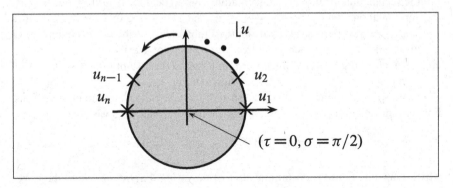

Fig. 3.14 One may also map the entire semi-infinite strip of the string diagram onto a unit disc via the transformation $u = (1 + i\,e^{\tau + i\sigma})/(1 - i\,e^{\tau + i\sigma})$, with $u_i = (1 + iz_i)/(1 - iz_i), z_i = e^{\tau_i}$

The transition amplitude A, involving $n - 2$ vertices, may be written in terms of propagators (see Appendix B of this chapter) connecting the vertices in question as[42]

$$A(\Lambda_1 k_1, \Lambda_2 k_2, \ldots, \Lambda_n k_n) = (g_0)^{n-2} \langle \Lambda_n\, k_n \text{ out} \,|\, \{\cdot\} \,|\, \Lambda_1 k_1 \text{ in} \rangle, \tag{3.5.3}$$

$$\{\cdot\} = V_{\Lambda_{n-1}}(k_{n-1})\, \Delta\, V_{\Lambda_{n-2}}(k_{n-2})\, \Delta \cdots \Delta\, V_{\Lambda_2}(k_2), \tag{3.5.4}$$

---

[42]As a standard notation, the variables Λ_i label the external particles.

where g_0 denotes an open string coupling constant, and from (B-3.5), the propagators Δ are given by

$$\Delta = \frac{1/\alpha'}{p^2 + M^2 - i\epsilon}. \tag{3.5.5}$$

The latter may be rewritten as

$$\Delta = i \int_{-\infty}^{0} d\beta\, e^{i\beta\alpha'(p^2+M^2-i\epsilon)} = i \int_{-\infty}^{0} d\beta\, e^{i\beta(H-i\epsilon)}, \tag{3.5.6}$$

where we have used the expression of the Hamiltonian in (B-3.3). Upon substituting the latter expression in (3.5.4) for $\{\cdot\}$, the latter may be rewritten as

$$\{\cdot\} = \delta_n \int_{-\infty}^{0} e^{\epsilon\alpha'\beta_{n-2}}\, d\beta_{n-2} \int_{-\infty}^{0} e^{\epsilon\alpha'\beta_{n-3}}\, d\beta_{n-3} \cdots \int_{-\infty}^{0} e^{\epsilon\alpha'\beta_2}\, d\beta_2$$

$$\times V_{\Lambda_{n-1}}(k_{n-1})$$

$$\times \Bigg[\exp(i\,\beta_{n-2}\,H)\, V_{\Lambda_{n-2}}(k_{n-2}) \exp(-i\,\beta_{n-2}\,H)$$

$$\times \exp(i\,(\beta_{n-2}+\beta_{n-3})\,H)\, V_{\Lambda_{n-2}}(k_{n-2}) \exp(-i\,(\beta_{n-2}+\beta_{n-3})\,H)$$

$$\vdots$$

$$\times \exp(i\,(\beta_{n-2}+\beta_{n-3}+\cdots\beta_2)\,H)\, V_{\Lambda_2}(k_2) \exp(-i\,(\beta_{n-2}+\beta_{n-3}+\cdots+\beta_2)\,H)\Bigg]$$

$$\times \exp\big(i\,(\beta_{n-2}+\beta_{n-3}+\cdots+\beta_3+\beta_2)\,H\big), \tag{3.5.7}$$

$\delta_n = (i)^{n-3}$ for $n \geq 3$, $\delta_2 = 1$.

We first make a change of variables from $\beta_2, \cdots \beta_{n-2}$ to $\tau_2, \cdots, \tau_{n-2}$, with the latter defined by

$$\tau_{n-2} = \beta_{n-2}, \quad \tau_{n-3} = \beta_{n-2}+\beta_{n-3}, \quad \ldots, \quad \tau_2 = \beta_{n-2}+\beta_{n-3}+\cdots+\beta_2. \tag{3.5.8}$$

Then we use the fact that the expression on the last line in (3.5.7) is nothing but

$$\exp\big(i\alpha'\tau_2(p^2+M^2)\big),$$

$(p^2 + M^2)|\Lambda_1 k_1 \text{in}\rangle = 0$, and for a particle-state in the spectrum of the string, this amounts to replacing $\exp\big(i\alpha'\tau_2(p^2+M^2)\big)$ by one. Finally we introduce the time translated operators, as given in (B-3.6)

$$\exp(i\tau H)\, V_\Lambda(k) \exp(-i\tau H) = V_\Lambda(k, \tau), \tag{3.5.9}$$

to obtain

$$A(\Lambda_1 k_1, \Lambda_2 k_2, \ldots, \Lambda_n k_n) = (g_0)^{n-2} \delta_n \int_{\mathcal{T}} d\tau_{n-2} \, d\tau_{n-3} \, \cdots \, d\tau_3 \, e^{\tau_2 \alpha' \epsilon} \, d\tau_2$$

$$\times \, \langle \Lambda_n k_n \text{ out} | \, V_{\Lambda_{n-1}}(k_{n-1}, \tau_{n-1} = 0) \, V_{\Lambda_{n-2}}(k_{n-2}, \tau_{n-2}) \, \cdots \, V_{\Lambda_2}(k_2, \tau_2) | \Lambda_1 k_1 \text{ in} \rangle, \tag{3.5.10}$$

$$\mathcal{T} : \; -\infty < \tau_2 < \tau_3 < \cdots < \tau_2 < \tau_{n-2} < 0. \tag{3.5.11}$$

The expression for the amplitude in (3.5.10) could have been written down directly as it indicates particles emissions (absorption) at the consecutive times $\tau_{n-2} \leq \cdots < \tau_2$ from the corresponding vertices.

Upon using the variable $z = e^\tau$ in (3.5.10), the amplitude takes the convenient form $(z_n = \infty, z_{n-1} = 1, z_1 = 0)$

$$A(\Lambda_1 k_1, \Lambda_2 k_2, \ldots, \Lambda_n k_n) = (g_0)^{n-2} \delta_n \int_{\mathcal{M}} \frac{dz_{n-2}}{z_{n-2}} \frac{dz_{n-3}}{z_{n-3}} \cdots \frac{dz_2}{z_2} (z_2)^{\alpha' \epsilon}$$

$$\times \, \langle \Lambda_n k_n \text{ out} | \, V_{\Lambda_{n-1}}(k_{n-1}, z_{n-1} = 1) \, V_{\Lambda_{n-2}}(k_{n-2}, z_{n-2}) \cdots V_{\Lambda_2}(k_2, z_2) | \Lambda_1 k_1 \text{ in} \rangle, \tag{3.5.12}$$

where (see also Problem 3.21)

$$\mathcal{M} : \; 0 < z_2 < \cdots < z_{n-3} < z_{n-2} < 1, \tag{3.5.13}$$

and a vertex operator $V_\Lambda(k_i, \tau_i)$ is now expressed in terms of the variable z_i.

Of particular interest is the so-called three-point-function

$$A(\Lambda_1 k_1, \Lambda_2 k_2, \Lambda_3 k_3) = g_0 \langle \Lambda_3 k_3 \text{ out} | \, V_{\Lambda_2}(k_2, z_2 = 1) | \Lambda_1 k_1 \text{ in} \rangle, \tag{3.5.14}$$

involving no integration to be carried out, and the one describing a scattering process such as of 2 particle $\rightarrow$ 2 particles

$$A(\Lambda_1 k_1, \Lambda_2 k_2, \Lambda_3 k_3, \Lambda_4 k_4) = (g_0)^2 \, i \int_0^1 \frac{dz_2}{z_2} \, z_2^{\alpha' \epsilon}$$

$$\times \, \langle \Lambda_4 k_4 \text{ out} | \, V_{\Lambda_3}(k_3, z_3 = 1) \, V_{\Lambda_2}(k_2, z_2) | \Lambda_1 k_1 \text{ in} \rangle, \tag{3.5.15}$$

with only one integration to be carried out.

To simplify the notation, it is customary to set $\ell = 1$, and hence $\alpha' = 1/2$. We may use (3.2.58) in a covariant notation, to express the string field for $\sigma = 0$ as

$$X^\mu(\tau) = x^\mu + p^\mu \tau + i \sum_{n=1}^{\infty} \left(\frac{\alpha^\mu(n)}{n} e^{-i n \tau} - \frac{\alpha^\mu(n)^\dagger}{n} e^{i n \tau} \right). \tag{3.5.16}$$

We define a vertex operator by the expression

$$V_\Lambda(k, \tau) = W_\Lambda(k, \tau) e^{i kX(\tau)}, \tag{3.5.17}$$

using a standard notation, where $X^\mu(\tau)$ is given (3.5.16). Here $W_\Lambda(k, \tau)$ depends on the state of an external particle.

In response to the translation of the position of a string : $X^\mu \to X^\mu + a^\mu$, the wave-function of an external particle of momentum k, develops the factor e^{ika}, as it should, which explains the presence of the factor $e^{i kX(\tau)}$. It is assumed to be defined as normal-ordered. That is, it is defined with creation operators set to the left of annihilation operators. To this end, let us spell out the normal-ordered expression for $e^{i kX}$, denoted by $: e^{i kX} :$.

Introducing the variable $z = e^\tau$ used in (3.5.12), we have

$$kX(z) = kx + kp \ln z + i \sum_{n=1}^{\infty} \left[\frac{k\alpha(n)}{n} z^{-in} - \frac{k\alpha(n)^\dagger}{n} z^{in} \right]. \tag{3.5.18}$$

The normal-ordered expression in question is then defined by

$$: \exp i kX(z) := \exp(k \cdot \mathcal{B}(z)) \exp[i (k \cdot x + k \cdot p \ln z)] \exp(k \cdot \mathcal{A}(z)), \tag{3.5.19}$$

$$\mathcal{A}^\mu(z) = -\sum_{n=1}^{\infty} \frac{\alpha^\mu(n)}{n} z^{-in}, \quad \mathcal{B}^\mu(z) = \sum_{n=1}^{\infty} \frac{\alpha^\mu(n)^\dagger}{n} z^{in}. \tag{3.5.20}$$

The state of a massless vector particle of momentum k and polarization specified by the vector $e_\lambda^\mu = (0, \mathbf{e}_\lambda)$, [see (3.2.175)], is given by

$$|\Lambda, k\rangle = e_\lambda \cdot \alpha(1)^\dagger |0; k\rangle = e \cdot \alpha(1)^\dagger |0; k\rangle, \quad \Lambda = \lambda, \tag{3.5.21}$$

suppressing the parameter λ, for the simplicity of the notation, and we have labeled the ground-state by the momentum k in a covariant description. The first factor of the vertex operator for such a particle, is then defined by

$$W_\Lambda(k, \tau) = e_{\mu\lambda} \frac{dX^\mu(\tau)}{d\tau}, \quad e_\lambda \cdot k = 0, \quad k^2 = 0, \tag{3.5.22}$$

where $dX^\mu/d\tau$ is the current. For three massless vector particles in (3.5.14), the so-called the three point function, is easily evaluated.

To the above end,

$$\frac{dX^{\mu}(\tau)}{d\tau}\bigg|_{\tau=0} = p^{\mu} + \sum_{n=1}^{\infty} [\alpha^{\mu}(n) + \alpha^{\mu}(n)^{\dagger}], \tag{3.5.23}$$

leading from (3.5.14), (3.5.17)–(3.5.22) to ($\tau_2 = 0, z_2 = 1$)

$$A(\Lambda_1 k_1, \Lambda_2 k_2, \Lambda_3 k_3) = g_0 \langle 0; k_3 | e_3 \alpha(1) [\cdot] e_1 \alpha(1)^{\dagger} | 0; k_1 \rangle \tag{3.5.24}$$

$$[\cdot] = [e_2 \cdot (p + \alpha(1) + \alpha(1)^{\dagger}) e^{k_2 \cdot \alpha(1)^{\dagger}} e^{ik_2 \cdot x} e^{-k_2 \cdot \alpha(1)}]. \tag{3.5.25}$$

To evaluate this, we use the relations $e_j \cdot k_j = 0$, for each $j = 1, 2, 3$, the covariant relation $[\alpha^{\mu}(1), \alpha^{\nu}(1)^{\dagger}] = \eta^{\mu\nu}$, and the fact that e^{ikx} denotes a momentum translation operator, then an elementary analysis gives[43]

$$A(\Lambda_1 k_1, \Lambda_2 k_2, \Lambda_3 k_3)$$
$$= g_0 (e_2 \cdot k_1 \ e_1 \cdot e_3 + e_3 \cdot k_2 \ e_2 \cdot e_1 + e_1 \cdot k_3 \ e_3 \cdot e_2 + k_1 \cdot e_2 \ k_2 \cdot e_3 \ k_3 \cdot e_1), \tag{3.5.26}$$

up to a momentum conserving delta function, with $k_1 + k_2 + k_3 = 0$, where, in the process of the derivation, we have finally chosen the direction of all the momenta such that momentum conservation reads as just given. Since the last term in (3.5.26) involves three powers of momenta, we may infer from dimensional analysis alone that its coefficient is suppressed by a factor α' relative to the other terms and hence is of higher order. The leading contribution to (3.5.26) may be then conveniently written as

$$e_{1\mu} e_{2\nu} e_{3\rho} A^{\mu\nu\rho}(k_1, k_2, k_3), \tag{3.5.27}$$

$$A^{\mu\nu\rho}(k_1, k_2, k_3) = \eta^{\mu\rho} k_1^{\nu} + \eta^{\mu\nu} k_2^{\rho} + \eta^{\nu\rho} k_3^{\mu}, \tag{3.5.28}$$

with $k_1 + k_2 + k_3 = 0$. It is easily verified, by the application of momentum conservation and the orthogonality relations $e_j k_j = 0$, for each $j = 1, 2, 3$, that this amplitude is anti-symmetric in the exchange of any pairs $(e_i, k^i) \leftrightarrow (e_j, k^j)$.[44] Hence symmetrization over the exchanges of such pairs give zero unless one considers a linear combination of such amplitudes with corresponding *anti-symmetric* coefficients thus leading naturally to a non-abelian gauge theory with anti-symmetric structure constants. Thus this three point-function contributes only in the non-abelian gauge theory case. This will be discussed again in Sect. 3.5.3, and applied in Sect. 3.7.

---

[43]See Problem 3.22.
[44]See Problem 3.23.

For historical reasons, the amplitude (3.5.15) is often applied to the scattering of four tachyons as the resulting amplitude, in particular, of scattering of four scalar particles coincides with the expression conjectured by Veneziano in the sixties, which was discussed in the introduction to this chapter, and has some desirable properties. The vertex operator for a tachyon state of momentum k, has the simplest possible structure given by

$$V_\Lambda(k, z) =: \exp i\, kX(z) :$$ (3.5.29)

i.e., $W_\Lambda(k, z) = 1$. We note from (3.2.94) that for the tachyon, $M^2 = -2$, for $\alpha' = 1/2$, i.e., $k^2 = 2$. We carry out a transformation of variable $(z)^i \to z$, which corresponds to so-called a Wick rotation in complex time plane.

In reference to (3.5.19), the following properties easily follow for a tachyon state,

$$\exp\left(-k_{2\mu} \sum_{n=1}^{\infty} \frac{\alpha^\mu(n)}{n} z^{-n}\right)|0; k_1\rangle = |0; k_1\rangle,$$ (3.5.30)

$$\exp[i\left(k_2 x - i k_2 p \ln z\right)]|0; k_1\rangle = \exp(k_1 k_2 \ln z) \exp\left(\frac{k_2^2}{2}\ln z\right)|0; k_1 + k_2\rangle,$$ (3.5.31)

where in writing the latter equation, we have used the elementary relation

$$e^{A+B} = e^A\, e^B\, e^{[B,A]/2},$$

for $[B, A]$ a c-number, the covariant generalization $[x^\mu, p^\nu] = i\, \eta^{\mu\nu}$, that $p\,|0; k_1\rangle = k_1|0; k_1\rangle$, and finally that x is the generator of momentum translation. We recall that $k_2^2 = 2$.

From (3.5.19), (3.5.20), (3.5.30), (3.5.31), the scattering amplitude (3.5.15) of four tachyons becomes

$$A(k_1, k_2, k_3, k_4) = (g_0)^2 \int_0^1 \frac{dz}{z}(z)^{k_1 k_2} z^{1-i\epsilon}$$

$$\times \left\langle 0; k_3 + k_4 \left| \exp\left[-\sum_{n=1}^{\infty} \frac{k_3\alpha(n)}{n}\right] \exp\left[\sum_{m=1}^{\infty} \frac{k_2\alpha(m)^\dagger}{m} z^m\right]\right| 0; k_1 + k_2\right\rangle.$$ (3.5.32)

Using the identity

$$\left\langle 0; k_3 + k_4 \left| \left[-\sum_{n=1}^{\infty} \frac{k_3\alpha(n)}{n}\right]\left[\sum_{m=1}^{\infty} \frac{k_2\alpha(m)^\dagger}{m} z^m\right]\right| 0; k_1 + k_2\right\rangle$$

$$= -k_2 k_3 \sum_{n=1}^{\infty} \frac{z^n}{n} \langle 0; k_3 + k_4|0; k_1 + k_2\rangle = k_2 k_3 \ln(1 - z)\, \langle 0; k_3 + k_4|0; k_1 + k_2\rangle,$$

(3.5.33)

after some labor, the following expression emerges

$$A(k_1, k_2, k_3, k_4) = (g_0)^2 \int_0^1 dz \, (z)^{k_1 k_2} (1-z)^{k_2 k_3} \langle 0; k_3 + k_4 | 0; k_1 + k_2 \rangle, \qquad (3.5.34)$$

where $\langle 0; k_3 + k_4 | 0; k_1 + k_2 \rangle$ expresses the conservation of momentum. Upon using Mandelstam variables and the identities following them ($k_i^2 = 2, i = 1, 2, 3, 4$)

$$s = -(k_1 + k_2)^2, \quad t = -(k_2 + k_3)^2, \quad k_1 k_2 = -\frac{s}{2} - 2 \quad k_2 k_3 = -\frac{t}{2} - 2, \qquad (3.5.35)$$

and the definition of the beta function

$$B(a, b) = \int_0^1 dz \, z^{a-1} (1-z)^{b-1} = \frac{\Gamma(a)\,\Gamma(b)}{\Gamma(a+b)}, \qquad (3.5.36)$$

the amplitude, up to a conserving delta function and a phase factor, becomes

$$A = (g_0)^2 B(-\frac{s}{2} - 1, -\frac{t}{2} - 1), \qquad (3.5.37)$$

which is the classic Veneziano amplitude.[45] In addition to the tower of string states contributing to this expression, leading to poles in the gamma functions in the s and t- channels. The pole, in the s-channel, for example, arising at $s = -2$, is the expected exchange of a tachyon in field theory.

3.5.2 Closed Bosonic Strings

The worldsheet of a closed string, does not have specific boundaries such as $\sigma = 0$, as in the open string case, and particle emission (absorption) is to be considered to occur from any σ value, as shown by the crosses in the string diagram of a closed string in Fig. 3.15: One may map the surface of the cylinder onto the surface of a unit sphere, as shown in Fig. 3.16, via the transformation

$$\tau = 2 \ln \left(\tan \frac{\theta}{2} \right), \quad 0 \le \theta \le \pi, \qquad (3.5.38)$$

$$x_1 = \cos(2\sigma) \sin \theta, \quad x_2 = \sin(2\sigma) \sin \theta, \quad x_3 = \cos \theta, \quad 0 \le \sigma \le \pi. \tag{3.5.39}$$

---

[45] See also the introduction to this chapter for additional details on the Veneziano amplitude.

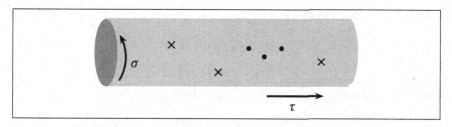

Fig. 3.15 Particle emission (absorption) is to be considered to occur from any points specified by the variable σ on the string diagram of a closed string. The latter is defined as the surface of the infinite cylinder generated by the closed string set up vertically straight, in a *circular shape* ($\bigcirc \rightarrow \bigcirc$), with τ labeled along a horizontal axis, from $\tau \rightarrow -\infty$ to $\tau \rightarrow \infty$

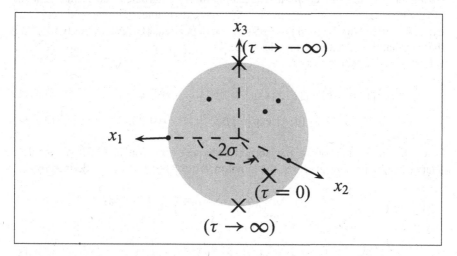

Fig. 3.16 One may map the surface of the cylinder onto the surface of a unit sphere, with transformation variables defined in (3.5.38), (3.5.39), where θ has its usual meaning and 2σ replaces the angle ϕ, in spherical coordinates

Needless to say, one cannot order points on the sphere and a permutation over vertex insertions may be then necessary.

In order to introduce a vertex for particle emission, let us consider, say, the graviton which is in the spectrum of a closed bosonic string. To this end, the state of graviton of momentum k, and described by a polarization tensor $\epsilon^{\mu\nu}$, [see (3.2.188), (3.2.189), (3.2.175)], may be written as [see (3.2.77), (3.2.78), (3.2.116), (3.2.117), (3.2.122)]

$$\epsilon_{\mu\nu}\, \alpha^{\mu}(1)\bar{\alpha}^{\nu}(1)|0; k\rangle, \qquad\qquad \text{(3.5.40)}$$

where $|0; k\rangle$, is the tachyonic ground-state,[46]

$$\epsilon^{\mu\nu} = \epsilon^{\nu\mu}, \quad \epsilon^{\mu 0} = 0, \quad \epsilon^{\mu}{}_{\mu} = 0, \quad k_{\mu}\epsilon^{\mu\nu} = 0. \tag{3.5.41}$$

A graviton vertex operator may be then defined by $V_\Lambda \propto \epsilon_{\mu\nu} \partial_\xi X^\mu \partial^\xi X^\nu \, e^{ikX}$, with $\xi = (\tau, \sigma)$, and hence simply, and more conveniently, as

$$V_\Lambda(\tau, \sigma) = \epsilon_{\mu\nu} \left(\partial_+ X^\mu \, \partial_- X^\nu \right) e^{ik\cdot X} = \epsilon_{\mu\nu} \left(\partial_+ X_L^\mu \, \partial_- X_R^\nu \right) e^{ik\cdot X}, \tag{3.5.42}$$

$\partial_\pm = (\partial_\tau \pm \partial_\sigma)/2$.

We note that for $\tau = 0$, we may write $k \cdot X(0, \sigma) = k \cdot x +$ an expression consisting of operators of creation and annihilation of particles and is independent of x, which we may denote by $k \cdot \widetilde{X}(0, \sigma)$. Also note that the coefficient of $\exp[i k \cdot X(0, \sigma)]$ in $V_\Lambda(0, \sigma)$ in (3.5.42) is independent of x. Accordingly by using the fact that the graviton is massless, i.e., $k^2 = 0$, we have for any constant β the following useful equations:

$$\exp[i \beta p^2] \exp[ik \cdot x] \exp[-i\beta p^2] = \exp[ik \cdot x], \tag{3.5.43}$$

$$\exp[i \beta p^2] V_\Lambda(0, \sigma) \exp[-i\beta p^2] = V_\Lambda(0, \sigma). \tag{3.5.44}$$

Using the expression of the normal ordered expression in (3.5.19), (3.5.20), adapted to the present situation, the graviton vertex operator may be defined by

$$V_\Lambda(k, \tau, \sigma) = \epsilon_{\mu\nu} \left(\frac{1}{2} p^\mu + \sum_{n \neq 0} \alpha^\mu(n) \, e^{-2in(\tau-\sigma)} \right) : e^{ik \cdot X_R(\tau-\sigma)} :$$

$$\times \left(\frac{1}{2} p^\mu + \sum_{n \neq 0} \tilde{\alpha}^\mu(n) \, e^{-2i\,n(\tau+\sigma)} \right) : e^{ik \cdot X_L(\tau+\sigma)} : . \tag{3.5.45}$$

Upon rewriting the latter as $V = \epsilon_{\mu\nu} V^{\mu\nu}$, we see that $V^{\mu\nu}$, is written as the product of a right- and left-vertex. Therefore up to polarization tensors, we may define a vertex operator in the closed string as a product of a right- and a left vertex operators expressed as

$$V_\Lambda(k, \tau - \sigma, \tau + \sigma) = V_{R\Lambda} \left(\frac{k}{2}, \tau - \sigma \right) V_{L\Lambda} \left(\frac{k}{2}, \tau + \sigma \right), \tag{3.5.46}$$

---

[46]The tachyonic state may be written as the product of a left and right-movers states.

and perform an integration over σ as an averaging over all σ, considering all of its possible values, and obtain a vertex operator which is independent of σ :

$$V_\Lambda(k, \tau) = \frac{1}{\pi} \int_0^\pi d\sigma \, V_{R\Lambda}\left(\frac{k}{2}, \tau - \sigma\right) V_{L\Lambda}\left(\frac{k}{2}, \tau + \sigma\right). \tag{3.5.47}$$

From (3.5.10), (3.5.11), we may directly write

$$A(\Lambda_1 k_1, \Lambda_2 k_2, \ldots, \Lambda_n k_n) = (g_c)^{n-2} \delta_n \int_{\mathscr{T}} d\tau_{n-2} \, d\tau_{n-3} \, \ldots \, d\tau_3 \, e^{\tau 2\alpha'\epsilon} \, d\tau_2$$

$$\times \langle \Lambda_n k_n \text{ out} | V_{\Lambda_{n-1}}(k_{n-1}, \tau_{n-1} = 0) \, V_{\Lambda_{n-2}}(k_{n-2}, \tau_{n-2}) \ldots V_{\Lambda_2}(k_2, \tau_2) | \Lambda_1 k_1 \text{ in}\rangle, \tag{3.5.48}$$

$$\mathscr{T} : -\infty < \tau_2 < \tau_3 < \cdots < \tau_2 < \tau_{n-2} < 0, \tag{3.5.49}$$

where a vertex $V_\Lambda(k, \tau)$ is defined in (3.5.47). Here g_c is a closed string coupling constant.

Upon introducing the variable $z = e^{2(\tau + i\sigma)}$, with $r = e^{2\tau}$, $\phi = 2\sigma$, we have

$$d\tau \, d\sigma = \frac{1}{4} \frac{dr \, d\phi}{r} = \frac{1}{4} \frac{r \, dr \, d\phi}{r^2} = \frac{1}{4} \frac{d^2 z}{|z|^2}. \tag{3.5.50}$$

From (3.5.47)–(3.5.50), we obtain the explicit expression

$$A(\Lambda_1 k_1, \Lambda_2 k_2, \ldots, \Lambda_n k_n) = \bar{\delta}_n \left(\frac{g_c}{4\pi}\right)^{n-2} \int_{\mathscr{B}} \frac{d^2 z_{n-2}}{|z_{n-2}|^2} \frac{d^2 z_{n-3}}{|z_{n-3}|^2} \cdots \frac{d^2 z_2}{|z_2|^2}$$

$$\times \langle \Lambda_n k_n \text{ out} | V_{\Lambda_{n-1}}(k_{n-1}, \tau_{n-1} = 0)$$

$$\times V_{\Lambda_{n-2}}(k_{n-2}, z_{n-2}, z_{n-2}^*) \ldots V_{\Lambda_2}(k_2, z_2, z_2^*) | \Lambda_1 k_1 \text{ in}\rangle, \tag{3.5.51}$$

where

$$\bar{\delta}_n = 4\pi (i)^{n-3} \quad \text{for} \quad n \geq 3, \quad \bar{\delta}_2 = 1, \tag{3.5.52}$$

and the vertices in (3.5.46), (3.5.48)

$$V_{\Lambda_{n-2}}(k_{n-2}, \tau_{n-2} - \sigma_{n-2}, \tau_{n-2} + \sigma_{n-2}) \ldots V_{\Lambda_2}(k_2, \tau_2 - \sigma_2, \tau_2 + \sigma_2),$$

are now expressed in terms of the variables $(z_{n-2}, z_{n-2}^*), \ldots, (z_2, z_2^*)$, respectively, where we have carried out a so-called Wick rotation $\tau \to i\tau$ so that

$$\exp[-2i(\tau - \sigma)] \to \exp[2(\tau + i\sigma)] = z$$

$$\exp[-2i(\tau + \sigma)] \to \exp[2(\tau - i\sigma)] = z^*,$$

in the right and left vertices, respectively. The domain of integration is given by

$$\mathscr{B} : 0 < |z_2| < \cdots < |z_{n-3}| < |z_{n-2}| < 1, \tag{3.5.53}$$

$$V_{\Lambda_{n-1}}(k_{n-1}, \tau_{n-1} = 0) = \frac{1}{\pi} \int_0^\pi d\sigma \; V_{R\Lambda_{n-1}}\left(\frac{k_{n-1}}{2}, 0 - \sigma\right) V_{L\Lambda_{n-1}}\left(\frac{k_{n-1}}{2}, 0 + \sigma\right). \tag{3.5.54}$$

For the three-graviton vertex, in particular, we have

$$A(\Lambda_1 k_1, \Lambda_2 k_2, \Lambda_3 k_3) = g_c \langle \Lambda_3 k_3 | V_{\Lambda_2}(k_2) | \Lambda_1 k_1 \rangle, \tag{3.5.55}$$

where

$$V_{\Lambda_2}(k_2) = \frac{1}{\pi} \int_0^\pi d\sigma \; V_{R\Lambda_2}\left(\frac{k_2}{2}, 0 - \sigma\right) V_{L\Lambda_2}\left(\frac{k_2}{2}, 0 + \sigma\right). \tag{3.5.56}$$

Moreover, *from* (3.5.44)–(3.5.46), and Appendix B, at the end of the chapter, we effectively have with $\alpha' \to 1/2, k_2^2 = 0$,

$$V_{R\Lambda_2}\left(\frac{k_2}{2}, 0 - \sigma\right)$$

$$= \exp[\frac{1}{2} i (0-\sigma)(p^2 + M_R^2)] \; V_{R\Lambda_2}\left(\frac{k_2}{2}, 0\right) \exp[-\frac{1}{2} i (0-\sigma)(p^2 + M_R^2)]$$

$$= \qquad \exp[-\frac{1}{2} i \sigma M_R^2] \; V_{R\Lambda_2}\left(\frac{k_2}{2}, 0\right) \exp[\frac{1}{2} i \; \sigma M_R^2], \tag{3.5.57}$$

as well as

$$V_{L\Lambda_2}\left(\frac{k_2}{2}, 0 + \sigma\right)$$

$$= \exp[\frac{1}{2} i (0+\sigma)(p^2 + M_L^2)] \; V_{L\Lambda_2}\left(\frac{k_2}{2}, 0\right) \exp[-\frac{1}{2} i (0+\sigma)(p^2 + M_L^2)]$$

$$= \qquad \exp[\frac{1}{2} i \sigma M_L^2] \; V_{L\Lambda_2}\left(\frac{k_2}{2}, 0\right) \exp[-\frac{1}{2} i \; \sigma M_L^2]. \tag{3.5.58}$$

Since M_L^2 and $V_{R\Lambda_2}$ commute, M_R^2 and $V_{L\Lambda_2}$ likewise commute, the above two equations lead to

$$V_{R\Lambda_2}\left(\frac{k_{n-1}}{2}, 0 - \sigma\right) V_{L\Lambda_2}\left(\frac{k_{n-1}}{2}, 0 + \sigma\right)$$

$$= \exp\left[-i\frac{1}{2}\sigma(M_R^2 - M_L^2)\right] V_{R\Lambda_2}\left(\frac{k_2}{2}, 0\right) V_{L\Lambda_2}\left(\frac{k_2}{2}, 0\right) \exp\left[i\frac{1}{2}\sigma (M_R^2 - M_L^2)\right],$$

$$(3.5.59)$$

For a particle-state $|\varphi\rangle$, in the spectrum of the string, we also have the constraint $[M_R^2 - M_L{}^2]|\varphi\rangle = 0$.

From (3.5.56) and (3.5.59), the three-graviton vertex is then simply given by

$$A(\Lambda_1 k_1, \Lambda_2 k_2, \Lambda_3 k_3) = g_c \langle \Lambda_3 k_3 | V_{\Lambda_2 R}\left(\frac{k_2}{2}, 0\right) V_{\Lambda_2 L}\left(\frac{k_2}{2}, 0\right) |\Lambda_1 k_1\rangle. \quad (3.5.60)$$

Upon comparison of this expression with the three-point function of three massless vector particles through (3.5.14)–(3.5.17), (3.5.27), (3.5.28), the expression of three-point function for three gravitons emerges as

$$g_c\, \epsilon^1_{\mu_1\mu_2}\, \epsilon^2_{\nu_1\nu_2}\, \epsilon^3_{\rho_1\rho_2}\, A^{\mu_1\nu_1\rho_1}\left(\frac{k_1}{2}, \frac{k_2}{2}, \frac{k_3}{2}\right) A^{\mu_2\nu_2\rho_2}\left(\frac{k_1}{2}, \frac{k_2}{2}, \frac{k_3}{2}\right), \quad (3.5.61)$$

$$A^{\mu\nu\rho}(k_1, k_2, k_3) = \eta^{\mu\rho} k_1^\nu + \eta^{\mu\nu} k_2^\rho + \eta^{\nu\rho} k_3^\mu, \quad (3.5.62)$$

with the direction of momenta finally chosen so that momentum conservation reads $k_1 + k_2 + k_3 = 0$, where we have retained only the terms that survive in the limit $\ell^2 \to 0$ implying that $A^{\mu\nu\rho}(k_1, k_2, k_3)$ must be linear in the momenta (see (3.5.28) and just below it). We recall that for simplicity of notation we have set the dimensional scale parameter $\ell^2 \to 1$ in the present and in the last subsection computations, and because of dimensional reasons the coefficient of the products of three momenta in $A^{\mu\nu\rho}(k_1, k_2, k_3)$ must be proportional to ℓ^2 since $\dim k = [\text{length}]^{-1}$ and is of *higher* order.

Considering the limit $\ell \to 0$ of (3.5.61), we may rewrite the corresponding expression in a more symmetrical manner. We Make use of momentum conservation: $k_1 + k_2 + k_3 = 0$, and the properties $k_i^\mu \epsilon_{\mu\nu} = 0$, $k_i^\nu \epsilon_{\mu\nu} = 0$, for $i = 1, 2, 3$. The expression for the three-point function in (3.5.61) equally holds true for $A^{\mu\nu\rho}(k_1, k_2, k_3) \to -A^{\mu\nu\rho}(k_3, k_1, k_2)$. Accordingly, we may rewrite the three-point function for three gravitons, with the polarization tensors labeling the gravitons, as

$$A(\varepsilon_1 k_1, \varepsilon_2 k_2, \varepsilon_3 k_3) = \epsilon^1_{\mu_1\mu_2}\, \epsilon^2_{\nu_1\nu_2}\, \epsilon^3_{\rho_1\rho_2}\, V_{\mu_1\mu_2, \nu_1\nu_2, \rho_1\rho_2}(k_1, k_2, k_3), \quad (3.5.63)$$

giving rise to a three-graviton vertex

$$V_{\mu_1\mu_2,\nu_1\nu_2,\rho_1\rho_2}(k_1,k_2,k_3) = (2\pi)^D \delta^{(D)}(k_1+k_2+k_3)\frac{gc}{8}$$

$$\times \text{Sym}\Big[A^{\mu_1\nu_1\rho_1}(k_1,k_2,k_3)\,A^{\mu_2\nu_2\rho_2}(k_1,k_2,k_3)$$

$$+ A^{\mu_1\nu_1\rho_1}(k_3,k_1,k_2)\,A^{\mu_2\nu_2\rho_2}(k_3,k_1,k_2)\Big], \qquad (3.5.64)$$

where $A^{\mu\nu\rho}(k_1,k_2,k_3)$ is defined in (3.5.62), and "Sym" stands for the symmetrization operation to be applied to the expression following it over $\mu_1 \leftrightarrow \mu_2$, $\nu_1 \leftrightarrow \nu_2$, $\rho_1 \leftrightarrow \rho_2$, as a consequence of the symmetric nature of the polarization tensors in their two indices. For completeness, we have also introduced a momentum conserving delta function factor in (3.5.64).

The above expression of the three-graviton vertex will turn up to be important when dealing with the connection between string theory and Einstein's theory of gravitation in Sect. 3.6.

3.5.3   Superstrings

Consider the open superstring in the NS sector. The basic fields are given in (3.2.58), (3.3.63), and for $(\tau,\sigma) = (0,0) \equiv (0)$, we may write

$$\psi^{\mu}(0) = \sqrt{2}\sum_{r=1/2,3/2,\dots}\big(d^{\mu}(r)+d^{\mu}(r)^{\dagger}\big) = \sqrt{2}\sum_{s=\pm 1/2,\pm 3/2,\dots} d^{\mu}(s), \qquad (3.5.65)$$

$$X^{\mu}(0) = x^{\mu}+i\sum_{m\neq 0}\frac{\alpha^{\mu}(m)}{m}, \qquad \frac{\partial X^{\mu}(0)}{\partial\tau} = \sum_{m=-\infty}^{\infty}\alpha(m), \qquad (3.5.66)$$

expressed in covariant notations, where $\partial X^{\mu}(0)/\partial\tau \equiv (\partial X^{\mu}(\tau,0)/\partial\tau)\big|_{\tau=0}$. As in the previous two subsections, we have set $\ell = 1$, for the simplicity of the notation.

A massless vector particle state with momentum k, and polarization, specified by a vector e_{λ}^{μ}, is given by (see (3.3.87), (3.2.175): $e_{\lambda}^{\mu}k_{\mu} = 0$, $e_{\lambda}^{0} = 0$, $e_{\lambda}^{i}e_{\lambda'}^{i} = \delta_{\lambda\lambda'}$)

$$e_{\lambda}\cdot d(1/2)^{\dagger}|\text{NS}\rangle|0;k\rangle = e\cdot d(1/2)^{\dagger}|k\rangle \qquad (3.5.67)$$

where we have simplified the notation in writing the last equality. We may then define a corresponding fermionic vertex as

$$\frac{1}{\sqrt{2}}\,e\cdot\psi(0)\,e^{ik\cdot X(0)}. \qquad (3.5.68)$$

Given this expression of the vertex, we may generate a bosonic vertex corresponding to the vector particle (a boson) by carrying the anti-commutation relation of the former vertex with a fermion-boson exchange operator[47]

$$G(r) = \sum_{n=-\infty}^{\infty} \alpha_\mu(n) \, d^\mu(r-n), \quad \text{for all} \quad r = \pm 1/2, \pm 3/2, \ldots, \quad (3.5.69)$$

as follows. First we note that

$$\left\{ G(r), \frac{1}{\sqrt{2}} e \cdot \psi(0) \, e^{ik \cdot X(0)} \right\}$$

$$= \frac{1}{\sqrt{2}} \left\{ G(r), e \cdot \psi(0) \right\} e^{ik \cdot X(0)} - \frac{1}{\sqrt{2}} e \cdot \psi(0) \left[G(r), e^{ik \cdot X(0)} \right]. \quad (3.5.70)$$

On the other hand,

$$\left\{ G(r), e \cdot \psi(0) \right\} = \sqrt{2} \, e_\nu \sum_{n=-\infty}^{\infty} \left(\sum_{s=\pm 1/2, \pm 3/2, \ldots} \alpha_\mu(n) \left\{ d^\mu(r-n), d^\nu(s) \right\} \right)$$

$$= \sqrt{2} \, e_\nu \sum_{n=-\infty}^{\infty} \left(\sum_{s=\pm 1/2, \pm 3/2, \ldots} \alpha_\mu(n) \, \eta^{\mu\nu} \, \delta(n, r-s) \right) = \sqrt{2} \, e \cdot \frac{\partial X(0)}{\partial \tau},$$

$$(3.5.71)$$

for all $r = \pm 1/2, \pm 3/2, \ldots$. Also,

$$[G(r), X^\mu(0)] = i \sum_{n=-\infty}^{\infty} \sum_{m \neq 0} \frac{1}{m} [\alpha_\nu(n), \alpha^\mu(m)] \, d^\nu(r-n)$$

$$= -i \sum_{s=\pm 1/2, \pm 3/2, \ldots} d^\mu(s) = -\frac{i}{\sqrt{2}} \psi^\mu(0), \quad (3.5.72)$$

for all $r = \pm 1/2, \pm 3/2, \ldots$. From (3.5.69), we then readily obtain

$$\left\{ G(r), \frac{1}{\sqrt{2}} e \cdot \psi(0) \, e^{ik \cdot X(0)} \right\} = \left(e \cdot \frac{\partial X(0)}{\partial \tau} - \frac{1}{2} e \cdot \psi(0) \, k \cdot \psi(0) \right) e^{ik \cdot X(0)},$$

$$(3.5.73)$$

for all r as above.

---

[47]This operator was introduced in [32]. See also [5].

An interesting application of the above vertex is to determine the three-point function involving three massless vector particles. The amplitude, as in the bosonic string case, now applied to the vertex in question, is given by

$$A(k_1 e_1, k_2 e_2, k_3 e_3) = \langle k_3 | \{ . \} | k_1 \rangle, \qquad (3.5.74)$$

$$\{ . \} = e_3 \cdot d(1/2) \left(e_2 \cdot \frac{\partial X(0)}{\partial \tau} - \frac{1}{2} \, e_2 \cdot \psi(0) \, k_2 \cdot \psi(0) \right) e^{i k_2 \cdot X(0)} \, e_1 \cdot d(1/2)^\dagger,$$
$$(3.5.75)$$

up to a coupling constant. By using the fact that $\alpha^{\,i}(n) | k \rangle = 0$, and the equality

$$e_2 \cdot \frac{\partial X(0)}{\partial \tau} = e_2 \cdot p + \sum_{n=1}^{\infty} \left(e_2 \cdot \alpha(n) + e_2 \cdot \alpha(n)^\dagger \right), \qquad (3.5.76)$$

it is easily seen, using in the process the fact that $k_2 \cdot e_2 = 0$, that only the $e_2 \cdot p$ term survives in the latter equation, and only the term $e^{i k_2 \cdot x}$ survives in the normal-ordered expression for $e^{i k_2 \cdot X(0)}$, when they are applied within the expression for the amplitude in (3.5.74), (3.5.75). Thus the expression for the amplitude reduces to the evaluation of

$$\langle k_3 | e_3 \cdot d(1/2) \Big(e_2 \cdot p - e_2 \cdot d(1/2) \, k_2 \cdot d(1/2)^\dagger$$
$$- e_2 \cdot d(1/2)^\dagger \, k_2 \cdot d(1/2) \Big) e^{i k_2 \cdot x} \, e_1 \cdot d(1/2)^\dagger | k_1 \rangle,$$
$$(3.5.77)$$

and the following final expression readily emerges for it[48]

$$A(k_1 e_1, k_2 e_2, k_3 e_3) = g \, e_{1 \mu_1} e_{2 \mu_2} e_{3 \mu_3} \big[\eta^{\mu_1 \mu_2} (k_1 - k_2)^{\mu_3} + \eta^{\mu_3 \mu_1} (k_3 - k_1)^{\mu_2}$$
$$+ \eta^{\mu_2 \mu_3} (k_2 - k_3)^{\mu_1} \big], \qquad (3.5.78)$$

up to a momentum conserving delta function, with $k_1 + k_2 + k_3 = 0$, where the direction of momenta were finally chosen so that momentum conservation reads as just given. We have also used the covariant anti-commutation relation $\{ d^\mu(r), d^\nu(r')^\dagger \} = \eta^{\mu\nu} \delta(r, r')$, the fact that $\exp[i k x]$ involves a momentum translation, and introduced a coupling parameter g.

---

[48]Note that this expression coincides with, or more precisely is proportional to, the one in (3.5.26)/(3.5.27), (3.5.28), in bosonic string theory, in the limit $\ell \to 0$ for the latter. See Problem 3.24. In the supersymmetric case, however, there is no term non-linear in the momenta in the corresponding three-point function and no limit $\ell \to 0$ is to be taken.

Remark Here we reach a point of importance. We easily check that the three-point function $A(k_1 e_1, k_2 e_2, k_3 e_3)$ is anti-symmetric in the exchange of any pairs of variables $(e_i, k^i) \leftrightarrow (e_j, k^j)$ for a theory meant to be dealing with bosons !. Thus the only way such a three-point function can contribute to a theory involving massless vector bosons is to consider linear combinations of such amplitudes with corresponding *anti*-symmetric coefficients. This is achieved by attaching an additional label (a Chan-Paton factor) to the polarization vectors $e^a, a = 1, \ldots, N$, introduce totally anti-symmetric constants f_{abc} and consider such a linear combination

$$\mathscr{A}(k_1 e_1, k_2 e_2, k_3 e_3) = e_{1\mu_1}^{a_1} e_{2\mu_2}^{a_2} e_{3\mu_3}^{a_3} V_{a_1 a_2 a_3}^{\mu_1 \mu_2 \mu_3}(k_1, k_2, k_3), \qquad (3.5.79)$$

giving rise to a three-vector-boson vertex

$$V_{a_1 a_2 a_3}^{\mu_1 \mu_2 \mu_3}(k_1, k_2, k_3) = (2\pi)^D \delta^D(k_1 + k_2 + k_3)$$

$$\times i g f_{a_1 a_2 a_3} \left[\eta^{\mu_1 \mu_2}(k_1 - k_2)^{\mu_3} + \eta^{\mu_3 \mu_1}(k_3 - k_1)^{\mu_2} + \eta^{\mu_2 \mu_3}(k_2 - k_3)^{\mu_1} \right], \tag{3.5.80}$$

where the overall phase factor "i" is introduced for convenience to ensure reality in x-space. The above three-point function $\mathscr{A}(k_1 e_1, k_2 e_2, k_3 e_3)$ for bosons, with necessarily totally anti-symmetric expansion coefficients f_{abc}, satisfies the Bose character of the vector particles as it should. Are we on the verge in rediscovering the Yang-Mills field theory from string theory, where the f_{abc} are so-called structure constants? This is the subject of Sect. 3.7. For completeness, we have also introduced a momentum conserving delta function factor in (3.5.80).

3.6   From String Theory to Einstein's Theory of Gravitation

We recall that string theory, for closed strings, gives rise to the graviton, a spin 2 massless particle, in their mass spectra. The graviton arises directly from string theory and was not, a priori, assumed to exist. String theory, among other things, also gave rise to a mathematical expression for the three-graviton vertex in (3.5.64), (3.5.62) for $\ell \to 0$. Given these two facts, that is a spin two massless particle in D dimensions, and the mathematical expression of the three-graviton vertex, as obtained in the just mentioned equation, would they lead to Einstein's theory of gravitation for $\ell \to 0$? The purpose of this section is involved, rather formally, with the simplest possible analysis of the underlying problem leaving a detailed dynamical analysis for a more advanced treatment. Let us see what happens. In this section all Greek indices go over from 0 to $D - 1$.

For the graviton description, we start from the first order formulation in Sect. 1.3.2, adapted readily to D dimensions. The corresponding action is given by [see (1.3.36)]

$$W = \frac{1}{\kappa^2} \int (dx) \left[\kappa \, \xi^{\mu\nu} (\partial_\mu \Gamma_{\rho\nu}{}^\rho - \partial_\rho \Gamma_{\mu\nu}{}^\rho) \right.$$
$$\left. + \eta^{\mu\nu} (\Gamma_{\lambda\rho}{}^\rho \Gamma_{\mu\nu}{}^\lambda - \Gamma_{\lambda\mu}{}^\rho \Gamma_{\rho\nu}{}^\lambda + \partial_\rho \Gamma_{\mu\nu}{}^\rho - \partial_\mu \Gamma_{\rho\nu}{}^\rho) \right], \tag{3.6.1}$$

where we have introduced a parameter κ, and its connection with the string coupling parameter will be made. The key equations obtained before, for spin 2, massless particles, now adapted to D dimensions of spacetime and with a parameter κ introduced, simply read [see (1.3.38)–(1.3.43)]

$$\Gamma_\mu{}^{\mu\beta} = \kappa \, \partial_\mu \, \xi^{\mu\beta} = \kappa \, \partial_\mu \, (h^{\mu\beta} - \frac{1}{2} \eta^{\mu\beta} h), \tag{3.6.2}$$

$$\Gamma_{\gamma\mu}{}^\mu = -\frac{\kappa}{D-2} \partial_\gamma \xi = \frac{\kappa}{2} \partial_\gamma h, \tag{3.6.3}$$

where

$$h_{\alpha\beta} = \xi_{\alpha\beta} - \frac{1}{D-2} \eta_{\alpha\beta} \xi, \qquad \xi_{\alpha\beta} = h_{\alpha\beta} - \frac{1}{2} \eta_{\alpha\beta} h, \tag{3.6.4}$$

$$\xi \equiv \xi^\mu{}_\mu, \qquad h^\alpha{}_\alpha \equiv h = -\frac{2}{D-2} \xi, \tag{3.6.5}$$

and we have used the fact that $\eta^\mu{}_\mu = D$,

$$\Gamma_{\alpha\beta\gamma} = \frac{\kappa}{2} (\partial_\alpha h_{\beta\gamma} + \partial_\beta h_{\alpha\gamma} - \partial_\gamma h_{\alpha\beta}). \tag{3.6.6}$$

Upon varying the action in (3.6.1), with respect to $\xi^{\mu\nu}$, and using, in the process, (3.6.5), (3.6.6), we obtain

$$-\Box \, \xi_{\sigma\lambda} + \partial_\rho \partial_\sigma \, \xi_\lambda{}^\rho + \partial_\rho \partial_\lambda \, \xi_\sigma{}^\rho - \eta_{\sigma\lambda} \partial_\alpha \partial_\beta \, \xi^{\alpha\beta} = 0, \tag{3.6.7}$$

$$\Box \, \xi + (D-2) \, \partial_\sigma \partial_\lambda \, \xi^{\sigma\lambda} = 0. \tag{3.6.8}$$

As in Sect. 1.3.2, we may use a gauge to transform the previous field to a field $\xi^{\mu\nu}$, using, for simplicity, the same notation, satisfying $\Box \, \xi^{\mu\nu} = 0$, $\partial_\mu \xi^{\mu\nu} = 0$, $\partial_\nu \xi^{\mu\nu} = 0$, $\xi^\mu{}_\mu = 0$, corresponding to $D(D-3)/2$ polarization states [see (3.2.189)], and $h_{\mu\nu}$ in the expression for $\Gamma_{\alpha\beta\gamma}$ in (3.6.6) becomes simply replaced by $\xi_{\mu\nu}$.

An important relation then follows from the expression for $\Gamma_{\alpha\beta\gamma}$ in (3.6.6), given by (see Problem 3.25)

$$-\frac{1}{\kappa}\,\xi^{\mu\nu}\left[\Gamma_{\lambda\rho}{}^{\rho}\Gamma_{\mu\nu}{}^{\lambda} - \Gamma_{\lambda\mu}{}^{\rho}\Gamma_{\rho\nu}{}^{\lambda}\right]$$

$$= -\frac{\kappa}{4}\,\xi^{\mu\nu}\left[\xi^{\lambda\rho}\,\partial_{\mu}\partial_{\nu}\,\xi_{\lambda\rho} + \partial_{\mu}\xi_{\lambda\rho}\,\partial^{\rho}\xi^{\lambda}{}_{\nu} + \partial_{\nu}\xi_{\lambda\rho}\,\partial^{\rho}\xi^{\lambda}{}_{\mu}\right], \tag{3.6.9}$$

up to a total derivative.

We recall the expression of the three-graviton vertex obtained in (3.5.64) from string theory

$$V^{\mu_1\mu_2,\nu_1\nu_2,\rho_1\rho_2}(k_1,k_2,k_3) = (2\pi)^D\delta^{(D)}(k_1+k_2+k_3)\,\frac{g_c}{8}\times$$

$$\text{Sym}\left[A^{\mu_1\nu_1\rho_1}(k_1,k_2,k_3)A^{\mu_2\nu_2\rho_2}(k_1,k_2,k_3)+A^{\mu_1\nu_1\rho_1}(k_3,k_1,k_2)A^{\mu_2\nu_2\rho_2}(k_3,k_1,k_2)\right], \tag{3.6.10}$$

$$A^{\mu\nu\rho}(k_1,k_2,k_3) = \eta^{\mu\rho}k_1^{\nu} + \eta^{\mu\nu}k_2^{\rho} + \eta^{\nu\rho}k_3^{\mu}, \tag{3.6.11}$$

where recall that "Sym" stands for the symmetrization operation to be applied to the expression following it over $\mu_1 \leftrightarrow \mu_2$, $\nu_1 \leftrightarrow \nu_2$, $\rho_1 \leftrightarrow \rho_2$.

As we will see below the vertex part $V_{\mu_1\mu_2,\nu_1\nu_2,\rho_1\rho_2}(k_1,k_2,k_3)$ may be obtained from the interaction Lagrangian density

$$\mathcal{L}_1(x) = -\frac{1}{\kappa}\,\xi^{\mu\nu}(x)\left[\Gamma_{\lambda\rho}{}^{\rho}(x)\Gamma_{\mu\nu}{}^{\lambda}(x) - \Gamma_{\lambda\mu}{}^{\rho}(x)\Gamma_{\rho\nu}{}^{\lambda}(x)\right], \tag{3.6.12}$$

provided we choose $\kappa = g_c/2$ as a coupling parameter, where κ is defined in D dimensional spacetime. The above expression is given in (3.6.9) involving the product of three fields and two derivatives, in conformity with the expression in (3.6.10) in x-space.

We introduce the Fourier transforms of the fields $\xi^{\mu\nu}(x)$

$$\xi^{\mu\nu}(x) = \int \frac{(dk)}{(2\pi)^D}\,e^{ikx}\,\xi^{\mu\nu}(k), \tag{3.6.13}$$

From (3.6.12), (3.6.9), we obtain,

$$\int (dx)\,\mathcal{L}_1(x) = \frac{\kappa}{4}\int (dx)\,e^{i(k_1+k_2+k_3)x}\,\frac{(dk_1)}{(2\pi)^D}\frac{(dk_2)}{(2\pi)^D}\frac{(dk_3)}{(2\pi)^D}$$

$$\times\left[k_{3\mu}\,k_{3\nu}\,\eta_{\sigma\lambda}\,\eta_{\gamma\rho} + k_{2\mu}\,k_{3\rho}\,\eta_{\sigma\lambda}\,\eta_{\gamma\nu} + k_{2\nu}\,k_{3\rho}\,\eta_{\sigma\lambda}\,\eta_{\gamma\mu}\right]\xi^{\mu\nu}(k_1)\,\xi^{\lambda\rho}(k_2)\,\xi^{\sigma\gamma}(k_3). \tag{3.6.14}$$

The three-graviton vertex $\widetilde{V}_{\mu_1\nu_1,\mu_2\nu_2,\mu_3\nu_3}(k_1,k_2,k_3)$, corresponding to the interaction Lagrangian density $\mathcal{L}_1$ in (3.6.12) is defined by:

$$\widetilde{V}_{\mu_1\mu_1,\nu_1\nu_2,\rho_1\rho_2}(k_1,k_2,k_3)$$

$$= (2\pi)^{(3D)}\frac{\delta}{\delta\xi^{\rho_1\rho_2}(k_3)}\frac{\delta}{\delta\xi^{\nu_1\nu_2}(k_2)}\frac{\delta}{\delta\xi^{\mu_1\mu_1}(k_1)}\int (dx)\,\mathcal{L}_1(x). \qquad (3.6.15)$$

With the functional derivative defined by

$$\frac{\delta\xi^{\mu\nu}(k_1)}{\delta\xi_{\mu_1\mu_2}(k_2)} = \frac{1}{2}\left[\eta^{\mu\mu_1}\eta^{\nu\mu_2}+\eta^{\mu\mu_2}\eta^{\nu\mu_1}\right]\delta^D(k_1-k_2), \qquad (3.6.16)$$

a straightforward application[49] of the functional differentiations in (3.6.15) to the expression in (3.6.14), shows that $\widetilde{V}^{\mu_1\mu_2,\nu_1\nu_2,\rho_1\rho_3}(k_1,k_2,k_3)$ is precisely given by $V^{\mu_1\mu_2,\nu_1\nu_2,\rho_1\rho_3}(k_1,k_2,k_3)$ in (3.6.10), with κ identified with the string coupling parameter through the relation $\kappa = g_c/2$.

The above analysis suggests to add the integral of the interaction term in (3.6.12),

$$\int (dx)\,\mathcal{L}_1(x) = -\frac{1}{\kappa}\int (dx)\,\xi^{\mu\nu}(x)\left[\Gamma_{\lambda\rho}{}^\rho(x)\Gamma_{\mu\nu}{}^\lambda(x)-\Gamma_{\lambda\mu}{}^\rho(x)\Gamma_{\rho\nu}{}^\lambda(x)\right], \qquad (3.6.17)$$

to the action in (3.6.1) as an iteration. Although the coefficient of $\eta^{\mu\nu}$, in the action in (3.6.1), contains two total-divergence terms they will not be neglected as we note that the coefficients of $-\kappa\,\xi^{\mu\nu}$ and of $\eta^{\mu\nu}$, of the resulting new action, are identical and may be combined. Most importantly, the fields $\xi^{\mu\nu}$ and $\Gamma_{\alpha\beta}{}^\gamma$ are now subject to the variation of the new action and hence are automatically modified. All told, the action becomes:

$$W = \frac{1}{\kappa^2}\int (dx)\,(\eta^{\mu\nu}-\kappa\,\xi^{\mu\nu})(\partial_\rho\Gamma_{\mu\nu}{}^\rho - \partial_\mu\Gamma_{\rho\nu}{}^\rho + \Gamma_{\lambda\rho}{}^\rho\Gamma_{\mu\nu}{}^\lambda - \Gamma_{\lambda\mu}{}^\rho\Gamma_{\rho\nu}{}^\lambda). \qquad (3.6.18)$$

We set

$$(\eta^{\mu\nu}-\kappa\,\xi^{\mu\nu}) \equiv \tilde{g}^{\mu\nu}, \qquad (3.6.19)$$

and introduce its inverse $\tilde{g}_{\mu\rho}$ as well as their determinants:

$$\tilde{g}_{\mu\rho}\tilde{g}^{\rho\nu} = \delta_\mu{}^\nu, \qquad \det[\tilde{g}_{\mu\rho}] \equiv \tilde{g}, \qquad \det[\tilde{g}^{\mu\rho}] = \frac{1}{\tilde{g}}, \qquad (3.6.20)$$

with $\tilde{g}$ not to be confused with the contraction of the indices in $\tilde{g}_\mu{}^\nu$.

---

[49]See Problem 3.26.

Finally we introduce a matrix with elements $g_{\mu\nu}$ defined by

$$g_{\mu\nu} = |\tilde{g}|^{-1/(D-2)}\tilde{g}_{\mu\nu}, \qquad \det[g_{\mu\nu}] \equiv g, \qquad (3.6.21)$$

from which

$$|g| = |\tilde{g}|^{-2/(D-2)}, \quad g_{\mu\rho}\tilde{g}^{\rho\nu} = |\tilde{g}|^{-1/(D-2)}\delta_\mu{}^\nu, \quad \tilde{g}^{\rho\mu} = |\tilde{g}|^{-1/(D-2)}g^{\rho\mu}.$$
$$(3.6.22)$$

That is,

$$\tilde{g}^{\mu\nu} = \sqrt{|g|}\,g^{\mu\nu}. \qquad (3.6.23)$$

Accordingly, the action in (3.6.18) may be rewritten as

$$W = \frac{1}{\kappa^2}\int (\mathrm{d}x)\sqrt{|g|}\,g^{\mu\nu}(\partial_\rho\Gamma_{\mu\nu}{}^\rho - \partial_\mu\Gamma_{\rho\nu}{}^\rho + \Gamma_{\lambda\rho}{}^\rho\Gamma_{\mu\nu}{}^\lambda - \Gamma_{\lambda\mu}{}^\rho\Gamma_{\rho\nu}{}^\lambda). \qquad (3.6.24)$$

We recognize the coefficient of $\sqrt{|g|}\,g^{\mu\nu}$, in the integrand, as the Ricci tensor $R_{\mu\nu}$. As mentioned above, now all the fields are subject to the variation of this new action, and lead to the Einstein equation $R_{\mu\nu} = 0$, with $\Gamma_{\alpha\beta}{}^\delta$ having the familiar expression. The analysis was based on the expression of the Lagrangian density of free gravitons, with action given in (3.6.1), and an interaction term which gives rise to the three-graviton vertex as obtained from *string theory*, followed by an iteration, and no principle of equivalence or general covariance, were used to obtain the expression for the effective action in (3.6.24). No attempt will be made, in this introductory presentation, however, to consider supergravity in the light of superstring theory.

3.7 From String Theory to the Yang and Mills-Field Theory

We have seen that string theory gives rise naturally to vector massless particles associated with modes of vibrations of some of the strings. Superstring theory, in particular, also gives rise to a mathematical expression for the three-vector-particle vertex for massless particles in (3.5.76). Would these two facts lead to the Yang-Mills field theory? The purpose of this section is involved, rather formally, with the simplest possible analysis of the underlying problem leaving a detailed dynamical analysis for a more advanced treatment. As in the previous section, all Greek indices go over from 0 to $D - 1$.

For the description of the massless vector particle, we start from a first order formulation of free massless vector particles in D dimensions.

As we have witnessed, in particular, in (3.5.80), in order for a three-point function, giving the interaction of massless vector bosons, to contribute, it necessitates to perform linear combinations of such three-point functions with anti-symmetric constants f_{abc}, consistent with Spin & Statistics.

The action of free massless vector particles may be defined in terms of two sets of fields $A_{a\mu}$, and $F_{a\mu\nu}$ as follows

$$W = \int (\mathrm{d}x) \left[-\frac{1}{2} F_{a\mu\nu} (\partial^\mu A_a^\nu - \partial^\nu A_a^\mu) + \frac{1}{4} F_{a\mu\nu} F_a^{\mu\nu} \right], \qquad (3.7.1)$$

in D dimensions, with a summation over a multiplicity index a understood.

The formal field equations resulting from the action are easily obtained to be

$$\partial^\mu F_{a\mu\nu} = 0, \qquad F_{a\mu\nu} = \partial_\mu A_{a\nu} - \partial_\nu A_{a\mu}. \qquad (3.7.2)$$

The *string theory* three-vector-particle vertex as obtained in (3.5.80) reads

$$V^{\mu_1\mu_2\mu_3}_{a_1 a_2 a_3}(k_1, k_2, k_3) = (2\pi)^D \delta^{(D)}(k_1 + k_2 + k_3)$$
$$\times \, \mathrm{i} g f_{a_1 a_2 a_3} \left[\eta^{\mu_1\mu_2}(k_1 - k_2)^{\mu_3} + \eta^{\mu_3\mu_1}(k_3 - k_1)^{\mu_2} + \eta^{\mu_2\mu_3}(k_2 - k_3)^{\mu_1} \right]. \qquad (3.7.3)$$

As we will see below, given the fields $A_{a\mu}$ and $F_{b\mu\nu}$, an interaction term involving the product of these fields which leads to the vertex part in (3.7.3) is simply given by

$$\mathscr{L}_1(x) = -\frac{1}{2} g f_{abc} F_{a\mu\nu}(x) A_b^\mu(x) A_c^\nu(x). \qquad (3.7.4)$$

The latter may be equivalently rewritten as

$$\mathscr{L}_1(x) = -g f_{abc} \partial_\mu A_{a\nu}(x) A_b^\mu(x) A_c^\nu(x), \qquad (3.7.5)$$

involving only *one* derivative in conformity with the expression in (3.7.3), in x-space.

The vertex part $\widetilde{V}^{\mu_1\mu_2\mu_3}_{a_1 a_2 a_3}(k_1, k_2, k_3)$ corresponding to this interaction is defined by

$$\widetilde{V}^{\mu_1\mu_2\mu_3}_{a_1 a_2 a_3}(k_1, k_2, k_3) = (2\pi)^{(3D)} \frac{\delta}{\delta A_{a_3\mu_3}(k_3)} \frac{\delta}{\delta A_{a_2\mu_2}(k_2)} \frac{\delta}{\delta A_{a_1\mu_1}(k_1)} \int (\mathrm{d}x) \, \mathscr{L}_1(x). \qquad (3.7.6)$$

We introducing the Fourier transform of the fields

$$A_{a\mu}(x) = \int \frac{(\mathrm{d}p)}{(2\pi)^D} \, \mathrm{e}^{\mathrm{i}px} A_{a\mu}(p), \qquad (3.7.7)$$

with functional derivatives of the fields defined by

$$\frac{\delta A_{a\mu}(p)}{\delta A_{b\nu}(k)} = \delta_{ab} \delta_\mu^{\ \nu} \delta^{(D)}(p - k). \qquad (3.7.8)$$

The expression for $\widetilde{V}^{\mu_1 \mu_2 \mu_3}_{a_1 a_2 a_3}(k_1, k_2, k_3)$ in (3.7.6), is then readily carried out. Clearly, it will contain six terms. Up to the $(2\pi)^D \delta^{(D)}(k_1 + k_2 + k_3)$ factor obtained by integrating over x, the first term in (3.7.6) is given by

$$- i g f_{abc} [k_{1\mu} \delta^{\mu_1}{}_\nu \delta_{aa_1}] [\eta^{\mu\mu_2} \delta_{ba_2}] [\eta^{\nu\mu_3} \delta_{ca_3}]. \quad (*)$$

A straightforward application of the functional derivatives in (3.7.6), by using, in the process, the anti-symmetry of f_{abc}, then gives precisely

$$\widetilde{V}^{\mu_1 \mu_2 \mu_3}_{a_1 a_2 a_3}(k_1, k_2, k_3) = (2\pi)^D \delta^D(k_1 + k_2 + k_3)$$

$$\times i g f_{a_1 a_2 a_3} \Big[\eta^{\mu_1 \mu_2}(k_1 - k_2)^{\mu_3} + \eta^{\mu_3 \mu_1}(k_3 - k_1)^{\mu_2} + \eta^{\mu_2 \mu_3}(k_2 - k_3)^{\mu_1} \Big].$$

$$(3.7.9)$$

For example, note that the term just given in (*) corresponds to the term $\eta^{\mu_3 \mu_1}(-k_1^{\mu_2})$ inside the square brackets in (3.7.9). The vertex part $\widetilde{V}^{\mu_1 \mu_2 \mu_3}_{a_1 a_2 a_3}(k_1, k_2, k_3)$ coincides with the one in (3.7.3)

The above analysis, in turn, suggests to add the integral of the interaction term in (3.7.4) to the action in (3.7.1) as an iteration. Here we note that the fields $F_{a\mu\nu}$ and $A_{a\nu}$ are now subject to the variation of the new action and hence are automatically modified. All told, the action becomes:

$$W = \int (\mathrm{d}x) \Big[-\frac{1}{2} F_{a\mu\nu} (\partial^\mu A_a^\nu - \partial^\nu A_a^\mu + g f_{abc} A_b^\mu A_c^\nu) + \frac{1}{4} F_{a\mu\nu} F_a^{\mu\nu} \Big]. \qquad (3.7.10)$$

The field equations that now follow are

$$\partial^\mu F_{c\mu\nu} = g F_{a\mu\nu} A_b^\mu f_{abc}, \qquad F_{a\mu\nu} = \partial_\mu A_{a\nu} - \partial_\nu A_{a\mu} + g f_{abc} A_{b\mu} A_{c\nu},$$

$$(3.7.11)$$

recognizing the Yang-Mills field theory equations,[50] where we have used, in the process, the anti-symmetry property of f_{abc}. The analysis was based on the expression of Lagrangian density of free gluons and an interaction term which gives rise to the three-vector-particle vertex as obtained from *string theory*, followed by an iteration. Obviously no non-abelian gauge invariance argument was used to obtain the expression for the effective action in (3.7.10).

---

[50]In earlier chapters we have conveniently denoted $F_{a\mu\nu}$ by $G_{a\mu\nu}$.

Appendix A: Summary of the Expressions for M^2 for Bosonic Strings

For open bosonic strings we have

$$\alpha' M^2 = \left(\sum_{m=1}^{\infty} \sum_{i=1}^{24} \alpha^i(m)^\dagger \alpha^i(m) - 1 \right), \qquad (\text{A-3.1})$$

as derived in (3.2.96), and for the closed ones

$$\frac{\alpha'}{4} M^2 = (N_R - 1) = (N_L - 1), \qquad (\text{A-3.2})$$

$$N_R = \sum_{m=1}^{\infty} \sum_{i=1}^{24} \alpha^i(m)^\dagger \alpha^i(m), \quad N_L = \sum_{m=1}^{\infty} \sum_{i=1}^{24} \bar{\alpha}^i(m)^\dagger \bar{\alpha}^i(m), \qquad (\text{A-3.3})$$

as derived in (3.2.125), (3.2.126).

If an extra spatial dimension, say, X^{25} is compactified into a circle of radius R, the expression for M^2 takes the form

$$\frac{\alpha'}{4} M^2 = \alpha' p_L^2 + (N_L - 1) = \alpha' p_R^2 + (N_R - 1), \qquad (\text{A-3.4})$$

as derived in (3.2.152)–(3.2.154), where p_R, p_L are defined in (3.2.145), $p^{25} = k/R$. The latter equation may be spelled out in detail as

$$N_L = \sum_{n=1}^{\infty} \sum_{i=1}^{23} \left(\bar{\alpha}^i(n)^\dagger \bar{\alpha}^i(n) + \bar{\alpha}(n)^\dagger \bar{\alpha}(n) \right), \qquad (\text{A-3.5})$$

$$N_R = \sum_{n=1}^{\infty} \sum_{i=1}^{23} \left(\alpha^i(n)^\dagger \alpha^i(n) + \alpha(n)^\dagger \alpha(n) \right), \qquad (\text{A-3.6})$$

$$M^2 = \frac{k^2}{R^2} + \frac{w^2 R^2}{\alpha'^2} + \frac{2}{\alpha'} (N_L + N_R - 2). \qquad (\text{A-3.7})$$

$$N_R - N_L = kw. \qquad (\text{A-3.8})$$

The zero-mass modes of the bosonic strings are as follows:

Oriented Bosonic String

Open String
A vector particle A^i with 24 number of states.

Closed String
A scalar (dilaton), with one state, an anti-symmetric tensor field B^{ij} (Kalb-Raymond field), with 276 number of states, and a symmetric traceless tensor field G^{ij}—the graviton, with 299 number of states. The total number of states being $1 + 276 + 299 = (24)^2$.

Unoriented Bosonic String

Open String (With Chan-Paton factors):
A vector particle, with $(24 \times N(N-1)/2)$ number of states, corresponding to gauge group SO(N).

Closed String
A scalar (dilaton), with 1 state, and a traceless symmetric tensor field (graviton), with 299 number of states.

Appendix B: Moving on a Worldsheet and Translations Operations

We derive the expressions for translation operators on the worldsheet for open and closed strings as we have encountered, for example, in bosonic string theory. These turn up to be quite important in scattering theory in Sect. 3.5.

Open Strings

The canonical conjugate momentum densities to the fields X^μ are given from the Lagrangian density in (3.2.55) to be $P^\mu = T\dot{X}^\mu$, from which the Hamiltonian is given by[51]

$$H = \int_0^\pi d\sigma \, (\dot{X} \cdot P - \mathscr{L}) = \frac{T}{2} \int_0^\pi d\sigma \, (\dot{X}^2 + X'^2) = \frac{1}{2} \sum_n \alpha^i(-n)\alpha^i(n),$$
(B-3.1)

where we have used the identity $\ell^2 \pi T = 1$ in (3.2.46), and (3.2.62). We make use of the expression for X^i, X^+ in (3.2.56), (3.2.58), respectively, that

$$X^{D-1} = \frac{(X^+ - X^-)}{\sqrt{2}}, \quad X^0 = \frac{(X^+ + X^-)}{\sqrt{2}},$$

and that $X'^+ = 0$, and hence

$$\partial_\alpha X^{D-1} \partial^\alpha X^{D-1} - \partial_\alpha X^0 \partial^\alpha X^0 = -\dot{X}^- \dot{X}^+ - \dot{X}^+ \dot{X}^-.$$
(B-3.2)

---

[51]With τ taken to be dimensionless, H is the Hamiltonian times a unit of time.

We also use the identities

$$\alpha^i(0) = \ell^2 p^i, \quad \alpha^\pm(0) = \ell^2 p^\pm, \quad \alpha^+(n) = 0, \quad \text{for } n \neq 0,$$

[see (3.2.56), (3.2.64)], the expression for $\partial_\tau X^-$ in (3.2.57), and finally that

$$\sum_{n \neq 0} \alpha^i(-n)\alpha^i(n) = 2 \left(\sum_{n=1}^\infty \alpha(n)^\dagger \alpha(n) - 1 \right),$$

[see (3.2.89)–(3.2.92), (3.2.96)], to obtain

$$H = \alpha'(p^2 + M^2). \tag{B-3.3}$$

Two things should be noted about this expression. From (3.2.66), we note that for a given particle state $|\varphi\rangle$, in the spectrum of the string, i.e., in particular, on the mass shell $(p^2 = -M^2)$,

$$(p^2 + M^2)|\varphi\rangle = 0. \tag{B-3.4}$$

If one compares this with the ordinary Klein-Gordon equation $(p^2 + m^2)|\varphi\rangle = 0$, which accounts for the exchange of only one particle of mass m via the corresponding propagator $(p^2 + m^2 - i\epsilon)^{-1}$, we see that in the string theory case, we have the equivalence of the exchange of an infinite number particles, in the the spectrum of the string, via a corresponding propagator (operator) given by

$$\Delta = \frac{1/\alpha'}{p^2 + M^2 - i\epsilon}. \tag{B-3.5}$$

We also note that the translation of an operator $\mathcal{O}$ through proper time τ may be defined by

$$\mathcal{O}(\tau) = \exp(i\tau\alpha'(p^2 + M^2))\,\mathcal{O}\exp(-i\tau\alpha'(p^2 + M^2)). \tag{B-3.6}$$

Closed Strings

A similar analysis as for an open string, gives the following expression for the Hamiltonian of a closed string[52]

$$H = \sum_n \left(\alpha(-n)^i \alpha(n)^i + \bar{\alpha}(-n)^i \bar{\alpha}(n)^i \right), \tag{B-3.7}$$

---

[52]Bailin and Love [5].

where $\alpha^i(n), \bar{\alpha}^i(n)$ are associated with right X_R, and left movers X_L, respectively. The relevant equations to consider now are (3.2.45), (3.2.48), (3.2.70), (3.2.71), (3.2.75), (A-3.2), (A-3.3). Here one exploits the independence of the right and left movers. A straightforward analysis, following the derivation of the corresponding case for open strings above in Problem 3.12, gives with

$$M^2 = M_R^2 + M_L^2,$$

the expression

$$H = \alpha' \left[\frac{1}{2}(p^2 + M_R^2) + \frac{1}{2}(p^2 + M_L^2) \right], \tag{B-3.8}$$

$$\frac{\alpha'}{2} M_R^2 = \sum_{n=1}^{\infty} \alpha^i(n)^\dagger \alpha^i(n) - 1, \qquad \frac{\alpha'}{2} M_L^2 = \sum_{n=1}^{\infty} \bar{\alpha}^i(n)^\dagger \bar{\alpha}^i(n) - 1. \tag{B-3.9}$$

Here for a particle state $|\varphi\rangle$, in the spectrum of the string, we have, according to (A-3.2), the constraint (on mass-shell)

$$(M_R^2 - M_L^2 |\varphi\rangle = 0. \tag{B-3.10}$$

Of particular interest is the translation of right and left movers operators, which, in particular and by definition, are functions of $\tau \mp \sigma$, respectively, defined as follows

$$\mathcal{O}_{R/L}(\tau \mp \sigma)$$

$$= \exp\left(i(\tau \mp \sigma)\alpha'(p^2 + M_{R/L}^2) \right) \mathcal{O}_{R/L}(0) \exp\left(-i(\tau \mp \sigma)\alpha'(p^2 + M_{R/L}^2) \right). \tag{B-3.11}$$

Appendix C: Summary of the Expressions for M^2 for Superstrings

∇ Open Superstrings

□ NS-*Sector*

$$\alpha' M^2 = \sum_{m=1}^{\infty} \alpha^i(m)^\dagger \alpha^i(m) \quad + \sum_{r=1/2,3/2,\ldots} r\, d^i(r)^\dagger d^i(r) \quad - \frac{1}{2}, \tag{C-3.1}$$

derived in (3.3.86)/(3.3.105). □

□ R-*Sector*

$$\alpha' M^2 = \sum_{m=1}^{\infty} \left(\alpha^i(m)^\dagger \alpha^i(m) + m \, d^i(m)^\dagger d^i(m) \right), \tag{C-3.2}$$

derived in (3.3.88). □

∇ Closed Superstrings

□ (N, N)-*Sector*

$$\frac{\alpha'}{4} M^2 = N_L - \frac{1}{2} = N_R - \frac{1}{2}, \tag{C-3.3}$$

$$N_L = \sum_{m=1}^{\infty} \bar{\alpha}^i(m)^\dagger \bar{\alpha}^i(m) \quad + \sum_{r=1/2,3/2,\ldots} r \, \bar{b}^i(r)^\dagger \bar{b}^i(r), \tag{C-3.4}$$

$$N_R = \sum_{m=1}^{\infty} \alpha^i(m)^\dagger \alpha^i(m) \quad + \sum_{r=1/2,3/2,\ldots} r \, b^i(r)^\dagger b^i(r), \tag{C-3.5}$$

derived in (3.3.123). □

□ (NS$_L$, R$_R$)-*Sector*

$$\frac{\alpha'}{4} M^2 = N_L - \frac{1}{2} = N_R, \tag{C-3.6}$$

$$N_L = \sum_{m=1}^{\infty} \bar{\alpha}^i(m)^\dagger \bar{\alpha}^i(m) \quad + \sum_{r=1/2,3/2,\ldots} r \, \bar{b}^i(r)^\dagger \bar{b}^i(r), \tag{C-3.7}$$

$$N_R = \sum_{m=1}^{\infty} \left(\alpha^i(m)^\dagger \alpha^i(m) + m \, d^i(m)^\dagger d^i(m) \right), \tag{C-3.8}$$

derived in (3.3.124). □

□ (R$_L$, NS$_R$)-*Sector*

$$\frac{\alpha'}{4} M^2 = N_L = N_R - \frac{1}{2}, \tag{C-3.9}$$

$$N_L = \sum_{m=1}^{\infty} \left(\bar{\alpha}^i(m)^\dagger \bar{\alpha}^i(m) + m \, \bar{d}^i(m)^\dagger \bar{d}^i(m) \right), \tag{C-3.10}$$

$$N_R = \sum_{m=1}^{\infty} \alpha^i(m)^\dagger \alpha^i(m) \quad + \sum_{r=1/2,3/2,\dots} r\, b^i(r)^\dagger b^i(r), \tag{C-3.11}$$

derived in (3.3.126). $\square$

$\square$ (R_L, R_R)-*Sector*

$$\frac{\alpha'}{4} M^2 = N_L = N_R, \tag{C-3.12}$$

$$N_L = \sum_{m=1}^{\infty} \left(\bar\alpha^i(m)^\dagger \bar\alpha^i(m) + m\, \bar d^i(m)^\dagger \bar d^i(m) \right), \tag{C-3.13}$$

$$N_R = \sum_{m=1}^{\infty} \left(\alpha^i(m)^\dagger \alpha^i(m) + m\, d^i(m)^\dagger d^i(m) \right), \tag{C-3.14}$$

derived in (3.3.141). $\square$

∇ The Heterotic String (Closed Superstring)

$\square$ R_R-*Sector*

$$\frac{\alpha'}{4} M^2 = \alpha'\, \mathbf{p}_L^2 + N_L - 1 = N_R, \tag{C-3.15}$$

$$N_L = \sum_{m=1}^{\infty} \left(\sum_{i=1}^{8} \bar\alpha^i(m)^\dagger \bar\alpha^i(m) + \sum_{I=10}^{25} \bar\alpha^I(m)^\dagger \bar\alpha^I(m) \right), \tag{C-3.16}$$

$$\widetilde N_R = \sum_{m=1}^{\infty} \sum_{i=1}^{8} \left(\alpha^i(m)^\dagger \alpha^i(m) + m\, d^i(m)^\dagger d^i(m) \right), \tag{C-3.17}$$

where

$$\mathbf{p}_L = (p_L^{10}, p_L^{11}, \dots, p_L^{25}), \tag{C-3.18}$$

from the first equalities in (3.2.153)/(A-3.4), for the compactified bosonic part of the string corresponding to the left-mover, and the expression in the second equality in (3.3.124) corresponding to the right-mover.

The compactification is along the dimensions specified by the index I taking the values $(10, 11, \dots, 25)$. $\square$

☐ NS $_R$-*Sector*

$$\frac{\alpha'}{4} M^2 = \alpha' \mathbf{p}_L^2 + N_L - 1 = N_R - \frac{1}{2}, \tag{C-3.19}$$

$$N_L = \sum_{m=1}^{\infty} \left(\sum_{i=1}^{8} \bar{\alpha}^i(m)^\dagger \bar{\alpha}^i(m) + \sum_{I=10}^{25} \bar{\alpha}^I(m)^\dagger \bar{\alpha}^I(m) \right), \tag{C-3.20}$$

$$N_R = \sum_{m=1}^{\infty} \sum_{i=1}^{8} \alpha^i(m)^\dagger \alpha^i(m) \quad + \sum_{r=1/2,3/2,...} \sum_{i=1}^{8} r\, b^i(r)^\dagger b^i(r), \tag{C-3.21}$$

where

$$\mathbf{p}_L = (p_L^{10}, p_L^{11}, \ldots, p_L^{25}), \tag{C-3.22}$$

from the first equalities in (3.2.153)/(A-3.4), for the compactified bosonic part of the string corresponding to the left-mover, and the expression in the second equality in (3.3.126) corresponding to the right-mover. The compactification is along the dimensions specified by the index I taking the values $(10, 11, \ldots, 25)$. ☐

Problems

3.1 Show that the response to the infinitesimal variation $\delta\tau = \tau - \tau' = \lambda(\tau)$ for infinitesimal λ, gives rise to the variations $\delta X^\mu(\tau) = X^\mu(\tau) - X'^\mu(\tau)$ and $\delta a(\tau) = a(\tau) - a'(\tau)$ as shown in (3.1.10) and (3.1.11), respectively.

3.2 Establish the result in (3.2.20).

3.3 In reference to (3.2.33), consider the equation

$$\begin{pmatrix} \hat{h}_{00} & \hat{h}_{01} \\ \hat{h}_{10} & \hat{h}_{11} \end{pmatrix} = \begin{pmatrix} a & b \\ c & d \end{pmatrix} \begin{pmatrix} h_{00} & h_{01} \\ h_{10} & h_{11} \end{pmatrix} \begin{pmatrix} a & c \\ b & d \end{pmatrix}$$

with the first matrix on the right-hand side being the transpose of the third one. Show that for $h_{00} = 0$ or $h_{00} > 0$ or $h_{00} < 0$, one may diagonalize the matrix on the left-hand side of the above equation to the form

$$\begin{pmatrix} \hat{h}_{00} & \hat{h}_{01} \\ \hat{h}_{10} & \hat{h}_{11} \end{pmatrix} = e^\phi \begin{pmatrix} -1 & 0 \\ 0 & 1 \end{pmatrix}.$$

3.4 Show that the wave equation $(\partial_\sigma^2 - \partial_\tau^2) X^i(\tau, \sigma) = 0$, may be rewritten as $\partial_{\tau-\sigma} \partial_{\tau+\sigma} X^i(\tau, \sigma) = 0$, with a general solution of the form $X^i(\tau, \sigma) = X_R^i(\tau -$

$\sigma) + X_L^i(\tau + \sigma)$. Interpret these solutions to justify the subscripts R/L attached to them.

3.5 Provide the details leading to (3.2.82).

3.6 In reference to (3.2.137), show that upon expressing the generators as

$$\Lambda = \begin{pmatrix} A & B \\ C & D \end{pmatrix}$$

written in terms of $(N/2) \times (N/2)$ sub-matrices, then the latter satisfy the conditions $D = -A^{\mathsf{T}}$, $B = B^{\mathsf{T}}$, $C = C^{\mathsf{T}}$, and that the independent components of Λ are $N(N+1)/2$.

3.7 Use the expressions in (3.2.146), (3.2.147) and the corresponding ones for the $X^i, i = 1, \ldots, 23$, in (3.2.76), to derive (3.2.148).

3.8 Derive the constraint in (3.2.152).

3.9 In reference to Sect. 3.2.9, dealing with the massless vector field, derive the expressions of the vacuum-to-vacuum transition amplitude $\langle\, 0_+|0_-\rangle$, and of the propagator. Finally establish the positivity of the formalism, i.e., show that $|\langle\, 0_+|0_-\rangle|^2 < 1$.

3.10 In reference to Sect. 3.2.9, dealing with the massless symmetric tensor field, derive the expressions of the vacuum-to-vacuum transition amplitude $\langle\, 0_+|0_-\rangle$, and of the propagator.

3.11 In reference to Sect. 3.2.9, dealing with the massless anti-symmetric tensor field, derive the expressions of the vacuum-to-vacuum transition amplitude $\langle\, 0_+|0_-\rangle$, and of the propagator.

3.12 Derive the expression for the Hamiltonian for open strings in (B-3.3) following the steps given in the text.

3.13 Show that a Majorana condition of a spinor in two dimensions is as stated in (3.3.11) corresponding to the reality condition of the spinor.

3.14 Prove the Fierz identity in two dimensions:

$$\delta_{ab}\,\delta_{cd} = \frac{1}{2}\,\delta_{ad}\,\delta_{cb} - \frac{1}{2}\,(\rho^\mu)_{ad}\,(\rho_\mu)_{cb} + \frac{1}{2}\,(\rho^5)_{ad}\,(\rho^5)_{cb}.$$

3.15 By considering ϵ to be ξ-dependent, derive the variational relation given in (3.3.38), thus obtaining the expression for the supercurrent in (3.3.39).

3.16 Derive the constraints in (3.3.47) using, in the process, (3.3.48) and (3.3.49).

3.17 Use the expression for the energy-momentum tensor $T_{\alpha\beta}$ in (3.3.35) to show that: (1) $T_{00}^{\mathrm{F}} + T_{01}^{\mathrm{F}} = 0$, leads to the constraint in (3.3.54), (2) $T_{00}^{\mathrm{F}} - T_{01}^{\mathrm{F}} = 0$, leads to the constraint in (3.3.55), and (3) $T_{01} = 0$ leads to the one in (3.3.56).

3.18 Show that the zero mode part, that is the $\exp[-in(\tau + \sigma)]$—independent part, of (3.3.54) for open strings is as given in (3.3.66) and (3.3.68) for the R and NS boundary conditions.

3.19 Show that the condition $\int_0^\pi d\sigma \, T_{01} = 0$ leads to the constraints in (3.3.77), (3.3.78) for closed superstrings with R, NS boundary conditions, respectively.

3.20 In reference to Sect. 3.3.8, dealing with the massless Rarita-Schwinger field, derive the expressions of the vacuum-to-vacuum transition amplitude $\langle 0_+ | 0_- \rangle$, and of the propagator.

3.21 Consider the transformation $W = (a\rho + b)/(c\rho + d)$ for any real parameters a, b, c, d such that $ad - bc = 1$, $\rho = e^\tau e^{i\sigma}$ as introduced in Sect. 3.5.1. (1) Show that W maps the upper complex ρ-plane into itself and the real line into itself. (2) Since three of the real parameters a, b, c, d are arbitrary, one may fix their values by fixing, in turn, the values of any three of the real variables $z_1, z_2, \ldots, z_{n-1}, z_n$ encountered in Sect. 3.5.1. What were the values given for z_1, z_{n-1}, z_n, in the text? and for what values of $\tau_1, \tau_{n-1}, \tau_n$ these correspond?

3.22 Use the expression for the three-point function in (3.5.24), (3.5.25), involving three massless vector bosons, to derive the corresponding expression in (3.5.26).

3.23 Show that the three-point function in (3.5.27)/(3.5.28), with $\ell \to 0$ is anti-symmetric in the exchange of any pairs $(e_i, k^i) \leftrightarrow (e_j, k^j)$.

3.24 Show that the three-point function in supersymmetric string theory involving three vector particles in (3.5.74) is simply proportional to the corresponding one obtained in bosonic string theory in (3.5.26)/(3.5.27),(3.5.28), for $\ell \to 0$.

3.25 Derive the equality in (3.6.9) which is given up to a total derivative.

3.26 Show that the vertex parts

$$V^{\mu_1\mu_2,\nu_1\nu_2,\rho_1\rho_2}(k_1, k_2, k_3), \quad \widetilde{V}^{\mu_1\mu_2,\nu_1\nu_2,\rho_1\rho_2}(k_1, k_2, k_3),$$

in (3.6.10), (3.6.15), respectively, are identical with the coupling κ identified with the string parameter through the relation $\kappa = g_c/2$.

Recommended Reading

Bailin, D., & Love, A. (1994). *Supersymmetric gauge field theory and string theory*. Bristol: Institute of Physics Publishing,

Becker, K, Becker, M., & Schwarz, J. H. (2006). *String theory and M-theory: A modern introduction*. Cambridge: Cambridge University Press.

Dine, M. (2007). *Supersymmetry and string theory: Beyond the standard model*. Cambridge: Cambridge University Press.

Elizalde, E. (1995). *Ten applications of spectral zeta functions*. Berlin: Springer.

Elizalde, E., Odintsov, S. D., Romeo, A, Bytsenko, A. A., & Zirbini, S. (1994). *Zeta regularization techniques with applications*. Singapore: World Scientific.

Manoukian, E. B. (2012). All the fundamental massless bosonic fields in bosonic string theory. *Fortschritte der Physik, 60*, 329–336.

Manoukian, E. B. (2012). All the fundamental massless bosonic fields in superstring theory. *Fortschritte der Physik, 60*, 337–344.

Manoukian, E. B. (2012). All the fundamental massless fermion fields in superstring theory. *Journal of Modern Physics, 3*, 682–685.

Manoukian, E. B. (2016). *Quantum field theory I: Foundations and abelian and non-abelian gauge theories*. Dordrecht: Springer.

Polchinski, J. (2005). *Superstring theory* (Vols. I and II). Cambridge: Cambridge University Press.

Zwiebach, B. (2009). *A first course in string theory*. Cambridge: Cambridge University Press.

References

1. Adesi, V. B., & Zerbini, S. (2002). Analytic continuation of the Hurwitz Zeta function with physical application. *Journal of Mathematical Physics, 43*, 3759–3765.
2. Aharoni, O., et al. (2008). $N = 6$ superconformal Chern-Simons matter theories, M2-Branes and their gravity duals. *JHEP, 0810*, 091.
3. Amati, D., Ciafaloni, M., & Veneziano, G. (1987). Superstring collisions at Planckian energies. *Physics Letters B, 197*, 81–88.
4. Amati, D., Ciafaloni, M., & Veneziano, G. (1988). Classical and quantum effects from Planckian energy superstring collisions. *International Journal of Modern Physics, 3*, 1615–1661.
5. Bailin, D., & Love, A. (1994). *Supersymmetric gauge field theory and string theory*. Bristol: Institute of Physics Publishing.
6. Bousso, R. (2002). The holographic principle. *Reviews of Modern Physics, 74*, 825–874.
7. Cohen, H. (2007). *Number theory. volume II: Analytic and modern tools*. New York: Springer.
8. Cremmer, E., Julia, B., & Scherk, J. (1978). Supergravity theory in eleven-dimensions. *Physics Letters B, 76*, 409–412.
9. Di Vecchia, P. (2008). The birth of string theory. In M. Gasperini & J. Maharana (Eds.), *String theory and fundamental interactions: Gabriele Veneziano and theoretical physics: Historical and contemporary perspectives*. Lecture notes in physics (Vol. 737, pp. 59–118). Berlin: Springer.
10. Duff, M. J. (1996). M-theory (The theory formerly known as superstrings). *International Journal of Modern Physics A, 11*, 5623–5642. hep–th/9608117.
11. Elizalde, E. (1995). *Ten physical applications of spectral zeta functions*. Berlin: Springer.
12. Elizalde, E., Odintsov, S. D., Romeo, A., Bytsenko, A. A., & Zerbini, S. (1994). *Zeta regularization techniques with applications*. Singapore: World Scientific.
13. Gliozzi, F., Scherk, J., & Olive, D. (1976). Supergravity and the spinor dual model. *Physics Letters B, 65*, 282–286.
14. Gliozzi, F., Scherk, J., & Olive, D. (1977). Supersymmetry, supergravity theories and the dual spinorl model. *Nuclear Physics B, 22*, 253–290.
15. Goto, T. (1971). Relativistic quantum mechanics of one-dimensional mechanical continuum and subsidiary condition of dual resonance model. *Progress in Theoretical Physics, 46*, 1560–1569.
16. Green, M. B., & Schwarz, J. H. (1981). Supersymmetrical dual string theory. *Nuclear Physics B 181*, 502–530.
17. Green, M. B., & Schwarz, J. H. (1982). Supersymmetrical string theories. *Physics Letters B, 109*, 444–448.

18. Green, M. B., Schwarz, J. H., & Witten, E. (1987). *Superstring theory* (Vols.1 and 2). Cambridge: Cambridge University Press.
19. Gross, D. J., Harvey, J. A., Martinec, E. J., & Rhom, R. (1985). Heterotic string theory (I). The free heterotic string. *Nuclear Physics B, 256*, 253–284.
20. Gross, D. J., Harvey, J. A., Martinec, E. J., & Rohm, R. (1985). The heterotic string. *Physical Reviews Letters, 54*, 502–505.
21. Gubser, S. S., Klebanov, I. R., & Polyakov, A. M. (1998). Gauge theory correlations from non-critical string theory. *Physics Letters B, 428*, 105–114. (hep-th/9802150).
22. Horowitz, G., Lowe, D. A., & Maldacena, J. (1996). Statistical entropy of non-extremal four dimensional black holes and U-duality. *Physical Review Letters, 77*, 430–433.
23. Horowitz, G. T. (2005). Spacetime in string theory. *New Journal of Physics, 7*(1–13), 201.
24. Lovelace, C., & Squires, E. (1970). Veneziano theory. *Proceedings of the Royal Society of London A, 318*, 321–353.
25. Maldacena, J. (1998). The large N limit of superconformal theories and gravitation. *Advances in Theoretical and Mathematical Physics, 2*, 231–252. (hep-th/9711200).
26. Manoukian, E. B. (2012). All the fundamental massless bosonic fields in bosonic string theory. *Fortschritte der Physik, 60*, 329–336.
27. Manoukian, E. B. (2012). All the fundamental bosonic massless fields in superstring theory. *Fortschritte der Physik, 60*, 337–344.
28. Manoukian, E. B. (2012). All the fundamental massless fermion fields in superstring theory: A rigorous analysis. *Journal of Modern Physics, 3*, 1027–1030.
29. Manoukian, E. B. (2016). *Quantum field theory I: Foundations and abelian and non-abelian gauge theories*. Dordrecht: Springer.
30. Nambu, Y. (1969). In *Proceedings of International Conference on Symmetries and Quark Models* (p. 269). New York: Wayne State University, Gordon and Breach.
31. Nambu, Y. (1970). Duality and electrodynamics. In T. Eguchi & K. Nishijima (Eds.), *Broken Symmetry: Selected Papers of Y. Nambu*. Singapore: World Scientific, 1995.
32. Neveu, A., & Schwarz, J. H. (1971). Factorizable dual model of pions. *Nuclear Physics B, 31*, 86–112.
33. Nielsen, H. (1970). *International Conference on High Energy Physics*, Kiev Conference, Kiev.
34. Polchinski, J. (2005). *Superstring theory* (Vols. I and II). Cambridge: Cambridge University Press.
35. Raymond, P. (1971). Dual theory for free fermions. *Physics Review D, 3*, 2415–2418.
36. Scherk, J., & Schwarz, J. H. (1974). Dual models for non-hadrons. *Nuclear Physics B, 81*, 118–144.
37. Schwarz, J. H. (1997). Lectures on superstrings and M-theory. *Nuclear Physics B - Proceedings Supplements, 55*, 1–32, hep–th/9607201.
38. Strominger, A., & Fava, G. (1996). Microscopic origin of the 'Bekenstein-Hawking Entropy'. *Physics Letters B, 379*, 99–104.
39. Susskind, L. (1970). Dual symmetric theory of hadrons. I. *Nuovo Cimento A, 69*, 457–496.
40. Susskind, L. (1995). The world as a hologram. *Journal of Mathematical Physics, 36*, 6377–6396.
41. 't Hooft, G. (1987). Can spacetime be probed below string size? *Physics Letters B, 198*, 61–63.
42. 't Hooft, G. (1995). Black holes and the dimensional reduction of spacetime. In U. Lindström (Ed.), *The Oskar Klein Centenary, 19–21 Sept. Stockholm, Sweden(1994)*. Worldscientific. See also: "Dimensional Reduction in Quantum gravity", Utrecht preprint THU-93/26 (gr-qg/9310026).
43. Thorn, C. B. (1991). Reformulating string theory with the $1/N$ expansion. In *International A. D. Sakharov Conference on Physics*, Moscow, pp. 447–454. (hep-th/9405069).
44. Townsend, P. K. (1995). The eleven-dimensional supermembrane revisited. *Physics Letters B, 350*, 184–187. hep–th/9501068.

45. Veneziano, G. (1968). Construction of a crossing-symmetric Regge-behaved amplitude for linearly rising trajectories. *Nuovo Cimento A, 57*, 190–197.
46. Witten, E. (1995). String theory dynamics in various dimensions. *Nuclear Physics B, 443*, 85–126. hep–th/9503124.
47. Witten, E. (1998). Anti-de-Sitter space and holography. *Advances in Theoretical and Mathematical Physics, 2*, 253–291. (hep-th/9802150).
48. Yoneya, T. (1974). Connection of dual models to electrodynamics and gravidynamics. *Progress in Theoretical Physics, 51*, 1907–1920.

General Appendices

Appendix I
The Gamma Matrices in Various Dimensions

In this appendix, we summarize some of the representations of the gamma matrices γ^μ in $D = 2, 4, 10$ dimensional spacetimes, satisfying the anti-commutation relations:[1]

$$\{\gamma^\mu, \gamma^\nu\} = -2\eta^{\mu\nu}, \quad \eta^{\mu\nu} = \text{diag}[-1, 1, \ldots, 1], \quad \mu = 0, 1, \ldots, (D-1). \quad \text{(I.1)}$$

Four Dimensional Spacetime

Some of the properties of the gamma matrices, based on their anti-commutations relations in (I.1) in 4D are given here in Box I.1.

---

[1]It should be noted that no gamma matrix may be chosen to be the identity matrix I without making the other gamma matrices non-zero. The reason for this is very simple. If, for example γ^μ, for a fixed μ, is chosen to be I, then for any other gamma matrix γ^ν, for $\nu \neq \mu$, the anti-commutation relation (I.1) implies that $2I\gamma^\nu = 0$, i.e., *all* the other gamma matrices γ^ν, for $\nu \neq \mu$, are *zero*. The latter, in turn, is inconsistent with (I.1) as the anti-commutator of the zero matrix with itself is zero.

© Springer International Publishing Switzerland 2016
E.B. Manoukian, *Quantum Field Theory II*, Graduate Texts in Physics,
DOI 10.1007/978-3-319-33852-1

Box I.1: Some properties of the gamma matrices

$$\gamma^{\mu}\gamma^{\nu} = -\eta^{\mu\nu}I + (1/2)[\gamma^{\mu},\gamma^{\nu}], \quad \text{Tr}[\gamma^{\mu}] = 0,$$

$$(\gamma^{0})^{2} = I, \quad (\gamma^{i})^{2} = -I, \quad i = 1,2,3.$$

$$\eta_{\mu\nu}\gamma^{\mu}\gamma^{\nu} = -4I,$$

$$\eta_{\mu\nu}\gamma^{\mu}(\gamma^{\sigma})\gamma^{\nu} = 2\gamma^{\sigma},$$

$$\eta_{\mu\nu}\gamma^{\mu}(\gamma^{\sigma}\gamma^{\lambda})\gamma^{\nu} = 4\eta^{\sigma\lambda},$$

$$\eta_{\mu\nu}\gamma^{\mu}(\gamma^{\sigma}\gamma^{\lambda}\gamma^{\rho})\gamma^{\nu} = 2\gamma^{\rho}\gamma^{\lambda}\gamma^{\sigma},$$

$$\left[\gamma^{\mu},\left[\gamma^{\sigma},\gamma^{\rho}\right]\right] = 4\left(\gamma^{\sigma}\eta^{\mu\rho} - \gamma^{\rho}\eta^{\mu\sigma}\right),$$

$$\text{Tr}[\gamma^{\mu}\gamma^{\nu}] = -4\eta^{\mu\nu},$$

$$\text{Tr}[\gamma^{\alpha}\gamma^{\beta}\gamma^{\mu}\gamma^{\nu}] = 4(\eta^{\alpha\beta}\eta^{\mu\nu} - \eta^{\alpha\mu}\eta^{\beta\nu} + \eta^{\alpha\nu}\eta^{\beta\mu}),$$

$$\text{Tr}[\gamma^{5}\gamma^{\alpha}\gamma^{\beta}\gamma^{\mu}\gamma^{\nu}] = -4i\varepsilon^{\alpha\beta\mu\nu},$$

$\varepsilon^{\alpha\beta\mu\nu}$ totally anti-symmetric with $\varepsilon^{0123} = +1.$

$\text{Tr}[\text{odd number of } \gamma\text{'s}] = 0.$

$$\gamma^{5} = i\gamma^{0}\gamma^{1}\gamma^{2}\gamma^{3}, \quad \text{Tr}[\gamma^{5}] = 0, \quad \text{Tr}[\gamma^{5}\gamma^{\mu}] = 0,$$

$$(\gamma^{5})^{2} = I, \quad \{\gamma^{5},\gamma^{\mu}\} = 0,$$

$$(\gamma^{\mu}a_{\mu})^{2} = -I[\mathbf{a}^{2} - (a^{0})^{2}], \quad (\boldsymbol{\gamma}\cdot\mathbf{a})^{2} = -I\mathbf{a}^{2},$$

$$\mathbf{a} = (a_{1},a_{2},a_{3}), \quad a_{0} = -a^{0}, \quad a_{i} = a^{i}, \quad i = 1,2,3.$$

Now we introduce various representations of the gamma matrices in 4D.

◊   Dirac representation:

$$\gamma^{0} = \begin{pmatrix} I & 0 \\ 0 & -I \end{pmatrix}, \quad \gamma^{i} = \begin{pmatrix} \mathbf{0} & \sigma^{i} \\ -\sigma^{i} & \mathbf{0} \end{pmatrix}, \quad i = 1,2,3, \quad \gamma^{5} \equiv i\gamma^{0}\gamma^{1}\gamma^{2}\gamma^{3} = \begin{pmatrix} 0 & I \\ I & 0 \end{pmatrix}.$$

◊   Chiral Representation:

$$\gamma^{0} = \begin{pmatrix} 0 & -I \\ -I & 0 \end{pmatrix}, \quad \gamma^{i} = \begin{pmatrix} \mathbf{0} & \sigma^{i} \\ -\sigma^{i} & \mathbf{0} \end{pmatrix}, \quad i = 1,2,3, \quad \gamma^{5} = \begin{pmatrix} I & 0 \\ 0 & -I \end{pmatrix}.$$

◊ Majorana representation:

$$\gamma^0 = \begin{pmatrix} 0 & \sigma^2 \\ \sigma^2 & 0 \end{pmatrix}, \quad \gamma^1 = \begin{pmatrix} i\sigma^3 & 0 \\ 0 & i\sigma^3 \end{pmatrix}, \quad \gamma^2 = \begin{pmatrix} 0 & -\sigma^2 \\ \sigma^2 & 0 \end{pmatrix},$$

$$\gamma^3 = \begin{pmatrix} -i\sigma^1 & 0 \\ 0 & -i\sigma^1 \end{pmatrix}, \quad \gamma^5 = \begin{pmatrix} \sigma^2 & 0 \\ 0 & -\sigma^2 \end{pmatrix}.$$

Two Dimensional Spacetime

$$\gamma^0 = \begin{pmatrix} 1 & 0 \\ 0 & -1 \end{pmatrix}, \quad \gamma^1 = \begin{pmatrix} 0 & 1 \\ -1 & 0 \end{pmatrix}, \quad \gamma^0\gamma^1 = \begin{pmatrix} 0 & 1 \\ 1 & 0 \end{pmatrix}.$$

Ten Dimensional Spacetime

$$\gamma^0 = \begin{pmatrix} 0 & -iI_{16} \\ iI_{16} & 0 \end{pmatrix}, \quad \gamma^i = \begin{pmatrix} 0 & -\Gamma^i \\ -\Gamma^i & 0 \end{pmatrix}, \quad i = 1, \ldots, 9,$$

$$\gamma_c = (-i)\gamma^0\gamma^1 \ldots \gamma^9 = \begin{pmatrix} I_{16} & 0 \\ 0 & -I_{16} \end{pmatrix},$$

where

$$\begin{aligned}
\Gamma^1 &= i\sigma^2 \times \sigma^2 \times \sigma^2 \times \sigma^2, & \Gamma^2 &= iI \times \sigma^1 \times \sigma^2 \times \sigma^2, \\
\Gamma^3 &= iI \times \sigma^3 \times \sigma^2 \times \sigma^2, & \Gamma^4 &= i\sigma^1 \times \sigma^2 \times I \times \sigma^2, \\
\Gamma^5 &= i\sigma^3 \times \sigma^2 \times I \times \sigma^2, & \Gamma^6 &= i\sigma^2 \times I \times \sigma^1 \times \sigma^2, \\
\Gamma^7 &= i\sigma^2 \times I \times \sigma^3 \times \sigma^2, & \Gamma^8 &= iI \times I \times I \times \sigma^1, & \Gamma^9 &= iI \times I \times I \times \sigma^3.
\end{aligned}$$

Appendix II
Some Basic Fields in 4D

Under a homogeneous Lorentz transformation,

$$x'^{\mu} = \Lambda^{\mu}{}_{\nu} x^{\nu}, \quad \partial'_{\mu} = \Lambda_{\mu}{}^{\nu} \partial_{\nu}, \tag{II.1}$$

$$\Lambda^{\mu}{}_{\rho} \eta_{\mu\nu} \Lambda^{\nu}{}_{\lambda} = \eta_{\rho\lambda}, \quad (\Lambda^{-1})^{\nu}{}_{\mu} = \Lambda_{\mu}{}^{\nu}, \quad [\eta_{\rho\lambda}] = \text{diag}\,[-1, 1, 1, 1]. \tag{II.2}$$

We note that the first equality in (II.2) rewritten in matrix form reads

$$\Lambda^{\mathsf{T}} \eta \Lambda = \eta, \tag{II.3}$$

and since $\det \eta = -1$, we may infer that for the transformations in (II.1) that

$$\det \Lambda = +1. \tag{II.4}$$

It is not -1 as this transformation includes neither time nor space reflections. From this and the fact that $\partial x'^{\mu}/\partial x^{\nu} = \Lambda^{\mu}{}_{\nu}$, we learn that the Jacobian of the transformation $x' \to x$ is given by $\det \Lambda = 1$. This establishes the Lorentz invariance of the volume element in Minkowski spacetime:

$$(dx) = (dx'), \quad \text{where } (dx) \equiv dx^0 \, dx^1 \, dx^2 \, dx^3. \tag{II.5}$$

This together with the definition of a Lorentz scalar $\Phi(x)$ by the condition

$$\Phi'(x') = \Phi(x), \tag{II.6}$$

under the above Lorentz transformations, guarantees the invariance of integrals of the form

$$\mathscr{A} = \int (dx)\, \Phi(x), \tag{II.7}$$

© Springer International Publishing Switzerland 2016
E.B. Manoukian, *Quantum Field Theory II*, Graduate Texts in Physics,
DOI 10.1007/978-3-319-33852-1

such as the action integral, and lead to the development of Lorentz invariant theories. In Chap. 2, we have the more ambitious programme of developing supersymmetric invariant actions involving superfields.

Under an infinitesimal transformation

$$\Lambda^{\mu}{}_{\nu} = \delta^{\mu}{}_{\nu} + \delta\omega^{\mu}{}_{\nu}, \quad \delta\omega^{\mu\nu} = -\delta\omega^{\nu\mu}. \tag{II.8}$$

For explicit expressions of $\Lambda^{\mu}{}_{\nu}$ and $\delta\omega^{\nu\mu}$, see, e.g., [1], Chap. 16.

The Langrangian density of a Hermitian scalar field interacting with an external source K is given by

$$\mathcal{L}(x) = -\frac{1}{2}\,\partial_{\mu}\varphi(x)\partial^{\mu}\varphi(x) - \frac{m^2}{2}\,\varphi^2(x) + K(x)\varphi(x), \tag{II.9}$$

and the vacuum-to-vacuum transition amplitude is given by

$$\langle\, 0_+\,|\,0_-\rangle = \exp\!\left[\frac{\mathrm{i}}{2}\int(\mathrm{d}x)(\mathrm{d}x')K(x)\Delta_+(x-x')K(x')\right], \tag{II.10}$$

from which the propagator is obtained by functional differentiations as follows:

$$\frac{\delta}{\delta K(x')}(-\mathrm{i})\frac{\delta}{\delta K(x)}\langle\, 0_+\,|\,0_-\rangle\Big|_{K=0} = \Delta_+(x-x') = \mathrm{i}\langle\, 0\,|\,\bigl(\varphi(x)\varphi^{\dagger}(x')\bigr)_+\,|\,0\rangle, \tag{II.11}$$

where

$$\Delta_+(x-x') = \int\frac{(\mathrm{d}p)}{(2\pi)^4}\frac{\mathrm{e}^{\mathrm{i}p\,(x-x')}}{p^2 + m^2 - \mathrm{i}\epsilon}, \tag{II.12}$$

and we have used the notation $\langle\, 0\,|\,.\,|\,0\,\rangle$ for $\langle\, 0_+\,|\,.\,|\,0_-\rangle$ for $K = 0$.

Of particular interest is the functional (path) integral expression in terms of classical fields given by

$$\int(\mathscr{D}\phi)\exp\mathrm{i}\left[\int(\mathrm{d}x)(\mathrm{d}x')\bigl(-\frac{1}{2}\,\phi(x)M(x,x')\phi(x')\bigr) + \int(\mathrm{d}x)\,\phi(x)K(x)\right] \tag{II.13}$$

$$= \langle\, 0_+\,|\,0_-\rangle\int(\mathscr{D}\phi)\exp\mathrm{i}\left[\int(\mathrm{d}x)(\mathrm{d}x')\bigl(-\frac{1}{2}\,\phi(x)M(x,x')\phi(x')\bigr)\right], \tag{II.14}$$

$$M(x,x') = [-\Box + m^2]\delta^{(4)}(x-x'), \tag{II.15}$$

$$M^{-1}(x,x') = \frac{1}{[-\Box + m^2 - \mathrm{i}\epsilon]}\,\delta^{(4)}(x-x') = \Delta_+(x-x'). \tag{II.16}$$

$$\int M(x,x')\,(dx')\,M^{-1}(x',y) = \delta^{(4)}(x-y). \tag{II.17}$$

From (II.13)–(II.17), we may, in a compact matrix notation in spacetime variables, write the useful expression

$$\frac{1}{\sqrt{\det M}}\exp\left[\frac{i}{2}KM^{-1}K\right] = \int(\mathscr{D}\phi)\exp\left[i\left(-\frac{1}{2}\phi M\phi + \phi K\right)\right]. \tag{II.18}$$

For spin 1/2, the Dirac equation is given by

$$\left[\frac{\gamma^\mu\partial_\mu}{i} + m\right]\psi(x) = 0, \tag{II.19}$$

which may be obtained from the Lagrangian density

$$\mathscr{L}(x) = -\frac{1}{2i}\left(\overline{\psi}(x)\gamma^\mu\partial_\mu\psi(x) - (\partial^\mu\overline{\psi})(x)\psi(x)\right) - m\overline{\psi}(x)\psi(x). \tag{II.20}$$

Under a homogeneous Lorentz transformation, which may include a 3D rotation, the Dirac equation reads

$$\left[\frac{\gamma^\nu\partial'_\nu}{i} + m\right]K\psi(x) = 0, \qquad K\psi(x) = \psi'(x'), \quad x' = \Lambda x, \tag{II.21}$$

where the matrix K satisfies the relations

$$K^\dagger\gamma^0 K = \gamma^0, \qquad \Lambda^\mu{}_\nu\gamma^\nu = K^{-1}\gamma^\mu K. \tag{II.22}$$

For infinitesimal transformations, as given in (II.8), we may set

$$K \simeq I + \frac{i}{2}\delta\omega^{\mu\nu} S_{\mu\nu}, \tag{II.23}$$

where $S_{\mu\nu}$ is to be determined. By substituting this expression in (II.22), we obtain

$$[S^{\lambda\mu}, \gamma^\nu] = i\left(\eta^{\lambda\nu}\gamma^\mu - \eta^{\mu\nu}\gamma^\lambda\right), \tag{II.24}$$

whose solution is $S^{\lambda\mu} = i[\gamma^\lambda, \gamma^\mu]/4$, and we may write

$$K \simeq I + \frac{i}{2}\delta\omega^{\mu\nu} S_{\mu\nu}, \quad S^{\lambda\mu} = \frac{i}{4}[\gamma^\lambda, \gamma^\mu]. \tag{II.25}$$

In the presence of an external electromagnetic field, the Dirac equation reads

$$\left[\gamma^\mu \left(\frac{\partial_\mu}{i} - e A_\mu(x)\right) + m\right] \psi(x) = 0, \tag{II.26}$$

from which one obtains the equation $(\overline{\psi} = \psi^\dagger \gamma^0)$ for the charge conjugate field $\psi^\mathscr{C}$:

$$\left[\gamma^\mu \left(\frac{\partial_\mu}{i} + e A_\mu(x)\right) + m\right] \psi^\mathscr{C}(x) = 0, \quad \psi^\mathscr{C}(x) = \mathscr{C}\,\overline{\psi}^{\mathsf{T}}(x), \tag{II.27}$$

with opposite sign of the charge e, where

$$\mathscr{C} = i\gamma^2 \gamma^0. \tag{II.28}$$

The Lagrangian density of a (Hermitian) massless vector (the photon), may be taken as

$$\mathscr{L} = -\frac{1}{4} F_{\mu\nu} F^{\mu\nu}, \qquad F^{\mu\nu} = \partial^\mu V^\nu - \partial^\nu V^\mu. \tag{II.29}$$

For a photon with momentum k^μ, $k^2 = 0$, we may introduce polarization vectors $e_\lambda^\mu = (0, \mathbf{e}_\lambda)$, with $\lambda = 1, 2$, such that $k_\mu e_\lambda^\mu = 0$, $e_\lambda^{\mu*} e_{\lambda'\mu} = \delta_{\lambda\lambda'}$, satisfying the completeness relation

$$\eta^{\mu\nu} = \frac{k^\mu \underline{k}^\nu + \underline{k}^\mu k^\nu}{k\,\underline{k}} + \sum_{\lambda=\pm 1} e_\lambda^\mu e_\lambda^{\nu*}, \qquad k = (k^0, \mathbf{k}),\ \underline{k} = (k^0, -\mathbf{k}) \tag{II.30}$$

For completeness we also give the Lagrangian densities of massless spin 3/2 field (the gravitino), as well as of a (Hermitian) spin 2 field (the graviton).

For the 3/2 field, we may take $(\gamma\partial \equiv \gamma^\mu \partial_\mu,\ \overset{\leftrightarrow}{\partial} \equiv \overset{\rightarrow}{\partial} - \overset{\leftarrow}{\partial}\,)$

$$\mathscr{L} = -\frac{1}{2i} \overline{\psi}_\mu \left(\eta^{\mu\nu} \gamma \overset{\leftrightarrow}{\partial} - (\gamma^\mu \overset{\leftrightarrow}{\partial}{}^\nu + \gamma^\nu \overset{\leftrightarrow}{\partial}{}^\mu) + \gamma^\mu(-\gamma \overset{\leftrightarrow}{\partial})\gamma^\nu\right)\psi_\nu, \tag{II.31}$$

while for the spin 2, we may take

$$\mathscr{L} = -\frac{1}{2} \partial^\sigma U_{\mu\nu} \, \partial_\sigma U^{\mu\nu} + \partial_\mu U^{\mu\nu} \, \partial_\sigma U^\sigma{}_\nu - \partial_\sigma U^{\sigma\mu} \, \partial_\mu U + \frac{1}{2} \partial^\mu U \, \partial_\mu U, \tag{II.32}$$

where $U = U^\mu{}_\mu.$[1]

---

[1] For details on all of the above see Sect. 4.7 of Vol. I [2].

References

1. Manoukian, E. B. (2006). *Quantum theory: A wide spectrum.* AA Dordrecht: Springer.
2. Manoukian, E. B. (2016). *Quantum field theory I: Foundations and abelian and non-abelian gauge theories.* Dordrecht: Springer.

Solutions to the Problems

Chapter 1

1.1. For the Minkowski metric $\eta_{\alpha\beta}$, $\partial_\nu \eta_{\alpha\beta} = 0$, $\eta_{\alpha\beta} = e_\alpha{}^\mu e_{\beta\mu}$. Also, by mere relabeling of dummy indices, we may write

$$\Gamma_{\nu\sigma}{}^\mu e_\alpha{}^\sigma e_{\beta\mu} = \Gamma_{\nu\mu}{}^\sigma e_\alpha{}^\mu e_{\beta\sigma}.$$

Accordingly, we obviously have

$$\partial_\nu(e_\alpha{}^\mu e_{\beta\mu}) + \Gamma_{\nu\sigma}{}^\mu e_\alpha{}^\sigma e_{\beta\mu} - \Gamma_{\nu\mu}{}^\sigma e_\alpha{}^\mu e_{\beta\sigma} = 0.$$

Upon expanding the latter equation we obtain

$$e_{\beta\mu}(\partial_\nu e_\alpha{}^\mu + \Gamma_{\nu\sigma}{}^\mu e_\alpha{}^\sigma) = -e_\alpha{}^\mu(\partial_\nu e_{\beta\mu} - \Gamma_{\nu\mu}{}^\sigma e_{\beta\sigma}),$$

or

$$e_{\beta\mu} \nabla_\nu e_\alpha{}^\mu = -e_\alpha{}^\mu \nabla_\nu e_{\beta\mu}.$$

The equality in question then follows upon rewriting the right-hand side of the above equation as

$$-e_\alpha{}^\mu \nabla_\nu e_{\beta\mu} = -e_\alpha{}^\mu \nabla_\nu e_\beta{}^\gamma g_{\gamma\mu} = -e_{\alpha\gamma} \nabla_\nu e_\beta{}^\gamma \equiv -e_{\alpha\mu} \nabla_\nu e_\beta{}^\mu,$$

using the fact that $\nabla_\nu g_{\gamma\mu} = 0$. Equivalently, we may write $\nabla_\nu e_\alpha{}^\mu e_{\beta\mu} = 0$.

1.2. (i) This is a special case of (1.1.40) obtained by contracting μ_1 and μ in the latter. We provide, however, a direct demonstration of this. Using the facts that $\nabla_\beta \xi^\alpha$ is a mixed tensor, $\nabla_\alpha \xi^\alpha$ is a scalar, and that $\Gamma_{\sigma\lambda}{}^\kappa$ is

© Springer International Publishing Switzerland 2016
E.B. Manoukian, *Quantum Field Theory II*, Graduate Texts in Physics,
DOI 10.1007/978-3-319-33852-1

symmetric in (σ, λ), give

$$\nabla_\alpha \left(\nabla_\beta \, \xi^\alpha\right) = \partial_\alpha \, \partial_\beta \, \xi^\alpha + \left(\partial_\alpha \Gamma_{\sigma\beta}{}^\alpha\right)\xi^\sigma + \Gamma_{\beta\sigma}{}^\lambda \partial_\lambda \xi^\sigma - \Gamma_{\beta\sigma}{}^\lambda \partial_\lambda \xi^\sigma$$
$$- \Gamma_{\beta\alpha}{}^\lambda \Gamma_{\sigma\lambda}{}^\alpha \xi^\sigma + \Gamma_{\alpha\sigma}{}^\alpha \partial_\beta \xi^\sigma + \Gamma_{\alpha\lambda}{}^\alpha \Gamma_{\sigma\beta}{}^\lambda \xi^\sigma,$$

$$\nabla_\beta(\nabla_\alpha \xi^\alpha) = \partial_\beta \partial_\alpha \xi^\alpha + (\partial_\beta \Gamma_{\sigma\alpha}{}^\alpha)\xi^\sigma + \Gamma_{\alpha\sigma}{}^\alpha \partial_\beta \xi^\sigma,$$

which upon subtraction gives the identity in (i).

(ii) $\nabla_\nu \, h^{\alpha\beta}$ is a mixed tensor field, $\nabla_\alpha h^{\alpha\beta}$ is a vector field. In a local Lorentz coordinate system at the point x in question, the connection vanishes but not its derivatives. Accordingly, we may write in the latter coordinate system:

$$\nabla_\alpha \nabla_\nu h^{\alpha\beta} \to \partial_\alpha \partial_\nu h^{\alpha\beta} + \partial_\alpha \left(\Gamma_{\sigma\nu}{}^\alpha h^{\sigma\beta} + \Gamma_{\nu\sigma}{}^\beta h^{\alpha\sigma}\right)$$

$$= \partial_\alpha \partial_\nu h^{\alpha\beta} + \partial_\alpha \Gamma_{\sigma\nu}{}^\alpha \, h^{\sigma\beta} + \partial_\alpha \Gamma_{\nu\sigma}{}^\beta \, h^{\alpha\sigma},$$

$$\nabla_\nu \nabla_\alpha h^{\alpha\beta} \to \partial_\nu \partial_\alpha h^{\alpha\beta} + \partial_\nu \left(\Gamma_{\sigma\alpha}{}^\alpha h^{\sigma\beta} + \Gamma_{\alpha\sigma}{}^\beta h^{\alpha\sigma}\right),$$

$$= \partial_\nu \partial_\alpha h^{\alpha\beta} + + \partial_\nu \Gamma_{\sigma\alpha}{}^\alpha \, h^{\sigma\beta} + \partial_\nu \Gamma_{\alpha\sigma}{}^\beta \, h^{\alpha\sigma},$$

from which

$$[\nabla_\alpha, \nabla_\nu] \, h^{\alpha\beta} \to \left(\partial_\alpha \Gamma_{\sigma\nu}{}^\alpha - \partial_\nu \Gamma_{\sigma\alpha}{}^\alpha\right)h^{\sigma\beta} + \left(\partial_\alpha \Gamma_{\sigma\nu}{}^\beta - \partial_\nu \Gamma_{\alpha\sigma}{}^\beta\right)h^{\alpha\sigma}.$$

From (1.1.50), (1.1.39), we recognize the coefficient of $h^{\sigma\beta}$ in the first term on the right-hand side, as $R_{\nu\sigma}$, and the coefficient of $h^{\alpha\sigma}$ as $R^\beta{}_{\sigma\alpha\nu}$ at the point x in such a coordinate system. As a tensor equation the statement of the problem holds true in a general coordinate system as well.

1.3. (i) According (A-1.9), for a matrix A: $\delta \det A = \det A \, \text{Tr}[(A^{-1})\delta A]$. Hence for $A = [g_{\mu\nu}]$, $\delta g = g g^{\alpha\beta} \delta g_{\alpha\beta}$, since $[g^{\alpha\beta}]$ is the inverse of $[g_{\mu\nu}]$. The first equality then follows by multiplying the latter by minus, and noting that $\delta\sqrt{-g} = (1/2)\delta(-g)/\sqrt{-g}$. On the other hand $g^{\alpha\gamma} g_{\gamma\beta} = \delta^\alpha{}_\beta$ which upon taking the variation of the latter gives, in particular, $g^{\alpha\beta}\delta g_{\alpha\beta} = -g_{\alpha\beta}\delta g^{\alpha\beta}$, implying the second equality.

(ii) The first equality in (i) above implies the first equality in (ii). On the other hand, we explicitly have $\Gamma_{\mu\sigma}{}^\sigma = (1/2)g^{\alpha\beta}\partial_\mu g_{\alpha\beta}$ which gives the second equality.

1.4. (i) The vanishing of the covariant derivative of the (inverse of the) metric $\nabla_\mu g^{\alpha\beta} = 0$, gives $\Gamma_{\mu\sigma}{}^\alpha g^{\sigma\beta} = -(\partial_\mu g^{\alpha\beta} + \Gamma_{\mu\sigma}{}^\beta g^{\sigma\alpha})$. Upon setting $\beta = \mu$, summing over it, and using part (ii) in Problem 1.3 establish the equality. Part (ii) is the content of part (ii) in Problem 1.3.

1.5. (i) $\sqrt{-g}\,\nabla_\mu \xi^\mu = \sqrt{-g}\,(\partial_\mu \xi^\mu + \Gamma_{\mu\sigma}{}^\mu \xi^\sigma)$ and the result follows upon using part (ii) of Problem 1.4.

(ii) Recall that $\partial_\nu \phi$ is a vector field, we may write $\sqrt{-g}\,g^{\mu\nu}\partial_\mu\phi\,\partial_\nu\phi = \sqrt{-g}\,\partial_\mu\phi\partial^\mu\phi$. Partial integration implies that the latter is equivalent to $-(\partial_\mu\sqrt{-g})\phi\,\partial^\mu\phi - \sqrt{-g}\,\phi\partial_\mu\partial^\mu\phi$. The result again follows from part (ii) of Problem 1.3, and the expression of the covariant derivative of a vector field.

(iii) The results follows from the application of part (i) of Problem 1.3 for $\sqrt{-g}$ followed by the variation of $g^{\mu\nu}$ (the coefficient of $\partial_\mu\phi\partial_\nu\phi$), keeping ϕ fixed as required.

1.6. From the just mentioned equation, the following transformation law of $\delta\Gamma_{\varrho\nu}{}^\gamma$ follows to be

$$\delta\Gamma'_{\varrho\nu}{}^\gamma = \frac{\partial x'^\gamma}{\partial x^\sigma}\frac{\partial x^\mu}{\partial x'^\varrho}\frac{\partial x^\kappa}{\partial x'^\nu}\,\delta\Gamma_{\mu\kappa}{}^\sigma,$$

which is the transformation rule of a tensor.

1.7. (i) The variation of the Riemann tensor is explicitly given by

$$\delta R^\mu{}_{\nu\rho\sigma} = \partial_\rho\delta\Gamma_{\nu\sigma}{}^\mu - \partial_\sigma\,\delta\Gamma_{\nu\rho}{}^\mu + \delta\Gamma_{\rho\kappa}{}^\mu\Gamma_{\nu\sigma}{}^\kappa + \Gamma_{\rho\kappa}{}^\mu\delta\Gamma_{\nu\sigma}{}^\kappa$$
$$-\delta\Gamma_{\sigma\kappa}{}^\mu\Gamma_{\nu\rho}{}^\kappa - \Gamma_{\sigma\kappa}{}^\mu\delta\Gamma_{\nu\rho}{}^\kappa.$$

In a local lorentz coordinate system, the connection vanishes *locally*, but *not* its variation since the latter is a tensor. Accordingly, in a local Lorentz coordinate system, at the point x in question, we may write $\delta R^\mu{}_{\nu\rho\sigma} = \partial_\rho\delta\Gamma_{\nu\sigma}{}^\mu - \partial_\sigma\,\delta\Gamma_{\nu\rho}{}^\mu$. Since $\delta\Gamma_{\nu\sigma}{}^\mu$, $\delta\Gamma_{\nu\rho}{}^\mu$ are tensors, a tensorial equation holding in every coordinate is obtained by replacing the partial derivatives by their covariant counterparts. This gives (i).

Part (ii) follows by writing $R_{\mu\nu\rho\sigma} = g_{\mu\lambda}R^\lambda{}_{\nu\rho\sigma}$ and using part (i).

Part (iii) is obtained by making a contraction between ρ and μ in (i). In a local Lorentz coordinate system, the first derivative of the metric is, locally, zero at the point in question, and we may write

$$\delta\Gamma_{\mu\nu}{}^\rho = \frac{1}{2}g^{\rho\sigma}(\partial_\mu\delta g_{\nu\sigma} + \partial_\nu\delta g_{\mu\sigma} - \partial_\sigma\delta g_{\mu\nu}).$$

Accordingly, in every coordinate system, $\delta\Gamma_{\mu\nu}{}^\rho$ has the structure as stated in the problem with covariant derivatives replacing partial derivatives, thus establishing part (iv).

1.8. Upon using the expression for $\delta R_{\mu\nu}$ in Problem 1.7, the expression for $g^{\mu\nu}\Gamma_{\mu\nu}{}^{\sigma}$ in part (i) of Problem 1.4, and the definition of a covariant derivative, we get

$$\sqrt{-g}\, g^{\mu\nu}\delta R_{\mu\nu} = \sqrt{-g}\, g^{\mu\nu}\left(\partial_{\rho}\delta\Gamma_{\mu\nu}{}^{\rho} - \Gamma_{\rho\mu}{}^{\sigma}\delta\Gamma_{\sigma\nu}{}^{\rho} - \Gamma_{\rho\nu}{}^{\sigma}\delta\Gamma_{\sigma\mu}{}^{\rho}\right.$$
$$\left. + \Gamma_{\rho\sigma}{}^{\rho}\delta\Gamma_{\mu\nu}{}^{\sigma}\right) - \sqrt{-g}\, g^{\mu\nu}\,\partial_{\nu}\delta\Gamma_{\mu\rho}{}^{\rho} - \partial_{\nu}(\sqrt{-g}\, g^{\mu\nu})\delta\Gamma_{\mu\rho}{}^{\rho}.$$

The last two terms constitute the total derivative $-\partial_{\nu}(\sqrt{-g}\, g^{\mu\nu}\delta\Gamma_{\mu\rho}{}^{\rho})$. On the other hand, by partial integration the sum of the remaining terms is equivalent to

$$-\sqrt{-g}\,(\partial_{\rho}g^{\nu\sigma} + \Gamma_{\rho\mu}{}^{\sigma}g^{\mu\nu} + \Gamma_{\rho\mu}{}^{\sigma}g^{\mu\nu})\delta\Gamma_{\nu\sigma}{}^{\rho} \equiv -\sqrt{-g}\,(\nabla_{\rho}g^{\nu\sigma})\delta\Gamma_{\nu\sigma}{}^{\rho},$$

which vanishes, since $\nabla_{\rho}g^{\mu\sigma} = 0$ [see (1.1.46)].

1.9. Using part (i) of Problems 1.3, and 1.8, we have

$$\delta(\sqrt{-g}\, g^{\mu\nu}R_{\mu\nu}) = \sqrt{-g}\left(-\frac{1}{2}R\, g_{\mu\nu}\,\delta g^{\mu\nu} + R_{\mu\nu}\,\delta g^{\mu\nu} + g^{\mu\nu}\delta R_{\mu\nu}\right)$$
$$= \sqrt{-g}\left(R_{\mu\nu} - \frac{1}{2}g_{\mu\nu}R\right)\delta g^{\mu\nu}.$$

1.10. (i) Note that: $\nabla_{\mu}T^{\mu_1\cdots\mu_k\mu_{k+1}\cdots\mu_n}$
$$= \partial_{\mu}T^{\mu_1\cdots\mu_k\mu_{k+1}\cdots\mu_n} + \Gamma_{\mu\sigma}{}^{\mu_1}T^{\sigma\mu_2\cdots\mu_n} + \cdots + \Gamma_{\mu\sigma}{}^{\mu_n}T^{\mu_1\mu_2\cdots\sigma}.$$
On the other hand,

$$\sqrt{-g}\,\partial_{\mu}T^{\mu_1\cdots\mu_n}S_{\mu_1\cdots\mu_k}{}^{\mu}{}_{\mu_{k+1}\cdots\mu_n} = -(\sqrt{-g}\,\Gamma_{\mu k}{}^{\kappa})T^{\mu_1\cdots\mu_n}S_{\mu_1\cdots\mu_k}{}^{\mu}{}_{\mu_{k+1}\cdots\mu_n}$$
$$- \sqrt{-g}\,T^{\mu_1\cdots\mu_n}\partial_{\mu}S_{\mu_1\cdots\mu_k}{}^{\mu}{}_{\mu_{k+1}\cdots\mu_n}, \;(*)$$

up to a total derivative, where we have used part (ii) of Problem 1.3. By mere relabeling of dummy indices, one may also write

$$\Gamma_{\mu k}{}^{\kappa}\,T^{\mu_1\cdots\mu_n}S_{\mu_1\cdots\mu_k}{}^{\mu}{}_{\mu_{k+1}\cdots\mu_n} = \Gamma_{\mu\sigma}{}^{\mu}\,T^{\mu_1\cdots\mu_n}S_{\mu_1\cdots\mu_k}{}^{\sigma}{}_{\mu_{k+1}\cdots\mu_n},$$

in reference to the first term on the right-hand side of $(*)$ above, also

$$\Gamma_{\mu\sigma}{}^{\mu_1}T^{\sigma\mu_2\cdots\mu_n}S_{\mu_1\cdots\mu_k}{}^{\mu}{}_{\mu_{k+1}\cdots\mu_n} = T^{\mu_1\mu_2\cdots\mu_n}\Gamma_{\mu\mu_1}{}^{\sigma}S_{\sigma\cdots\mu_k}{}^{\mu}{}_{\mu_{k+1}\cdots\mu_n},$$

in reference to the second term on the right-hand side of the expression for $\nabla_{\mu}T^{\mu_1\cdots\mu_k\mu_{k+1}\cdots\mu_n}$ above, and similarly for the other indices

$\mu_2, \ldots, \mu_n$. Accordingly, the left-hand side of (i) in the statement of the problem, may be rewritten, up to a total derivative, as

$$- \sqrt{-g}\, T^{\mu_1 \cdots \mu_n} (\partial_\mu S_{\mu_1 \ldots \mu_k}{}^\mu{}_{\mu_{k+1} \ldots \mu_n} - \cdots + \Gamma_{\mu\sigma}{}^\mu S_{\mu_1 \ldots \mu_k}{}^\sigma{}_{\mu_{k+1} \ldots \mu_n}$$
$$- \Gamma_{\mu\mu_n}{}^\sigma S_{\mu_1 \ldots \mu_k}{}^\mu{}_{\mu_{k+1} \ldots \mu_{n-1}\sigma}) = -\sqrt{-g}\, T^{\mu_1 \cdots \mu_n} \nabla_\mu S_{\mu_1 \ldots \mu_k}{}^\mu{}_{\mu_{k+1} \ldots \mu_n},$$

establishing the first equality.

(ii) The derivation of the second one is essentially the same except the following term $\sqrt{-g}\,\Gamma_{\mu\sigma}{}^\sigma T^{\mu_1 \cdots \mu_k \mu \mu_{k+1} \cdots \mu_n} S_{\mu_1 \ldots \mu_k \mu_k \mu_{k+1} \ldots \mu_n}$ cancels out in the process, as is easily verified upon partial integration. Part (iii) is a special case of (i).

1.11. From part (iii) of Problem 1.10, we may rewrite the integrand of the action W_{matter} as: $-(1/2)\sqrt{-g}\,(-\phi\, g^{\mu\nu}\nabla_\mu(\partial_\nu\phi) + m^2\phi^2)$, leading to the field equation

$$(\Box - m^2)\,\phi = 0, \qquad \Box\phi = g^{\mu\nu}\nabla_\mu(\partial_\nu\phi) \equiv g^{\mu\nu}\nabla_\mu\nabla_\nu\,\phi.$$

(ii) From part (iii) of Problem 1.5, we obtain

$$T_{\mu\nu} = \partial_\mu\phi\,\partial_\nu\phi - \frac{1}{2}\,g_{\mu\nu}(\partial^\rho\phi\,\partial_\rho\phi + m^2\phi^2).$$

(iii) With suitable relabeling of dummy indices, we explicitly have

$$\nabla_\mu T^{\mu\nu} = (\Box\phi - m^2\phi)\,\partial^\nu\phi + g^{\sigma\nu}(\nabla_\mu\partial_\sigma\phi - \nabla_\sigma\partial_\mu\phi)\,\partial^\mu\phi.$$

But $\nabla_\mu\partial_\sigma\phi = \partial_\mu\partial_\sigma\phi - \Gamma_{\mu\sigma}{}^\lambda\partial_\lambda\phi = \nabla_\sigma\partial_\mu\phi$, leading to

$$\nabla_\mu T^{\mu\nu} = (\Box\phi - m^2\phi)\,\partial^\nu\phi,$$

which vanishes as a consequence of the field equation. Due to symmetry in (μ, ν), this also holds for $\nabla_\nu T^{\mu\nu} = 0$.

1.12. $\delta\mathcal{L}$ is worked out to be

$$\left(\partial^\mu\Lambda^\nu + \partial^\nu\Lambda^\mu - \eta^{\mu\nu}\partial.\Lambda\right)\left(\partial_\mu\Gamma_{\rho\nu}{}^\rho - \partial_\rho\Gamma_{\mu\nu}{}^\rho\right) + \xi^{\mu\nu}\left(\partial_\mu\partial_\rho\partial_\nu\Lambda^\rho - \partial_\rho\partial_\mu\partial_\nu\Lambda^\rho\right)$$
$$+ \eta^{\mu\nu}\left(\Gamma_{\mu\nu}{}^\lambda\partial_\lambda\partial_\rho\Lambda^\rho + \Gamma_{\lambda\rho}{}^\rho\partial_\mu\partial_\nu\Lambda^\lambda - \Gamma_{\rho\nu}{}^\lambda\partial_\lambda\partial_\mu\Lambda^\rho - \Gamma_{\lambda\mu}{}^\rho\partial_\rho\partial_\nu\Lambda^\lambda\right)$$
$$= -\left(\partial^\mu\Gamma_{\mu\lambda}{}^\lambda - \partial^\lambda\Gamma_\mu{}^\mu{}_\lambda\right)\partial_\rho\Lambda^\rho + \left(\partial^\mu\Gamma_{\mu\lambda}{}^\lambda - \partial^\lambda\Gamma_\mu{}^\mu{}_\lambda\right)\partial_\rho\Lambda^\rho = 0,$$

up to total derivatives, after cancelation of various terms.

1.13. Let us start from the right-hand side of (1.4.4) which in detail reads

$$(g_{\mu\lambda} + h_{\mu\lambda})\partial_\nu\Lambda^\lambda + (g_{\nu\lambda} + h_{\nu\lambda})\partial_\mu\Lambda^\lambda$$
$$+ \Lambda^\lambda\left(\partial_\lambda h_{\mu\nu} + (g_{\mu\sigma} + h_{\mu\sigma})\Gamma_{\nu\lambda}{}^\sigma + (g_{\nu\sigma} + h_{\nu\sigma})\Gamma_{\mu\lambda}{}^\sigma - h_{\sigma\nu}\Gamma_{\mu\lambda}{}^\sigma - h_{\sigma\mu}\Gamma_{\nu\lambda}{}^\sigma\right).$$

Note that the terms involving $h_{..} \Gamma_{..}{}^{\cdot}$ in the second line cancel out, while the corresponding ones with $g_{..}$ give $\partial_\lambda g_{\mu\nu}$ on account of the fact that $\nabla_\lambda g_{\mu\nu} = 0$. This establishes the statement of the problem. A more direct way of seeing this is that (1.4.3) as a tensor equation and must hold with the partial derivatives in it replaced by covariant ones. After having done this, make use of the fact that $\nabla_\lambda g_{\mu\nu} = 0$ to obtain (1.4.4).

1.14. $\Gamma_{\mu\nu}{}^\sigma = (1/2) g^{\sigma\lambda} \left(\partial_\mu g_{\nu\lambda} + \partial_\nu g_{\mu\lambda} - \partial_\lambda g_{\mu\nu} \right)$. Now for $g_{\mu\nu} \to g_{\mu\nu} + h_{\mu\nu}$, $\delta g^{\mu\nu} = -h^{\mu\nu} + h^{\mu\lambda} h_\lambda{}^\nu$, with the latter given in (1.4.6),

$$\delta\Gamma_{\mu\nu}{}^\sigma = -\frac{1}{2} \left(h^{\sigma\lambda} - h^{\sigma\rho} h_\rho{}^\lambda \right) \left(\partial_\mu g_{\nu\lambda} + \partial_\nu g_{\mu\lambda} - \partial_\lambda g_{\mu\nu} \right)$$

$$+ \frac{1}{2} g^{\sigma\lambda} \left(\partial_\mu h_{\nu\lambda} + \partial_\nu h_{\mu\lambda} - \partial_\lambda h_{\mu\nu} \right) - \frac{1}{2} h^{\sigma\lambda} \left(\partial_\mu h_{\nu\lambda} + \partial_\nu h_{\mu\lambda} - \partial_\lambda h_{\mu\nu} \right).$$

Although $\Gamma_{\mu\nu}{}^\sigma$ is not a tensor, $\delta\Gamma_{\mu\nu}{}^\sigma$ is a tensor (see Problem 1.6). Accordingly, in a local Lorentz coordinate system, at the point in question, $\partial_\lambda g_{\mu\nu} = 0$. Hence, in a general coordinate system we may simply replace partial derivatives by covariant ones to obtain

$$\delta\Gamma_{\mu\nu}{}^\sigma = \frac{1}{2} \left(\nabla_\mu h_\nu{}^\sigma + \nabla_\nu h_\mu{}^\sigma - \nabla^\sigma h_{\mu\nu} \right) - \frac{1}{2} h^{\sigma\lambda} \left(\nabla_\mu h_{\nu\lambda} + \nabla_\nu h_{\mu\lambda} - \nabla_\lambda h_{\mu\nu} \right).$$

1.15. For a given function f: $\delta(1/f) = -(1/f^2) \delta f$. Hence from Problem 1.3,(i), $\delta(1/\sqrt{-g}) = (1/2\sqrt{-g}) g_{\rho\kappa} \delta g^{\rho\kappa}$, from which the statement of the problem follows.

1.16. With $\eta^{\mu\nu\lambda\sigma} = \varepsilon^{\mu\nu\lambda\sigma}/\sqrt{-g}$, $\eta_{\mu\nu\lambda\sigma} = \sqrt{-g}\, \varepsilon_{\mu\nu\lambda\sigma}$, we have the identity

$$\eta^{\mu\nu\rho\sigma} \eta_{\lambda\kappa\gamma\epsilon} = -\sum_{P[\mu\nu\rho\sigma]} \mathrm{sgn}_P\, \delta^\mu{}_\lambda \delta^\nu{}_\kappa \delta^\rho{}_\gamma \delta^\sigma{}_\epsilon,$$

where $\sum_{P[\mu\nu\rho\sigma]} \mathrm{sgn}_P$ stands for a summation over all permutations of $\{\mu\nu\rho\sigma\}$ with corresponding signs attached. Upon writing

$$R_{\mu\nu\rho\sigma} R_{\lambda\kappa\gamma\epsilon}\, \eta^{\mu\nu\lambda\kappa}\, \eta^{\rho\sigma\gamma\epsilon} = R_{\mu\nu}{}^{\rho\sigma} R_{\lambda\kappa}{}^{\gamma\epsilon}\, \eta^{\mu\nu\lambda\kappa}\, \eta_{\rho\sigma\gamma\epsilon},$$

the previous identity leads to the expression given in the problem.

1.17. The Bianchi identity (1.1.44) reads

$$\nabla_\lambda R_{\mu\nu\rho\sigma} + \nabla_\rho R_{\mu\nu\sigma\lambda} + \nabla_\sigma R_{\mu\nu\lambda\rho} = 0.$$

Upon multiplying the latter by $\eta^{\rho\sigma\alpha\lambda}$ and conveniently relabeling the indices gives:

$$0 = [\eta^{\rho\sigma\alpha\lambda} + \eta^{\lambda\rho\alpha\sigma} + \eta^{\sigma\lambda\alpha\rho}] \nabla_\lambda R_{\mu\nu\rho\sigma} = 3\,\eta^{\rho\sigma\alpha\lambda} \nabla_\lambda R_{\mu\nu\rho\sigma},$$

by finally using the totally anti-symmetric nature of $\eta^{\rho\sigma\alpha\lambda}$.

1.18. Since no indices in $\eta^{\mu\nu\alpha\beta}$, due to its complete antisymmetry, may be equal, it is sufficient to establish the above result for η^{0123}. In detail

$$\nabla_\lambda \eta^{0123} = \varepsilon^{0123} \partial_\lambda \frac{1}{\sqrt{-g}} + \Gamma_{\lambda 0}{}^0 \eta^{0123} + \Gamma_{\lambda 1}{}^1 \eta^{0123} + \Gamma_{\lambda 2}{}^2 \eta^{0123} + \Gamma_{\lambda 3}{}^3 \eta^{0123}$$

$$= \varepsilon^{0123} \left(\partial_\lambda \frac{1}{\sqrt{-g}} + \frac{1}{\sqrt{-g}} \Gamma_{\lambda\sigma}{}^\sigma \right).$$

From Problem 1.15 and Problem 1.3 (ii): $\partial_\lambda(1/\sqrt{-g}) = -(1/\sqrt{-g})\Gamma_{\lambda\sigma}{}^\sigma$, and the statement in the problem follows.

1.19. Starting from (1.9.10), we have by an elementary iteration procedure

$$\chi(\gamma(\bar{s})) \simeq \left[1 + (i) \int_{\underline{s}}^{\bar{s}} ds \, \dot{\gamma}^a(s) \, A_a(\gamma(s)) \right] \chi(\gamma(\underline{s})),$$

$$\vdots$$

$$\chi(\gamma(\bar{s})) = \left[1 + \sum_{n \geq 1} (i)^n \int_{\underline{s}}^{\bar{s}} ds_n \int_{\underline{s}}^{s_n} ds_{n-1} \cdots \int_{\underline{s}}^{s_2} ds_1 \right.$$

$$\left. \times \dot{\gamma}^{a_n}(s_n) A_{a_n}(\gamma(s_n)) \ldots \dot{\gamma}^{a_1}(s_1) A_{a_1}(\gamma(s_1)) \right] \chi(\gamma(\underline{s})).$$

Upon using the path ordering notation

$$\mathscr{P}\left(A_{a_{i_1}}(\gamma(s_{i_1})) \ldots A_{a_{i_k}}(\gamma(s_{i_k})) \right) = A_{a_1}(\gamma(s_1)) \ldots A_{a_k}(\gamma(s_k)),$$

for $s_1 \geq \cdots \geq s_k$, where $\{i_1, \ldots, i_k\}$ is any permutation of $1, \ldots, k$, we obtain

$$\chi(\gamma(\bar{s})) = \mathscr{P}\left(\sum_{n \geq 0} \frac{(i)^n}{n!} \left[\int_{\underline{s}}^{\bar{s}} ds \, \dot{\gamma}^a(s) A_a(\gamma(s)) \right]^n \right) \chi(\gamma(\underline{s})),$$

with $n!$ corresponding to the $n!$ possible permutations of the indices $\{1, 2, \ldots, n\}$, for a given n. The above leads to the expression of the holonomy in (1.9.12).

Chapter 2

2.1. From the second identity in (2.1.9), we have $\gamma^0 K = (\gamma^0 K^{-1})^\dagger$. Therefore

$$\bar{\epsilon} K \gamma^\mu K^{-1} \epsilon = \epsilon^\dagger (\gamma^0 K^{-1})^\dagger \gamma^\mu K^{-1} \epsilon = (K^{-1}\epsilon)^\dagger \gamma^0 \gamma^\mu K^{-1} \epsilon = \overline{(K^{-1}\epsilon)} \gamma^\mu (K^{-1}\epsilon).$$

2.2. From (2.1.10), (2.1.11),

$$x'' = \Lambda'\left(\Lambda x + \frac{i}{2}\bar{\epsilon}\gamma K\epsilon - b\right) + \frac{i}{2}\bar{\epsilon}'\gamma K'\left(K\theta + \epsilon\right) - b',$$

which upon using the identity $\Lambda'\gamma = K'^{-1}\gamma K'$, given in (2.1.9), and by collecting terms, the transformation rule in (2.1.14) emerges. The transformation rule in (2.1.15) simply follows by replacing θ' by $K\theta + \epsilon$ in $\theta'' = K'\theta' + \epsilon'$.

2.3. (i) The first identity was derived in (2.2.18). The second follows upon taking the adjoint of the first, multiplying by γ^0 from the right, and using the properties $\{\gamma^5, \gamma^\sigma\} = 0$, $(\gamma^\mu)^\dagger\gamma^0 = \gamma^0\gamma^\mu$. The fourth follows upon taking the adjoint of the third and multiplying from the right by γ^0. The third follows by choosing, in turn,

$$(A_1 = \gamma^5\gamma^\mu, A_2 = I), \quad (A_1 = \gamma^5, A_2 = \gamma^\mu),$$

in (2.2.16), and adding the results.

(ii) The first follows by multiplying the first identity in (2.2.22) by $(\bar{\theta}\gamma^5)_a$. The second follows by multiplying the first in (2.2.22) by $(\bar{\theta}\gamma^5\gamma^\mu)_a$ and using the identity $\bar{\theta}\gamma^\mu\theta = 0$ in (2.2.11). The third follows upon multiplying the third identity in (2.2.22) by $(\bar{\theta}\gamma^5)_a$. The third one also follows by multiplying the fourth identity in (2.2.22) by $(\gamma^5\theta)_a$. The fourth one follows by multiplying the third identity in (2.2.22) by $(\bar{\theta}\gamma^5\gamma^\sigma)_a$, using the identity $\gamma^\sigma\gamma^\mu = [\gamma^\sigma, \gamma^\mu]/2 - \eta^{\mu\nu}$, and, from the last equality in (2.2.11), that $\bar{\theta}\gamma^5[\gamma^\sigma, \gamma^\mu]\theta = 0$. Finally the fifth one follows upon multiplying the first one in (2.2.22) by $\bar{\theta}_a$.

(iii) These easily follow upon multiplying (2.2.21), in turn, by $\bar{\theta}\theta$, $\bar{\theta}\gamma^5\theta$, $\bar{\theta}\gamma^5\gamma^\sigma\theta$ and making use of the equalities in (2.2.23), (2.2.24).

2.4. (i) This identity follows simply upon multiplying (2.2.21) by θ_c, using the identities in (2.2.22), anti-symmetrizing with respect to the indices a, b, c, and finally using the Fierz identity (A-2.1) in Chap. 2. (ii) This identity follows by multiplying (2.2.21) by $\theta_c\theta_d$, using the identities in (2.2.25), (2.2.26), anti-symmetrizing with respect to the indices a, b, c, d, and finally using the Fierz identity (A-2.2).

2.5. Multiply the classic Fierz identity (A-2.3) by $(\gamma^5)_{a'a}\mathcal{C}_{dd'}$ to obtain

$$(\gamma^5\gamma^\mu)_{a'b}(\gamma_\mu\mathcal{C})_{cd'} = -(\gamma^5\mathcal{C})_{a'd'}\delta_{cb} - \frac{1}{2}(\gamma^5\gamma^\mu\mathcal{C})_{a'd'}(\gamma_\mu)_{cb}$$

$$-\frac{1}{2}(\gamma^\mu\mathcal{C})_{a'd'}(\gamma^5\gamma_\mu)_{cb} + \mathcal{C}_{a'd'}(\gamma^5)_{cb}.$$

Make a copy of this equation by making the replacements: $a' \to a$, $b \to d$, $d' \to k$, and another copy, this time making replacements:

$$a' \to k, b \to d, d' \to a,$$

and add the two resulting equations, using, in the process:

$(\gamma^5 \gamma^\mu \mathscr{C})_{ka} = -(\gamma^5 \gamma^\mu \mathscr{C})_{ak}, (\gamma^5 \mathscr{C})_{ka} = -(\gamma^5 \mathscr{C})_{ak}, (\gamma^\mu \mathscr{C})_{ka} = (\gamma^\mu \mathscr{C})_{ak},$
and the identity in question follows.

2.6. Using the reality of the matrices, γ^0, $\mathscr{C}$, the identity $\{\gamma^0, \mathscr{C}\} = 0$, and the property $\mathscr{C}^T = \mathscr{C}^{-1} = -\mathscr{C}$, we obtain

$$(\overline{\psi}_+)_c = (\overline{\psi}_c + (\gamma^0)_{ca}(\mathscr{C})_{ab}(\psi)_k(\gamma^0)_{kb})/2,$$

which upon multiplying by $\mathscr{C}_{dc}$, gives $\mathscr{C}\overline{\psi}_+^T = (\mathscr{C}\overline{\psi}^T + \psi)/2 = \psi_+$. Similarly,

$$(\overline{\psi}_-)_c = \mathrm{i}(\overline{\psi}_c - (\gamma^0)_{ca}(\mathscr{C})_{ab}(\psi)_k(\gamma^0)_{kb})/2,$$

which upon multiplying by $\mathscr{C}_{dc}$, gives $\mathscr{C}\overline{\psi}_-^T = \mathrm{i}(\mathscr{C}\overline{\psi}^T - \psi)/2 = \psi_-$.

2.7. Using the definition $\theta_b = \mathscr{C}_{bk}\overline{\theta}_k$ gives $(\partial/\partial\overline{\theta}_a)\mathscr{C}_{bk}\overline{\theta}_k = \mathscr{C}_{ba}$. Also $\mathscr{C}^{-1}\gamma^\mu \mathscr{C} = -(\gamma^\mu)^T$, $\mathscr{C}^T = -\mathscr{C}$ imply that $(\gamma^\mu \mathscr{C})_{ab} = (\gamma^\mu \mathscr{C})_{ba}$. Hence from the definition of D_a in (2.3.9), the anti-commutator in (2.3.10) follows after carrying one differentiation with respect to $\overline{\theta}_a$ as spelled out above. The anti-commutator in (2.3.11) emerges from the definition $\overline{D}_c = -D_b \mathscr{C}_{bc}^{-1}$, and by multiplying the first one by $-\mathscr{C}_{bc}^{-1}$ giving $\{D_a, \overline{D}_c\} = \mathrm{i}(\gamma^\mu)_{ac} \partial_\mu$.

2.8. Using the property $(\mathscr{C}\gamma^\mu)_{ab} = (\mathscr{C}\gamma^\mu)_{ba}$, we have

$$\overline{D}\gamma^\mu D = (\mathscr{C}\gamma^\mu)_{ab}\{D_a, D_b\}/2 = (\mathrm{i}/2)\,\mathrm{Tr}\,(\gamma^\mu \gamma^\sigma)\partial_\sigma = -2\,\mathrm{i}\,\partial^\mu,$$

where we have used the anti-commutator in (2.3.10), $\mathscr{C} = -\mathscr{C}^{-1}$. Similarly, $\overline{D}[\gamma^\mu, \gamma^\nu]D = (\mathscr{C}[\gamma^\mu, \gamma^\nu])_{ab}\{D_a, D_b\}/2 = 0$, as it gives rise to the trace of an odd number of gamma matrices.

2.9. From (2.2.14) $\mathscr{C}^{-1}B\,\mathscr{C} = B^T$, which implies that $(\mathscr{C}B)_{ab} = -(\mathscr{C}B)_{ba}$. The anti-commutator in (2.3.10) implies that $D_a D_b = -D_b D_a - \mathrm{i}(\gamma^\mu \mathscr{C})_{ab}\partial_\mu$. Hence

$$\overline{D}BD\,D_a = D_c(\mathscr{C}B)_{cd}D_d\,D_a = -D_c(\mathscr{C}B)_{cd}[D_a D_d + \mathrm{i}(\gamma^\mu \mathscr{C})_{da}\partial_\mu]$$

$$= [D_a D_c + \mathrm{i}(\gamma^\mu \mathscr{C})_{ca}\partial_\mu](\mathscr{C}BD)_c - \mathrm{i}\,D_c(\mathscr{C}B\gamma^\mu \mathscr{C})_{ca}\partial_\mu,$$

which is equal to $D_a(\overline{D}BD) - 2\,\mathrm{i}\,(\gamma^\mu BD)_a \partial_\mu$, where we have used the property $(\mathscr{C}B\gamma^\mu \mathscr{C})_{ca} = (\gamma^\mu B)_{ac}$ on account that $\mathscr{C} = -\mathscr{C}^{-1}$

2.10. • (2.3.18): Simply choose $B = I$ in (2.3.13) to obtain this identity.
   • (2.3.19): Multiply (2.3.17) by $(\gamma^5)_{ba}$, and choose $A = I$. Add the resulting equation to the one in (2.3.17), with $A = \gamma^5$, b replaced by a, and simplify to obtain this identity.
   • (2.3.20): Choose $B = \gamma^5$ in (2.3.13) to obtain

$$\overline{D}\gamma^5 D\,D_a = D_a \overline{D}\gamma^5 D - 2\,\mathrm{i}(\gamma^\mu \gamma^5 D)_a \partial_\mu = D_a \overline{D}\gamma^5 D + 2\,\mathrm{i}(\gamma^5 \gamma^\mu D)_a \partial_\mu,$$

and then use the identity established in (2.3.19) to replace $\overline{D}\gamma^5DD_a$ by $-\overline{D}D(\gamma^5D)_a$.

- (2.3.21): Multiply (2.3.20) by $(\overline{D}\gamma^5)_a$ to obtain

$$(\overline{D}\gamma^5D)^2 = -\overline{D}_b\overline{D}DDD_b - 2i(\overline{D}\gamma^\mu D)\partial_\mu,$$

and then replace $\overline{D}DDD_b$ by the right-hand side of the identity established in (2.3.18), with a replaced by b, and simplify to obtain this identity.

- (2.3.22): Choose $A = \gamma^5\gamma^\sigma$ in (2.3.17) with b replaced by a, and add the resulting equation to the one obtained from (2.3.17) multiplied by $(\gamma^5\gamma^\sigma)_{ab}$, and $A = I$, and simplify to obtain this identity.
- (2.3.23): Choose $A = \gamma^5\gamma^\sigma$ in (2.3.17), then use the result just established in (2.3.22) to replace $(\overline{D}\gamma^5\gamma^\sigma D)D_a$ by the expression on the right-hand side of the latter equation and simplify.

2.11. From (2.1.10), $\partial x'/\partial x = \Lambda$, $\partial x'/\partial\theta = -i\bar{\epsilon}\gamma K/2$. On the other hand, from (2.4.13), $\partial\theta'/\partial x = 0$, $\partial\theta'/\partial\theta = K$, which lead to the expression of the matrix in question.

2.12. We carry out an expansion of the logarithm as follows:

$$\ln[I - (I - M)] = -\sum_{n\geq1}\begin{pmatrix} I-\Lambda & -T \\ 0 & I-K \end{pmatrix}^n \Big/ n,$$

$$\begin{pmatrix} I-\Lambda & -T \\ 0 & I-K \end{pmatrix}\begin{pmatrix} I-\Lambda & -T \\ 0 & I-K \end{pmatrix} = \begin{pmatrix} (I-\Lambda)^2 & -[(I-\Lambda)T+T(I-K)] \\ 0 & (I-K)^2 \end{pmatrix},$$

$$\begin{pmatrix} I-\Lambda & -T \\ 0 & I-K \end{pmatrix}^n = \begin{pmatrix} (I-\Lambda)^n & C_n \\ 0 & (I-K)^n \end{pmatrix},$$

$C_n = -[(I-\Lambda)^{n-1}T + (I-\Lambda)^{n-2}T(I-K) + \ldots + T(I-K)^{n-1}]$, which corresponds to the expression in (2.4.20).

2.13. We may write

$$\begin{pmatrix} C & \eta \\ \xi & D \end{pmatrix} = \begin{pmatrix} C & 0 \\ \xi & I \end{pmatrix}\begin{pmatrix} I & C^{-1}\eta \\ 0 & D-\xi C^{-1}\eta \end{pmatrix},$$

Sdet of the first matrix on the right-hand side of the above equation, is $\det C/\det I$, while for the second one is $\det I/\det(D - \xi C^{-1}\eta)$, and the stated result follows upon their multiplication.

2.14. Upon adding (2.3.18), (2.3.19), and dividing by 2 gives:

$$\overline{D}D^RD_a = (1/2)D_a\overline{D}D - (1/2)\overline{D}D(\gamma^5D)_a - i(\gamma^\mu D)_a\partial_\mu.$$

On the other hand, (2.3.20) gives: $\overline{D}D(\gamma\,^5D)_a = -D_a\overline{D}\gamma\,^5D - 2i(\gamma\,^5\gamma^\mu D)_a\partial_\mu$. From these two equations, the statement of the problem follows upon multiplication by $(1-\gamma\,^5)/2$ and using the property $\{\gamma\,^5, \gamma^\mu\} = 0$.

2.15. (i) From (2.2.5), (2.2.6) $\overline{\theta}\xi_L = -\theta^T\mathscr{C}^{-1}[(1-\gamma\,^5)/2]\xi$, $\overline{\theta}\xi_R = -\theta^T\mathscr{C}^{-1}[(1+\gamma\,^5)/2]\xi$,

$$\mathscr{C}^{-1}[(1-\gamma\,^5)/2] = \begin{pmatrix} 0 & 0 \\ 0 & -i\sigma^2 \end{pmatrix}, \qquad \mathscr{C}^{-1}[(1+\gamma\,^5)/2] = \begin{pmatrix} i\sigma^2 & 0 \\ 0 & 0 \end{pmatrix}.$$

The first two results in (i) then immediately follow. From Box 2.1 in the beginning of Sect. 2.6,

$$\overline{\theta}\gamma^\mu\gamma^\nu\theta = -\overline{\theta}\theta\,\eta^{\mu\nu}, \qquad \overline{\theta}\gamma^\mu\gamma^\nu\gamma\,^5\theta = -\overline{\theta}\gamma\,^5\theta\,\eta^{\mu\nu},$$

which when combined lead to the third result in (i). The last one in (i) follows by adding $0 = \overline{\theta}\gamma^\mu\theta$ to $\overline{\theta}\gamma\,^5\gamma^\mu\theta$ and using $\overline{\theta} = -\theta^T\mathscr{C}^{-1}$. Using the fact that $\{\gamma\,^5, \gamma\,^0\} = 0$, we have

$$(\overline{\theta}\xi_L)^\dagger = \overline{\xi}\theta_R = -\xi\mathscr{C}^{-1}\frac{1}{2}(1+\gamma\,^5)\theta,$$

and the first statement in (ii) then follows upon using the definition $\mathscr{C}^{-1}\theta = \overline{\theta}^T$, $[\mathscr{C}^{-1}, \gamma\,^5] = 0$, and the anti-commutativity of the components of $\overline{\theta}$ and of ξ. Similarly, using the facts that $(\gamma^\mu)^\dagger\gamma\,^0 = \gamma\,^0(\gamma^\mu)$, $\{\gamma\,^5, \gamma\,^0\} = 0$, we have

$$(\overline{\theta}\gamma^\mu[(1-\gamma\,^5)/2]\xi)^\dagger = \overline{\xi}\gamma^\mu\theta_L = -\xi^T(\mathscr{C}^{-1}\gamma^\mu[(1-\gamma\,^5)/2]\mathscr{C})\mathscr{C}^{-1}\theta,$$

$$\mathscr{C}^{-1}\gamma^\mu\mathscr{C} = -(\gamma^\mu)^T.$$

The last equality then follows from the identity $\{\gamma^\mu, \gamma\,^5\} = 0$, and the anti-commutativity of the spinor field components.

2.16. Using the expression of Λ in (2.6.97), and part (ii) of the previous Problem, together with the identities: $(\overline{\theta}\gamma\,^5\gamma^\mu\theta)^\dagger = \overline{\theta}\gamma\,^5\gamma^\mu\theta$, $(\overline{\theta}\gamma\,^5\theta)^\dagger = -\overline{\theta}\gamma\,^5\theta$, lead directly to the expression given in the problem.

2.17. It is easily verified that $\mathscr{V}^\rho(x,\theta)$ in (2.6.61), may be rewritten as

$$e^{i[\overline{\theta}\gamma\,^5\gamma^\mu\theta\partial_\mu]/4}\left[V^\rho(x) + \frac{i}{\sqrt{2}}\overline{\theta}\gamma\,^\rho\chi(x) + \overline{\theta}\gamma\,^5\gamma_\lambda\theta\left(A^{\lambda\rho}(x) - \frac{i}{4}\partial^\lambda V^\rho(x)\right)\right.$$

$$\left. + \overline{\theta}\gamma\,^5\theta\,\overline{\theta}\left(B^\rho(x) + \frac{1}{4\sqrt{2}}\gamma^\mu\gamma\,^\rho\partial_\mu\chi(x)\right)\right], \quad (*)$$

and note that the quadratic term $(i\overline{\theta}\gamma\,^5\gamma^\mu\theta\partial_\mu/4)^2/2$ coming from the exponential generates also a term $-\Box V^\rho/32$ from the θ-independent term V^ρ

within the square brackets in (*) which combines with the $-(i/4)(i/4)\partial^\lambda V^\rho$ coming from the term, next to $A^{\lambda\rho}$, within the round brackets in (*), to give the $-(i/8)\partial^\lambda V^\rho$ term next to $A^{\lambda\rho}$, within the round brackets in the last term in (2.6.61). Also note that $\overline{\theta}\gamma^{\mu}\theta\,\overline{\theta}\gamma^\rho = \overline{\theta}\gamma^5\theta\,\overline{\theta}\gamma^\mu\gamma^\rho$, as obtained from the last identity in (2.2.22). On the other hand, since the exponential term represents the translation operator of the argument x^μ of the component fields by $i\overline{\theta}\gamma^5\gamma^\mu\theta/4$, our expression for $\mathscr{V}^\rho(x,\theta)$, in the Wess-Zumino supergauge, becomes simply as given in (2.6.66).

2.18. Let us work it out for the abelian case first. We have: $\varepsilon^{\mu\nu\alpha\beta}F_{\mu\nu}F_{\alpha\beta} = 4\,\varepsilon^{\mu\nu\alpha\beta}(\partial_\mu V_\nu\partial_\alpha V_\beta) = 4\,\varepsilon^{\mu\nu\alpha\beta}[\partial_\mu(V_\nu\partial_\alpha V_\beta) - V_\nu(\partial_\mu\partial_\alpha)V_\beta]$. The second term within the square brackets gives zero since it is symmetric in $(\mu\alpha)$ while $\varepsilon^{\mu\nu\alpha\beta}$ is anti-symmetric in interchanging these two indices. This establishes the statement of the problem for the abelian case. For the non-abelian case, we have

$$\varepsilon^{\mu\nu\alpha\beta}G_{A\mu\nu}G_{A\alpha\beta} = 4\varepsilon^{\mu\nu\alpha\beta}\left(\partial_\mu V_{A\nu} + \frac{g}{2}f_{ABC}V_{B\mu}V_{C\nu}\right)\left(\partial_\alpha V_{A\beta} + \frac{g}{2}f_{ADE}V_{D\alpha}V_{E\beta}\right),$$

where note the symmetry in the interchange $(\mu\nu) \leftrightarrow (\alpha\beta)$. Upon using the equalities $\varepsilon^{\mu\nu\alpha\beta}f_{ABC} = \varepsilon^{\mu\beta\alpha\nu}f_{CBA} = \varepsilon^{\mu\nu\beta\alpha}f_{ACB}$, the above equation becomes

$$4\,\varepsilon^{\mu\nu\alpha\beta}\left(\partial_\mu\left(V_{A\nu}\partial_\alpha V_{A\beta}\right) - V_{A\nu}(\partial_\mu\partial_\alpha)V_{A\beta}\right)$$

$$+\frac{g}{3}f_{ABC}\partial_\mu\left(V_{A\nu}V_{B\alpha}V_{C\beta}\right) + \frac{g^2}{4}f_{ABC}f_{ADE}V_{B\mu}V_{C\nu}V_{D\alpha}V_{E\beta}\right).$$

Only the last term needs to be considered. The last factor multiplying g^2, may be rewritten as

$$-2\,\varepsilon^{\mu\nu\alpha\beta}\mathrm{Tr}\left([V_\mu, V_\nu][V_\alpha, V_\beta]\right),$$

where we have used the normalization $\mathrm{Tr}(t_A t_A') = \delta_{AA'}/2$, and that

$$[V_\mu, V_\nu] = i\,t_A f_{ABC}V_{B\mu}V_{C\nu}.$$

Using the anti-symmetry property of $\varepsilon^{\mu\nu\alpha\beta}$ under the exchange of its indices, and by simply relabeling them, the last term is then equal to

$$g^2\varepsilon^{\mu\nu\alpha\beta}f_{ABC}f_{ADE}V_{B\mu}V_{C\nu}V_{D\alpha}V_{E\beta} = -8\,g^2\varepsilon^{\mu\nu\alpha\beta}\mathrm{Tr}[V_\mu V_\nu V_\alpha V_\beta].$$

The latter is obviously zero, since, for example, the trace factor does not change for $(\mu,\nu,\alpha,\beta) \rightarrow (\nu,\alpha,\beta,\mu)$, while $\varepsilon^{\mu\nu\alpha\beta} \rightarrow -\varepsilon^{\nu\alpha\beta\mu}$. This establishes the statement of the problem in the non-abelian case as well.

2.19. The part of the Lagrangian density in (2.10.1) depending on the Majorana field ψ may be rewritten as

$$\frac{1}{2}\psi_a\left[\mathscr{C}^{-1}\left(\frac{\gamma\partial}{i}+m\right)\right]_{ab}\psi_b+\bar{\eta}_a\psi_a+\sqrt{2}\lambda\left[\varphi_1\,\mathscr{C}^{-1}+i\,\varphi_2\,(\mathscr{C}^{-1}\gamma^{\,5})\right]_{ab}\}\psi_a\psi_b.$$

We recall the anti-symmetry of the charge conjugation matrix $\mathscr{C}$, its commutativity with γ^5, the identity $\mathscr{C}^{-1}\gamma^\mu\,\mathscr{C}=-(\gamma^\mu)^\top$, and the anti-commutativity of ψ_a, ψ_b. Integrating by parts, (2.10.4) follows from the action principle by multiplying, in the process, the resulting equation by $(-i\gamma\partial+m)^{-1}\mathscr{C}$, and finally taking the vacuum expectation value. In particular, in the absence of interaction, we recognize the familiar equation of a spinor field in the presence of an external source $\eta=\mathscr{C}\bar{\eta}^\top$.

2.20. From (2.10.4), $\langle\,(\varphi_1(x')\overline{\psi}_a(x'')\psi_a(x))_+\,\rangle\big|_1$ is equal to

$$(-2\sqrt{2}\lambda)\left(\frac{i\gamma\partial+m}{-\Box+m^2}\right)_{bc}\langle\,(\varphi_1(x')\varphi_1(x)\overline{\psi}_a(x'')\psi_c(x))_+\rangle\big|_0$$

$$=-2\sqrt{2}\,\lambda\left(\frac{i\gamma\partial+m}{-\Box+m^2}\right)_{ac}\{(-i)\Delta_+(x'-x)\,i\,S_{+ca}(x-x'')\}$$

$$=-2\sqrt{2}\,\lambda\int(dz)\Delta_+(x'-z)S_{+ca}(z-x'')\left(\frac{i\gamma\partial+m}{-\Box+m^2}\right)_{ac}\delta^{(4)}(x-z)$$

$$=-2\sqrt{2}\,\lambda\int(dz)\Delta_+(x'-z)S_{+ca}(z-x'')S_{+ac}(x-z).$$

By Fourier transform, and taking the trace, the above integral, now applied to $\langle\,(\varphi_1(x')\overline{\psi}_a(x)\psi_a(x))_+\,\rangle\big|_1$, takes the form

$$4\int(dz)\frac{(dk)(dk_1)(dk_2)}{(2\pi)^{12}}\frac{e^{ik(x'-z)}}{k^2+m^2}\frac{e^{ik_1(z-x)}}{k_1^2+m^2}\frac{e^{ik_2(x-z)}}{k_2^2+m^2}(-k_1k_2+m^2).$$

Upon using the identity

$$(-k_1k_2+m^2)=(1/2)\{[(k_1-k_2)^2+m^2]-[k_1^2+m^2]-[k_2^2+m^2]+3m^2\},$$

integrating over z, and k, we readily obtain the three expressions given in (2.10.16) upon multiplying by $(-2\sqrt{2}\,\lambda)$.

2.21. We use the fact that on may write $\gamma^\alpha\gamma^5=(i/3!)\,\varepsilon^{\,\alpha\rho\omega\lambda}\,\gamma_\rho\gamma_\omega\gamma_\lambda$, and the identity $(\varepsilon^{\,0123}=+1,\ \varepsilon_{0123}=-1)$

$$\varepsilon_{\alpha\beta\mu\nu}\,\varepsilon^{\alpha\rho\omega\lambda}=-\,\delta^\rho{}_\beta(\delta^\omega{}_\mu\delta^\lambda{}_\nu-\delta^\omega{}_\nu\delta^\lambda{}_\mu)-\delta^\rho{}_\mu(\delta^\omega{}_\nu\delta^\lambda{}_\beta-\delta^\omega{}_\beta\delta^\lambda{}_\nu)$$

$$-\,\delta^\rho{}_\nu\,(\delta^\omega{}_\beta\delta^\lambda{}_\mu-\delta^\omega{}_\mu\delta^\lambda{}_\beta).$$

Upon substituting the above two equalities in the above expression given in the statement of the problem, and using the elementary anti-commutativity properties of the gamma matrices, we readily recover our earlier expression for the Lagrangian density in (2.15.1) by using now the Latin alphabet for Lorentz indices.

2.22. By definition, $[\mathscr{D}_\mu, \mathscr{D}_\nu]$ follows directly to be given by

$$\frac{1}{8}[\gamma^a, \gamma^b]\left(\partial_\nu(\omega_\mu)_{ab} - \partial_\mu(\omega_\nu)_{ab}\right) - \frac{1}{4}(\omega_\nu)_{cd}(\omega_\mu)_{ab}[S^{ab}, S^{cd}],$$

where $S^{ab} = i[\gamma^a, \gamma^b]/4$ provides a representation of homogeneous Lorentz transformation (spin). That is, from (4.2.10) in Chap. 4 of Vol. I, it satisfies the commutation relation

$$[S^{ab}, S^{cd}] = i\left(\eta^{ac}S^{bd} - \eta^{bc}S^{ad} + \eta^{bd}S^{ac} - \eta^{ad}S^{bc}\right).$$

Upon replacing this in the previous equation and relabeling some of the indices, the statement of the problem follows.

Chapter 3

3.1. For infinitesimal $\lambda(\tau)$, then $\tau - \tau' = \lambda(\tau)$ implies that $d\tau'/d\tau = 1 - \dot{\lambda}$, $\dot{\lambda} \equiv d\lambda/d\tau$. Using (3.1.9), we have $a'(\tau') = \left(1 + \dot{\lambda}(\tau)\right)a(\tau)$, or $a'(\tau) = a(\tau) + \lambda(\tau)\dot{a}(\tau) + a(\tau)\dot{\lambda}(\tau)$. From which (3.1.11) follows. Similarly, the relation $X'^\mu(\tau') = X^\mu(\tau)$, in reference to the worldline, leads to (3.1.10).

3.2. From (3.2.16) we note that $(\dot{X} \cdot X')^2 - \dot{X} \cdot \dot{X} \, X' \cdot X' = h_{01} h_{10} - h_{00} h_{11} = -h$, which appears under the square root in (3.2.8). By using this together (3.2.17), (3.2.19), and (3.2.20) immediately follows.

3.3. One may explicitly write

$$\hat{h}_{01} = h_{00} \, ac + h_{01}(bc + ad) + h_{11} \, bd,$$

$$\hat{h}_{00} = h_{00} \, a^2 + 2h_{01} \, ab + h_{11} \, b^2,$$

$$\hat{h}_{11} = h_{00} \, c^2 + 2h_{01}cd + h_{11} \, d^2,$$

For $h_{00} = 0$, and hence $h = -h_{01}^2 \neq 0$, choose $d = b \neq 0$, $a = -(1 + b^2 h_{11})/2b \, h_{01}$, $c = (1 - b^2 h_{11})/2b \, h_{01}$ giving $[\hat{h}_{\alpha\beta}] = \text{diag}[-1, 1]$.

For $h_{00} > 0$, choose $a = h_{01}/\sqrt{-h}$, $b = -h_{00}/\sqrt{-h}$, $c = \pm 1$, $d = 0$. This gives $[\hat{h}_{\alpha\beta}] = (h_{00}) \, \text{diag}[-1, 1]$.

For $h_{00} < 0$, choose $c = h_{01}/\sqrt{-h}$, $d = -h_{00}/\sqrt{-h}$, $a = \pm 1$, $b = 0$. This gives $[\hat{h}_{\alpha\beta}] = (-h_{00}) \, \text{diag}[-1, 1]$.

That is, we may write $[\hat{h}_{\alpha\beta}] = e^{\phi} \operatorname{diag}[-1, 1]$, with e^{ϕ} defining a positive function.

3.4. The chain rule $\partial_{\sigma} = \partial_{\sigma}(\tau - \sigma)/\partial_{\tau-\sigma} + \partial_{\sigma}(\tau + \sigma)/\partial_{\tau+\sigma}$, implies that $\partial_{\sigma} = -\partial/\partial_{\tau-\sigma} + \partial/\partial_{\tau+\sigma}$. Similarly $\partial_{\tau} = \partial/\partial_{\tau-\sigma} + \partial/\partial_{\tau+\sigma}$. These give

$$(\partial_{\sigma})^2 - (\partial_{\tau})^2 = -4\,\partial_{\tau-\sigma}\partial_{\tau+\sigma},$$

from which the wave equation becomes $\partial_{\tau-\sigma}\partial_{\tau+\sigma}X^i(\tau, \sigma) = 0$. Using the facts that

$$\tau = (\tau-\sigma)/2 + (\tau+\sigma)/2, \quad \sigma = -(\tau-\sigma)/2 + (\tau+\sigma)/2,$$

give, from the chain rule,

$$\partial_{\tau-\sigma} = (1/2)(\partial/\partial_{\tau} - \partial/\partial_{\sigma}), \quad \partial_{\tau+\sigma} = (1/2)(\partial/\partial_{\tau} + \partial/\partial_{\sigma}).$$

Hence $\partial_{\tau\pm\sigma}(\tau \mp \sigma) = 0$, as expected, from which the newly obtained wave equation implies the structure of the solutions mentioned in the problem. We note that for $\sigma \to (\sigma + \Delta\sigma)$, with $\Delta\sigma > 0$, the arguments

$$(\tau \mp \sigma) \to (\tau \mp (\sigma + \Delta\sigma)) = ((\tau \mp \Delta\sigma) \mp \sigma),$$

and $\tau > \tau - \Delta\sigma, \tau < \tau + \Delta\sigma$, with τ corresponding, respectively, to the future and the past in the evolution process. This justifies the subscripts R/L attached to these solutions as right- and left-movers, respectively.

3.5. Using the integral $\int_0^{\pi} d\sigma\, e^{-2iN\sigma} = \pi\delta(N, 0)$ for integer N, the expression in (3.2.81) gives

$$\int_0^{\pi} d\sigma\, \partial_{\tau}X^- = \ell^2 p^- \pi.$$

On the other hand, the explicit expressions in (3.2.79), (3.2.80) lead to

$$\frac{1}{2}\int_0^{\pi} d\sigma\, [(\partial_{\sigma}X^i)^2 + (\partial_{\tau}X^i)^2]$$

$$= \pi\,\ell^2\left(\frac{1}{2}[\alpha^i(0)+\bar{\alpha}^i(0)]^2 + \sum_{m\neq 0}[\alpha^i(-m)\alpha^i(m) + \bar{\alpha}^i(-m)\bar{\alpha}^i(m)]\right).$$

where recall from (3.2.75) that $\alpha^i(0) = \bar{\alpha}^i(0)$. Upon comparing the latter two integrals with the result obtained by integrating the equality in (3.2.48) over σ from 0 to π gives (3.2.82).

3.6. The identity in (3.2.136) is explicitly given by

$$\begin{pmatrix} 0 & I \\ -I & 0 \end{pmatrix} \begin{pmatrix} A & B \\ C & D \end{pmatrix} = - \begin{pmatrix} A^{\mathsf{T}} & C^{\mathsf{T}} \\ B^{\mathsf{T}} & D^{\mathsf{T}} \end{pmatrix} \begin{pmatrix} 0 & I \\ -I & 0 \end{pmatrix},$$

which leads to the stated conditions. Since no restrictions are set on the matrix A, and the elements of the matrix D are obtained from those of A, the number of independent components of the generators Λ are:
$(N^2/4) + N(N/2+1)/4 + N(N/2+1)/4 = N(N+1)/2$, where $N(N/2+1)/4$ denotes the number of independent elements of the matrix B or of C.

3.7. The expressions in (3.2.146), (3.2.147), may be rewritten as

$$\partial_\tau X = \ell \sum_n \left[\bar{\alpha}(n) e^{-2in(\tau+\sigma)} + \alpha(n) e^{-2in(\tau-\sigma)} \right], \qquad \ell p^{25} = \bar{\alpha}(0) + \alpha(0),$$

$$\partial_\sigma X = \ell \sum_n \left[\bar{\alpha}(n) e^{-2in(\tau+\sigma)} - \alpha(n) e^{-2in(\tau-\sigma)} \right], \quad 2Rw = \ell \left(\bar{\alpha}(0) - \alpha(0) \right),$$

where we have, in the process, used (3.2.140), (3.2.141). An elementary integration over σ, as defined below, and the identity $2\ell^2 \left(\bar{\alpha}^2(0) + \alpha^2(0) \right) = 4R^2 w^2 + \ell^4 (p^{25})^2$, together give

$$\frac{1}{2} \int_0^\pi d\sigma \left[(\partial_\sigma X)^2 + (\partial_\tau X)^2 \right]$$

$$= \pi \frac{1}{2} (4R^2 w^2 + \ell^4 (p^{25})^2) + \ell^2 \pi \sum_{m \neq 0} \left[\alpha(-m)\alpha(m) + \bar{\alpha}(-m)\bar{\alpha}(m) \right].$$

Now we use (3.2.48), the integral $\int_0^\pi d\sigma \partial_\tau X^- = \ell^2 \pi p^-$, and add the contribution of the X^i obtained in Problem 3.5, now for $i = 1, \ldots, 23$, to the above equation, to obtain (3.2.148).

3.8. In reference to the constraint in (3.2.109), we have from (3.2.146), (3.2.147) for the $X^{25} \equiv X$ contribution,

$$\frac{1}{\pi} \int_0^\pi d\sigma \partial_\tau X \partial_\sigma X = 2R\omega \ell^2 p^{25} - \ell^2 \sum_{n \neq 0} \left(\alpha(-n)\alpha(n) - \bar{\alpha}(-n)\bar{\alpha}(n) \right),$$

Adding the contribution of the X^i, which is $-\ell^2$ times the expression on the extreme left-hand side of (3.2.110), now for $i = 1, .., 23$, and setting the sum equal to zero, lead to the constraint in (3.2.152).

3.9. We note that if constraints are imposed on the external source, thus changing the right-hand side of (3.2.173), for example, by imposing a conservation law, the complete expression of a propagator does not follow. Since no constraints were imposed on J^μ, we may vary its components independently. Upon taking the vacuum expectation values $\langle 0_+ | \cdot | 0_- \rangle$ of (3.2.173) and (3.2.174), setting

$\langle 0_+|A^\mu(x)|0_-\rangle = (-i\delta/\delta J_\mu(x))\,\langle 0_+|0_-\rangle$, and functionally integrating with respect to the external source, we get

$$\langle 0_+|0_-\rangle = \exp\left[\frac{i}{2}\int (dx)J_\mu(x)D_+^{\mu\nu}(x-x')J_\nu(x')\right],$$

where $D_+^{\mu\nu}(x-x') = \int (dp)/(2\pi)^D\, e^{ik(x-x')}D_+^{\mu\nu}(k)$ is the propagator,

$$D_+^{ij}(k) = \frac{1}{k^2 - i\epsilon}\left(\delta^{ij} - \frac{k^ik^j}{k^2}\right),\quad D_+^{00}(k) = -\frac{1}{k^2},\quad D_+^{0i} = 0.$$

Clearly $D_+^{00}(k)$ gives rise to a phase to $\langle 0_+|0_-\rangle$. Upon using the identity

$$i\left[\frac{1}{k^2 - i\epsilon} - \frac{1}{k^2 + i\epsilon}\right] = -\frac{\pi}{|\mathbf{k}|}[\delta(k^0 - |\mathbf{k}|) + \delta(k^0 + |\mathbf{k}|)],$$

and the expansion $\pi^{ij} = \sum_{\lambda=1}^{D-2} e_\lambda^i e_\lambda^j$ [see (3.2.175)], then give

$$|\langle 0_+|0_-\rangle|^2 = \exp\left[-\int \left(d^{D-1}\mathbf{k}/(2|\mathbf{k}|(2\pi)^{D-1})\right)|J(\lambda,k)|^2\right] < 1,$$

where $J(\lambda,k) = e_\lambda^i J^i(k)$, $k^0 = |\mathbf{k}|$, establishing the positivity of the formalism.

3.10. We take the vacuum expectation values of (3.2.183), (3.2.184), and make use of the equation

$$\langle 0_+|h^{\mu\nu}(x)|0_-\rangle = (-i\delta/\delta T_{\mu\nu}(x))\,\langle 0_+|0_-\rangle,$$

where we note that no constraints were imposed on the external source and hence all of its components may be varied independently. Upon integrating with respect to the external source we obtain,

$$\langle 0_+|0_-\rangle = \exp\left[\frac{i}{2}\int (dx)(dx')\,T_{\mu\nu}(x)\Delta_+^{\mu\nu,\lambda\sigma}(x-x')T_{\lambda\sigma}(x')\right],\quad \text{where}$$

$$\Delta_+^{\mu\nu,\lambda\sigma}(x-x') = \int [(dk)/(2\pi)^D]\,e^{ik(x-x')}\Delta_+^{\mu\nu,\lambda\sigma}(k),\quad \Delta_+^{00,0i} = \Delta_+^{0i,00} = 0,$$

$$\Delta_+^{00,00}(k) = \frac{D-3}{D-2}\frac{(k^2)}{|\mathbf{k}|^4},\quad \Delta_+^{00,ij}(k) = \frac{2}{D-2}\frac{1}{|\mathbf{k}|^2}\pi^{ij},\quad \Delta_+^{0i,0k}(k) = -\frac{1}{2|\mathbf{k}|^2}\pi^{ik},$$

$$\Delta_+^{ij,k\ell}(k) = \frac{1}{k^2 - i\epsilon}\frac{1}{D-2}\left[\frac{D-2}{2}(\pi^{ik}\pi^{j\ell} + \pi^{i\ell}\pi^{jk}) - \pi^{ij}\pi^{k\ell}\right],\quad \epsilon \to +0,$$

$\Delta_+^{ij,00}(k) = (2/[(D-2)|\mathbf{k}|^2])\,\pi^{ij}$, and $\pi^{ij} = \delta^{ij} - k^ik^j/|\mathbf{k}|^2 = \sum_{\lambda=1}^{D-2} e_\lambda^i e_\lambda^j$. Clearly, $\Delta_+^{00,00}$, $\Delta_+^{0i,0k}$, $\Delta_+^{ij,00}$, $\Delta_+^{00,ij}$, provide phase factors to $\langle 0_+|0_-\rangle$. We use the following identity

$$\frac{1}{D-2}\left[\frac{D-2}{2}(\pi^{ik}\pi^{j\ell}+\pi^{i\ell}\pi^{jk})-\pi^{ij}\pi^{k\ell}\right]=\sum_{\lambda,\lambda'=1}^{D-2}\epsilon^{ij}(\lambda,\lambda')\,\epsilon^{k\ell}(\lambda,\lambda'),$$

where $\epsilon^{ij}(\lambda,\lambda')$ is defined in (3.2.188). We note that the independent degrees of freedom now is easily obtained from

$$\sum_{\lambda,\lambda'=1}^{D-2}\epsilon^{ij}(\lambda,\lambda')\,\epsilon^{ij}(\lambda,\lambda')=\frac{1}{D-2}\left[\frac{D-2}{2}(\pi^{ii}\pi^{jj}+\pi^{ij}\pi^{ji})-\pi^{ij}\pi^{ij}\right]$$

to be simply $D(D-3)/2$ since $\pi^{ii}=D-2$, $\pi^{ij}\,\pi^{ij}=\pi^{ij}\delta^{ij}=D-2$. Finally the vacuum persistence probability is given by

$$|\langle\,0_+|0_-\rangle|^2=\exp\left[-\int\sum_{\lambda,\lambda'=1}^{D-2}\frac{\mathrm{d}^{D-1}\mathbf{k}}{2|\mathbf{k}|(2\pi)^{D-1}}\,|T(\lambda,\lambda',k)|^2\right]<1,$$

where $T(\lambda,\lambda',k)$ is defined in (3.2.187), $k^0=|\mathbf{k}|$, establishing the positivity of the formalism.

3.11. As before no constraints are set on the external source and hence all of its components may be varied independently. We take the vacuum expectation values of (3.2.197), (3.2.198), and setting $\langle\,0_+|A^{\mu\nu}(x)|0_-\rangle=(-\mathrm{i}\delta/\delta J_{\mu\nu}(x))\,\langle 0_+|0_-\rangle$. Upon integrating with respect to the external source we obtain,

$$\langle 0_+|0_-\rangle=\exp[\frac{\mathrm{i}}{2}\int(\mathrm{d}x)(\mathrm{d}x')\,T_{\mu\nu}(x)\tilde{\Delta}_+^{\mu\nu,\lambda\sigma}(x-x')T_{\lambda\sigma}(x')],$$

$$\tilde{\Delta}_+^{\mu\nu,\sigma\lambda}(x-x')=\int\frac{(\mathrm{d}k)}{(2\pi)^D}\,\mathrm{e}^{\mathrm{i}k(x-x')}\,\tilde{\Delta}_+^{\mu\nu,\sigma\lambda}(k),$$

$$\tilde{\Delta}_+^{00,00}=0,\quad\tilde{\Delta}_+^{00,0i}=\tilde{\Delta}_+^{0i,00}=0,\quad\tilde{\Delta}_+^{ij,00}=0,$$

$$\tilde{\Delta}_+^{0i,0j}(k)=-\frac{1}{|\mathbf{k}|^2}\,\pi^{ij},\quad\tilde{\Delta}_+^{ij,k\ell}(k)=\frac{1}{k^2-\mathrm{i}\epsilon}\,\frac{(\pi^{ik}\pi^{j\ell}-\pi^{i\ell}\pi^{jk})}{2}.$$

Clearly, $\tilde{\Delta}_+^{0i,0j}(k)$ gives rise to a phase factor to $\langle 0_+|0_-\rangle$. Upon using the identity

$$\frac{(\pi^{ik}\pi^{j\ell}-\pi^{i\ell}\pi^{jk})}{2}=\sum_{\lambda,\lambda'=1}^{D-2}\varepsilon^{ij}(\lambda,\lambda')\varepsilon^{k\ell}(\lambda,\lambda'),$$

where $\varepsilon^{ij}(\lambda, \lambda')$ is defined in (3.2.200). The number of independent polarization states are given, as before, to be

$$\sum_{\lambda,\lambda'=1}^{D-2} \varepsilon^{ij}(\lambda, \lambda') \varepsilon^{ij}(\lambda, \lambda') = \frac{1}{2}\left[\pi^{ii}\pi^{jj} - \pi^{ij}\pi^{ji}\right] = \frac{1}{2}(D-2)(D-3).$$

The vacuum persistence probability emerges as

$$|\langle 0_+|0_-\rangle|^2 = \exp\left[-\int \sum_{\lambda,\lambda'=1}^{D-2} \frac{d^{D-1}\mathbf{k}}{2|\mathbf{k}|(2\pi)^{D-1}} |J(\lambda, \lambda', k)|^2\right] < 1,$$

with $J(\lambda, \lambda', k)$ defined in (3.2.200), $k^0 = |\mathbf{k}|$, thus establishing the positivity of the formalism.

3.12. The Hamiltonian density is given by $\mathcal{H} = P \cdot \dot{X} - \mathcal{L}$, where

$$\mathcal{L} = -(T/2)(\partial_\alpha X \cdot \partial^\alpha X),$$

is the Lagrangian density, and $P^\mu = T\dot{X}^\mu$. These give $\mathcal{H} = (T/2)(\dot{X}^2 + X'^2)$. Upon using $X^\pm = (X^0 \pm X^{D-1})/\sqrt{2}$, we may rewrite $\mathcal{H}$ as

$$\mathcal{H} = \frac{T}{2}(\dot{X}^i\dot{X}^i + X'^iX'^i) - T(\dot{X}^-\dot{X}^+ + X'^-X'^+).$$

From (3.2.62), we obtain,

$$\int_0^\pi d\sigma(\dot{X}^i\dot{X}^i + X'^iX'^i) = \pi\ell^2 \sum_n \alpha^i(-n)\alpha^i(n),$$

where we recall from (3.2.61) that $\alpha^i(0) = \ell p^i$. On the other hand, (3.2.56) implies that $\dot{X}^+ = \ell^2 p^+$, $X'^+ = 0$. From (3.2.64), we then obtain $\int_0^\pi d\sigma (\dot{X}^-\dot{X}^+ + X'^-X'^+) = \pi\ell^4 p^+ p^-$. Using the identities in (3.2.46), relating T, ℓ^2, α', the identity $p^2 = p^i p^i - 2p^+ p^-$, and the explicit expression of M^2 in (A-3.1), the expression for $H = \int_0^\pi d\sigma \mathcal{H}$ in (B-3.3) follows.

3.13. The two dimensional Dirac equation in (3.3.10), in the presence of an electromagnetic coupling, reads $[\rho^\alpha(\partial_\alpha/i - eA_\alpha) + m]\psi = 0$. Upon taking the complex conjugate of this equation and using the fact that the Dirac matrices are pure imaginary, give $[\rho^\alpha(\partial_\alpha/i + eA_\alpha) + m]\psi^* = 0$, from which we infer that $\psi_C = \psi^*$, for the charge conjugate spinor.

3.14. A quick way of establishing this is to set on general grounds

$$\delta_{ab}\delta_{cd} = B\,\delta_{ad}\delta_{cb} + C\,(\rho^\mu)_{ad}(\rho_\mu)_{cb} + D\,(\rho^5)_{ad}(\rho^5)_{cb}.$$

The coefficients B, C, D, are readily obtained by considering specific matrix elements, e.g., $a = b = c = d = 1; a = d = 1, b = c = 2; a = b = 1, c = d = 2$.

3.15. The transformation rules in (3.3.37) imply, after straightforward manipulations, that $\delta(\partial^\alpha X^\mu \partial_\alpha X_\mu) = \sqrt{2}\,\partial^\alpha(\bar{\epsilon}\psi^\mu \partial_\alpha X_\mu) - \sqrt{2}\,\bar{\epsilon}\psi^\mu \Box X_\mu$,

$$\frac{1}{i}\delta(\bar{\psi}^\mu \rho^\alpha \partial_\alpha \psi_\mu) = \frac{1}{\sqrt{2}}\,\partial_\alpha(\bar{\epsilon}\rho^\beta \rho^\alpha \psi_\mu \partial_\beta X^\mu) + \sqrt{2}\,\bar{\epsilon}\psi^\mu \Box X_\mu$$
$$- \sqrt{2}(\partial_\alpha \bar{\epsilon})(\rho^\beta \rho^\alpha \psi^\mu \partial_\beta X_\mu),$$

from which (3.3.38) follows, up to a total derivative, where, in the process of the derivation, we have used the first identity in (3.3.16), and $\rho^\beta \rho^\alpha \partial_\beta \partial_\alpha X^\mu = -\Box X^\mu$.

3.16. Using the definitions of the ρ^μ, ρ^5 matrices in (3.3.9), the definition of light-cone variables, the light-cone gauge property $X^+ = x^+ + \ell^2 p^+ \tau$ in (3.2.56), (3.3.45), and the definitions $\psi_{R/L} = [(I \pm \rho^5)/2]\psi$, we may write the expression for J^0 in (3.3.46) as

$$J^0 = \frac{1}{2}\rho^\beta \rho^0 \psi^i \partial_\beta X^i - \frac{1}{2}\ell^2 p^+ \psi^- - \frac{1}{2}\rho^\beta \rho^0 \psi^+ \partial_\beta X^-$$
$$= \frac{1}{2}\left[\begin{pmatrix} \psi_R^i(\partial_0 - \partial_1)X_R^i \\ \psi_L^i(\partial_0 + \partial_1)X_L^i \end{pmatrix} - \ell^2 p^+ \begin{pmatrix} \psi_R^- \\ \psi_L^- \end{pmatrix} - \begin{pmatrix} \psi_R^+(\partial_0 - \partial_1)X_R^- \\ \psi_L^+(\partial_0 + \partial_1)X_L^- \end{pmatrix} \right].$$

The constraints then follow by using the definitions $\partial_\pm = (\partial_0 \pm \partial_1)/2$, and setting the above equation equal to zero. Also recall that $\partial_\pm X_{R/L}(\tau \mp \sigma) = 0$.

3.17. Using the explicit expressions of the matrices ρ^0, ρ^1 in (3.3.9), it easily follows that for the fermionic parts

$$T_{00}^F = -\frac{i}{2}\psi_R^\mu \partial_- \psi_{\mu R} - \frac{i}{2}\psi_L^\mu \partial_- \psi_{\mu L},$$

$$T_{01}^F = \frac{i}{2}\psi_R^\mu \partial_- \psi_{\mu R} + \frac{i}{2}\psi_L^\mu \partial_- \psi_{\mu L}.$$

By invoking the boundary conditions satisfied by the spinor ψ^μ in (3.3.52), the above equations give

$$T_{00}^F + T_{01}^F = i\left(\psi_L^i \partial_+ \psi_L^i - \psi_L^+ \partial_+ \psi_L^- \right),$$

$$T_{00}^F - T_{01}^F = i\left(\psi_R^i \partial_- \psi_R^i - \psi_R^+ \partial_+ \psi_R^- \right),$$

$$T_{00}^B + T_{01}^B = 2\left(\partial_+ X_L^i \partial_+ X_L^i - \ell^2 p^+ \partial_+ X_L^- \right)$$

$$T_{00}^B - T_{01}^B = 2\left(\partial_- X_R^i \partial_- X_R^i - \ell^2 p^+ \partial_- X_R^- \right).$$

For the bosonic part we have $T_{00}^B = [\partial_0 X^\mu \partial_0 X_\mu + \partial_1 X^\mu \partial_1 X_\mu]/2$, $T_{01}^B = \partial_0 X^\mu \partial_1 X_\mu$. The three constraints mentioned in the problem immediately follow upon setting $[T_{00} + T_{01}] = 0$, $[T_{00} - T_{01}] = 0$, $T_{01} = ([T_{00} + T_{01}] - [T_{00} - T_{01}])/2 = 0$.

3.18. For the R boundary condition, $\partial_+ X_L^-$, $\partial_+ \psi_L^-$ have the general structures

$$\partial_+ X_L^- = \frac{1}{2} \ell^2 p^- + \frac{\ell}{2} \sum_{n \neq 0} A^-(n) \, e^{-in(\tau+\sigma)},$$

$$\partial_+ \psi_L^- = -\frac{i}{\sqrt{2}} \ell \sum_n \chi^-(n) \, n \, e^{-in(\tau+\sigma)}.$$

The zero mode part, i.e., the $e^{-in(\tau+\sigma)}$-independent part, of the right-hand side of the first equation is $\ell^2 p^-/2$, while for the second one, it is zero. The zero mode part of $\partial_+ X_L^i \partial_+ X_L^i + i \psi_L^i \partial_+ \psi_L^i/2$, is clearly as given on the right-hand of (3.3.66). The statement in the problem then follows from the application of (3.3.54). The demonstration for the NS boundary condition is almost identical.

3.19. Using the facts that $\partial_\alpha \psi^+ = 0$ and $\partial_1 X^+ = 0$ [see (3.3.51), (3.3.52)], we have, with $T_{01} = T_{01}^F + T_{01}^B$, the following explicit expressions: (see also Problem 3.17)

$$T_{01}^F = -\frac{i}{2} (\psi_L^+ \partial_+ \psi_L^- - \psi_R^+ \partial_- \psi_R^-) + \frac{i}{2} (\psi_L^i \partial_+ \psi_L^i - \psi_R^i \partial_- \psi_R^i),$$
$$T_{01}^B = -\ell^2 p^+ \partial_1 X^- + \partial_0 X^i \partial_1 X^i.$$

consistent with (3.3.56). For the R boundary condition, obviously the σ-integrals of $\partial_+ \psi_L^-$ and $\partial_1 X^-$ are both zero on account that $\int_0^\pi d\sigma \, e^{-2in\sigma} = 0$, for non-zero integer n. On the other-hand, for the NS boundary condition, the same reason gives zero for the σ- integral of $\partial_1 X^-$ and, $\psi^+ = 0$, in this case. On the other hand as a consequence of the orthogonality relation

$$\int_0^\pi d\sigma \, e^{-2i(n-m)\sigma} = \pi \, \delta_{n,m},$$

the remaining terms in T_{01} readily give the constraints in (3.3.77), and (3.3.78), by finally invoking the boundary condition $T_{01} = 0$ holding true as a special case of (3.3.35).

3.20. Since no constraints were imposed on the external sources $\overline{K}^\mu, K^\mu$, we may vary each of their components independently. Upon taking the vacuum expectation values $\langle 0_- | . | 0_+ \rangle$ of (3.3.177), (3.3.178), and setting $\langle 0_+ | \psi_a^\mu(x) | 0_- \rangle = (-i\delta/\delta \overline{K}_{\mu a}(x)) \langle 0_+ | 0_- \rangle$, and integrating with respect to the sources, we obtain

$$\langle 0_+|0_-\rangle = \exp\left[\,i\int (dx)(dx')\,\overline{K}_{\mu a}(x)\Delta_{+\,ab}^{\mu\nu}(x-x')K_{vb}(x')\right],$$

where the Rarita-Schwinger propagator $\Delta_{+\,ab}^{\mu\nu}(x-x')$ in 10 dimensions is given by $\Delta_{+\,ab}^{\mu\nu}(x-x') = \int[(dp)/(2\pi)^{10}]\,e^{ip\,(x-x')}\,\Delta_{+\,ab}^{\mu\nu}(p)$, and $\Delta_{+\,ab}^{\mu\nu}(p)$ is explicitly worked out to be $(\gamma\cdot p = \gamma^{\mu}p_{\mu})$

$$\Delta_+^{ij}(p) = \frac{(-\gamma\cdot p)}{p^2 - i\epsilon}\alpha^{ij}(p), \qquad \Delta_+^{00}(p) = \frac{7}{8}\frac{(\gamma\cdot p)}{\mathbf{p}^2},$$

$$\Delta_+^{0i}(p) = +\frac{1}{8\mathbf{p}^2}\left(p^i + \gamma^{}\!{}^{j}p^{j}\,\gamma^{i}\right)\gamma^0, \qquad \Delta_+^{i0}(p) = -\frac{1}{8\mathbf{p}^2}\left(p^i + \gamma^{}\!{}^{j}p^{j}\,\gamma^{i}\right)\gamma^0,$$

$$\alpha^{ij} = \left(\delta^{ij} - \frac{p^i p^j}{\mathbf{p}^2}\right) + \frac{1}{8}\left(\delta^{ik} - \frac{p^i p^k}{\mathbf{p}^2}\right)\gamma^k\gamma^\ell\left(\delta^{\ell j} - \frac{p^\ell p^j}{\mathbf{p}^2}\right).$$

Clearly only $\Delta_+^{ij}(p)$ propagates.

3.21. Upon setting $\rho = e^{\tau}e^{i\sigma}$, with $W = W_R + i\,W_{\mathrm{Im}}$, $ad - bc = 1$, and $0 \le \sigma \le \pi$, we obtain

$$W_{\mathrm{Re}} = \left[(ac\,e^{2\tau} + bd) + e^{\tau}(ad + bc)\cos\sigma\right]/D,$$

$$W_{\mathrm{Im}} = \left[e^{\tau}\sin\sigma\right]/D \ge 0$$

$$D = (c^2\,e^{2\tau} + d^2) + 2\,cd\,e^{\tau}\cos\sigma \ge (c\,e^{\tau} - d)^2,$$

which establish the facts that W maps the upper complex ρ-plane into itself and the real line into itself. Since any three of the real variables $z_1, z_2, \ldots, z_{n-1}, z_n$, encountered in Sect. 3.5.1, may be chosen at will, on account that three of real parameters a, b, c, d, such that $ad - bc = 1$, are arbitrary, a natural choice, for a scattering process, is $\tau_1 \to -\infty$, $\tau_n \to +\infty$, which correspond to $z_1 = 0$, $z_n = \infty$. Finally, z_{n-1} was chosen to be 1, corresponding to $\tau = 0$, obtaining the following restriction on the variables

$$0 = z_1 < z_2 < \cdots < z_{n-3} < z_{n-2} < 1 = z_{n-1} < z_n = \infty,$$

as appearing in (3.5.13).

3.22. This involves three terms:

$$e_2\cdot k_1\,\langle 0; k_3\,|\,e_3\cdot\alpha(1)\,e^{k_2\cdot\alpha(1)^{\dagger}}\,e^{-k_2\cdot\alpha(1)}\,e_1\cdot\alpha(1)^{\dagger}\,|\,0; k_1 + k_2\rangle$$

$$= (e_2\cdot k_1\,e_1\cdot e_3 + e_2\cdot k_1\,e_3\cdot k_2\,e_1\cdot k_3)\,\langle 0; k_3\,|\,0; k_1 + k_2\rangle,$$

$$\langle 0; k_3 \,|\, e_3 \cdot \alpha(1) \; e_2 \cdot \alpha(1) \; e^{k_2 \cdot \alpha(1)^\dagger} \, e^{-k_2 \cdot \alpha(1)} \; e_1 \cdot \alpha(1)^\dagger \,|\, 0; k_1 + k_2 \rangle$$
$$= e_3 \cdot k_2 \; e_2 \cdot e_1 \, \langle 0; k_3 | 0; k_1 + k_2 \rangle,$$

$$\langle 0; k_3 \,|\, e_3 \cdot \alpha(1) \; e_2 \cdot \alpha(1)^\dagger \; e^{k_2 \cdot \alpha(1)^\dagger} \, e^{-k_2 \cdot \alpha(1)} \; e_1 \cdot \alpha(1)^\dagger \,|\, 0; k_1 + k_2 \rangle$$
$$= e_1 \cdot k_3 \; e_3 \cdot e_2 \, \langle 0; k_3 \,|\, 0; k_1 + k_2 \rangle,$$

where we have used, in the process, the fact that $p^\mu |0; k\rangle = k^\mu |0; k\rangle$, $k_i^2 = 0$, that x generates momentum translation, and that $k_1 + k_2 + k_3 = 0$, and, in the process of derivation, the directions of momenta were finally chosen such that the conservation of momenta reads as just given. Adding these three terms lead immediately to the expression in (3.5.26).

3.23. The three-point function in question, up to an overall coupling parameter, may be rewritten as

$$e_{1\mu} \, e_{2\nu} \, e_{3\varrho} \left(\eta^{\mu\varrho} k_1^\nu + \eta^{\mu\nu} k_2^\varrho + \eta^{\nu\varrho} k_3^\mu \right). \qquad (*)$$

It is sufficient to establish the anti-symmetry property of, say, under the exchange $(e_1, k_1) \leftrightarrow (e_2, k_2)$. Momentum conservation allows us to write, respectively, in reference to these corresponding momenta: $k_1 = -k_2 - k_3$, $k_2 = -k_1 - k_3$, $k_3 = -k_1 - k_2$. Using the properties $e_i \cdot k_i = 0, i = 1, 2, 3$, the above three-point function may be rewritten as

$$- e_{1\mu} \, e_{2\nu} \, e_{3\varrho} \left(\eta^{\mu\varrho} k_3^\nu + \eta^{\mu\nu} k_1^\varrho + \eta^{\nu\varrho} k_2^\mu \right), \qquad (**)$$

which upon the exchange $(e_1, k_1) \leftrightarrow (e_2, k_2)$ it is transformed to

$$- e_{2\mu} \, e_{1\nu} \, e_{3\varrho} \left(\eta^{\mu\varrho} k_3^\nu + \eta^{\mu\nu} k_2^\varrho + \eta^{\nu\varrho} k_1^\mu \right).$$

By a mere relabeling of the Lorentz indices, note that this is just the initial three-point function of opposite sign.

3.24. The three-point function, in question, in bosonic string theory, with $\ell \to 0$, up to an overall coupling parameter, as in the previous problem, is given by

$$e_{1\mu} \, e_{2\nu} \, e_{3\varrho} \left(\eta^{\mu\varrho} k_1^\nu + \eta^{\mu\nu} k_2^\varrho + \eta^{\nu\varrho} k_3^\mu \right).$$

Using momentum conservation: $k_1 + k_2 + k_3 = 0$, we may rewrite k_1, k_2, k_3, respectively, within the brackets above as

$$k_1 = -\frac{1}{2}(k_2 + k_3 - k_1), \quad k_2 = -\frac{1}{2}(k_1 + k_3 - k_2), \quad k_3 = -\frac{1}{2}(k_1 + k_2 - k_3).$$

Upon using the properties

$$e_i \cdot k_i = 0, \; i = 1, 2, 3,$$

the above three-point function simply becomes

$$-\frac{1}{2} e_{1\mu} e_{2\nu} e_{3\varrho} \left(\eta^{\mu\varrho} (k_3 - k_1)^\nu + \eta^{\mu\nu} (k_1 - k_2)^\varrho + \eta^{\nu\varrho} (k_2 - k_3)^\mu \right),$$

and the proportionality of the latter to the superstring expression in (3.5.74) is now evident. If the reader has gone through solution of the previous problem, then note that the equality of the two equations (*), (**) in it imply they are also both equal to $[(*) + (**)]/2$, which is the equation derived above.

3.25. The equality follows by using, in the process, the basic equalities:

$$\xi^{\mu\nu} \partial_\rho \xi_{\mu\lambda} \, \partial^\rho \xi^\lambda{}_\nu = \frac{1}{2} \partial_\rho (\xi^{\mu\nu} \xi_{\mu\lambda} \partial^\rho \xi^\lambda{}_\nu),$$

$$\xi^{\mu\nu} \partial_\mu \xi^{\lambda\rho} \, \partial_\nu \xi_{\lambda\rho} = \partial_\mu (\xi^{\mu\nu} \xi^{\lambda\rho} \partial_\nu \xi_{\lambda\rho}) - \xi^{\mu\nu} \xi^{\lambda\rho} \partial_\mu \partial_\nu \xi_{\lambda\rho},$$

$$\xi^{\mu\nu} \partial_\lambda \xi_\mu{}^\rho \, \partial_\rho \xi_\nu{}^\lambda = \partial_\lambda (\xi^{\mu\nu} \xi_\mu{}^\rho \partial_\rho \xi_\nu{}^\lambda) - \xi^{\mu\nu} \partial_\nu \xi_{\rho\lambda} \partial^\lambda \xi_\mu{}^\rho,$$

where, for convenience, we have *relabeled* some of the indices in writing down these equations. The last two equalities easily follow. The first one in detail is given by

$$\xi^{\mu\nu} \partial_\rho \xi_{\mu\lambda} \, \partial^\rho \xi^\lambda{}_\nu = \partial_\rho \left(\xi^{\mu\nu} \xi_{\mu\lambda} \partial^\rho \xi^\lambda{}_\nu \right) - \partial_\rho \xi^{\mu\nu} \xi_{\mu\lambda} \partial^\rho \xi^\lambda{}_\nu$$

$$= \partial_\rho \left(\xi^{\mu\nu} \xi_{\mu\lambda} \partial^\rho \xi^\lambda{}_\nu \right) - \xi^{\mu\nu} \partial_\rho \xi_{\mu\lambda} \partial^\rho \xi^\lambda{}_\nu$$

where in writing the last term, on the extreme right-hand side, we have exchanged the indices $\lambda \leftrightarrow \nu$, and when this last term is brought to the left-hand side of the equality, the result follows.

3.26. The functional derivatives in (3.6.15) as applied to the integral in (3.6.14) is explicitly given by

$$(2\pi)^D \delta^{(D)}(k_1 + k_2 + k_3) \frac{\kappa}{4} \mathrm{Sym}[\,\cdot\,],$$

where $[\,\cdot\,]$ is equal to

$$\left[k_{3\mu_1} k_{3\mu_2} \eta_{\nu_1 \varrho_1} \eta_{\nu_2 \varrho_2} + k_{3\nu_1} k_{3\nu_2} \eta_{\mu_1 \varrho_1} \eta_{\mu_2 \varrho_2} \right]$$

$$+ \left[k_{2\varrho_1} k_{2\varrho_2} \eta_{\mu_1 \nu_1} \eta_{\mu_2 \nu_2} + k_{2\mu_1} k_{2\mu_2} \eta_{\nu_1 \varrho_1} \eta_{\nu_2 \varrho_2} \right]$$

$$+ \left[k_{1\nu_1} k_{1\nu_2} \eta_{\mu_1 \varrho_1} \eta_{\mu_2 \varrho_2} + k_{1\varrho_1} k_{1\varrho_2} \eta_{\mu_1 \nu_1} \eta_{\mu_2 \nu_2} \right]$$

$$+ \left[k_{3\mu_1} k_{2\varrho_2} \eta_{\nu_1 \varrho_1} \eta_{\mu_2 \nu_2} + k_{2\varrho_1} k_{3\mu_2} \eta_{\mu_1 \nu_1} \eta_{\nu_2 \varrho_2} \right]$$

$$+ \left[k_{3\nu_1} k_{2\mu_2} \eta_{\mu_1 \varrho_1} \eta_{\nu_2 \varrho_2} + k_{2\mu_1} k_{3\nu_2} \eta_{\nu_1 \varrho_1} \eta_{\mu_2 \varrho_2} \right]$$

$$+ \left[k_{1\nu_1} k_{2\varrho_2} \eta_{\mu_1 \varrho_1} \eta_{\mu_2 \nu_2} + k_{2\varrho_1} k_{1\nu_2} \eta_{\mu_1 \nu_1} \eta_{\mu_2 \varrho_2} \right]$$

$$+ \left[k_{1\varrho_1} k_{2\mu_2} \eta_{\mu_1 \nu_1} \eta_{\nu_2 \varrho_2} + k_{2\mu_1} k_{1\varrho_2} \eta_{\nu_1 \varrho_1} \eta_{\mu_2 \nu_2} \right]$$

$$+ \left[k_{1\nu_1} k_{3\mu_2} \eta_{\mu_1 \varrho_1} \eta_{\nu_2 \varrho_2} + k_{3\mu_1} k_{1\nu_2} \eta_{\nu_1 \varrho_1} \eta_{\mu_2 \varrho_2} \right]$$

$$+ \left[k_{1\varrho_1} k_{3\nu_2} \eta_{\mu_1 \nu_1} \eta_{\mu_2 \varrho_2} + k_{3\nu_1} k_{1\varrho_2} \eta_{\mu_1 \varrho_1} \eta_{\mu_2 \nu_2} \right],$$

where recall that "Sym" stands for the symmetrization operation to be applied to the expression following it over $\mu_1 \leftrightarrow \mu_2$, $\nu_1 \leftrightarrow \nu_2$, $\varrho_1 \leftrightarrow \varrho_2$. This gives precisely the vertex part in (3.6.10) with κ identified with the string coupling parameter through $\kappa = g_c/2$.

Index

© Springer International Publishing Switzerland 2016
E.B. Manoukian, *Quantum Field Theory II*, Graduate Texts in Physics,
DOI 10.1007/978-3-319-33852-1

Printed in the United States
By Bookmasters